NUMERICAL ANALYSIS FOR APPLIED SCIENCE

PURE AND APPLIED MATHEMATICS

A Wiley-Interscience Series of Texts, Monographs, and Tracts

A complete list of the titles in this series appears at the end of this volume.

NUMERICAL ANALYSIS FOR APPLIED SCIENCE

MYRON B. ALLEN III
ELI L. ISAACSON

Department of Mathematics
University of Wyoming

A Wiley-Interscience Publication

JOHN WILEY & SONS, INC.

New York • Chichester • Weinheim • Brisbane • Singapore • Toronto

This text is printed on acid-free paper.

Library of Congress Cataloging in Publication Data:

Allen, Myron B., 1954–
Numerical analysis for applied science / Myron B. Allen III and Eli L. Isaacson.
p. cm. — (Wiley-Interscience series of pure and applied mathematics)
"A Wiley-Interscience publication."
Includes bibliographical references and index.
ISBN 0-471-55266-6 (acid-free paper)
1. Numerical analysis. I. Isaacson, Eli L. II. Title.
III. Series.
QA297.A53 1997
519.4—dc21 97-16688

Printed in the United States of America

10 9 8 7 6 5 4 3 2 1

CONTENTS

PREFACE

Once in a while you get shown the light
In the strangest of places, if you look at it right.
Robert Hunter

We intend this book to serve as a first graduate-level text for applied mathematicians, scientists, and engineers. We hope that these students have had some exposure to numerics, but the book is self-contained enough to accommodate students with no numerical background. Students *should* know a computer programming language, though.

In writing the text, we have tried to adhere to three principles:

1. The book should cover a significant range of numerical methods now used in applications, especially in scientific computation involving differential equations.
2. The book should be appropriate for mathematics students interested in the theory behind the methods.
3. The book should also appeal to students who care less for rigorous theory than for the heuristics and practical aspects of the methods.

The first principle is a matter of taste. Our omissions may appall some readers; they include polynomial root finders, linear and nonlinear programming, digital filtering, and most topics in statistics. On the other hand, we have included topics that receive short shrift in many other texts at this level. Examples include

- Multidimensional interpolation, including interpolation on triangles.
- Quasi-Newton methods in several variables.
- A brief introduction to multigrid methods.
- Conjugate-gradient methods, including error estimates.
- Rigorous treatment of the QR method for eigenvalues.
- An introduction to adaptive methods for numerical integration and ordinary differential equations.

- A thorough treatment of multistep schemes for ordinary differential equations (ODEs).
- Consistency, stability, and convergence of finite-difference schemes for partial differential equations (PDEs).
- An introduction to finite-element methods, including basic convergence arguments and methods for time-dependent problems.

All of these topics are prominent in scientific applications.

The second and third principles conflict. Our strategy for addressing this conflict is threefold. First, most sections of the book have a "pyramid" structure. We begin with the *motivation and construction* of the methods, then discuss *practical considerations* associated with their implementation, then present rigorous *mathematical details*. Thus students in a "methods" course can concentrate on motivation, construction, and practical considerations, perhaps grazing from the mathematical details according to the instructor's tastes. Students in an "analysis" course should delve into the mathematical details *as well as* the practical considerations.

Second, we have included Chapter 0, "Some Useful Tools," which reviews essential notions from undergraduate analysis and linear algebra. Mathematics students should regard this chapter as a review; engineering and applied science students may profit from reading it thoroughly.

Third, at the end of each chapter are both theoretical and computational exercises. Engineers and applied scientists will probably concentrate on the computational exercises. Mathematicians should work a variety of both theoretical and computational problems. Numerical analysis without computation is a sterile enterprise.

The book's format allows instructors to use it in either of two modes. For a "methods" course, one can cover a significant set of topics in a single semester by covering the motivation, construction, and practical considerations. At the University of Wyoming, we teach such a course for graduate engineers and geophysicists. For an "analysis" course, one can construct a two- or three-semester sequence that involves proofs, computer exercises, and projects requiring written papers. At Wyoming, we offer a two-semester course along these lines for students in applied mathematics.

Most instructors will want to skip topics. The following remarks may help avoid infelicitous gaps:

- We typically start our courses with Chapter 1. Sections 1.2 and 1.3 (on polynomial interpolation) and 1.7 (on least squares) seem essential.
- Even if one has an aversion to direct methods for linear systems, it is worthwhile to discuss Sections 2.1 and 2.2. Also, the introduction to matrix norms and condition numbers in Sections 2.5 and 2.6 is central to much of numerical analysis.
- While Sections 3.1 through 3.4 contain the traditional core material on nonlinear equations, our experience suggests that engineering students profit from *some* coverage of the multidimensional methods discussed in Sections 3.6 and 3.7.

- Even in a proof-oriented course, one might reasonably leave some of the theory in Sections 4.3 and 4.4 for independent reading. Section 4.5, The Conjugate-Gradient Method, is independent of earlier sections in that chapter.
- Taste permitting, one can skip Chapter 5, Eigenvalue Problems, completely.
- One should cover Section 6.1 and at least some of Section 6.2, Newton-Cotes Formulas, in preparation for Chapter 7. Engineers use Gauss quadrature so often, and the basic theory is so elegant, that we seldom skip Section 6.4.
- We rarely cover Chapter 7 (on ODEs) completely. Still, in preparation for Chapter 8, one should cover at least the most basic material—through Euler methods—from Sections 7.1 and 7.2.
- While many first courses in numerics omit the treatment of PDEs, at least some coverage of Chapter 8 seems critical for virtually all of the students who take our courses.
- Chapter 9, on finite-element methods, emphasizes analysis at the expense of coding, since the latter seems to lie at the heart of most semester-length engineering courses on the subject. It is hard to get this far in a one-semester "methods" course.

We owe tremendous gratitude to many people, including former teachers and many remarkable colleagues too numerous to list. We thank the students and colleagues who graciously endured our drafts and uncovered an embarrassing number of errors. Especially helpful were the efforts of Marian Anghel, Damian Betebenner, Bryan Bornholdt, Derek Mitchum, Patrick O'Leary, Eun-Jae Park, Gamini Wickramage, and the amazingly keen-eyed Li Wu. (Errors undoubtedly remain; they are *our* fault.) The first author wishes to thank the College of Engineering and Mathematics at the University of Vermont, at which he wrote early drafts during a sabbatical year. Finally, we thank our wives, Adele Aldrich and Lynne Ipiña, to whom we dedicate the book. Their patience greatly exceeds that required to watch a book being written.

NUMERICAL ANALYSIS FOR APPLIED SCIENCE

Chapter 0

Some Useful Tools

0.1 Introduction

One aim of this book is to make a significant body of mathematics accessible to people in various disciplines, including engineering, geophysics, computer science, the physical sciences, and applied mathematics. People who have had substantial mathematical training enjoy a head start in this enterprise, since they are more likely to be familiar with ideas that, too often, receive little emphasis outside departments of mathematics. The purpose of this preliminary chapter is to "level the playing field" by reviewing mathematical notations and concepts used throughout the book.

We begin with some notation. A **set** is a collection of **elements**. If x is an element of the set S, we write $x \in S$ and say that x **belongs to** S. If every element of a set R also belongs to the set S, we say that R is a **subset** of S and write $R \subset S$. There are several ways to specify the elements of a set. One way is simply to list them:

$$R = \{2, 4, 6\}, \qquad S = \{2, 4, 6, 8, 10, \ldots\}.$$

Another is to give a rule for selecting elements from a previously defined set. For example,

$$R = \{x \in S : x \leq 6\}$$

denotes the set of all elements of S that are less than or equal to 6. If the statement $x \in S$ fails for all x, then S is the **empty set**, denoted as $\emptyset$.

The notation $x = y$ should be familiar enough, but two related notions are worth mentioning. By $x \leftarrow y$, we mean "assign the value held by the variable y to the variable x." Distinguishing between $x = y$ and $x \leftarrow y$ can seem pedantic until one recalls such barbarisms as "$k = k + 1$" that occur in Fortran and several other programming languages. Also, we use $x := y$ to indicate that x is defined to have the value y.

If R and S are sets, then $R \cup S$ is their **union**, which is the set containing all elements of R and all elements of S. The **intersection** $R \cap S$ is the set of all elements that belong to *both* R and S. If S_i is a set for each i belonging to some index set I, then

$$\bigcup_{i \in I} S_i, \qquad \bigcap_{i \in I} S_i$$

denote, respectively, the set containing all elements that belong to at least one of the sets S_i and the set containing just those elements that belong to every S_i. The **difference** $R \backslash S$ is the set of all elements of R that do not belong to S. If $S_1, S_2, \ldots, S_n$ are sets, then their **Cartesian product** $S_1 \times S_2 \times \cdots \times S_n$ is the set of all **ordered n-tuples** $(x_1, x_2, \ldots, x_n)$, where each $x_i \in S_i$. Two such n-tuples $(x_1, x_2, \ldots, x_n)$ and $(y_1, y_2, \ldots, y_n)$ are equal precisely when $x_1 = y_1,\ x_2 = y_2,\ \ldots,\ x_n = y_n$.

Among the most commonly occurring sets in this book are $\mathbb{R}$, the set of all real numbers; $\mathbb{C}$, the set of all complex numbers $x + iy$, where $x, y \in \mathbb{R}$ and $i^2 = -1$, and

$$\mathbb{R}^n := \underbrace{\mathbb{R} \times \mathbb{R} \times \cdots \times \mathbb{R}}_{n \text{ times}},$$

the set of all n-tuples $\mathbf{x} = (x_1, x_2, \ldots, x_n)$ of real numbers. We often write these n-tuples as **column vectors**:

$$\mathbf{x} = \begin{bmatrix} x_1 \\ x_2 \\ \vdots \\ x_n \end{bmatrix}.$$

$\mathbb{R}$ itself has several important types of subsets, including **open intervals**,

$$(a, b) := \left\{ x \in \mathbb{R} : a < x < b \right\};$$

closed intervals,

$$[a, b] := \left\{ x \in \mathbb{R} : a \leq x \leq b \right\};$$

and the **half-open** intervals

$$[a, b) := \left\{ x \in \mathbb{R} : a \leq x < b \right\}, \qquad (a, b] := \left\{ x \in \mathbb{R} : a < x \leq b \right\}.$$

To extend this notation, we sometimes use the symbol ∞ in a slightly abusive fashion:

$$\begin{aligned} (a, \infty) &:= \left\{ x \in \mathbb{R} : a < x \right\}, \\ (-\infty, b] &:= \left\{ x \in \mathbb{R} : x \leq b \right\}, \\ (-\infty, \infty) &:= \mathbb{R}, \end{aligned}$$

and so forth.

In specifying functions, we write $f: R \to S$. This graceful notation indicates that $f(x)$ is defined for every element x belonging to R, the **domain** of f, and that each such value $f(x)$ belongs to the set S, called the **codomain** of f. The codomain of f contains as a subset the set $f(R)$ of all images $f(x)$ of points x belonging to the domain R. We call $f(R)$ the **range** of f.

The notation $f: x \mapsto y$ indicates that $f(x) = y$, the domain and codomain of f being understood from context. Sometimes we write $x \mapsto y$ when the function itself, as well as its domain and codomain, are understood from context.

Throughout this book we assume that readers are familiar with the basics of calculus and linear algebra. However, it may be useful to review a few notions from these subjects. We devote the rest of this chapter to a summary of facts about bounded sets and normed vector spaces and some frequently used results from calculus.

0.2 Bounded Sets

In numerical analysis, sets of real numbers arise in many contexts. Examples include sequences of approximate values for some quantity, ranges of values for the errors in such approximations, and so forth. It is often important to estimate where these sets lie on the real number line — for example, to guarantee that the possible values for a numerical error lie in a small region around the origin. We say that a set $S \subset \mathbb{R}$ is **bounded above** if there exists a number $B \in \mathbb{R}$ such that $x \leq B$ for every $x \in S$. In this case, B is an **upper bound** for S. Similarly, S is **bounded below** if, for some $b \in \mathbb{R}$, $b \leq x$ for every $x \in S$. In this case, b is a **lower bound** for S. A **bounded** set is one that is bounded both above and below. Note that a set S is bounded if and only if there exists a number $M \in \mathbb{R}$ such that $|x| \leq M$ for every $x \in S$.

By extension, if $f: S \to \mathbb{R}$ is a function whose range $f(S)$ is bounded above, bounded below, or bounded, then we say that f is bounded above, bounded below, or bounded, respectively.

Most upper and lower bounds give imprecise information. For example, 17 is an upper bound for the set $S = (0, 2)$, but, as Figure 1 illustrates, the upper bound 2 tells us more about S. We call B_0 a **least upper bound** or **supremum** for $S \subset \mathbb{R}$ if B_0 is an upper bound for S and $B_0 \leq B$ whenever B is an upper bound for S. In this case, we write $B_0 = \sup S$. Similar reasoning applies to lower bounds: -109 is a lower bound for $(0, 2)$, but so is the more informative number 0. We call b_0 a **greatest lower bound** or **infimum** for $S \subset \mathbb{R}$ if b_0 is a lower bound for S and $b_0 \geq b$ whenever b is also a lower bound for S. We write $b_0 = \inf S$. The notations inf and sup have obvious extensions. For example, if $S_2 := \{(x, y) \in \mathbb{R}^2 : x^2 + y^2 = 1\}$ denotes the unit circle in $\mathbb{R}^2$ and $f: S_2 \to \mathbb{R}$ is a real-valued function defined on S_2, then

$$\sup_{S_2} f := \sup_{(x,y)\in S_2} f(x,y) := \sup\Big\{ f(x,y) \in \mathbb{R} \,:\, x^2 + y^2 = 1 \Big\}. \tag{0.2-1}$$

Shortly we discuss conditions under which this quantity exists.

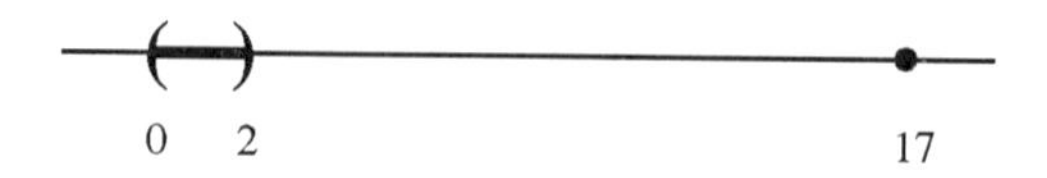

FIGURE 1. *The set $(0,2) \subset \mathbb{R}$ and two of its upper bounds.*

Not every set has a supremum or an infimum. For example, the set

$$\Big\{ \ldots, -2, -1, 0, 1, 2, \ldots \Big\}$$

of all **integers** has neither a supremum nor an infimum. The set

$$\Big\{ 1, 2, 3, \ldots \Big\}$$

of **natural numbers** has 1 as its infimum but has no supremum. One should take care to distinguish between $\sup S$ and $\inf S$ and the notions of maximum and minimum. By a **maximum** of a set $S \subset \mathbb{R}$ we mean an element $M \in S$ for which $x \leq M$ whenever $x \in S$, and we write $M = \max S$. Thus $\sup(0,2) = 2 = \sup[0,2] = \max[0,2]$, but $\max(0,2)$ does not exist. Similarly, an element $m \in S$ is a **minimum** of S if $m \leq x$ for every $x \in S$. Thus $\inf(0,2) = 0 = \inf[0,2] = \min[0,2]$, while $\min(0,2)$ does not exist. These examples illustrate the fact that sup and inf are more general notions than max and min: $\sup S = \max S$ when $\sup S \in S$, but $\sup S$ may exist even when $\max S$ does not. A corresponding statement holds for inf and min.

The following principle, which one can take as a defining characteristic of $\mathbb{R}$, confirms the fundamental importance of sup and inf:

LEAST-UPPER-BOUND PRINCIPLE. *If a nonempty subset of $\mathbb{R}$ is bounded above, then it has a least upper bound.*

R.P. Boas ([1], Section 2) gives an accessible introduction to this principle. Similarly, every nonempty subset of $\mathbb{R}$ that is bounded below has a greatest lower bound. For example,

$$\inf\Big\{ \tfrac{1}{2}, \tfrac{1}{3}, \tfrac{1}{4}, \ldots \Big\} = 0, \quad \sup(-\infty, 0) = 0, \quad \sup\Big\{ 2, 4, 6 \Big\} = 6.$$

The set $\{2, 4, 6, 8, 10, \ldots\}$, however, is not bounded above, and it has no least upper bound. The least-upper-bound principle ensures that $\sup_{S_2} f$, defined in Equation (0.2-1), exists whenever the set of real numbers

$$\Big\{ f(x,y) \in \mathbb{R} \,:\, (x,y) \in S_2 \Big\}$$

is bounded above. However, without knowing more about f, we cannot guarantee the existence of a point $(x, y) \in S_2$ where f attains the value $\sup_{S_2} f$.

Let us turn to the multidimensional sets $\mathbb{R}^n$. Which subsets of $\mathbb{R}^n$ are bounded? Here we generally have no linear order analogous to the relation $\leq$ on which to base a definition of boundedness. Instead, we rely on the idea of distance, which is familiar from geometry:

DEFINITION. *The* **Euclidean length** *of* $\mathbf{x} = (x_1, x_2, \ldots, x_n) \in \mathbb{R}^n$ *is*

$$||\mathbf{x}||_2 := \sqrt{x_1^2 + x_2^2 + \cdots + x_n^2}.$$

The **Euclidean distance** *between two points* $\mathbf{x}, \mathbf{y} \in \mathbb{R}^n$ *is the Euclidean length of their difference,* $||\mathbf{y} - \mathbf{x}||_2$.

Given a point $\mathbf{x} \in \mathbb{R}^n$ and a positive real number r, we call the set of all points in $\mathbb{R}^n$ whose Euclidean distance from $\mathbf{x}$ is less than r the **ball** of radius r about $\mathbf{x}$. We denote this set as $\mathcal{B}_r(\mathbf{x})$. Figure 2 depicts such a set in $\mathbb{R}^2$. A set $S \subset \mathbb{R}^n$ is **bounded** if it is a subset of some ball having finite radius in $\mathbb{R}^n$. Observe that, if $x \in \mathbb{R} = \mathbb{R}^1$, then $\mathcal{B}_r(x) = (x - r, x + r)$. One easily checks that a subset of $\mathbb{R}$ is bounded in this sense if and only if it is bounded above and below.

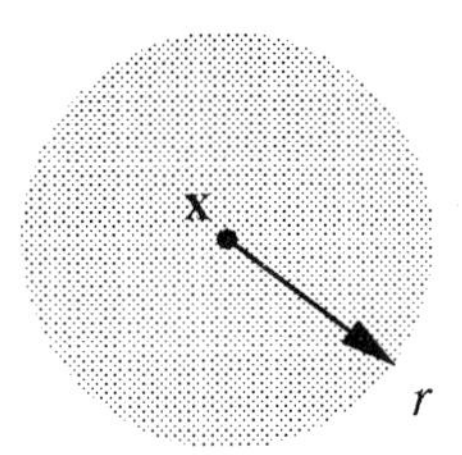

FIGURE 2. *The ball* $\mathcal{B}_r(\mathbf{x})$ *of radius* r *about the point* $\mathbf{x} \in \mathbb{R}^2$.

Other structural aspects of $\mathbb{R}^n$ also prove useful. Let $S \subset \mathbb{R}^n$. A point $\mathbf{x} \in S$ is an **interior point** of S if there is *some* ball $\mathcal{B}_r(\mathbf{x})$, possibly having very small radius r, such that $\mathcal{B}_r(\mathbf{x}) \subset S$. In Figure 3, the point $\mathbf{a}$ is an interior point of S, but $\mathbf{b}$ and $\mathbf{c}$ are not. A point $\mathbf{x} \in \mathbb{R}^n$ (not necessarily belonging to S) is a **limit point** of S if *every* ball $\mathcal{B}_r(\mathbf{x})$, no matter what its radius, contains at least one element of S distinct from $\mathbf{x}$. In Figure 4, $\mathbf{a}$ and $\mathbf{b}$ are limit points of S, but $\mathbf{c}$ is not. If every element of S is an interior point, then we call S an **open** set. If S contains all of its limit points, then we say that S is a **closed** set. The definitions are by no means mutually exclusive: $\mathbb{R}^n$ itself is both open and closed.

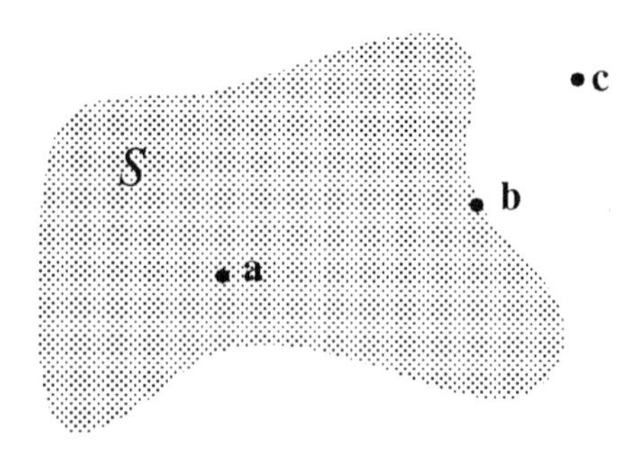

FIGURE 3. *A set $S \subset \mathbb{R}^2$, showing an interior point* **a** *and two points* **b**, **c** *that are not interior points.*

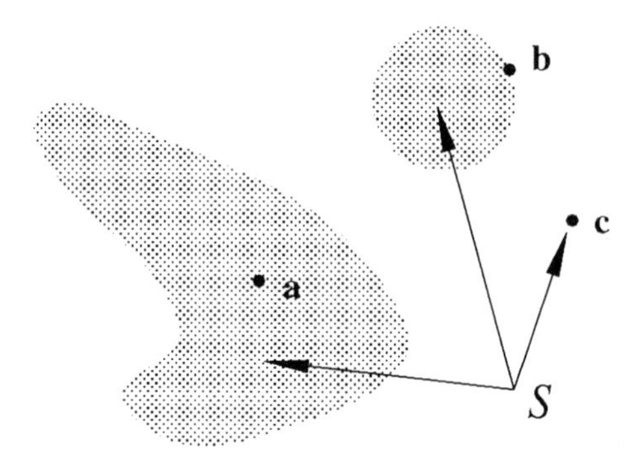

FIGURE 4. *A set $S \subset \mathbb{R}^2$, along with two limit points* **a** *and* **b** *and a point* **c** *that is not a limit point of S.*

Finally, a subset of $\mathbb{R}^n$ that is both closed and bounded is **compact**.[1] Thus the following subsets of $\mathbb{R}^2$ are compact:

$$[0,1] \times [0,1], \quad \Big\{(0,0),(0,\pi),(1,-\pi)\Big\}, \quad S_2 = \Big\{\mathbf{x} \in \mathbb{R}^2 : ||\mathbf{x}||_2 = 1\Big\},$$

while the sets

$$(0,1) \times (0,1), \quad \mathcal{B}_1(\mathbf{0}), \quad \Big\{(0,0),(1,1),(2,2),\ldots\Big\}$$

[1]This characterization of compactness is not the most general one, but it suffices for $\mathbb{R}^n$. The customary approach begins with a more technical and less intuitive definition of compactness, proving as a theorem the assertion that a subset of $\mathbb{R}^n$ is compact if and only if it is closed and bounded. Boas ([1], Section 7) supplies details.

are not. Compact sets in $\mathbb{R}^n$ have several interesting properties, one of which is especially useful in numerical analysis:

THEOREM 0.1. *If $S \subset \mathbb{R}^n$ is nonempty and compact and $f: S \to \mathbb{R}$ is a continuous function, then there are points $\mathbf{a}, \mathbf{b} \in S$ for which $f(\mathbf{a})$ and $f(\mathbf{b})$ are the minimum and maximum, respectively, of the set $f(S)$.*

For a proof, see Boas ([1], Section 7).

This theorem partially settles an issue raised earlier: If f is a continuous, real-valued function defined on the unit circle S_2, then there is at least one point $(x, y) \in S_2$ where f takes the value $\sup_{S_2} f$ defined in Equation (0.2-1). By considering the function $-f$, one can also show that f takes the value $\inf_{S_2} f$ at some point in S_2. Both of these statements hold just as well in $\mathbb{R}^n$, where $S_2 := \{\mathbf{x} \in \mathbb{R}^n : \|\mathbf{x}\|_2 = 1\}$. We use this generalization in the next section.

0.3 Normed Vector Spaces

Vector Spaces

In numerical analysis, vector spaces are ubiquitous:

DEFINITION. *A set $\mathcal{V}$ is a* **vector space** *over $\mathbb{R}$ if there are two operations,* **addition** *(+) and* **scalar multiplication**, *that obey the following rules for any $x, y, z \in \mathcal{V}$ and $a, b \in \mathbb{R}$:*

(i) *$x + y \in \mathcal{V}$ and $ax \in \mathcal{V}$; in other words, $\mathcal{V}$ is* **closed** *under addition and scalar multiplication.*

(ii) $x + y = y + x$.

(iii) $x + (y + z) = (x + y) + z$.

(iv) *There is a unique vector $0 \in \mathcal{V}$ such that $x + 0 = x$ for all $x \in \mathcal{V}$.*

(v) *For any $x \in \mathcal{V}$, there is a unique vector $-x \in \mathcal{V}$ such that $-x + x = 0$.*

(vi) $1x = x$.

(vii) $a(bx) = (ab)x$.

(viii) $a(x + y) = ax + ay$.

(ix) $(a + b)x = ax + bx$.

In this case we sometimes refer to $\mathbb{R}$ as the field of **scalars**. *The elements of $\mathcal{V}$ are* **vectors**. *A set $\mathcal{U}$ is a* **subspace** *of $\mathcal{V}$ if every element of $\mathcal{U}$ belongs to $\mathcal{V}$ and $\mathcal{U}$ is a vector space under the operations that it inherits from $\mathcal{V}$. Analogous definitions hold for vector spaces over $\mathbb{C}$.*

We denote the scalar multiple ax by juxtaposing the scalar a and the vector x. In most cases of interest in this book, the algebraic properties of addition and scalar multiplication are obvious from the definitions of the two operations, and the main issue is whether $\mathcal{V}$ is closed under these two operations.

Among the common examples of vector spaces are the finite-dimensional **Euclidean spaces** $\mathbb{R}^n$, with their familiar rules of addition and scalar multiplication:

$$\mathbf{x}+\mathbf{y} = \begin{bmatrix} x_1 \\ \vdots \\ x_n \end{bmatrix} + \begin{bmatrix} y_1 \\ \vdots \\ y_n \end{bmatrix} := \begin{bmatrix} x_1 + y_1 \\ \vdots \\ x_n + y_n \end{bmatrix};$$

$$a\mathbf{x} = a \begin{bmatrix} x_1 \\ \vdots \\ x_n \end{bmatrix} := \begin{bmatrix} ax_1 \\ \vdots \\ ax_n \end{bmatrix}.$$

In this vector space, the zero vector is $\mathbf{0}$, the array that has 0 as each of its n entries. The real line $\mathbb{R}$ is the simplest Euclidean space.

Various sets of functions constitute another important class of vector spaces. For example, if $S \subset \mathbb{R}$, then $C^k(S)$ signifies the vector space of all functions $f: S \to \mathbb{R}$ for which f and its derivatives $f', f'', \ldots, f^{(k)}$ through order k are continuous on the set S. By extension of this notation, $C^\infty(S)$ denotes the vector space of functions that have continuous derivatives of all orders on S. On all of these spaces we define addition and scalar multiplication pointwise:

$$(f+g)(x) := f(x) + g(x); \qquad (af)(x) := a\, f(x).$$

Here, the vector 0 is the function that assigns the number 0 to all arguments x. A slightly more general function space is $L^2(S)$. Although the rigorous definition of this space involves some technicalities, for our purposes it suffices to think of $L^2(S)$ as the set of all functions $f: S \to \mathbb{R}$ for which $\int_S f^2(x)\, dx$ exists and is finite. Readers curious about the technicalities may consult Rudin ([2], Chapter 11).

A third class of vector spaces consists of the sets $\mathbb{R}^{m\times n}$ of real $m \times n$ matrices. Our notational convention is to a use sans-serif capital letter, such as A, to signify the matrix whose entry in row i, column j is the number denoted by the corresponding lowercase symbol $a_{i,j}$. If C and D are two such matrices, then

$$\begin{aligned} \mathsf{C}+\mathsf{D} &= \begin{bmatrix} c_{1,1} & \cdots & c_{1,n} \\ \vdots & & \vdots \\ c_{m,1} & \cdots & c_{m,n} \end{bmatrix} + \begin{bmatrix} d_{1,1} & \cdots & d_{1,n} \\ \vdots & & \vdots \\ d_{m,1} & \cdots & d_{m,n} \end{bmatrix} \\ &:= \begin{bmatrix} c_{1,1}+d_{1,1} & \cdots & c_{1,n}+d_{1,n} \\ \vdots & & \vdots \\ c_{m,1}+d_{m,1} & \cdots & c_{m,n}+d_{m,n} \end{bmatrix}, \\ a\mathsf{C} &:= \begin{bmatrix} ac_{1,1} & \cdots & ac_{1,n} \\ \vdots & & \vdots \\ ac_{m,1} & \cdots & ac_{m,n} \end{bmatrix}, \end{aligned}$$

The additive identity in $\mathbb{R}^{m\times n}$ is the $m \times n$ matrix $\mathsf{0}$ all of whose entries are 0.

Finally, the set $\{\mathbf{0}\}$ is trivially a vector space.

One can use addition and scalar multiplication to construct subspaces.

DEFINITION. *If $\mathcal{V}$ is a real vector space, a* **linear combination** *of the vectors $x_1, x_2, \ldots, x_n \in \mathcal{V}$ is a vector of the form $c_1x_1 + c_2x_2 + \cdots + c_nx_n$, where $c_1, c_2, \ldots, c_n \in \mathbb{R}$. If $S \subset \mathcal{V}$, the* **span** *of S, denoted* $\operatorname{span} S$, *is the set of all linear combinations of vectors belonging to S. If $\mathcal{U} = \operatorname{span} S$, then S* **spans** *$\mathcal{U}$.*

It is easy to show that $\operatorname{span} S$ is a subspace of $\mathcal{V}$ whenever $S \subset \mathcal{V}$.

DEFINITION. *If $\mathcal{V}$ is a vector space, then a set $S \subset \mathcal{V}$ is* **linearly independent** *if no vector $x \in S$ belongs to* $\operatorname{span}(S\backslash\{x\})$, *that is, no vector in S is a linear combination of the other vectors in S. Otherwise, S is* **linearly dependent**.

One can regard a linearly independent set as containing minimal information needed to determine its span:

DEFINITION. *A subset S of a vector space $\mathcal{V}$ is a* **basis** *for $\mathcal{V}$ if S is linearly independent and* $\operatorname{span} S = \mathcal{V}$.

It is a basic theorem of linear algebra that, whenever two finite sets S_1 and S_2 are bases for a vector space $\mathcal{V}$, S_1 and S_2 have the same number of elements (see [3], Section 2.3) We call this number the **dimension** of $\mathcal{V}$. For example,

$\mathbb{R}^n$ has the **standard basis** $\{\mathbf{e}_1, \mathbf{e}_2, \ldots, \mathbf{e}_n\}$, where

$$\mathbf{e}_1 := \begin{bmatrix} 1 \\ 0 \\ \vdots \\ 0 \end{bmatrix}, \quad \mathbf{e}_2 := \begin{bmatrix} 0 \\ 1 \\ \vdots \\ 0 \end{bmatrix}, \quad \ldots, \quad \mathbf{e}_n := \begin{bmatrix} 0 \\ 0 \\ \vdots \\ 1 \end{bmatrix}.$$

If $\mathcal{V}$ has a basis containing finitely many vectors, then we say that $\mathcal{V}$ is **finite-dimensional**. If not, then $\mathcal{V}$ is **infinite-dimensional**.

Matrices as Linear Operators

Given matrices $\mathsf{A} \in \mathbb{R}^{m \times n}$ and $\mathsf{B} \in \mathbb{R}^{n \times p}$, one can compute their **matrix product**

$$\begin{aligned} \mathsf{AB} &= \begin{bmatrix} a_{1,1} & \cdots & a_{1,n} \\ \vdots & & \vdots \\ a_{m,1} & \cdots & a_{m,n} \end{bmatrix} \begin{bmatrix} b_{1,1} & \cdots & b_{1,p} \\ \vdots & & \vdots \\ b_{n,1} & \cdots & b_{n,p} \end{bmatrix} \\ &= \begin{bmatrix} c_{1,1} & \cdots & c_{1,p} \\ \vdots & & \vdots \\ c_{m,1} & \cdots & c_{m,p} \end{bmatrix}, \end{aligned}$$

where

$$c_{i,j} = \sum_{k=1}^{n} a_{i,k} b_{k,j}.$$

If we identify vectors in $\mathbb{R}^n$ with matrices in $\mathbb{R}^{n \times 1}$, then the product of an $m \times n$ real matrix with a vector in $\mathbb{R}^n$ is a vector in $\mathbb{R}^m$:

$$\begin{bmatrix} a_{1,1} & \cdots & a_{1,n} \\ \vdots & & \vdots \\ a_{m,1} & \cdots & a_{m,n} \end{bmatrix} \begin{bmatrix} x_1 \\ \vdots \\ x_n \end{bmatrix} = \begin{bmatrix} b_1 \\ \vdots \\ b_m \end{bmatrix},$$

where $b_i = a_{i,1}x_1 + \cdots + a_{i,n}x_n$. In this way, any $m \times n$ real matrix acts as a mapping $\mathsf{A}: \mathbb{R}^n \to \mathbb{R}^m$. It is easy to check that this mapping is a **linear operator** or **linear transformation**, that is, that it satisfies the following properties: For any $\mathbf{x}, \mathbf{y} \in \mathbb{R}^n$ and any $c \in \mathbb{R}$,

(i) $\mathsf{A}(\mathbf{x} + \mathbf{y}) = \mathsf{A}\mathbf{x} + \mathsf{A}\mathbf{y}$.

(ii) $\mathsf{A}(c\mathbf{x}) = c(\mathsf{A}\mathbf{x})$.

In this context, the **identity matrix** in $\mathbb{R}^{n\times n}$ plays a special role. This matrix has the form

$$\mathsf{I} = \begin{bmatrix} 1 & 0 & \cdots & 0 \\ 0 & 1 & \cdots & 0 \\ \vdots & \vdots & \ddots & \vdots \\ 0 & 0 & \cdots & 1 \end{bmatrix}.$$

It is easy to verify that $\mathsf{IA} = \mathsf{A}$ for every matrix $\mathsf{A} \in \mathbb{R}^{n\times m}$ and that $\mathsf{AI} = \mathsf{A}$ for every matrix $\mathsf{A} \in \mathbb{R}^{m\times n}$.

Frequently we are given a matrix $\mathsf{A} \in \mathbb{R}^{n\times n}$ and a vector $\mathbf{b} \in \mathbb{R}^n$ and would like to find a vector $\mathbf{x} \in \mathbb{R}^n$ such that $\mathsf{A}\mathbf{x} = \mathbf{b}$.

DEFINITION. *The matrix* $\mathsf{A} \in \mathbb{R}^{n\times n}$ *is* **nonsingular** *if, for any* $\mathbf{b} \in \mathbb{R}^n$, *there exists a unique vector* $\mathbf{x} \in \mathbb{R}^n$ *such that* $\mathsf{A}\mathbf{x} = \mathbf{b}$. *Otherwise,* A *is* **singular**.

If A is singular, then the equation $\mathsf{A}\mathbf{x} = \mathbf{b}$ may have no solutions $\mathbf{x}$, or solutions may exist but not be unique. There are several equivalent characterizations of these notions. In the next theorem, $\det \mathsf{A}$ denotes the determinant of the matrix $\mathsf{A} \in \mathbb{R}^{n\times n}$. Strang ([3], Chapter 4) reviews the definition of this quantity.

THEOREM 0.2. *If* $\mathsf{A} \in \mathbb{R}^{n\times n}$, *then the following statements are equivalent:*

(*i*) A *is nonsingular.*

(*ii*) $\det \mathsf{A} \neq 0$.

(*iii*) *If* $\mathsf{A}\mathbf{x} = \mathbf{0}$, *then* $\mathbf{x} = \mathbf{0}$.

(*iv*) *There is a unique matrix* $\mathsf{A}^{-1} \in \mathbb{R}^{n\times n}$ *such that* $\mathsf{AA}^{-1} = \mathsf{A}^{-1}\mathsf{A} = \mathsf{I}$.

The matrix A^{-1} in part (*iv*) is the **inverse** of A, and its existence means that A is **invertible**. For proof of the theorem, see Strang ([3], Chapter 2).

Suppose that $\mathsf{A} \in \mathbb{R}^{m\times n}$, and denote its (i,j)th entry by $a_{i,j}$. The **transpose** of A, denoted $\mathsf{A}^\top$, is the matrix in $\mathbb{R}^{n\times m}$ whose entry in the (i,j)th position is $a_{j,i}$. A matrix A is **symmetric** when $\mathsf{A}^\top = \mathsf{A}$. This equation guarantees that A is square and that $a_{i,j} = a_{j,i}$. The transpose of a column vector $\mathbf{v} \in \mathbb{R}^m$ is a row vector,

$$\mathbf{v}^\top = (v_1, v_2, \cdots, v_m),$$

which we also say is in $\mathbb{R}^m$. One easily shows that $(\mathsf{AB})^\top = \mathsf{B}^\top\mathsf{A}^\top$.

Norms

In analyzing errors associated with numerical approximations, we often estimate the "lengths" of vectors or the "distances" between pairs of vectors. The following concept captures the notion of length in settings even more general than $\mathbb{R}^n$:

DEFINITION. *A* **norm** *on a vector space* $\mathcal{V}$ *is a function* $\|\cdot\|: \mathcal{V} \to \mathbb{R}$ *that satisfies the following conditions for any* $x, y \in \mathcal{V}$ *and* $a \in \mathbb{R}$*:*

(i) $\|x\| \geq 0$, *and* $\|x\| = 0$ *if and only if* $x = 0$.

(ii) $\|ax\| = |a|\,\|x\|$.

(iii) $\|x + y\| \leq \|x\| + \|y\|$.

If such a function exists, then $\mathcal{V}$ *is a* **normed vector space**.

The third condition is the **triangle inequality**. From it there follows an alternative version:

PROPOSITION 0.3. *If* $\|\cdot\|$ *is a norm on a vector space* $\mathcal{V}$*, then, for any* $x, y \in \mathcal{V}$,

$$\big|\,\|x\| - \|y\|\,\big| \leq \|x - y\|. \tag{0.3-1}$$

PROOF: By the triangle inequality,

$$\|x\| = \|(x - y) + y\| \leq \|x - y\| + \|y\|,$$

so $\|x\| - \|y\| \leq \|x - y\|$. Interchanging x and y gives $\|y\| - \|x\| \leq \|y - x\| = \|x - y\|$, and the two results together imply the inequality (0.3-1). ∎

The prototypical norm is the absolute value function $|\cdot|: \mathbb{R} \to \mathbb{R}$. This familiar function has many extensions to $\mathbb{R}^n$, three of which are defined for $\mathbf{x} = (x_1, x_2, \ldots, x_n) \in \mathbb{R}^n$ as follows:

$$\begin{aligned}
\|\mathbf{x}\|_1 &:= |x_1| + |x_2| + \cdots + |x_n|, \\
\|\mathbf{x}\|_2 &:= \sqrt{x_1^2 + x_2^2 + \cdots + x_n^2}, \\
\|\mathbf{x}\|_\infty &:= \max_{1 \leq i \leq n} |x_i|.
\end{aligned}$$

By using properties of $|\cdot|$, one easily verifies that $\|\cdot\|_1$ and $\|\cdot\|_\infty$ satisfy the conditions to be norms. The function $\|\cdot\|_2$ is just the Euclidean length introduced earlier, and for this function the first two properties of norms follow from corresponding facts for $|\cdot|$. We review below an argument establishing the triangle inequality for $||\cdot||_2$.

Analogous norms exist for function spaces. Consider $C^k([a,b])$, the vector space of all real-valued functions defined on the bounded, closed interval $[a,b] \subset \mathbb{R}$ whose derivatives through order k are continuous. For $f \in C^k([a,b])$,

$$\begin{aligned} ||f||_1 &:= \int_a^b |f(x)|\,dx, \\ ||f||_2 &:= \left[\int_a^b f^2(x)\,dx\right]^{1/2}, \\ ||f||_\infty &:= \sup_{x\in[a,b]} |f(x)|. \end{aligned}$$

It is relatively straightforward to show that $||\cdot||_1$ and $||\cdot||_\infty$ satisfy the properties required to be a norm. For $||\cdot||_2$, proving the triangle inequality requires slightly more work, which we undertake shortly.

It is also possible to construct norms for the spaces $\mathbb{R}^{m\times n}$. We explore this idea in Chapter 3.

An interpretation in terms of length is natural for the norm $||\cdot||_2$ on $\mathbb{R}^n$, which is just the Euclidean length function. For the norms $||\cdot||_1$ and $||\cdot||_\infty$ on $\mathbb{R}^n$ the interpretation may be slightly less familiar. Figure 1 illustrates the **unit spheres**

$$\begin{aligned} S_1 &= \left\{\mathbf{x} \in \mathbb{R}^2 : ||\mathbf{x}||_1 = 1\right\}, \\ S_2 &= \left\{\mathbf{x} \in \mathbb{R}^2 : ||\mathbf{x}||_2 = 1\right\}, \qquad (0.3\text{-}2) \\ S_\infty &= \left\{\mathbf{x} \in \mathbb{R}^2 : ||\mathbf{x}||_\infty = 1\right\}, \end{aligned}$$

in $\mathbb{R}^2$. Each unit sphere consists of all those vectors whose length, measured in the appropriate norm, is 1.

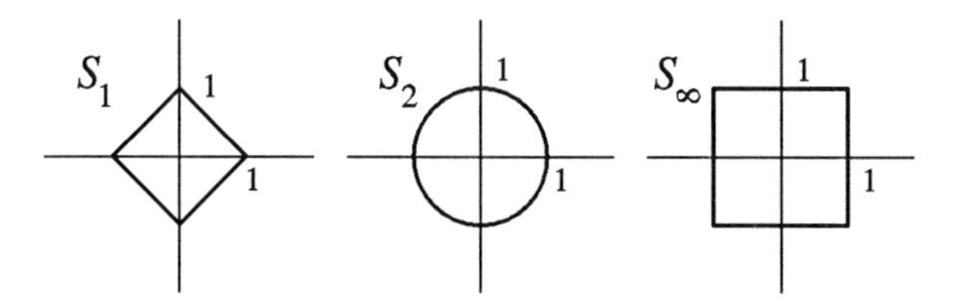

FIGURE 1. *The unit spheres S_1, S_2, and S_∞ in $\mathbb{R}^2$.*

In the function spaces $C^k([a,b])$, a norm typically assigns to a given function f some quantity whose interpretation as a length is more abstract. For example, $||f||_1$ is the average value of $|f|$ over $[a, b]$, multiplied by the length $|b-a|$ of the interval. Similarly, $||f||_2$ essentially gives the "root-mean-square" average of f over $[a, b]$, again multiplied by $|b - a|$. Finally, $||f||_\infty$ measures the largest excursion that f takes from the x-axis, as Figure 2 illustrates.

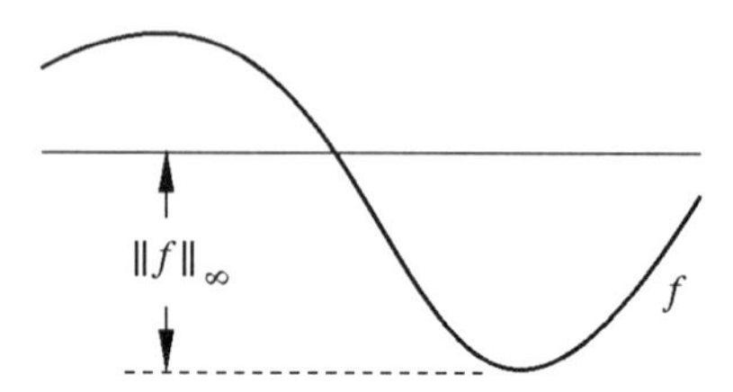

FIGURE 2. *Geometric interpretation of* $||f||_\infty$ *as a measure of the largest excursion that* f *takes from the* x*-axis.*

Viewing the length of a vector as its distance from 0 leads to another geometric idea: The distance between two vectors x and y in a normed vector space is the norm of their difference, $||y - x||$. Figure 3 illustrates this idea for two vectors using the Euclidean length $|| \cdot ||_2$ in $\mathbb{R}^2$, where the interpretation corresponds to familiar concepts in plane geometry. By abstracting this geometric notion to other norms and to vector spaces other than $\mathbb{R}^n$, we establish a useful means of measuring how close an approximation — whether to an n-tuple of numbers or to a function — lies to an exact answer.

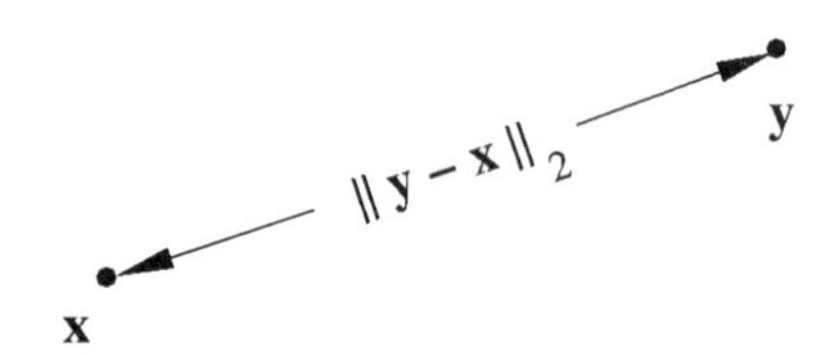

FIGURE 3. *The distance* $||\mathbf{y} - \mathbf{x}||_2$ *between two vectors* $\mathbf{x}, \mathbf{y} \in \mathbb{R}^2$.

Inner Products

In many vector spaces of interest in numerical analysis there is yet another level of geometric structure:

DEFINITION. *If* $\mathcal{V}$ *is a vector space, a function* $\langle \cdot, \cdot \rangle: \mathcal{V} \times \mathcal{V} \to \mathbb{R}$ *is an* **inner product** *on* $\mathcal{V}$ *if, for all* $x, y, z \in \mathcal{V}$,

(i) $\langle x, x \rangle \geq 0$, *and* $\langle x, x \rangle = 0$ *only if* $x = 0$ (**positive definiteness**).

(ii) $\langle x, y \rangle = \langle y, x \rangle$ (**symmetry**).

(iii) $\langle x, ay + bz \rangle = a\langle x, y \rangle + b\langle x, z \rangle$ *for any* $a, b \in \mathbb{R}$ (**linearity**).

If such a function exists, then $\mathcal{V}$ is an **inner-product space**.

The ordinary dot product on $\mathbb{R}^n$ is an inner product: If $\mathbf{x}, \mathbf{y} \in \mathbb{R}^n$, then

$$\langle \mathbf{x}, \mathbf{y} \rangle := \mathbf{x} \cdot \mathbf{y} = \begin{bmatrix} x_1 \\ \vdots \\ x_n \end{bmatrix} \cdot \begin{bmatrix} y_1 \\ \vdots \\ y_n \end{bmatrix} = x_1 y_1 + \cdots + x_n y_n.$$

The notation for matrix transposes allows us to write the dot product $\mathbf{u} \cdot \mathbf{v}$ as $\mathbf{u}^{\top}\mathbf{v}$, using the rules for array multiplication.

Each of the function spaces $C^k([a,b])$ also possesses an inner product, defined for two functions f, g as follows:

$$\langle f, g \rangle := \int_a^b f(x)g(x)\,dx.$$

The extra geometry associated with inner-product spaces stems from the following concept:

DEFINITION. *Two vectors $x, y \in \mathcal{V}$ are* **orthogonal** *if* $\langle x, y \rangle = 0$.

When $\mathcal{V} = \mathbb{R}^n$ and $\langle \cdot, \cdot \rangle$ is the ordinary dot product, this definition of orthogonality coincides with the usual notion of perpendicularity. In the function spaces $C^k([a,b])$ and in most other examples of inner-product spaces, the picture is more abstract, but the geometric analogy remains just as profitable.

Any inner-product space is a normed vector space, the natural norm being defined (and denoted) by analogy with the Euclidean length:

$$||x||_2 := \sqrt{\langle x, x \rangle}.$$

This definition includes the norms $||\cdot||_2$ defined on the vector spaces $\mathbb{R}^n$ and $C^k([a,b])$. To show that $||\cdot||_2$ indeed defines a norm, we must establish the triangle inequality. The argument hinges on the following fact:

THEOREM 0.4. (CAUCHY-SCHWARZ INEQUALITY). *If $\mathcal{V}$ is an inner-product space with inner product $\langle \cdot, \cdot \rangle$, then, for any $x, y \in \mathcal{V}$,*

$$|\langle x, y \rangle| \leq ||x||_2 ||y||_2. \tag{0.3-3}$$

PROOF: If $y = 0$, then both sides of the inequality (0.3-3) vanish, and the theorem is true trivially. Assume that $y \neq 0$. In this case, for any $r \in \mathbb{R}$, positive definiteness of the inner product implies that

$$0 \leq \langle x + ry, x + ry \rangle = ||x||_2^2 + 2r\langle x, y \rangle + r^2 ||y||_2^2.$$

The expression on the right is quadratic in r, and the fact that it is nonnegative implies that the discriminant $4\langle x,y\rangle^2 - 4\|x\|_2^2\|y\|_2^2 \le 0$. The inequality (0.3-3) follows. ∎

To prove the triangle inequality for $\|\cdot\|_2$, one simply observes that

$$(\|x\|_2 + \|y\|_2)^2 = \|x\|_2^2 + 2\|x\|_2\|y\|_2 + \|y\|_2^2.$$

The Cauchy-Schwarz inequality guarantees that the middle term on the right side of this identity is at least as large as $\langle x,y\rangle$, so we have

$$(\|x\|_2 + \|y\|_2)^2 \ge \|x\|_2^2 + 2\langle x,y\rangle + \|y\|_2^2 = \|x+y\|_2^2.$$

Taking square roots completes the argument.

The connection between norms and orthogonality in inner-product spaces allows us to specify a particularly useful type of basis:

DEFINITION. *A basis S for an inner-product space $\mathcal{V}$ is an* **orthonormal basis** *if the following conditions hold:*

(i) *Whenever $x, y \in S$, and $x \ne y$, $\langle x,y\rangle = 0$.*

(ii) *For every $x \in S$, $\|x\|_2 = 1$.*

When $\mathcal{V}$ is a finite-dimensional inner-product space, one can always construct an orthonormal basis from an arbitrary basis for $\mathcal{V}$ using an algorithm known as the Gram-Schmidt procedure. We refer to Strang ([3], Section 3.3) for details.

Norm Equivalence

While one can define infinitely many norms on $\mathbb{R}^n$, the structures that they impose are essentially the same, in a sense defined below. We devote the rest of this section to a discussion of this remarkable fact, which does not hold for normed vector spaces in general. We begin with the following general property of norms.

LEMMA 0.5. *Let $\mathcal{V}$ be a vector space over $\mathbb{R}$.*

(i) *Any norm $\|\cdot\|: \mathcal{V} \to \mathbb{R}$ is uniformly continuous.*

(ii) *In the special case $\mathcal{V} = \mathbb{R}^n$, any norm $\|\cdot\|: \mathbb{R}^n \to \mathbb{R}$ is uniformly continuous with respect to the Euclidean norm $\|\cdot\|_2$.*

PROOF: To prove (i) we must show that, for any $\epsilon > 0$, there exists a number $\delta > 0$ such that, whenever the vectors $\mathbf{x}, \mathbf{y} \in \mathcal{V}$ satisfy $\|\mathbf{x} - \mathbf{y}\| < \delta$, $\big|\|\mathbf{x}\| - \|\mathbf{y}\|\big| < \epsilon$. By the version (0.3-1) of the triangle inequality, we can simply choose $\delta = \epsilon$.

To establish (ii), let $\{\mathbf{e}_1, \mathbf{e}_2, \ldots, \mathbf{e}_n\}$ be the standard basis for $\mathbb{R}^n$, and suppose that $\epsilon > 0$ is given. For $\mathbf{x} = x_1\mathbf{e}_1 + x_2\mathbf{e}_2 + \cdots + x_n\mathbf{e}_n$ and $\mathbf{y} = y_1\mathbf{e}_1 + y_2\mathbf{e}_2 + \cdots + y_n\mathbf{e}_n$, we have

$$
\begin{aligned}
\big|\|\mathbf{x}\| - \|\mathbf{y}\|\big| \leq \|\mathbf{x} - \mathbf{y}\| &= \|(x_1 - y_1)\mathbf{e}_1 + \cdots + (x_n - y_n)\mathbf{e}_n\| \\
&\leq |x_1 - y_1|\,\|\mathbf{e}_1\| + \cdots + |x_n - y_n|\,\|\mathbf{e}_n\| \\
&\leq \left(\sum_{i=1}^{n} |x_i - y_i|^2\right)^{1/2} \underbrace{\left(\sum_{i=1}^{n} \|\mathbf{e}_i\|^2\right)^{1/2}}_{M} \\
&= M\,\|\mathbf{x} - \mathbf{y}\|_2,
\end{aligned}
$$

the number M being independent of $\mathbf{x}$ and $\mathbf{y}$. The third inequality in this chain follows from the Cauchy-Schwarz inequality. Choosing $\delta = \epsilon/M$ guarantees that $\big|\|\mathbf{x}\| - \|\mathbf{y}\|\big| < \epsilon$ whenever $\|\mathbf{x} - \mathbf{y}\|_2 < \delta$. ∎

The crucial question for norm equivalence is whether inequalities derived using one norm $\|\cdot\|_{\mathrm{I}}$ can be converted to analogous inequalities expressed in a different norm $\|\cdot\|_{\mathrm{II}}$.

DEFINITION. *Let $\|\cdot\|_{\mathrm{I}}$ and $\|\cdot\|_{\mathrm{II}}$ be norms on a vector space $\mathcal{V}$. Then $\|\cdot\|_{\mathrm{I}}$ and $\|\cdot\|_{\mathrm{II}}$ are* **equivalent** *if there exist constants $m, M > 0$ such that*

$$
m\|\mathbf{x}\|_{\mathrm{I}} \leq \|\mathbf{x}\|_{\mathrm{II}} \leq M\|\mathbf{x}\|_{\mathrm{I}} \tag{0.3-4}
$$

for all $\mathbf{x} \in \mathcal{V}$. If this relationship holds, then we write $\|\cdot\|_{\mathrm{I}} \approx \|\cdot\|_{\mathrm{II}}$.

PROPOSITION 0.6. *The relation $\approx$ of norm equivalence is an* **equivalence relation**, *that is,*

(i) *The relation is* **reflexive**: $\|\cdot\| \approx \|\cdot\|$.

(ii) *The relation is* **symmetric**: $\|\cdot\|_{\mathrm{I}} \approx \|\cdot\|_{\mathrm{II}}$ *implies* $\|\cdot\|_{\mathrm{II}} \approx \|\cdot\|_{\mathrm{I}}$.

(iii) *The relation is* **transitive**: *If* $\|\cdot\|_{\mathrm{I}} \approx \|\cdot\|_{\mathrm{II}}$ *and* $\|\cdot\|_{\mathrm{II}} \approx \|\cdot\|_{\mathrm{III}}$, *then* $\|\cdot\|_{\mathrm{I}} \approx \|\cdot\|_{\mathrm{III}}$.

We leave the proof for Problem 3. Symmetry implies that one can reverse the roles of the two norms in the inequalities (0.3-4), possibly using different values for the constants m and M.

Here is the main result:

THEOREM 0.7. *All norms on $\mathbb{R}^n$ are equivalent.*

PROOF: It suffices to show that any norm on $\mathbb{R}^n$ is equivalent to $||\cdot||_2$ by finding appropriate constants m and M, as stipulated in (0.3-4). So, let $||\cdot||$ be any norm on $\mathbb{R}^n$. Notice that the unit sphere S_2 defined in Equation (0.3-2) is compact, that is, it is closed and bounded in $\mathbb{R}^n$. Moreover, the function $||\cdot||$ is continuous with respect to $||\cdot||_2$ on S_2 by part (*ii*) of Lemma 0.5. From these two facts Theorem 0.1 it follows that $||\cdot||$ attains maximum and minimum values at some points $\mathbf{x}_{\text{max}}$ and $\mathbf{x}_{\text{min}}$, respectively, on S_2. This means that, for any $\mathbf{x} \in S_2$,

$$||\mathbf{x}_{\text{min}}|| \leq ||\mathbf{x}|| \leq ||\mathbf{x}_{\text{max}}||.$$

We claim that we can choose $m = ||\mathbf{x}_{\text{min}}||$ and $M = ||\mathbf{x}_{\text{max}}||$.

First, notice that $||\mathbf{x}_{\text{min}}|| > 0$ since $\mathbf{x}_{\text{min}} \neq \mathbf{0}$. Next, select an arbitrary vector $\mathbf{x} \in \mathbb{R}^n$. If $\mathbf{x} = \mathbf{0}$, then the claim is trivially true. Otherwise, $\mathbf{x}/||\mathbf{x}||_2 \in S_2$, which implies that

$$||\mathbf{x}_{\text{min}}|| \leq \left\| \mathbf{x}/||\mathbf{x}||_2 \right\| \leq ||\mathbf{x}_{\text{max}}||.$$

Multiplying these inequalities through by $||\mathbf{x}||_2$ establishes the claim and hence the theorem. ■

0.4 Results from Calculus

We conclude this chapter with a review of basic results from calculus, leading to several versions of the Taylor theorem. We begin with four familiar facts.

THEOREM 0.8 (INTERMEDIATE VALUE THEOREM). *Let $f \in C^0([a,b])$, and suppose that $f(a) < c < f(b)$. Then there exists a point $\zeta \in (a,b)$ such that $f(\zeta) = c$.*

Rudin ([2], Chapter 4) gives a proof.

THEOREM 0.9. *If $f, g \in C^0([a,b])$ and $f(x) \leq g(x)$ for every $x \in [a,b]$, then*

$$\int_a^b f(x)\,dx \leq \int_a^b g(x)\,dx.$$

For a proof, see Rudin ([2], Chapter 6). This theorem has a useful corollary, which serves as a continuous analog of the triangle inequality:

COROLLARY 0.10. *If $f \in C^0([a,b])$, then*

$$\left| \int_a^b f(x)\, dx \right| \leq \int_a^b |f(x)|\, dx.$$

PROOF: Letting $g(x) = |f(x)|$ in Theorem 0.9 gives the inequality

$$\int_a^b f(x)\, dx \leq \int_a^b |f(x)|\, dx.$$

Now replace f by $-f$ to prove that

$$-\int_a^b |f(x)|\, dx \leq \int_a^b f(x)\, dx.$$

■

THEOREM 0.11 (FUNDAMENTAL THEOREM OF CALCULUS). *If $f \in C^1([a,b])$ and $x \in [a,b]$, then*

$$f(x) = f(a) + \int_a^x f'(t)\, dt.$$

Again, Rudin ([2], Chapter 6) gives a proof.

THEOREM 0.12 (INTEGRATION BY PARTS). *If $g, h \in C^1([a,b])$, then*

$$\int_a^b g(x)h'(x)\, dx = g(x)h(x)\Big|_a^b - \int_a^b g'(x)h(x)\, dx, \tag{0.4-1}$$

where

$$g(x)h(x)\Big|_a^b := g(b)h(b) - g(a)h(a).$$

In lieu of a formal proof, which one can find in Rudin ([2], Chapter 6), we mention that Equation (0.4-1) follows directly from Theorem 0.11 and the product rule $(gh)' = g'h + gh'$ for differentiation.

We now have the tools needed to prove the following:

THEOREM 0.13 (TAYLOR). *Let $f \in C^{n+1}([a,b])$ for some $n \geq 0$, and let $c, x \in [a,b]$. There is a point ζ, lying strictly between c and x (unless $c = x$, in which case $\zeta = c = x$), such that*

$$\begin{aligned} f(x) &= \underbrace{f(c) + \frac{1}{1!}f'(c)(x-c) + \cdots + \frac{1}{n!}f^{(n)}(c)(x-c)^n}_{T_n(x-c)} \\ &\quad + \underbrace{\frac{1}{(n+1)!}f^{(n+1)}(\zeta)(x-c)^{n+1}}_{R_{n+1}(x,c)}. \end{aligned} \tag{0.4-2}$$

Before giving the proof, it is worthwhile to comment on this theorem. The idea is to approximate f near a point c, where we have information about the values of f and its first few derivatives. The **Taylor polynomial** $T_n(x-c)$ in Equation (0.4-2) is a polynomial of degree at most n in the expression $x-c$, which we regard as a small parameter. If we neglect the **remainder** $R_{n+1}(x,c)$, then we can view $T_n(x-c)$ as a polynomial approximation to $f(x)$ that is valid "close to" c.

The success of this idea depends upon whether R_{n+1} is small. One difficulty here is the fact that ζ, while guaranteed to exist, remains unknown except for the stipulation that it lies between c and x. To circumvent this problem, observe that

$$|R_{n+1}(x,c)| \leq \underbrace{\sup_{y\in[a,b]} \left|f^{(n+1)}(y)\right|}_{M_{n+1}} |x-c|^{n+1},$$

the constant M_{n+1} being independent of ζ and hence of the choice of x. This estimate shows, heuristically speaking, that R_{n+1} shrinks at least as fast as $(x-c)^{n+1}$, which grows smaller either as $x \rightarrow c$ or as the allowable order $n+1$ of differentiation increases, provided M_{n+1} is bounded as $n \rightarrow \infty$. To express succinctly the rate at which R_{n+1} shrinks with the small parameter $(x-c)^{n+1}$, we write $R_{n+1} = \mathcal{O}((x-c)^{n+1})$.

The notation $\mathcal{O}(\cdot)$ appears in so many contexts that it warrants a formal definition:

DEFINITION. *Let $\alpha(\epsilon)$ and $\beta(\epsilon)$ depend on some parameter ϵ. The notation $\alpha(\epsilon) = \mathcal{O}(\beta(\epsilon))$ as $\epsilon \rightarrow 0$ means there exist positive constants M and $\epsilon_{\max}$ such that $|\alpha(\epsilon)| \leq M|\beta(\epsilon)|$ whenever $0 < |\epsilon| < \epsilon_{\max}$. Similarly, $\alpha(\epsilon) = \mathcal{O}(\beta(\epsilon))$ as $\epsilon \rightarrow \infty$ if there exist positive constants M and $\epsilon_{\min}$ such that $|\alpha(\epsilon)| \leq M|\beta(\epsilon)|$ whenever $\epsilon > \epsilon_{\min}$.*

Whether $\epsilon \rightarrow 0$ or $\epsilon \rightarrow \infty$ is often clear from context, and in these cases we typically omit explicit mention of the limits. This notation uses the symbol $=$ in an unusual way. For example, the definition implies the following:

(i) *If $\alpha(\epsilon) = \mathcal{O}(\gamma(\epsilon))$ and $\beta(\epsilon) = \mathcal{O}(\gamma(\epsilon))$, then*

$$\alpha(\epsilon) \pm \beta(\epsilon) = \mathcal{O}(\gamma(\epsilon)). \tag{0.4-3a}$$

(ii) *If $0 < p < q$ and $\alpha(\epsilon) = \mathcal{O}(\epsilon^q)$ as $\epsilon \rightarrow 0$, then*

$$\alpha(\epsilon) = \mathcal{O}(\epsilon^p) \quad \textit{as} \quad \epsilon \rightarrow 0. \tag{0.4-3b}$$

(iii) *If $0 < p < q$ and $\alpha(\epsilon) = \mathcal{O}(\epsilon^p)$ as $\epsilon \rightarrow \infty$, then*

$$\alpha(\epsilon) = \mathcal{O}(\epsilon^q) \quad \textit{as} \quad \epsilon \rightarrow \infty. \tag{0.4-3c}$$

Problem 6 asks for proofs.

PROOF OF THEOREM 0.13: Assume that $x \neq c$, the case $x = c$ being trivial. According to Theorem 0.11,

$$f(x) = f(c) + \int_c^x f'(t)\,dt.$$

If $n = 0$, let $T_n(x-c) = f(c)$. If $n \geq 1$, integrate by parts, using $g = f'$ and $h'(t) = 1$ (or $h(t) = -(x-t)$) in Theorem 0.12 to get

$$f(x) = f(c) + f'(c)(x-c) + \int_c^x (x-t)f''(t)\,dt.$$

Continue to integrate by parts in this way, using $g(t) = f^{(k)}(t)$ and $h'(t) = -(x-t)^{k-1}$ at the kth stage, until the allowable derivatives of f are exhausted. We then have

$$f(x) = T_n(x-c) + \int_c^x \frac{(x-t)^n}{n!} f^{(n+1)}(t)\,dt.$$

It remains to show that the integral on the right of this identity equals R_{n+1}, as defined in Equation (0.4-2). We argue for the case when $c < x$, the case $c > x$ being similar. Call

$$m := \inf_{t\in[c,x]} f^{(n+1)}(t), \qquad M := \sup_{t\in[c,x]} f^{(n+1)}(t).$$

Theorem 0.9 yields the inequalities

$$m\int_c^x \frac{(x-t)^n}{n!}\,dt \leq \int_c^x \frac{(x-t)^n}{n!} f^{(n+1)}(t)\,dt \leq M\int_c^x \frac{(x-t)^n}{n!}\,dt.$$

Computing the integrals on the left and right and rearranging gives

$$m \leq \frac{(n+1)!}{(x-c)^{n+1}} \int_c^x \frac{(x-t)^n}{n!} f^{(n+1)}(t)\,dt \leq M.$$

But $f^{(n+1)}$ is continuous, so the intermediate value theorem guarantees that there is some point $\zeta \in (c, x)$ such that

$$f^{(n+1)}(\zeta) = \frac{(n+1)!}{(x-c)^{n+1}} \int_c^x \frac{(x-t)^n}{n!} f^{(n+1)}(t)\,dt.$$

Solving this identity for the integral shows that it is identical to R_{n+1}. ∎

The Taylor theorem admits two special cases important enough to have their own names.

THEOREM 0.14 (MEAN VALUE THEOREM). *If $f \in C^1([a,b])$, then there is a point $\zeta \in (a,b)$ such that*

$$f'(\zeta) = \frac{f(b) - f(a)}{b - a}.$$

This theorem is just the Taylor theorem for the case $n = 0$. It guarantees the existence of a point $\zeta \in (a,b)$ where the derivative of f equals the average slope of f over $[a,b]$, as shown in Figure 1.

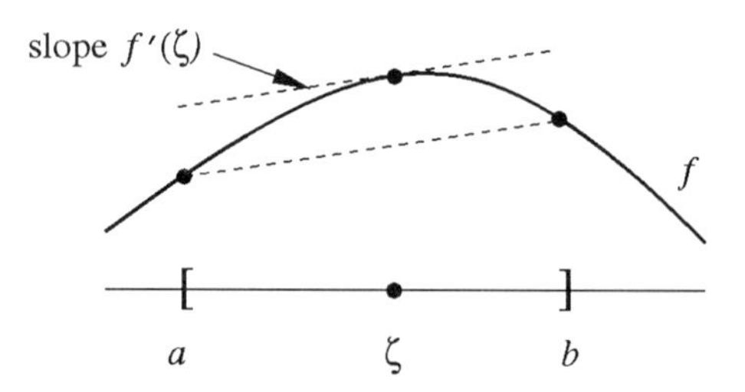

FIGURE 1. *Graphic example of the mean value theorem. At the point ζ, the value of f' equals the average slope of f over the interval $[a,b]$.*

THEOREM 0.15 (ROLLE). *If $f \in C^1([a,b])$ has zeros at a and b, then there is a point $\zeta \in (a,b)$ where $f'(\zeta) = 0$.*

This is the mean value theorem for the case $f(a) = f(b) = 0$.

The Taylor theorem has extensions to functions of several real variables. Instead of introducing the most general statement of the theorem, we examine two useful particular cases.

DEFINITION. *Let $\Omega \subset \mathbb{R}^n$ be an open set, with $f: \Omega \to \mathbb{R}$. We say that $f \in C^1(\Omega)$ if f is continuous at each point $\mathbf{x} = (x_1, x_2, \ldots, x_n) \in \Omega$ and each of the partial derivatives $\partial f/\partial x_1, \partial f/\partial x_2, \ldots, \partial f/\partial x_n$ exists and is continuous at each $\mathbf{x} \in \Omega$. The vector-valued function $\nabla f: \Omega \to \mathbb{R}^n$ defined by*

$$\nabla f(\mathbf{x}) := \left(\frac{\partial f}{\partial x_1}(\mathbf{x}), \frac{\partial f}{\partial x_2}(\mathbf{x}), \ldots, \frac{\partial f}{\partial x_n}(\mathbf{x}) \right)$$

is the **gradient** *of f.*

The first extension of the Taylor theorem is the following:

THEOREM 0.16. *Let $f \in C^1(\Omega)$, and suppose that* $\mathbf{c}, \mathbf{x} \in \Omega$ *and that the line segment connecting* $\mathbf{c}$ *and* $\mathbf{x}$ *lies entirely in* Ω. *Then there is a point* $\boldsymbol{\zeta}$ *lying on that line segment such that*

$$f(\mathbf{x}) = f(\mathbf{c}) + \nabla f(\boldsymbol{\zeta}) \cdot (\mathbf{x} - \mathbf{c}).$$

Figure 2 illustrates the theorem. Think of the line segment connecting $\mathbf{c}$ and $\mathbf{x}$ as an analog of the interval (c, x) in the one-dimensional Theorem 0.13.

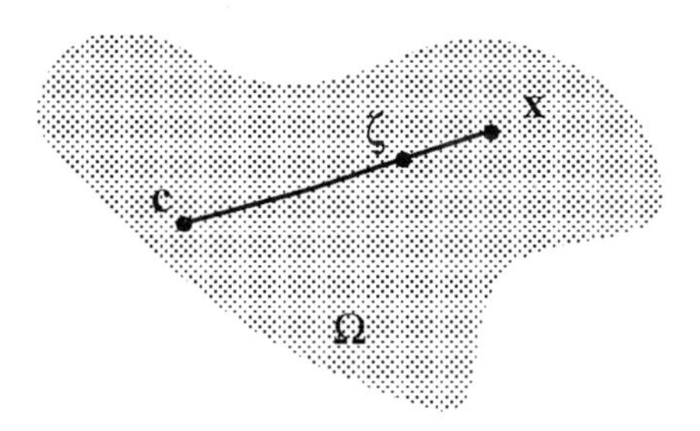

FIGURE 2. *An open set* $\Omega \subset \mathbb{R}^2$, *with the points* $\mathbf{c}$, $\mathbf{x}$, *and* $\boldsymbol{\zeta}$ *referred to in Theorem 0.16.*

PROOF: Define a function $\phi: [0,1] \to \mathbb{R}$ by setting $\phi(t) := f(\mathbf{c} + t(\mathbf{x} - \mathbf{c}))$. By the chain rule,

$$\begin{aligned} \phi'(t) &= \sum_{i=1}^{n} \frac{\partial f}{\partial x_i}(\mathbf{c} + t(\mathbf{x} - \mathbf{c})) \frac{d}{dt} [c_i + t(x_i - c_i)] \\ &= \sum_{i=1}^{n} \frac{\partial f}{\partial x_i}(\mathbf{c} + t(\mathbf{x} - \mathbf{c}))\,(x_i - c_i), \end{aligned}$$

the continuity of the individual terms in the sum guaranteeing that $\phi \in C^1([0,1])$. The mean value theorem yields a point $\zeta \in (0,1)$ such that $\phi(1) = \phi(0) + \phi'(\zeta)$. By the definition of ϕ, we therefore have

$$f(\mathbf{x}) = f(\mathbf{c}) + \nabla f\,(\mathbf{c} + \zeta(\mathbf{x} - \mathbf{c})) \cdot (\mathbf{x} - \mathbf{c}).$$

The vector $\boldsymbol{\zeta} := \mathbf{c} + \zeta(\mathbf{x} - \mathbf{c})$, which lies on the line segment between $\mathbf{c}$ and $\mathbf{x}$, is the desired point. ∎

To carry the Taylor expansion for $f: \Omega \to \mathbb{R}$ one term further, it is necessary to introduce more notation. We say that $f \in C^2(\Omega)$ if $f \in C^1(\Omega)$ and each of the second partial derivatives $\partial^2 f / \partial x_i \partial x_j$, $i, j = 1, 2, \ldots, n$, exists and is continuous at every $\mathbf{x} \in \Omega$. The matrix $\mathsf{H}_f(\mathbf{x}) \in \mathbb{R}^{n \times n}$ whose (i,j)th entry is

$$\frac{\partial^2 f}{\partial x_i \partial x_j}(\mathbf{x})$$

is the **Hessian matrix** of f at $\mathbf{x}$. The continuity of the second partial derivatives guarantees that $\partial^2 f/\partial x_i \partial x_j = \partial^2 f/\partial x_j \partial x_i$.

THEOREM 0.17. *Let $f \in C^2(\Omega)$, and suppose that $\mathbf{c}, \mathbf{x} \in \Omega$ and that the line segment connecting $\mathbf{c}$ and $\mathbf{x}$ lies entirely in Ω. Then there exists a point ζ lying on that line segment such that*

$$f(\mathbf{x}) = f(\mathbf{c}) + \nabla f(\mathbf{c}) \cdot (\mathbf{x} - \mathbf{c}) + \tfrac{1}{2}(\mathbf{x} - \mathbf{c}) \cdot \mathsf{H}_f(\zeta)(\mathbf{x} - \mathbf{c}).$$

The proof, which uses the one-dimensional Taylor expansion through order 2, is the subject of Problem 7.

While it is possible to extend the Taylor expansion for functions $f: \Omega \to \mathbb{R}$ to any order, depending upon the smoothness of f, we do not use expansions past the second order. One can also prove analogs of the Taylor theorem for vector-valued functions, a task that we postpone until Chapter 4.

0.5 Problems

PROBLEM 1. For each of the following subsets of $\mathbb{R}$, determine the least upper bound and greatest lower bound, if they exist.

(A) $(0,1) \cup (2,2.1) \cup (3,3.01) \cup (4,4.001) \cup \cdots$.

(B) $\left\{1, -1, \frac{1}{2}, -\frac{1}{2}, \frac{1}{3}, -\frac{1}{3}, \ldots\right\}$.

(C) $\left\{\exp(-x^2) \in \mathbb{R} \,:\, x \in \mathbb{R}\right\}$.

PROBLEM 2. Prove that $\|\cdot\|_1$, $\|\cdot\|_2$, and $\|\cdot\|_\infty$ are norms on $\mathbb{R}^n$.

PROBLEM 3. Show that norm equivalence ($\|\cdot\|_{\mathrm{I}} \approx \|\cdot\|_{\mathrm{II}}$) is an equivalence relation.

PROBLEM 4. Show that, for any $\mathbf{x} \in \mathbb{R}^n$,

(A) $\|\mathbf{x}\|_\infty \le \|\mathbf{x}\|_2 \le \sqrt{n}\,\|\mathbf{x}\|_\infty$.

(B) $\sqrt{1/n}\,\|\mathbf{x}\|_1 \le \|\mathbf{x}\|_2 \le \|\mathbf{x}\|_1$.

(C) $\|\mathbf{x}\|_\infty \le \|\mathbf{x}\|_1 \le n\|\mathbf{x}\|_\infty$.

Also show that these inequalities are sharp, in the sense that each inequality becomes an equality for some appropriate nonzero vector $\mathbf{x}$.

PROBLEM 5. With respect to a given norm $||\cdot||$, a sequence $\{\mathbf{x}_k\}$ of vectors in $\mathbb{R}^n$ **converges** to $\mathbf{x} \in \mathbb{R}^n$ (written $\mathbf{x}_k \to \mathbf{x}$) under the following condition: For any $\epsilon > 0$, there is a number $N > 0$ such that $||\mathbf{x}_k - \mathbf{x}|| < \epsilon$ whenever $k > N$. Let $||\cdot||_{\text{I}}$ and $||\cdot||_{\text{II}}$ be two norms on $\mathbb{R}^n$. Show that $\mathbf{x}_k \to \mathbf{x}$ with respect to $||\cdot||_{\text{I}}$ if and only if $\mathbf{x}_k \to \mathbf{x}$ with respect to $||\cdot||_{\text{II}}$.

PROBLEM 6. Prove the statements (0.4-3).

PROBLEM 7. Prove Theorem 0.17.

PROBLEM 8. Prove that, if $||\cdot||$ is a norm on $\mathbb{R}^n$, then the unit sphere $S := \{\mathbf{x} \in \mathbb{R}^n : ||\mathbf{x}|| = 1\}$ is compact.

PROBLEM 9. Let $\mathsf{A} \in \mathbb{R}^{n\times n}$, and let $||\cdot||$ be a norm on $\mathbb{R}^n$. Prove that the linear map defined by $\mathbf{x} \mapsto \mathsf{A}\mathbf{x}$ is uniformly continuous with respect to $||\cdot||$.

0.6 References

1. R.P. Boas, *A Primer of Real Functions* (3rd ed.), Carus Mathematical Monograph 13, Mathematical Association of America, Washington, DC, 1981.

2. W. Rudin, *Principles of Mathematical Analysis* (3rd ed.), McGraw-Hill, New York, 1976.

3. G. Strang, *Linear Algebra and Its Applications* (2nd ed.), Academic Press, New York, 1980.

Chapter 1

Approximation of Functions

1.1 Introduction

A fundamental task of numerical analysis is to approximate functions about which one has incomplete information. For example, one may know the values of a function f at two points x_1 and x_2 and want an estimate of $f(y)$, where y lies between x_1 and x_2. Or, one may know the values of f at a discrete set $\{x_0, x_1, \ldots, x_N\}$ of points in some interval $[a, b]$ and seek an approximate value of $\int_a^b f(x)\,dx$. In such applications it is helpful to construct an **approximating function** $\hat{f}$ to use as a surrogate for f. This chapter concerns methods for constructing such approximating functions and for analyzing their properties.

To say anything significant about how well $\hat{f}$ approximates f requires qualitative information about f, such as its continuity, differentiability, and so forth. We explore the connections between these properties and the effectiveness of the approximating function $\hat{f}$ later. For now, we ask that $\hat{f}$ satisfy three conditions. First, it should be easy to compute. Second, such basic properties of $\hat{f}$ as its continuity, differentiability, and integrability should be well understood. Third, $\hat{f}$ should be "close" to f in some sense. Here, the idea is to view f and $\hat{f}$ as elements in some normed vector space and then to ask whether the distance $||f - \hat{f}||$ is small.

This chapter explores methods for constructing $\hat{f}$. The methods differ in the information about f that they require and in qualitative features of the approximation $\hat{f}$. Also, we encounter differences in the norms $||\cdot||$ in which it is most natural to measure the distance between f and $\hat{f}$. We begin with polynomial interpolation, a scheme that is more useful as a foundation for other methods than as a direct approximation technique. We then investigate piecewise polynomial interpolation, Hermite interpolation, and interpolation in two space dimensions. We finish the chapter by discussing three somewhat specialized methods: cubic splines, least squares, and trigonometric interpolation.

1.2 Polynomial Interpolation

Motivation and Construction

Suppose that we know values of a function f at a set $\{x_0, x_1, \ldots, x_N\}$ of distinct points in an interval $[a, b]$. Assume that $x_0 = a$ and $x_N = b$ and that we have labeled the points so that $x_0 < x_1 < \cdots < x_N$, as Figure 1 illustrates. We call the abscissae x_i **nodes** and denote the known ordinates $f(x_i)$ by y_i. In many applications the following problem arises: Construct an approximating function $\hat{f}: [a, b] \to \mathbb{R}$ such that $\hat{f}(x_i) = y_i$ for each of the index values $i = 0, 1, \ldots, N$. In other words, find a function that passes through the known points on the graph of f and that is defined throughout the interval $[a, b]$. This is the **interpolation** problem.

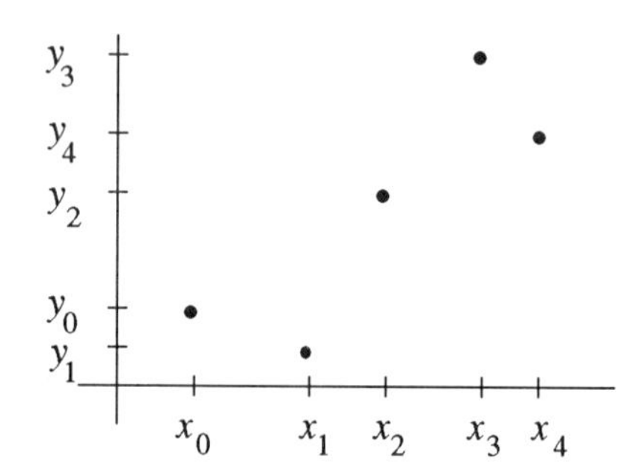

FIGURE 1. *A grid on $[a, b]$ with known points (x_i, y_i) on the graph of a function f.*

One solution to this problem is to pass a polynomial

$$\hat{f}(x) = a_0 + a_1 x + a_2 x^2 + \cdots + a_N x^N$$

through the known points (x_i, y_i). Polynomials are simple to work with, and their properties are well known. The computational task is to determine values for the coefficients $a_0, a_1, \ldots, a_N$ so that the graph of $\hat{f}$ indeed passes through the known points.

Observe that the number $N + 1$ of coefficients a_i equals the number of points x_i at which values of f are known. In other words, if an interpolant of this form exists, then there is a match between the number of unknowns (in this case, coefficients of $\hat{f}$) and the number of conditions available to determine them. In Section 1.7 we discuss a different approach that seems to relax this requirement.

A brute-force method for determining the coefficients a_i of the approximating function $\hat{f}$ is to write the system of equations implied by the conditions and then to solve them. The equations are as follows:

$$\begin{aligned} \hat{f}(x_0) &= a_0 + a_1 x_0 + a_2 x_0^2 + \cdots + a_N x_0^N &= y_0, \\ \hat{f}(x_1) &= a_0 + a_1 x_1 + a_2 x_1^2 + \cdots + a_N x_1^N &= y_1, \\ &\vdots \\ \hat{f}(x_N) &= a_0 + a_1 x_N + a_2 x_N^2 + \cdots + a_N x_N^N &= y_N. \end{aligned}$$

This set of equations is linear in the unknowns $a_0, a_1, \ldots, a_N$. It has a unique solution vector $(a_0, a_1, \ldots, a_N)$ provided that the nodes $x_0, x_1, \ldots, x_N$ are distinct. (We prove an equivalent fact later in this section.) However, solving the system can be tedious, especially when $N > 3$. Besides, the solution obtained in this way yields a collection of $N + 1$ numerical values affording little insight into the structure of this problem, let alone more general ones.

Instead, we decompose $\hat{f}$ into parts, each of which solves a simpler problem. Consider the simple interpolation problem in which one of the known ordinates, y_i, has the value 1, while the other ordinates are all 0. The solution to this problem is easy to construct: Being a polynomial L_i of degree at most N, with zeros at $x_0, \ldots, x_{i-1}, x_{i+1}, \ldots, x_N$, it must be some multiple of the polynomial

$$(x - x_0) \cdots (x - x_{i-1})(x - x_{i+1}) \cdots (x - x_N).$$

Now impose the requirement that $L_i(x_i) = 1$. The expression

$$L_i(x) = \frac{(x - x_0) \cdots (x - x_{i-1})(x - x_{i+1}) \cdots (x - x_N)}{(x_i - x_0) \cdots (x_i - x_{i-1})(x_i - x_{i+1}) \cdots (x_i - x_N)} \tag{1.2-1}$$

takes the value 1 at $x = x_i$. Therefore, this polynomial solves the simple interpolation problem where $y_i = 1$ and all other ordinates $y_j = 0$. For later reference, observe that

$$L_i(x) = \frac{\omega_N(x)}{(x - x_i)\omega_N'(x_i)}, \tag{1.2-2a}$$

where

$$\omega_N(x) := (x - x_0)(x - x_1) \cdots (x - x_N). \tag{1.2-2b}$$

This simple problem leads to a convenient solution $\hat{f}$ to the interpolation problem involving arbitrary ordinates $y_0, y_1, \ldots, y_N$. Since each of the polynomials $L_0, L_1, \ldots, L_N$ vanishes at all but one of the points x_j, we can scale them and add the results to get a solution to the more general problem. The resulting polynomial $\hat{f}$ has the form

$$\hat{f}(x) = \sum_{i=0}^{N} y_i L_i(x). \tag{1.2-3}$$

It should be clear that $\hat{f}(x_i) = y_i$, and $\hat{f}$ must be a polynomial of degree no greater than N, since it is a linear combination of such polynomials.

Figure 2 shows an example for the case $N = 2$, in which $x_0 = 1$, $x_1 = 3$, and $x_2 = 4$. The upper part of the figure shows graphs of the basis functions L_0, L_1, and L_2, each of which is a quadratic polynomial having the value 1 at its associated node and having a zero at each of the other two nodes. Specifically,

$$\begin{aligned} L_0(x) &= \frac{(x-3)(x-4)}{(-2)(-3)}, \\ L_1(x) &= \frac{(x-1)(x-4)}{2(-1)}, \\ L_2(x) &= \frac{(x-1)(x-3)}{(3)(1)}. \end{aligned}$$

The lower part of Figure 2 shows the linear combination $\hat{f} = -1L_0 + \frac{1}{2}L_1 + 0L_2$, which is the quadratic polynomial passing through the points $(x_0, y_0) = (1, -1)$, $(x_1, y_1) = (3, \frac{1}{2})$, and $(x_2, y_2) = (4, 0)$.

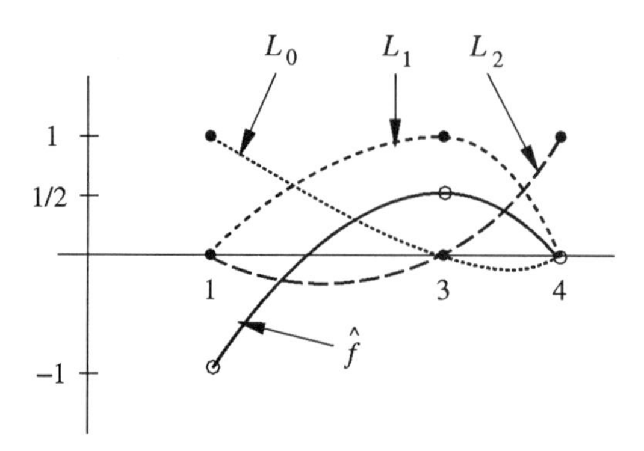

FIGURE 2. *Basis functions L_i and interpolant $\hat{f}$ for a sample problem involving quadratic interpolation.*

The expression in Equation (1.2-3) has two appealing features. First, each of the functions L_i is easy to remember or to reconstruct, regardless of the location or number of nodes x_i. Second, given the functions L_i, it is trivial to determine the coefficients that multiply them: The coefficients are simply the given ordinates y_i.

It is profitable to think of f and its **interpolant** $\hat{f}$ in the setting of a normed vector space. Equation (1.2-3) suggests that the functions L_i serve as basis vectors for some subspace of the functions with which we are concerned. The advantage of this more abstract point of view is that it allows us to estimate the distance $\|f-\hat{f}\|$ between the original function and its interpolant, measured in some norm $\|\cdot\|$. For example, if we are interested in keeping the interpolant close to f in a pointwise sense in the interval $[a,b]$, we might use the norm $\|\cdot\|_\infty$ defined in Section 0.3. This norm measures the maximum excursion that $\hat{f}$ takes from f in $[a,b]$, as Figure 3 depicts.

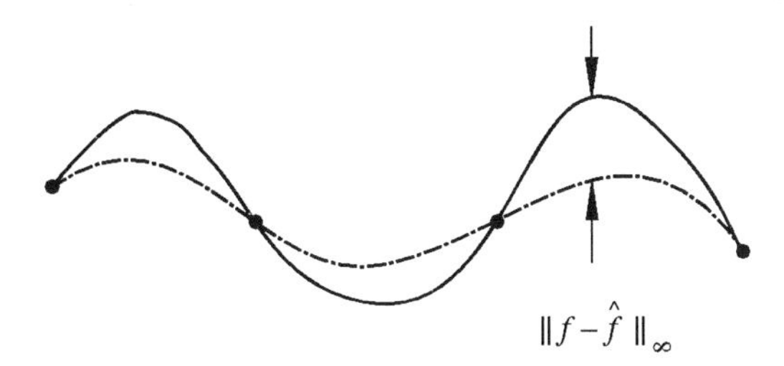

FIGURE 3. *Geometric meaning of* $\|f-\hat{f}\|_\infty$.

Practical Considerations

The procedure just discussed is called **standard polynomial interpolation**. It uses the **Lagrange interpolating polynomials** L_i to construct the interpolant $\hat{f}$. There are two theoretical aspects of this procedure that are important in applications. One is existence and uniqueness of the interpolant; the other is an estimate of $\|f-\hat{f}\|$. We now introduce these considerations and discuss their implications, postponing rigorous proofs for the end of the section.

Consider a collection $\{x_0, x_1, \ldots, x_N\}$ of nodes in an interval $[a,b]$ on the real line, chosen to satisfy the following definition:

DEFINITION. *A finite subset* $\Delta := \{x_0, x_1, \ldots, x_N\}$ *of the interval* $[a,b]$, *labeled so that* $a = x_0 < x_1 < \cdots < x_N = b$, *is a* **grid** *on* $[a,b]$. *The* **mesh size** *of* Δ *is*

$$h := \max_{1\le i\le N}(x_i - x_{i-1}).$$

One might expect more accurate interpolation of a given function f on fine grids — those having small mesh size — than on coarse grids — those having large mesh size. Shortly we investigate the extent to which this heuristic is reliable.

If we know the values $f(x_i) = y_i$ of f at the nodes $x_0, x_1, \ldots, x_N$ of a grid, then the distinctness of the nodes guarantees the existence of a unique interpolating polynomial $\hat{f}$ having degree at most N. We prove this fact later. The function $\hat{f}(x) = y_0 L_0(x) + y_1 L_1(x) + \cdots + y_N L_N(x)$ is clearly a polynomial of degree at most N (each of the functions L_i is), and $\hat{f}(x_i) = y_i$ (each function L_j vanishes at all of the nodes except x_j, where $L_j(x_j) = 1$). Thus existence and uniqueness of $\hat{f}$ hinge on the existence and uniqueness of the basis functions $L_0, L_1, \ldots, L_N$. Intuitively, we are on firm ground here, since we have used the $N + 1$ conditions

$$L_i(x_j) = \begin{cases} 1 & \text{if } i = j \\ 0 & \text{if } i \neq j \end{cases}$$

to determine the $N + 1$ degrees of freedom associated with each function L_i.

The connection between accuracy and mesh size is more complicated. To assess the accuracy of an interpolating scheme, imagine interpolating a *known* function f, for which the interpolation error $f - \hat{f}$ is computable. Then we can estimate this error in terms of generic properties of f and the mesh size h. The estimate depends on how "smooth" f is — that is, how many orders of continuous derivatives it possesses. In fact, if $f \in C^{N+1}([a, b])$ and $\hat{f}$ is a polynomial of degree N that interpolates f on a grid on $[a, b]$ having mesh size h, then

$$\|f - \hat{f}\|_\infty \leq \frac{\|f^{(N+1)}\|_\infty}{4(N+1)} h^{N+1}. \tag{1.2-4}$$

Here, $f^{(N+1)}$ denotes the $(N + 1)$st derivative of f. We prove this assertion later.

One enticing — but incomplete — way to interpret the estimate (1.2-4) is as a statement about how fast the interpolation error shrinks as we reduce the mesh size h. One might hope to abbreviate the statement of the estimate as follows:

$$\|f - \hat{f}\|_\infty = \mathcal{O}(h^{N+1}).$$

This view naively interprets the inequality (1.2-4) as saying that the interpolation error shrinks at least as fast as h^{N+1}, leading one to expect the error to shrink faster when the degree N of the interpolant $\hat{f}$ is large.

This interpretation is incorrect. It ignores the factor $\|f^{(N+1)}\|_\infty$ in the error estimate (1.2-4). When higher derivatives of f behave no worse than f itself, this factor causes no problems. This circumstance occurs, for example, when f itself is a polynomial, in which case derivatives of f are just polynomials of lower degree. (In fact, the inequality (1.2-4) implies that the interpolation error vanishes when f is itself a polynomial of degree N or less — a fact that is hardly astonishing.) However, polynomials are by no means typical in this respect. Many functions have derivatives that behave progressively worse, in the sense that $\|f^{(N+1)}\|_\infty$ grows, as the order of differentiation increases. In these cases, taking N larger (and hence h smaller) can be disastrous.

A classic example shows how the factor $\|f^{(N+1)}\|_\infty$ can foil attempts at high-degree interpolation. Consider the function $f(x) = 1/(1+25x^2)$, which possesses derivatives of all orders on $[-1,1]$. Problem 1 asks for various polynomial interpolants for this function, one of which looks like the highly oscillatory interpolant $\hat{f}$ shown in Figure 4. The occurrence of such surprisingly large excursions of $\hat{f}$ from f is known as the **Runge phenomenon**. While there are ways to minimize this phenomenon, as Problem 1 examines, the example serves a cautionary purpose. High-degree polynomial interpolation is risky! Section 1.3 discusses one way to circumvent the use of high-degree polynomial interpolation without abandoning the use of fine grids.

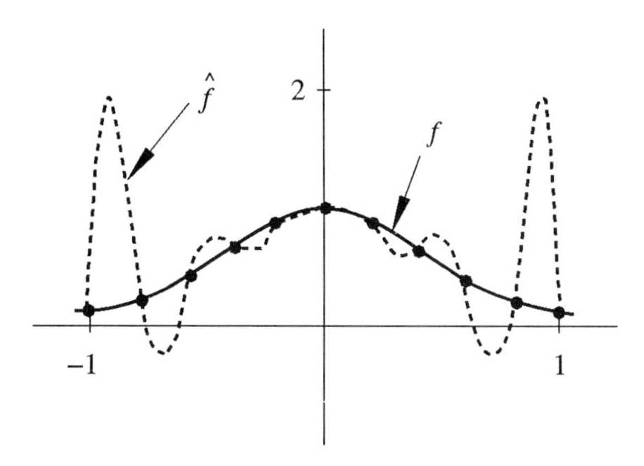

FIGURE 4. *The Runge phenomenon for standard polynomial interpolation of $f(x) = 1/(1+25x^2)$ on a grid containing 11 uniformly spaced points on $[-1,1]$.*

Mathematical Details

Underlying the considerations just discussed are rigorous arguments and a useful theoretical framework. There are many possible approaches to the problem of interpolating a function f on a given grid. Here are three examples of forms for the interpolant $\hat{f}$:

$$a_0 + a_1 x + a_2 x^2 + \cdots + a_N x^N \qquad \textbf{(polynomial)},$$

$$a_{-M} e^{-iMx} + a_{-(M-1)} e^{-i(M-1)x} + \cdots + a_M e^{iMx} \qquad \textbf{(trigonometric)},$$

$$\frac{a_0 + \cdots + a_k x^k}{a_{k+1} + \cdots + a_{N+1} x^{N-k}} \qquad \textbf{(rational)}.$$

Among the plethora of possibilities, let us identify a particularly tractable class: An interpolation problem is **linear** if the interpolant is a linear combination of known basis functions $\varphi_0, \varphi_1, \ldots, \varphi_N$, that is, if $\hat{f}$ has the form

$$\hat{f}(x) = a_0 \varphi_0(x) + a_1 \varphi_1(x) + \cdots + a_N \varphi_N(x).$$

Polynomial and trigonometric interpolation problems are linear, while rational interpolation problems typically are not.

In the case of polynomial interpolation on an interval $[a, b]$, we seek an interpolant $\hat{f}$ that belongs to the set of all polynomials of degree less than or equal to N on $[a, b]$. We denote this set by $\Pi_N([a, b])$. It is an easy exercise to show that $\Pi_N([a, b])$ is a vector space having dimension $N + 1$. This fact is useful in proving the basic existence and uniqueness theorem:

THEOREM 1.1. *Given a grid* $\Delta = \{x_0, x_1, \ldots, x_N\}$ *on* $[a, b]$ *and a collection* $\{y_0, y_1, \ldots, y_N\}$ *of ordinates, there is a unique polynomial* $\hat{f} \in \Pi_N([a, b])$ *such that* $\hat{f}(x_i) = y_i$ *for* $i = 0, 1, \ldots, N$.

PROOF: First we show uniqueness. Suppose that $\hat{f}_1, \hat{f}_2 \in \Pi_N([a, b])$ satisfy $\hat{f}_1(x_i) = \hat{f}_2(x_i) = y_i$, for $i = 0, 1, \ldots, N$. Then $\hat{f}_1 - \hat{f}_2 \in \Pi_N([a, b])$, since $\Pi_N([a, b])$ is a vector space, and consequently $\hat{f}_1 - \hat{f}_2$ is a polynomial of degree at most N having the $N+1$ zeros $x_0, x_1, \ldots, x_N$. This is possible only if $\hat{f}_1 - \hat{f}_2$ is identically zero; therefore $\hat{f}_1 = \hat{f}_2$.

To prove existence, we refer to the explicit expression (1.2-1). ■

We turn now to error estimates. The Rolle theorem (Theorem 0.15) is a crucial ingredient in the proof of the following lemma.

LEMMA 1.2. *Suppose that* $f \in C^{N+1}([a, b])$, *and let* $\hat{f} \in \Pi_N([a, b])$ *interpolate* f *on a grid* $\Delta = \{x_0, x_1, \ldots, x_N\}$ *on* $[a, b]$. *For any point* $x \in [a, b]$, *there exists a point* $\zeta \in (a, b)$ *such that*

$$f(x) - \hat{f}(x) = \frac{\omega_N(x) f^{(N+1)}(\zeta)}{(N+1)!}. \qquad (1.2\text{-}5)$$

(Recall the definition of the function ω_N in Equation (1.2-2b).)

PROOF: In the case that x is one of the nodes x_i in the grid Δ, $f(x)-\hat{f}(x)=0$, and Equation (1.2-5) follows from the fact that $\omega_N(x_i)=0$. When x is not a node, $\omega(x)\neq 0$, and we must argue differently. Define a new function $F:[a,b]\to\mathbb{R}$ by the formula

$$F(t)=f(t)-\hat{f}(t)-[f(x)-\hat{f}(x)]\frac{\omega_N(t)}{\omega_N(x)}.$$

Three properties of F are easy to check. First, $F\in C^{N+1}([a,b])$. Second, $F(x_i)=0$ for each node x_i. Third, $F(x)=0$. Thus F has $N+2$ zeros on $[a,b]$, and in any subinterval of $[a,b]$ bounded by two adjacent zeros F satisfies the hypotheses of the Rolle theorem. Therefore the function F' has at least one zero in each of these subintervals, as illustrated for the case $N=2$ in Figure 5. In other words, F' has at least $N+1$ zeros on (a,b). We apply similar reasoning to F' to deduce that F'' has at least N zeros on (a,b), and so forth, finally concluding that $F^{(N+1)}$ has at least one zero $\zeta\in[a,b]$. Thus,

$$0=f^{(N+1)}(\zeta)-\hat{f}^{(N+1)}(\zeta)-\frac{f(x)-\hat{f}(x)}{\omega_N(x)}\omega_N^{(N+1)}(\zeta). \tag{1.2-6}$$

But $\hat{f}^{(N+1)}(\zeta)=0$, since $\hat{f}$, a polynomial of degree at most N, has a vanishing $(N+1)$st derivative. Also, it is easy to check that

$$\omega_N^{(N+1)}(\zeta)=(N+1)!.$$

Incorporating these results and rearranging Equation (1.2-6) yields Equation (1.2-5). ∎

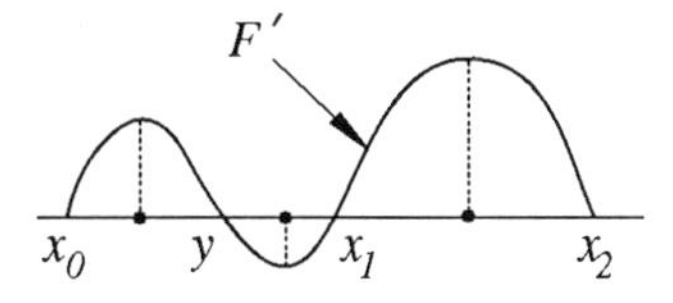

FIGURE 5. *Roots of F' guaranteed by the Rolle theorem.*

The estimate (1.2-4) is a simplified version of this lemma. Specifically, by Problem 2,

$$|\omega_N(x)|=|(x-x_0)(x-x_1)\cdots(x-x_N)|\leq N!h^{N+1}/4. \tag{1.2-7}$$

Using this result, we can convert Equation (1.2-5) to an estimate for $\|f-\hat{f}\|_\infty$. The upshot is the following theorem.

THEOREM 1.3. *Suppose that $f \in C^{N+1}([a,b])$, and let $\hat{f}$ be a polynomial of degree N that interpolates f on a grid on $[a,b]$ having mesh size h. Then*

$$\|f - \hat{f}\|_\infty \leq \frac{\|f^{(N+1)}\|_\infty}{4(N+1)} h^{N+1}. \tag{1.2-4}$$

Further Remarks

Knowing how well polynomial interpolants $\hat{f}$ approximate functions f, one might naturally ask how well derivatives of $\hat{f}$ approximate derivatives of f. (We examine the corresponding question for integrals of $\hat{f}$ and f in Chapter 6.) Problem 6 asks for a proof of the following fact:

PROPOSITION 1.4. *Let $f \in C^{N+1}([a,b])$, and let Δ be a grid of mesh size h on $[a,b]$. If $\hat{f}$ is the standard polynomial interpolant of degree at most N for f on Δ, then*

$$\|f' - \hat{f}'\|_\infty \leq \|f^{(N+1)}\|_\infty h^N. \tag{1.2-8}$$

In other words, we lose one power of h in the interpolation error estimate when we differentiate f and its standard polynomial interpolant.

PROOF (SKETCH): Observe that $f' - \hat{f}'$ has N zeros $\tilde{x}_1, \tilde{x}_2, \ldots, \tilde{x}_N$ in (a,b), located so that each $\tilde{x}_i \in (x_{i-1}, x_i)$. Now mimic the argument used in establishing Theorem 1.2 to estimate $f'(y) - \hat{f}'(y)$ for any $y \in [a,b]$ that is distinct from the points $\tilde{x}_1, \tilde{x}_2, \ldots, \tilde{x}_N$. ∎

This proposition suggests a conjecture:

THEOREM 1.5. *If f and $\hat{f}$ satisfy the hypotheses of the last proposition, then for each order $k = 1, 2, \ldots, N$ there exists a constant C, independent of h, such that*

$$\|f^{(k)} - \hat{f}^{(k)}\|_\infty \leq C\|f^{(N+1)}\|_\infty h^{N+1-k}.$$

(The constant C may be different for different values of k and N.) Thus with each differentiation we lose a power of h in the error estimates. Indeed, such estimates hold:

PROOF (SKETCH): Use the Rolle theorem repeatedly to locate zeros of higher derivatives of $f - \hat{f}$; then reason as for the cases $k = 0$ and 1. ∎

We close this section with a few words about the Runge phenomenon. The example shown in Figure 4 is closely related to one discussed by Runge [11]. Underlying this example is a surprising theorem stating, in essence, that it is possible to defeat any polynomial interpolating strategy based on high-degree interpolants. In particular, suppose that we select, in advance, a sequence

$\{\Delta_1, \Delta_2, \Delta_3, \ldots\}$ of grids on $[a, b]$ having $2, 3, 4, \ldots$ nodes, respectively. Then there exists a continuous function $f: [a, b] \to \mathbb{R}$ such that, if $\hat{f}_m$ denotes the standard polynomial interpolant of degree at most m for f on Δ_m, then $\|f - \hat{f}_m\|_\infty \to \infty$. Cheney ([2], p. 215) proves this fact. In practice, however, one can often find grids for which $\hat{f}_m$ approximates f quite well, so long as m is not large. Part of Problem 1 calls for the use of special grids, involving zeros of Chebyshev polynomials, to suppress the Runge phenomenon. Still, Runge's observations reveal an inherent difficulty in polynomial interpolation — one whose resolution we discuss next.

1.3 Piecewise Polynomial Interpolation

Motivation and Construction

The Runge phenomenon is an affront to numerical economy: There is no point in doing more work unless it yields better results. In the present context, more work corresponds to the use of higher-degree interpolating polynomials and hence more terms to evaluate in Equation (1.2-1). In this section, we examine a way to incorporate many points (x_i, y_i) into the interpolation of a function f without using high-degree polynomials.

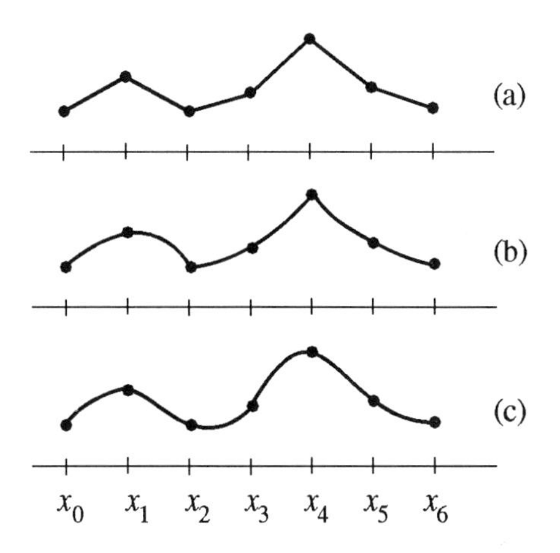

FIGURE 1. *Piecewise polynomial interpolants of degree at most (a) $n = 1$, (b) $n = 2$, and (c) $n = 3$.*

The idea is to fix a maximum degree n of polynomials to be used in the interpolation. Then, for a given grid Δ on $[a, b]$, patch together segments of polynomials on appropriately chosen subintervals of $[a, b]$. The result is a *single* interpolant $\hat{f}$ that coincides with different polynomials of degree at most n on different subintervals. This scheme of **piecewise polynomial interpolation** allows the total number N of points in the grid to become large, without courting the potentially disastrous effects associated with the factor $\|f^{(N+1)}\|_\infty$ in the error estimate (1.2-4).

(To simplify the discussion, we use the term "degree n" to describe polynomials of degree *at most* n in this section and the next. Thus, for example, the word "quadratic" refers to polynomials having exact degree zero, one, or two, the first two cases being degenerate instances of the third.)

To illustrate the case $n = 1$, consider a grid $\Delta = \{x_0, x_1, \ldots, x_N\}$ on $[a, b]$ and a set $\{y_0, y_1, \ldots, y_N\}$ of ordinates corresponding to values of f at nodes. Each pair $(x_i, y_i), (x_{i+1}, y_{i+1})$ of adjacent known points on the graph of f defines a polynomial arc of degree $n = 1$ (that is, a line segment) over the subinterval $[x_i, x_{i+1}]$. The result, known as **piecewise linear interpolation**, is the "connect-the-dots" interpolant $\hat{f}$ shown in Figure 1(a).

The case $n = 2$ is slightly more complicated. In this case, triples

$$(x_i, y_i), (x_{i+1}, y_{i+1}), (x_{i+2}, y_{i+2})$$

of adjacent points on the graph of f define quadratic arcs over subintervals of the form $[x_i, x_{i+2}]$. The tricky part of this construction is to ensure that the resulting interpolant $\hat{f}$ has a unique value for any $x \in [a, b]$. Figure 1(b) shows such an interpolant. Observe that $\hat{f}$ in this figure consists of three quadratic pieces, each of which is defined over one of the intervals $[x_0, x_2]$, $[x_2, x_4]$, $[x_4, x_6]$. Attempting to make $\hat{f}$ coincide with a single quadratic arc over the interval $[x_1, x_3]$ *and* with a single quadratic arc over $[x_0, x_2]$ would almost surely produce an interpolant whose values at points $x \in (x_1, x_2)$ would not be unique. Figure 2 depicts this situation.

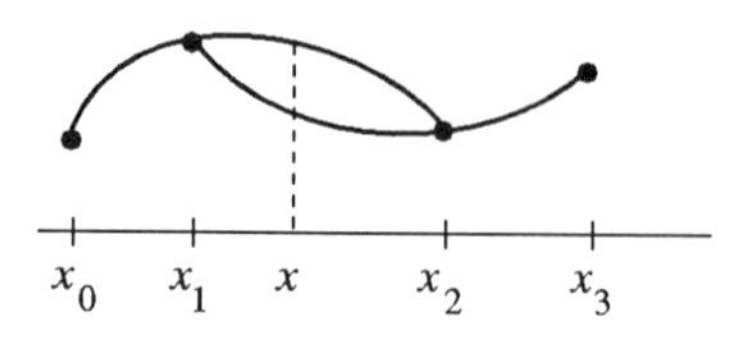

FIGURE 2. *Conflicting definitions of $\hat{f}$ in the interval $[x_1, x_2]$, arising from a failure to partition the global interval $[a, b]$ into elements before constructing the interpolant.*

To avoid this problem, we partition $[a, b]$ into subintervals

$$[x_0, x_2], \quad [x_2, x_4], \quad \ldots, \quad [x_{N-2}, x_N]$$

before constructing the piecewise quadratic interpolant. We call these subintervals **elements**; they are the largest subintervals of $[a, b]$ in which $\hat{f}$ coincides a priori with polynomial arcs of fixed maximal degree. (The polynomial with which $\hat{f}$ agrees in one element might be the same as the one with which $\hat{f}$ agrees in an adjacent element, but such occurrences are fortuitous.) This construction has an interesting consequence: To get an integer number of elements in piecewise quadratic interpolation, the index N of the rightmost node in the grid must be even.

We construct piecewise cubic interpolants similarly. First, partition $[a, b]$ into elements $[x_0, x_3], [x_3, x_6], \ldots, [x_{N-3}, x_N]$. This partitioning demands that N be an integer divisible by 3. Then the four known points

$$(x_i, y_i), \quad (x_{i+1}, y_{i+1}), \quad (x_{i+2}, y_{i+2}), \quad (x_{i+3}, y_{i+3})$$

on the graph of f in each element $[x_i, x_{i+3}]$ define a cubic arc over the element. The result is an interpolant $\hat{f}$ like the one plotted in Figure 1(c).

All of the graphs in Figure 1 are continuous. Allowing adjacent elements $[x_i, x_{i+n}]$, $[x_{i+n}, x_{i+2n}]$ to share the data (x_{i+n}, y_{i+n}) guarantees this fact. However, we have done nothing to guarantee any higher order of continuity at the element boundaries. In general, our interpolants will not be differentiable there, as the "corners" on the graphs in Figure 1 suggest. Therefore, in the global sense, piecewise polynomial interpolants constructed according to the procedures just described typically belong to $C^0([a, b])$ but not to $C^k([a, b])$ for any integer $k > 0$.

Practical Considerations

There are three issues that we need to settle to guarantee that piecewise polynomial interpolation is a useful technique. One is the existence and uniqueness of the interpolants. Another is the construction of basis functions. The third is a discussion of error estimates.

The existence and uniqueness of piecewise polynomial interpolants is a straightforward consequence of Theorem 1.1, which guarantees that the values of $\hat{f}$ at x_k and x_{k+1} uniquely determine $\hat{f}$ on the interval $[x_k, x_{k+1}]$.

The most useful bases for piecewise polynomial interpolation are the **nodal** (or **cardinal**) bases. For the grid $\Delta = \{x_0, x_1, \ldots, x_N\}$, the nodal basis for piecewise polynomial interpolation of any appropriate degree is a collection $\{\ell_0, \ell_1, \ldots, \ell_N\}$ of functions satisfying the conditions

$$\ell_i(x_j) = \begin{cases} 1, & \text{if } i = j, \\ 0, & \text{if } i \neq j. \end{cases} \tag{1.3-1}$$

While these conditions are similar to those used in Lagrange polynomial interpolation, they differ in that they implicitly require each function ℓ_j to be a piecewise polynomial of degree at most n on $[a, b]$.

Consider the case $n = 1$. The basis function ℓ_3, for example, has the value 1 at x_3 and vanishes at all other nodes $x_0, x_1, x_2, x_4, x_5, \ldots$. Between nodes, the graph of ℓ_3 consists of line segments. Explicitly,

$$\ell_3(x) = \begin{cases} \dfrac{x - x_2}{x_3 - x_2}, & \text{if } x \in [x_2, x_3], \\[2ex] \dfrac{x - x_4}{x_3 - x_4}, & \text{if } x \in [x_3, x_4], \\[2ex] 0, & \text{otherwise.} \end{cases}$$

The graph of ℓ_3 over the entire interval $[a, b]$ looks like the tent-shaped function drawn in Figure 3. In particular, ℓ_3 vanishes outside the two-element interval $[x_2, x_4]$ surrounding x_3. Figure 3 also depicts the graph of ℓ_0, showing that basis functions associated with the end nodes of a grid take nonzero values over only *one* element. Some people, believing that these basis functions resemble hats, call the set $\{\ell_0, \ell_1, \ldots, \ell_N\}$ defined in this way the **chapeau basis** for piecewise linear interpolation.

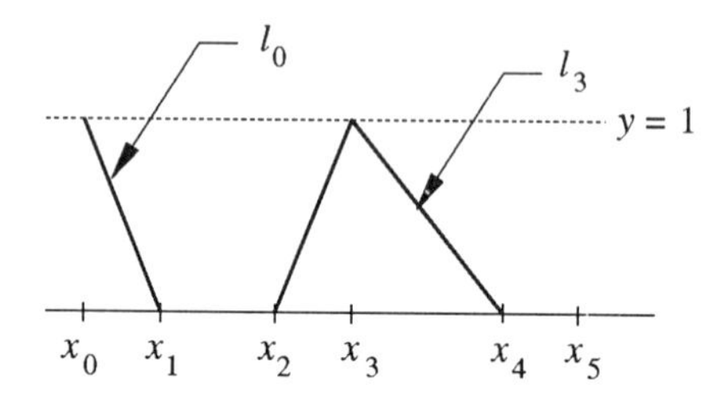

FIGURE 3. *Piecewise linear basis functions ℓ_0 and ℓ_3.*

Now consider the case $n = 3$. The condition (1.3-1) implies that there are two types of basis functions: one associated with element boundaries, the other associated with nodes belonging to the interiors of elements. The piecewise cubic basis function ℓ_3, sketched in Figure 4, exemplifies the first type. It has the following definition:

$$\ell_3(x) = \begin{cases} \dfrac{(x - x_0)(x - x_1)(x - x_2)}{(x_3 - x_0)(x_3 - x_1)(x_3 - x_2)}, & \text{if } x \in [x_0, x_3], \\[2ex] \dfrac{(x - x_4)(x - x_5)(x - x_6)}{(x_3 - x_4)(x_3 - x_5)(x_3 - x_6)}, & \text{if } x \in [x_3, x_6], \\[2ex] 0, & \text{otherwise.} \end{cases}$$

Notice that ℓ_3 takes nonzero values over the two elements $[x_0, x_3]$ and $[x_3, x_6]$. The function ℓ_6, being associated with the element boundary x_6, also takes nonzero values over two adjacent elements, as Figure 4 indicates.

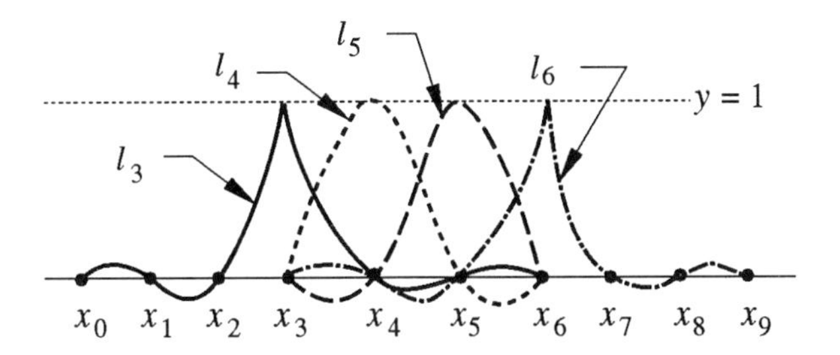

FIGURE 4. *The four piecewise Lagrange cubic functions ℓ_3, ℓ_4, ℓ_5, and ℓ_6 that take nonzero values over the element $[x_3, x_6]$.*

Figure 4 also shows basis functions ℓ_4 and ℓ_5, which belong to the second type. The explicit definition of ℓ_4 is as follows:

$$\ell_4(x) = \begin{cases} \dfrac{(x-x_3)(x-x_5)(x-x_6)}{(x_4-x_3)(x_4-x_5)(x_4-x_6)}, & \text{if } x \in [x_3, x_6], \\ 0, & \text{otherwise.} \end{cases}$$

As with the Lagrange interpolating bases discussed in Section 1.2, the beauty of these **piecewise Lagrange interpolating bases** is that they yield interpolants whose forms are trivial to construct once we know the ordinates $y_0, y_1, \ldots, y_N$. In fact,

$$\hat{f}(x) = \sum_{i=0}^{N} y_i \ell_i(x), \tag{1.3-2}$$

regardless of the degree n of the polynomial pieces of $\hat{f}$. To compute $\hat{f}(x)$ is simple: First determine the element to which the point x belongs, then add the $n+1$ terms in the sum in Equation (1.3-2) involving nodes lying in that element. All other terms in the sum vanish, since they involve basis functions that vanish at x. We refer to interpolation schemes based on these basis functions as **piecewise Lagrange interpolation**.

Error estimates for piecewise Lagrange interpolation follow from estimates for the corresponding global interpolation. Since any piecewise Lagrange interpolant $\hat{f}$ of degree n is a "global" interpolant on each element $[x_i, x_{i+n}]$, Theorem 1.3 applies on each element. In particular, if the interpolated function $f \in C^{n+1}([a,b])$ and $x \in [x_i, x_{i+n}]$, then there is a point $\zeta \in (x_i, x_{i+n})$ for which

$$f(x) - \hat{f}(x) = \frac{\omega_n(x) f^{(n+1)}(\zeta)}{(n+1)!}, \tag{1.3-3}$$

where

$$\omega_n(x) = (x-x_i)(x-x_{i+1})\cdots(x-x_{i+n}).$$

We render the factor $\omega_n(x)$ more intelligible by estimating it in terms of the mesh size h. If $x \in [x_i, x_{i+n}]$, pick $\alpha \leq n$ such that $x = x_i + \alpha h$. Then

$$\begin{aligned} |\omega_n(x)| &= |(x_i + \alpha h - x_i)(x_i + \alpha h - x_{i+1}) \cdots (x_i + \alpha h - x_{i+n})| \\ &\leq |\alpha(\alpha - r_1)(\alpha - r_2) \cdots (\alpha - r_n)| h^{n+1}, \end{aligned}$$

where $r_k = (x_{i+k} - x_i)/h$. The factor $|\alpha(\alpha - r_1) \cdots (\alpha - r_n)|$ is bounded above by a positive constant C, depending on the piecewise degree n of the interpolation but not on the total number N of points in the grid. In particular, C is independent of the mesh size h, as is the order $n+1$ of the derivative appearing in Equation (1.3-3). Hence Equation (1.3-3) yields the estimate

$$|f(x) - \hat{f}(x)| \leq \frac{Ch^{n+1}}{(n+1)!} |f^{(n+1)}(\zeta)| = \mathcal{O}(h^{n+1}).$$

This interpolation error tends to zero as $h \to 0$ (and $N \to \infty$); however, n remains fixed. Therefore, in using the symbol $\mathcal{O}(h^{n+1})$, we do not mask unpleasant subtleties like those lurking in the estimate (1.2-4) for global polynomial interpolation.

This line of reasoning proves the following theorem:

THEOREM 1.6. *Let $f \in C^{n+1}([a, b])$, and let $\hat{f}$ be the piecewise Lagrange interpolant of degree at most n for f on a grid Δ on $[a, b]$ having mesh size h. Then there exists a constant $C > 0$, independent of h, such that*

$$\|f - \hat{f}\|_\infty \leq \frac{Ch^{n+1}}{(n+1)!} \|f^{(n+1)}\|_\infty.$$

The theorem promises qualitative improvement over the error estimates for global polynomial interpolation. To illustrate how dramatic the improvement can be, Figure 5 shows the graph of $f(x) = 1/(1 + 25x^2)$, used earlier to demonstrate the Runge phenomenon, along with its piecewise linear interpolant on a uniform grid with $N = 10$. Compare this plot with Figure 4 of Section 1.2.

Mathematical Details

In addition to the error estimates of Theorem 1.6, piecewise Lagrange interpolation admits a worthy generalization. We have so far examined schemes for producing piecewise polynomial interpolants that vary in their maximal degree n but not in their smoothness: All piecewise Lagrange interpolants on $[a, b]$ belong to $C^0([a, b])$. Denote the set of all such interpolants for a given maximal degree n and a given grid Δ by $\mathcal{M}_0^n(\Delta)$. The subscript 0 indicates the highest order of derivatives of functions in the set that are guaranteed to be continuous.

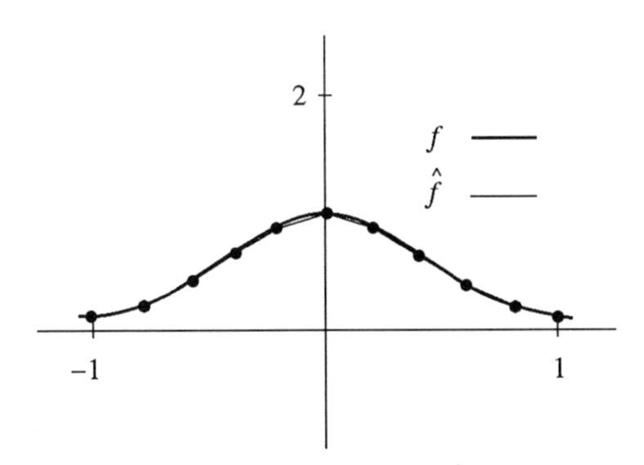

FIGURE 5. *The graph of $f(x) = 1/(1+25x^2)$ on $[-1,1]$, together with the graph of its piecewise linear interpolant on a uniform grid having 11 nodes.*

PROPOSITION 1.7. *$\mathcal{M}_0^n(\Delta)$ is a vector space.*

PROOF: (This is an exercise.) ∎

Indeed, $\mathcal{M}_0^n(\Delta)$ is the span of the piecewise Lagrange polynomial basis functions constructed above.

The generalization comes when we allow subscripts other than 0:

DEFINITION. *Given a grid Δ on $[a,b]$, $\mathcal{M}_k^n(\Delta)$ is the set of all functions in $C^k([a,b])$ whose restrictions to any element formed by the grid Δ are polynomials of maximal degree n.*

This definition remains somewhat vague, since it does not specify what might constitute "elements" for interpolants having higher orders of smoothness. Subsequent sections clarify this issue.

We end with a remark. By convention, the class $C^{-1}([a,b])$ contains all functions that possess at most finitely many discontinuities on $[a,b]$, all of which must be jump discontinuities. In other words, for any $\hat{f} \in C^{-1}([a,b])$, there are at most finitely many points $y \in [a,b]$ for which

$$\hat{f}(y-) = \lim_{x \to y-} \hat{f}(x) \neq \lim_{x \to y+} \hat{f}(x) = \hat{f}(y+),$$

and at each such point both of the one-sided limits $\hat{f}(y-)$ and $\hat{f}(y+)$ exist and are finite. Accordingly, the set $\mathcal{M}_{-1}^n(\Delta)$ is the set of all piecewise polynomials on $[a,b]$ that have degree n on any element formed by the grid Δ and that may have jump discontinuities at the element boundaries $x_i \in \Delta$. (In many applications, the actual values that a function $\hat{f} \in \mathcal{M}_{-1}^n(\Delta)$ takes at these element boundaries is immaterial. Unless circumstances demand otherwise, we do not distinguish, for instance, cases for which $\hat{f}(x-) = \hat{f}(x)$ from those for which $\hat{f}(x-) \neq \hat{f}(x)$, so long as the behavior of $\hat{f}$ in the interiors of elements is well defined.)

For example, $\mathcal{M}^0_{-1}(\Delta)$ denotes the set of all functions that are **piecewise constant** on the subintervals formed by $\Delta = \{x_0, x_1, \ldots, x_N\}$, as drawn in Figure 6. The set $\{\ell_i : i = 1, 2, \ldots, N\}$, with

$$\ell_i(x) = \begin{cases} 1, & \text{if } x_{i-1} \leq x < x_i, \\ 0, & \text{otherwise,} \end{cases}$$

serves as a basis for this space. By analogy with the spaces $\mathcal{M}^n_0(\Delta)$, one might expect an interpolation error estimate of the form $\|f - \hat{f}\|_\infty = \mathcal{O}(h)$ to hold for piecewise constant interpolation, provided $f \in C^1([a,b])$. Problem 5 asks for verification of this estimate.

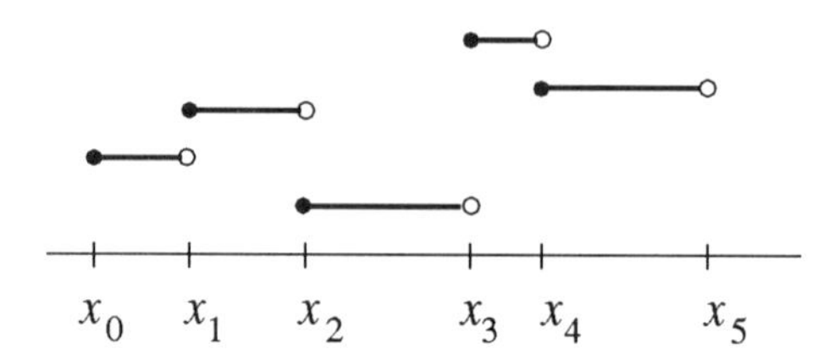

FIGURE 6. *A function in* $\mathcal{M}^0_{-1}(\Delta)$.

1.4 Hermite Interpolation

Motivation and Construction

It is sometimes useful to construct interpolants having higher orders of smoothness. For example, some applications call for globally differentiable interpolants. In other cases, we may know both function values and slopes at the nodes of a grid and seek interpolants that respect that knowledge. Still other applications call for approximations to functions that we know to be smooth.

This section discusses solutions to the following problem: Given a grid $\Delta = \{x_0, x_1, \ldots, x_N\}$ on an interval $[a,b]$, a set $\{y_0, y_1, \ldots, y_N\}$ of corresponding function values, and a set $\{y'_0, y'_1, \ldots, y'_N\}$ of corresponding slopes, find a continuously differentiable interpolant $\hat{f}$ such that

$$\hat{f}(x_i) = y_i, \qquad \frac{d\hat{f}}{dx}(x_i) = y'_i, \qquad i = 0, 1, \ldots, N. \tag{1.4-1}$$

The notation y_i' signifies a numerical value of slope at the abscissa x_i; the symbol "$'$" does *not* call for differentiating y_i, which is just a number. Polynomial interpolants that incorporate nodal values of derivatives in this way are called **Hermite interpolants**.

As with Lagrange polynomial interpolation, we begin by fitting a globally defined polynomial to the given data. In particular, we seek an interpolant of the form

$$\hat{f}(x) = \sum_{i=0}^{N} [y_i H_{0,i}(x) + y_i' H_{1,i}(x)], \tag{1.4-2}$$

in which the coefficients y_i, y_i' are precisely the given data. For this form to work, the set $\{H_{0,i}, H_{1,i} : i = 0, 1, \ldots, N\}$ must constitute a nodal basis, whose defining conditions are as follows:

$$H_{0,i}(x_j) = \begin{cases} 1, & \text{if } i = j, \\ 0, & \text{if } i \neq j, \end{cases} \qquad \frac{dH_{0,i}}{dx}(x_j) = 0, \tag{1.4-3}$$

for $i, j = 0, 1, \ldots, N$, and

$$H_{1,i}(x_j) = 0, \qquad \frac{dH_{1,i}}{dx}(x_j) = \begin{cases} 1, & \text{if } i = j, \\ 0, & \text{if } i \neq j, \end{cases} \tag{1.4-4}$$

for $i, j = 0, 1, \ldots, N$.

These $4(N+1)^2$ conditions suffice to determine $2N+2$ coefficients for each of $2N+2$ polynomials having degree at most $2N+1$. Problem 4 asks for proof that the functions

$$H_{0,i}(x) = \left[1 - 2(x - x_i)\frac{dL_i}{dx}(x_i)\right] L_i^2(x), \tag{1.4-5}$$

$$H_{1,i}(x) = (x - x_i)L_i^2(x) \tag{1.4-6}$$

satisfy these conditions. Here, L_i denotes the Lagrange basis function of degree N associated with the node x_i, defined in Equation (1.2-1). With these **Hermite interpolating basis functions**, the function $\hat{f}$ given in Equation (1.4-2) solves the interpolation problem (1.4-1).

One important lesson of the previous two sections is that piecewise polynomial interpolation has much to offer over global polynomial interpolation. Perhaps the most important version of piecewise Hermite interpolation is the cubic case. Following notation introduced in the previous section, the space of piecewise Hermite cubics on the grid $\Delta = \{x_0, x_1, \ldots, x_N\}$ is $\mathcal{M}_1^3(\Delta)$, which contains all functions $\hat{f} \in C^1([a,b])$ whose restrictions to any element of the grid are cubic polynomials. The elements in this case are subintervals of the form $[x_{i-1}, x_i]$, containing two nodes. The four data $y_{i-1}, y_{i-1}', y_i, y_i'$ give exactly the number of conditions needed to determine the four coefficients of a cubic polynomial on $[x_{i-1}, x_i]$.

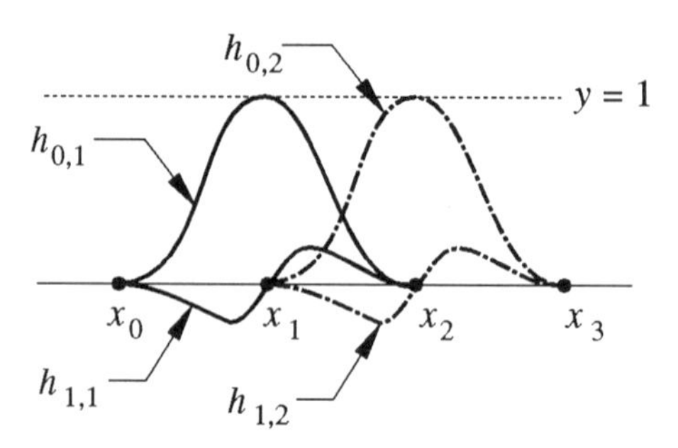

FIGURE 1. *Piecewise Hermite cubic basis functions associated with the nodes x_1 and x_2.*

We define a nodal basis for $\mathcal{M}_1^3(\Delta)$ using conditions similar to those used in global Hermite interpolation:

$$h_{0,i}(x_j) = \begin{cases} 1, & \text{if } i = j, \\ 0, & \text{if } i \neq j, \end{cases} \qquad \frac{dh_{0,i}}{dx}(x_j) = 0,$$

for $i, j = 0, 1, \ldots, N$, and

$$h_{1,i}(x_j) = 0, \qquad \frac{dh_{1,i}}{dx}(x_j) = \begin{cases} 1, & \text{if } i = j, \\ 0, & \text{if } i \neq j, \end{cases}$$

for $i, j = 0, 1, \ldots, N$. These equations assign to each node x_i two basis functions $h_{0,i}$ and $h_{1,i}$. The following equations give explicit formulas for these functions in the case where $x_i - x_{i-1} = x_{i+1} - x_i = h$:

$$h_{0,i}(x) = \begin{cases} h^{-3}(x - x_{i-1})^2[2(x_i - x) + h], & \text{if } x_{i-1} \leq x \leq x_i, \\ h^{-3}(x_{i+1} - x)^2[3h - 2(x_{i+1} - x)], & \text{if } x_i \leq x \leq x_{i+1}, \\ 0, & \text{otherwise;} \end{cases}$$

$$h_{1,i}(x) = \begin{cases} h^{-2}(x - x_{i-1})^2(x - x_i), & \text{if } x_{i-1} \leq x \leq x_i, \\ h^{-2}(x_{i+1} - x)^2(x - x_i), & \text{if } x_i \leq x \leq x_{i+1}, \\ 0, & \text{otherwise.} \end{cases}$$

As illustrated in Figure 1, each of these functions coincides with one nonzero cubic polynomial over the element $[x_{i-1}, x_i]$, with another nonzero cubic over $[x_i, x_{i+1}]$, and with zero outside these two subintervals. (The functions $h_{0,0}$, $h_{1,0}$, $h_{0,N}$, and $h_{1,N}$, being associated with the end nodes x_0 and x_N, are nonzero over only one element.)

Given this nodal basis, we solve the interpolation problem (1.4-1) by constructing the piecewise cubic, continuously differentiable function

$$\hat{f}(x) = \sum_{i=0}^{N} [y_i h_{0,i}(x) + y_i' h_{1,i}(x)].$$

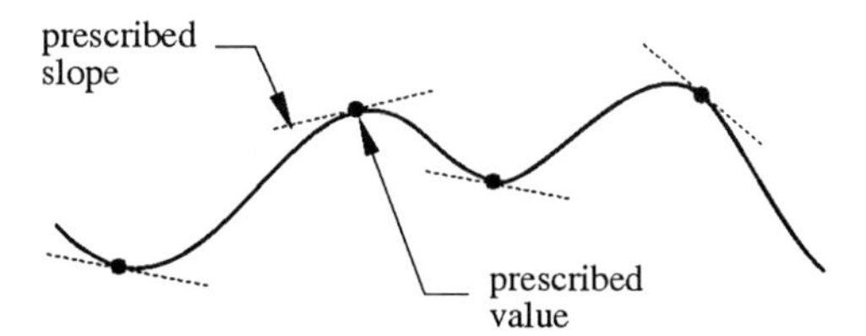

FIGURE 2. *Example of a piecewise Hermite cubic that interpolates prescribed nodal values and slopes.*

Figure 2 illustrates how $\hat{f}$ matches the prescribed function values y_i and slopes y_i' at each node x_i. This interpolant has a derivative that is well defined and continuous throughout $[a,b]$, namely,

$$\frac{d\hat{f}}{dx}(x) = \sum_{i=0}^{N} \left[y_i \frac{dh_{0,i}}{dx}(x) + y_i' \frac{dh_{1,i}}{dx}(x) \right].$$

Evaluating either of these sums at a typical point $x \in [a,b]$ involves calculating only those terms that are nonzero over the element $[x_{i-1}, x_i]$ to which x belongs. In this case, there are four such terms: two associated with the index $i-1$ and two associated with the index i.

As a concrete example, consider the piecewise Hermite cubic interpolant for $f(x) = \sin x$ on $[0, 2\pi]$, using the grid $\Delta = \{0, \pi/2, \pi, 3\pi/2, 2\pi\}$. At $x = 1$, the interpolant has the value

$$\hat{f}(1) = \sin(0)\, h_{0,0}(1) + \cos(0)\, h_{1,0}(1) + \sin(\pi/2)\, h_{0,1}(1) + \cos(\pi/2)\, h_{1,1}(1),$$

since the argument $x = 1$ lies in the subinterval $[x_0, x_1] = [0, \pi/2]$ formed by Δ. Because $\sin(0) = \cos(\pi/2) = 0$ and $\cos(0) = \sin(\pi/2) = 1$,

$$\begin{aligned}\hat{f}(1) = h_{1,0}(1) + h_{0,1}(1) &= \left(\frac{\pi}{2}\right)^{-2} \left(\frac{\pi}{2} - 1\right)^2 + \left(\frac{\pi}{2}\right)^{-3} \left[2\left(\frac{\pi}{2} - 1\right) + \frac{\pi}{2}\right] \\ &\simeq 0.831874.\end{aligned}$$

This value compares with the true value $\sin(1) \simeq 0.841487$.

Practical Considerations

We have already established the existence of Hermite interpolants by exhibiting polynomials that satisfy the interpolation constraints. Uniqueness follows by a zero-counting argument only slightly more sophisticated than the one used in the Lagrange case. We examine this argument in a general setting later in this section. For now, let us consider the questions of existence and uniqueness to be settled and focus on some practical aspects of piecewise Hermite cubic interpolation.

First, we consider error estimates. As with Lagrange interpolants, error estimates for the piecewise case are direct corollaries of estimates for global polynomial interpolation, which we examine rigorously below. The important result has a familiar form, bounding $\|f - \hat{f}\|_\infty$ in terms of a power of h and the norm of a corresponding derivative of f:

THEOREM 1.8. *Let $f \in C^4([a,b])$, and let $\Delta = \{x_0, x_1, \ldots, x_N\}$ be a grid on $[a,b]$ having mesh size h. Then the piecewise Hermite cubic function $\hat{f} \in \mathcal{M}_1^3(\Delta)$ that satisfies the interpolation problem (1.4-1), with $y_i = f(x_i)$ and $y_i' = f'(x_i)$ for $i = 0, 1, \ldots, N$, obeys the following error estimate:*

$$\|f - \hat{f}\|_\infty \leq C\|f^{(4)}\|_\infty h^4.$$

(Henceforth, unless we explicitly indicate the contrary, the symbol C denotes a positive constant, independent of the mesh size h, that can vary from estimate to estimate.) Thus piecewise Hermite cubic interpolation yields an error that is $\mathcal{O}(h^4)$, which is comparable to the error in piecewise Lagrange cubic interpolation.

When is piecewise Hermite cubic interpolation more appropriate than piecewise Lagrange cubic interpolation? This question calls for some judgment. Hermite interpolation may be more appropriate when one has numerical information about slopes at the nodes and when the intended application calls for a continuously differentiable interpolant. (Section 1.6 explores an approach that produces smooth interpolants without requiring numerical information about nodal slopes.) In a rough sense, piecewise Hermite cubics expend some of their interpolating power satisfying the constraints imposed by continuous differentiability. As a consequence, these interpolants demand about twice as much computational effort per node as piecewise Lagrange cubics.

Mathematical Details

Hermite polynomial interpolation admits more generality than the discussion so far suggests. Once we decide to match interpolants to prescribed nodal derivatives as well as function values, there is no a priori reason to limit the order of derivatives matched to 1 or, for that matter, to prescribe the same number of derivatives at all nodes. We may as well consider interpolation problems of the following form: Find a polynomial $\hat{f} : [a,b] \to \mathbb{R}$ such that

$$\hat{f}(x_i) = y_i^0, \quad \hat{f}'(x_i) = y_i^1, \quad \ldots, \quad \hat{f}^{(m_i)}(x_i) = y_i^{m_i}, \tag{1.4-7}$$

where $x_i, i = 0, 1, \ldots, N$, ranges over the nodes of a grid Δ. Here, we prescribe values of the first m_i derivatives of $\hat{f}$ at each node x_i, the integer m_i possibly varying from node to node.

In this case we expect to find $\hat{f} \in \Pi_d([a,b])$, where the degree

$$d = \sum_{i=0}^{N} (m_i + 1) - 1$$

guarantees that the number of interpolatory constraints matches the number of coefficients to be determined. Thus we must solve $d+1$ equations, given by the interpolatory constraints (1.4-3), for the $d+1$ coefficients of $\hat{f}$. These equations are linear in each coefficient: If $\hat{f}(x) = c_d x^d + c_{d-1}x^{d-1} + \cdots + c_0$, then the constraint $\hat{f}'(x_3) = y_3^1$, for example, has the explicit form

$$d x_3^{d-1} c_d + (d-1) x_3^{d-2} c_{d-1} + \cdots + c_1 = y_3^1.$$

The system of all such constraints has the form $\mathsf{A}\mathbf{c} = \mathbf{y}$, where A is a $(d+1) \times (d+1)$ matrix, $\mathbf{c}$ stands for the vector containing unknown coefficients $c_0, c_1, \ldots, c_d$, and $\mathbf{y}$ denotes the vector of prescribed values y_i^j. To show existence and uniqueness of the interpolant $\hat{f}$, we prove the following:

PROPOSITION 1.9. *The matrix* A *is nonsingular.*

The proof uses the fact that, if the number ξ is a zero of a polynomial p and $p(\xi) = p'(\xi) = \cdots = p^{(m-1)}(\xi) = 0$, then $p(x) = (x - \xi)^m q(x)$ for some other polynomial $q(x)$. The number m is the **multiplicity** of the zero ξ. We generalize this notion in Chapter 3.

PROOF: By Theorem 1.2, it suffices to show that the only solution to the system $\mathsf{A}\mathbf{c} = \mathbf{0}$ is the trivial solution, $\mathbf{c} = \mathbf{0}$. This system is equivalent to the following set of interpolatory constraints:

$$\hat{f}(x_i) = 0, \quad \hat{f}'(x_i) = 0, \quad \ldots, \quad \hat{f}^{(m_i)}(x_i) = 0, \qquad i = 0, 1, \ldots, N.$$

Thus, either $\hat{f}$ is the zero polynomial or else $\hat{f}$ has a zero of multiplicity $m_i + 1$ at each node x_i. In the latter case, $\hat{f}$ must be a multiple of the polynomial

$$(x - x_0)^{m_0+1}(x - x_1)^{m_1+1} \cdots (x - x_N)^{m_N+1}.$$

However, this polynomial has degree

$$\sum_{i=0}^{N} (m_i + 1) = d + 1,$$

and hence no nonzero polynomial in $\Pi_d([a,b])$ can be a multiple of it. The only remaining possibility is that $\hat{f}$ must be identically zero, that is, $c_0 = c_1 = \cdots = c_d = 0$. ∎

The greater generality afforded by the global interpolation constraints (1.4-7) translates, in a limited way, to greater generality in the piecewise case. For example, one can satisfy the interpolation constraints

$$\hat{f}(x_i) = y_i^0, \quad \hat{f}'(x_i) = y_i^1, \quad i = 0, 1, \ldots, N, \tag{1.4-8}$$

by constructing a piecewise quintic function whose coefficients over each element $[x_j, x_{j+2}]$, $j = 0, 2, 4, \ldots, N-2$, are chosen to satisfy the six equations of the form (1.4-4) for $i = j, j+1, j+2$. Alternatively, one can use a piecewise quintic to satisfy the interpolation constraints

$$\hat{f}(x_i) = y_i^0, \quad \hat{f}'(x_i) = y_i^1, \quad \hat{f}''(x_i) = y_i^2, \quad i = 0, 1, \ldots, N. \tag{1.4-9}$$

Here, $\hat{f}$ must be quintic over each element $[x_j, x_{j+1}]$, and one chooses its coefficients there to satisfy the six conditions of the form (1.4-9) applicable at the nodes x_j and x_{j+1}.

These observations notwithstanding, the most commonly used forms of piecewise Hermite interpolation employ only the values of $\hat{f}(x_i)$ and $\hat{f}'(x_i)$. It is for this case that we examine error estimates in detail. As with Lagrange interpolation, the estimates for the piecewise case follow from estimates for the global case, which the following theorem establishes.

THEOREM 1.10. *Given a function $f \in C^{2N+2}([a, b])$ and a grid*

$$\Delta = \left\{x_0, x_1, \ldots, x_N\right\}$$

on $[a, b]$, let $\hat{f} : [a, b] \to \mathbb{R}$ be the Hermite interpolant of f satisfying the equation

$$\hat{f}(x) = \sum_{i=0}^{N} [f(x_i) H_{0,i}(x) + f'(x_i) H_{1,i}(x)].$$

Then for any $x \in [a, b]$ there exists a point $\zeta \in (a, b)$ such that

$$f(x) - \hat{f}(x) = \frac{\omega_N^2(x)}{(2N+2)!} f^{(2N+2)}(\zeta).$$

PROOF (SKETCH): Define

$$F(t) = f(t) - \hat{f}(t) - [f(x) - \hat{f}(x)] \frac{\omega_N^2(t)}{\omega_N^2(x)}, \tag{1.4-10}$$

then reason as for the Lagrange case, noting that F' has $2N+2$ distinct zeros. Problem 7 asks for details. ∎

By the usual device of estimating $|\omega_N(x)|$, one can readily convert this theorem to an estimate having the form

$$\|f - \hat{f}\|_\infty \le C \|f^{(2N+2)}\|_\infty (b-a)^{2N+2}.$$

It is also possible to estimate the errors associated with derivatives of Hermite interpolants:

THEOREM 1.11. *With f and $\hat{f}$ as in Theorem 1.10,*

$$\|f^{(k)} - \hat{f}^{(k)}\|_\infty \leq C\|f^{(2N+2)}\|_\infty (b-a)^{2N+2-k}, \tag{1.4-11}$$

for $k = 0, 1, \ldots, 2N+1$.

By now the strategy should be cloying:

PROOF (SKETCH): For successively higher values of k, use the Rolle theorem to locate the zeros of $f^{(k)} - \hat{f}^{(k)}$. Then analyze a function analogous to the one defined in Equation (1.4-10) using repeated applications of the Rolle theorem. ∎

Error estimates for piecewise Hermite interpolation now fall out easily:

THEOREM 1.12. *Let $f \in C^{2n+2}([a,b])$, and let $\Delta = \{x_0, x_1, \ldots, x_N\}$ be a grid on $[a,b]$ having mesh size h, with N an integer multiple of n. If $\hat{f} : [a,b] \to \mathbb{R}$ is the piecewise Hermite interpolant of f having degree at most $2n+1$ on each of the elements $[x_0, x_n], [x_n, x_{2n}], \ldots, [x_{N-n}, x_N]$, then*

$$\|f^{(k)} - \hat{f}^{(k)}\|_\infty \leq C\|f^{(2n+2)}\|_\infty h^{2n+2-k}, \qquad k = 0, 1, \ldots, 2n+1. \tag{1.4-12}$$

(The value of C depends on n.)

PROOF: In the piecewise context, the factor $b - a$ appearing in the estimates (1.4-7) corresponds to the element length, and N corresponds to the number n of subintervals $[x_i, x_{i+1}]$ contained in an element. Element length, in turn, is bounded above by h in the piecewise cubic case ($n = 1$), where elements have the form $[x_i, x_{i+1}]$; by $2h$ in the piecewise quintic case ($n = 2$), where elements have the form $[x_i, x_{i+2}]$; and so forth. Therefore, by replacing the factor $(b-a)^{2N+2-k}$ in the estimate (1.4-11) by $(nh)^{2n+2-k}$, taking the norm $\|f^{(2n+2)}\|_\infty$ over the union of all elements, and absorbing all constant factors into C, we obtain the estimate (1.4-12). ∎

The error estimate given in Theorem 1.8 for piecewise Hermite cubic interpolation is just an instance of the estimate (1.4-12) for the case $n = 1$, $k = 0$.

1.5 Interpolation in Two Dimensions

When the function to be interpolated depends on several variables, there are many new avenues to explore. This section briefly introduces some of the possibilities for functions of two variables. The two main ideas that we investigate are tensor-product interpolation and piecewise linear interpolation on triangles.

Constructing Tensor-Product Interpolants

Tensor-product interpolation offers the most direct way to use one-dimensional results in multidimensional settings. The idea is to use one-dimensional interpolation in each coordinate direction, allowing the products of the interpolating functions to govern variations in directions oblique to the coordinate axes.

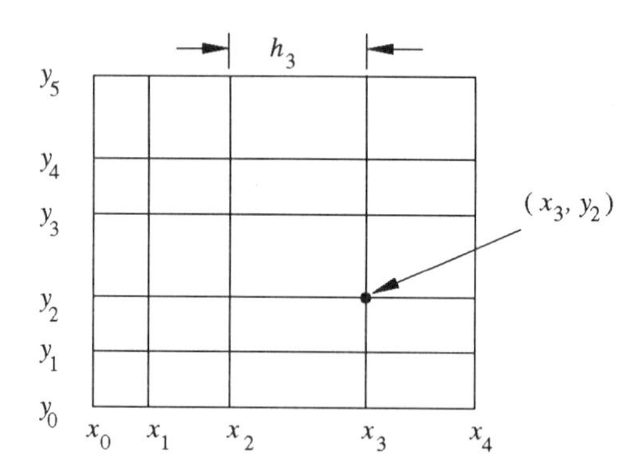

FIGURE 1. *The rectangle* $[a,b]\times[c,d]$, *along with the points* (x_i, y_j) *formed by grid-line intersections associated with grids* Δ_1 *and* Δ_2.

Central to tensor-product interpolation is the notion of Cartesian product, reviewed in Section 0.1. The Cartesian product of two intervals $[a,b]$ and $[c,d]$ is the rectangle $[a,b] \times [c,d]$ containing all points (x,y) for which $a \leq x \leq b$ and $c \leq y \leq d$, as drawn in Figure 1. Given grids $\Delta_1 = \{x_0, x_1, \ldots, x_M\}$ and $\Delta_2 = \{y_0, y_1, \ldots, y_N\}$ on $[a,b]$ and $[c,d]$, respectively, their Cartesian product is the set $\Delta_1 \times \Delta_2$ containing all of the points (x_i, y_j), where $x_i \in \Delta_1$ and $y_j \in \Delta_2$. Think of these points as intersections of the horizontal and vertical **grid lines** $x = x_i$ and $y = y_j$ in the rectangle $[a,b] \times [c,d]$, illustrated in Figure 1.

Given the grid $\Delta_1 \times \Delta_2$ on $[a,b] \times [c,d]$ and a collection $\{z_{i,j} : i = 0,1,\ldots,M;\ j = 0,1,\ldots,N\}$ of ordinates, we seek an interpolant $\hat{f} : [a,b] \times [c,d] \to \mathbb{R}$ such that $\hat{f}(x_i, y_j) = z_{i,j}$ for $i = 0,1,\ldots,M$ and $j = 0,\ldots,N$. It is also possible to demand that certain derivatives of $\hat{f}$ must agree with prescribed values, as with Hermite interpolation for functions of a single variable.

The first step is to construct a basis for the interpolation. This chore is easy: We use products of basis functions associated with Δ_1 and basis functions associated with Δ_2 as the two-dimensional basis functions.

DEFINITION. *Let* Δ_1 *and* Δ_2 *be grids on* $[a,b]$ *and* $[c,d]$, *respectively, and consider corresponding interpolation spaces* $\mathcal{M}(\Delta_1)$ *and* $\mathcal{N}(\Delta_2)$ *having bases*

$\{\varphi_0, \varphi_1, \ldots, \varphi_M\}$ *and* $\{\psi_0, \psi_1, \ldots, \psi_N\}$, *respectively. The* **tensor-product space** *is*

$$\mathcal{M}(\Delta_1) \otimes \mathcal{N}(\Delta_2) := \operatorname{span}\left\{\varphi_i\psi_j : i = 0, 1, \ldots, M;\ j = 0, 1, \ldots, N\right\},$$

where $(\varphi_i\psi_j)(x, y) := \varphi_i(x)\psi_j(y)$.

Thus $\mathcal{M}(\Delta_1) \otimes \mathcal{N}(\Delta_2)$ contains all linear combinations

$$\sum_{i=0}^{M}\sum_{j=0}^{N} c_{i,j}\varphi_i(x)\psi_j(y).$$

It is straightforward albeit slightly tedious to show that the set $\mathcal{M}(\Delta_1) \otimes \mathcal{N}(\Delta_2)$ is a vector space, that the set of products $\varphi_i\psi_j$ indeed constitutes a basis, and that the elements of $\mathcal{M}(\Delta_1) \otimes \mathcal{N}(\Delta_2)$ are independent of the particular bases used for the separate spaces $\mathcal{M}(\Delta_1)$ and $\mathcal{N}(\Delta_2)$. Problem 9 asks for details.

Perhaps the simplest example of a tensor-product interpolation space is $\mathcal{M}_0^1(\Delta_1) \otimes \mathcal{M}_0^1(\Delta_2)$, the space of **piecewise Lagrange bilinear** interpolants on $\Delta = \Delta_1 \times \Delta_2$. Functions in this space are piecewise linear along each line in the x- and y-directions, like the function graphed in Figure 2. If we denote the nodal bases for $\mathcal{M}_0^1(\Delta_1)$ and $\mathcal{M}_0^1(\Delta_2)$ by $\{\ell_0^1, \ell_1^1, \ldots, \ell_M^1\}$ and $\{\ell_0^2, \ell_1^2, \ldots, \ell_N^2\}$, respectively, then the tensor-product space has the basis

$$\left\{\ell_i^1\ell_j^2 : i = 0, 1, \ldots, M;\ j = 0, 1, \ldots, N\right\}.$$

A typical function $\ell_i^1\ell_j^2$ in this basis has a graph that looks like a pyramidal tent over the four-element rectangle $[x_{i-1}, x_{i+1}] \times [y_{j-1}, y_{j+1}]$, as depicted in Figure 3. Any cross-section of this graph along a line x = constant or y = constant consists of line segments. However, any other cross-section, such as that lying above the diagonal line connecting the corner points (x_{i-1}, y_{j-1}) and (x_{i+1}, y_{j+1}), consists of quadratic arcs. Given this basis, the solution to the two-dimensional interpolation problem on $\Delta_1 \times \Delta_2$ is the function

$$\hat{f}(x, y) = \sum_{i=0}^{M}\sum_{j=0}^{N} z_{i,j}\ell_i^1(x)\ell_j^2(y). \tag{1.5-1}$$

As a numerical example, take $[a, b] \times [c, d]$ to be the rectangle $[0, 4] \times [3, 7]$, and let $\Delta_1 = \{0, 1, 3, 4\}$ and $\Delta_2 = \{3, 6, 7\}$. Let $\hat{f}$ be the piecewise bilinear function on $\Delta_1 \times \Delta_2$ interpolating the data $z_{i,j} = \exp(x_i - y_j)$. To compute $\hat{f}(2, 5)$, note that $(2, 5) \in [x_1, x_2] \times [y_0, y_1]$. Evaluating the four corresponding nonzero terms in the sum (1.5-1) gives

$$\begin{aligned}\hat{f}(2, 5) &= z_{1,0}\ell_1^1(2)\ell_0^2(5) + z_{2,0}\ell_2^1(2)\ell_0^2(5) + z_{1,1}\ell_1^1(2)\ell_1^2(5) + z_{2,1}\ell_2^1(2)\ell_1^2(5) \\ &= e^{1-3} \cdot \tfrac{1}{2} \cdot \tfrac{1}{3} + e^{3-3} \cdot \tfrac{1}{2} \cdot \tfrac{1}{3} + e^{1-6} \cdot \tfrac{1}{2} \cdot \tfrac{2}{3} + e^{3-6} \cdot \tfrac{1}{2} \cdot \tfrac{2}{3} \\ &\simeq 0.208064.\end{aligned}$$

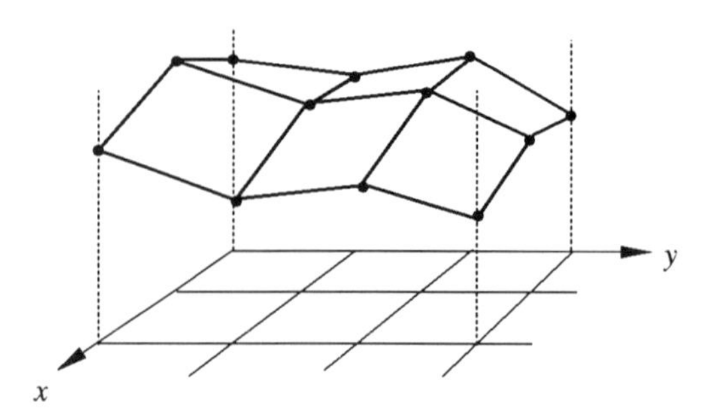

FIGURE 2. *Graph of a piecewise Lagrange bilinear function.*

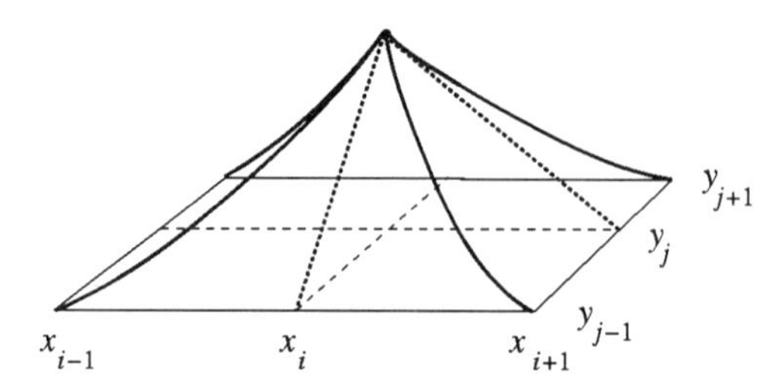

FIGURE 3. *Graph of a nodal basis function $\ell_i^1(x)\ell_j^2(y)$ for piecewise Lagrange bilinear interpolation.*

This result compares with the value $\exp(x-y) = e^{-3} \simeq 0.049787$; the interpolated answer in this case is wrong by a factor greater than 4. (Why is this interpolated value so inaccurate? How could we construct a more accurate tensor-product interpolant?)

For a somewhat more complicated instance of tensor-product interpolation, consider the **piecewise Hermite bicubic** interpolants. These functions belong to the space $\mathcal{M}_1^3(\Delta_1) \otimes \mathcal{M}_1^3(\Delta_2)$, which has a basis

$$\{h_{k,i}^1 h_{l,j}^2 \ : \ k,l = 0 \text{ or } 1; \ i = 0,1,\ldots,M; \ j = 0,1,\ldots,N\}.$$

This basis associates four functions $h_{k,i}^1 h_{l,j}^2$ with each node (x_i, y_j) in the grid. (It is a worthwhile exercise to compute the values of the derivatives $\partial/\partial x$, $\partial/\partial y$, and $\partial^2/\partial x \partial y$ of each of these functions at each node (x_p, y_q).) To interpolate a function $f : [a,b] \times [c,d] \to \mathbb{R}$ in this space, we therefore

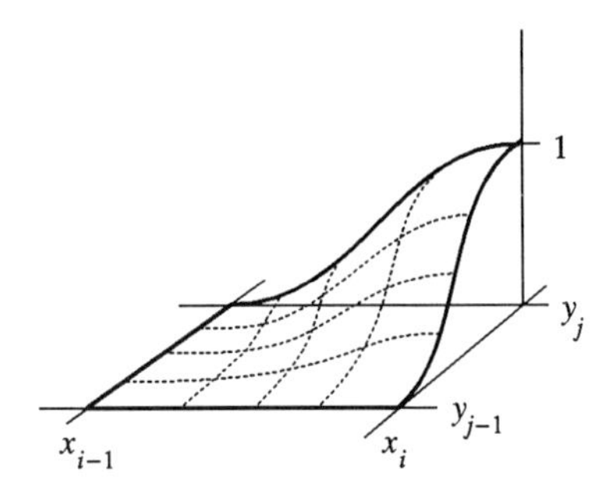

FIGURE 4. *A typical element for piecewise Hermite bicubic interpolation.*

must know four corresponding values

$$z_{i,j} = f(x_i, y_j), \qquad z_{i,j}^{(x)} = \frac{\partial f}{\partial x}(x_i, y_j),$$

$$z_{i,j}^{(y)} = \frac{\partial f}{\partial y}(x_i, y_j), \quad z_{i,j}^{(xy)} = \frac{\partial^2 f}{\partial x \partial y}(x_i, y_j).$$

Given data of this sort, the linear combination

$$\hat{f}(x,y) = \sum_{i=0}^{M}\sum_{j=0}^{N} \Big[z_{i,j} h^1_{0,i}(x) h^2_{0,j}(y) + z^{(x)}_{i,j} h^1_{1,i}(x) h^2_{0,j}(y) + z^{(y)}_{i,j} h^1_{0,i}(x) h^2_{1,j}(y) + z^{(xy)}_{i,j} h^1_{1,i}(x) h^2_{1,j}(y) \Big]$$

defines a continuously differentiable interpolant of f on the rectangle $[a,b] \times [c,d]$. To differentiate $\hat{f}$, we apply the appropriate derivative operators to each term in this sum. For example,

$$\frac{\partial \hat{f}}{\partial y}(x,y) = \sum_{i=0}^{M}\sum_{j=0}^{N} \left[z_{i,j} h^1_{0,i}(x) \frac{dh^2_{0,j}}{dy}(y) + z^{(x)}_{i,j} h^1_{1,i}(x) \frac{dh^2_{0,j}}{dy}(y) + z^{(y)}_{i,j} h^1_{0,i}(x) \frac{dh^2_{1,j}}{dy}(y) + z^{(xy)}_{i,j} h^1_{1,i}(x) \frac{dh^2_{1,j}}{dy}(y) \right].$$

To evaluate $\hat{f}(x,y)$ or any of its derivatives at a point (x,y) in the rectangle $[a,b] \times [c,d]$, it is typically necessary to compute 16 terms in this sum, namely, the four terms associated with each of the four corner nodes for the element $[x_{i-1}, x_i] \times [y_{j-1}, y_j]$ in which (x,y) lies, as shown in Figure 4.

The one-dimensional spaces $\mathcal{M}(\Delta_1)$ and $\mathcal{N}(\Delta_2)$ need not have the same piecewise polynomial degree. For example, try sketching some of the ramplike functions in the tensor-product space $\mathcal{M}^1_0(\Delta_1) \otimes \mathcal{M}^0_{-1}(\Delta_2)$.

Also, nothing prevents us from extending the tensor-product approach to the interpolation of functions $f(x_1, x_2, \ldots, x_d)$ of arbitrarily many variables. The main requirement is that the function f be defined on some hypercube $[a_1, b_1] \times [a_2, b_2] \times \cdots \times [a_d, b_d]$ in the d-dimensional Euclidean space $\mathbb{R}^d$. If this is the case, then we can construct a d-dimensional grid $\Delta_1 \times \Delta_2 \times \cdots \times \Delta_d$ using grids in each of the orthogonal coordinate directions $x_1, x_2, \ldots, x_d$. By associating with each one-dimensional grid Δ_k an interpolating space $\mathcal{M}_k(\Delta_k)$, we automatically have a tensor-product scheme, for which the space of interpolating functions is

$$\mathcal{M}_1(\Delta_1) \otimes \mathcal{M}_2(\Delta_2) \otimes \cdots \otimes \mathcal{M}_d(\Delta_d).$$

Error Estimates for Tensor-Product Schemes

Tensor-product interpolation schemes inherit error estimates from the interpolation schemes used in the individual coordinate directions. To show this for the two-dimensional case, we introduce the notion of **interpolatory projections**. An interpolatory projection is the mapping that associates with each function f its interpolant $\hat{f}$ in some predetermined interpolation space. For example, suppose $f \in C^2([a, b])$. The piecewise linear interpolant of f on a grid $\Delta = \{x_0, x_1, \ldots, x_N\}$ on the interval $[a, b]$ is the function $\hat{f} \in \mathcal{M}_0^1(\Delta) = \text{span}\{\ell_0, \ell_1, \ldots, \ell_N\}$ defined by the equation

$$\hat{f}(x) = \sum_{i=0}^{N} f(x_i)\ell_i(x).$$

The interpolatory projection in this case is the mapping $\pi : f \mapsto \hat{f}$.

When $f : [a, b] \times [c, d] \to \mathbb{R}$, the same notion applies to each of the arguments of f. For example, if f is twice continuously differentiable in x and in y, we might interpolate f by using piecewise bilinear functions defined over a two-dimensional grid $\Delta_1 \times \Delta_2$, as described above. The interpolatory projections $\pi_1 : C^2([a, b]) \to \mathcal{M}_0^1(\Delta_1)$ and $\pi_2 : C^2([c, d]) \to \mathcal{M}_0^1(\Delta_2)$ are then as follows:

$$(\pi_1 f)(x, y) := \sum_{i=0}^{M} f(x_i, y)\ell_i^1(x),$$

$$(\pi_2 f)(x, y) := \sum_{j=0}^{N} f(x, y_j)\ell_j^2(y).$$

Thus the tensor-product interpolant of f in $\mathcal{M}_0^1(\Delta_1) \otimes \mathcal{M}_0^1(\Delta_2)$ is the function

$\pi_1\pi_2 f$, where

$$(\pi_1\pi_2 f)(x,y) = (\pi_1(\pi_2 f))(x,y) = \sum_{i=0}^{M} (\pi_2 f)(x_i,y)\ell_i^1(x)$$

$$= \sum_{i=0}^{M}\sum_{j=0}^{N} f(x_i,y_j)\ell_i^1(x)\ell_j^2(y).$$

It should be clear that $\pi_2\pi_1 f = \pi_1\pi_2 f$. Recasting error estimates for one-dimensional piecewise polynomial interpolation in this new formalism, we find that $||f - \pi f||_\infty \le Ch^p$. Here, h signifies the mesh size of the grid, and C is a generic positive constant depending on f.

There is a connection between one-dimensional error estimates and the error estimates for the corresponding tensor-product scheme. In what follows, we use the notation $f(\cdot, y)$ to indicate that we vary the first argument of f, holding the second argument at a fixed value y, and vice versa for $f(x,\cdot)$.

THEOREM 1.13. *Let Δ_1 and Δ_2 be grids on $[a,b]$ and $[c,d]$, respectively, and suppose that Δ_1 has mesh size h_1 and Δ_2 has mesh size h_2. Let $\mathcal{M}(\Delta_1)$ and $\mathcal{N}(\Delta_2)$ be piecewise polynomial spaces with associated interpolatory projections π_1 and π_2, respectively. Suppose that $\mathcal{S}$ is a vector space of functions $f : [a,b]\times[c,d] \to \mathbb{R}$ such that $\pi_1, \pi_2 : \mathcal{S} \to \mathcal{S}$ and that we have interpolation error estimates of the form*

$$||f(\cdot,y) - \pi_1 f(\cdot,y)||_\infty \le Ch_1^p, \qquad ||f(x,\cdot) - \pi_2 f(x,\cdot)||_\infty \le Ch_2^q.$$

Then the tensor-product space $\mathcal{M}(\Delta_1)\otimes\mathcal{N}(\Delta_2)$ contains an interpolant $\pi_1\pi_2 f$ obeying the estimate

$$||f - \pi_1\pi_2 f||_\infty \le C\max\{h_1^p, h_2^q\}.$$

The constants C are independent of x and y. However, C can be different in different inequalities, in accordance with our convention. Also, if $h = \max\{h_1,h_2\}$ and $p,q \ge 0$, then

$$\max\{h^p, h^q\} = \mathcal{O}(h^{\min\{p,q\}}) \qquad \text{as} \quad h \to 0.$$

Hence the accuracy of the tensor-product interpolation scheme is limited by the accuracy of the one-dimensional scheme that has lower order.

PROOF (SKETCH): Apply the triangle inequality to the identity

$$f - \pi_1\pi_2 f = f - \pi_1 f \;+\; (\pi_1 f - \pi_1\pi_2 f) - (f - \pi_2 f) \;+\; f - \pi_2 f.$$

Problem 11 asks for details. ∎

One can easily generalize Theorem 1.13 to functions of d variables using mathematical induction.

Interpolation on Triangles: Background

One problem with tensor-product interpolation is its reliance on domains that are Cartesian products of intervals. Thus, in two dimensions we find ourselves restricted to rectangular domains of the form $[a,b] \times [c,d]$. There are ways of "jury-rigging" tensor-product schemes to handle nonrectangular domains. For example, one can use curvilinear coordinate systems, either globally (througout $[a,b] \times [c,d]$) or locally (by using a separate coordinate transformation on each element $[x_{i-1}, x_i] \times [y_{j-1}, y_j]$). However, it is perhaps more satisfying to construct interpolation schemes that have greater inherent geometric flexibility. In two dimensions, interpolation schemes based on triangles offer the simplest methods for accomplishing this task.

We begin by identifying domains that are amenable to decomposition into triangles:

DEFINITION. *A bounded, open, connected set $\Omega \subset \mathbb{R}^2$ is* **polygonal** *if it is simply connected and its boundary $\partial\Omega$ is a union of line segments. A* **triangular set** *is a polygonal set whose boundary is a triangle.*

This definition contains some technical verbiage, much of which we review in Section 0.2. To say that a bounded, open set in $\mathbb{R}^n$ is **connected** is to say that one can connect any two points in the set by a continuous path that lies entirely in the set. The term **simply connected** indicates that Ω has no "holes" or "islands." (Making this definition rigorous requires some work.) Figure 5 illustrates the idea.

The **boundary** $\partial\Omega$ of Ω is the set containing all limit points of Ω that are not also interior points of Ω. We denote by $\overline{\Omega}$ the **closure** of Ω, which is the union $\Omega \cup \partial\Omega$. Figure 6 illustrates these sets.

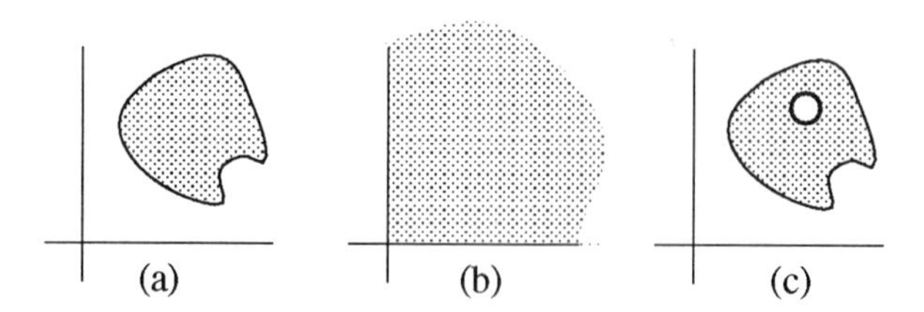

FIGURE 5. *(a) A bounded, simply connected set; (b) an unbounded set; (c) a set that is not simply connected.*

The definition of polygonal sets has a practical consequence. Figure 7(a) shows such a set. The requirement that the boundary $\partial\Omega$ be a union of line segments formally excludes domains having curved boundaries. However, as Figure 7(b) illustrates, it is often possible to approximate a nonpolygonal domain reasonably well by a polygonal one. Doing so raises the issue of how

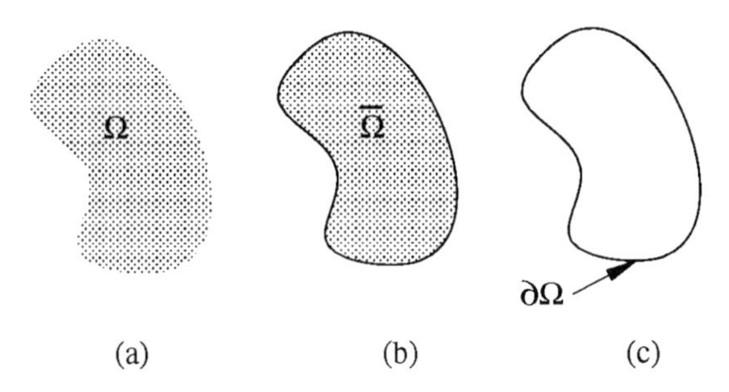

FIGURE 6. *(a) An open set Ω; (b) the set $\overline{\Omega}$ containing Ω and all of its limit points; (c) the boundary, $\partial\Omega$.*

good such an approximation can be — an issue that we do not explore here. Strang and Fix [13] give an introductory discussion of this problem.

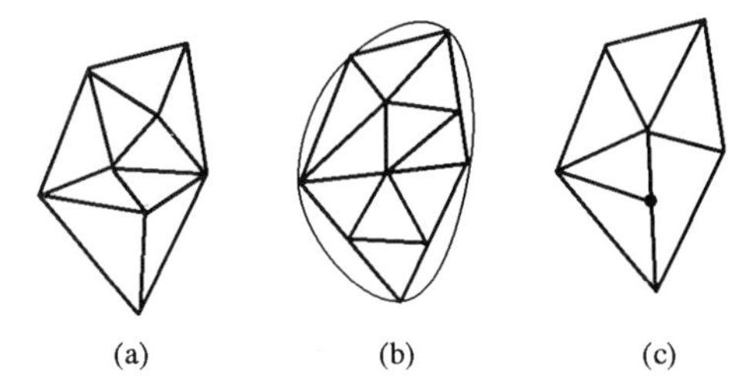

FIGURE 7. *(a) A polygonal set and a triangulation of it; (b) a nonpolygonal set with a polygonal approximation; (c) an invalid triangulation.*

Using these definitions, we partition polygonal sets into elements for piecewise polynomial interpolation.

DEFINITION. *Let Ω be a polygonal set. A* **triangulation** *of Ω is a decomposition of the closure $\overline{\Omega}$ into finitely many subsets,*

$$\overline{\Omega} = \bigcup_{e=1}^{E} \overline{\Omega}_e,$$

such that (i) each subset $\overline{\Omega}_e$ is the closure of a triangular set Ω_e and (ii) the intersection $\overline{\Omega}_e \cap \overline{\Omega}_f$ of any two of these subsets is either empty, a common vertex, or a common edge.

Figure 7(a) shows a triangulation of a polygonal set; Figure 7(c) illustrates a decomposition of the same set that is not a triangulation, since two of the subsets $\overline{\Omega}_e$ intersect in a forbidden way. Although we do not prove the fact here, every polygonal set has a trianglulation.

Construction of Linear Interpolants on Triangles

Given a triangulation of a polygonal set Ω, we interpolate functions $f : \overline{\Omega} \to \mathbb{R}$ by using the vertices of the triangular subsets Ω_e as nodes of a grid. The **elements** formed by this grid are the subsets $\overline{\Omega}_e$ in the triangulation. Thus the interpolation problem is as follows: Given a grid $\{\mathbf{x}_i = (x_i, y_i) : i = 0, 1, \ldots, N\}$ associated with a triangulation of Ω and a set $\{z_0, z_1, \ldots, z_N\}$ of corresponding ordinates, find a function $\hat{f} : \overline{\Omega} \to \mathbb{R}$ such that $\hat{f}(\mathbf{x}_i) = z_i$.

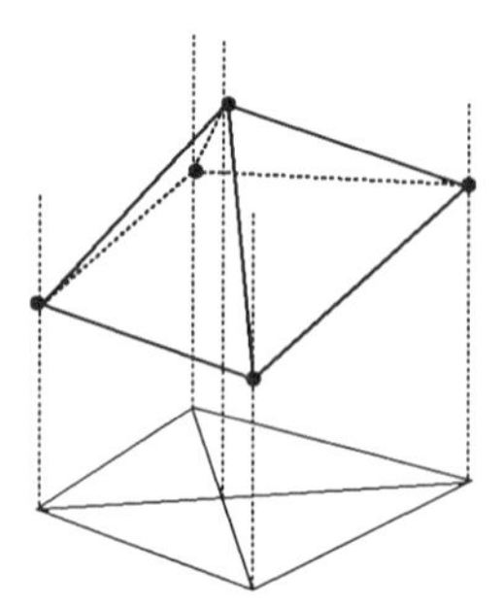

FIGURE 8. *Graph of a piecewise linear function $\hat{f}$ in two dimensions.*

The idea behind piecewise linear interpolation on triangles is to use the triples (x_i, y_i, z_i) associated with the vertices (x_i, y_i) of each triangle $\overline{\Omega}_e$ to define a plane segment over $\overline{\Omega}_e$. The interpolant $\hat{f}$ is then the function whose graph consists of these triangular plane segments, as Figure 8 illustrates.

The tricky part of this construction is to identify a nodal basis for the interpolation. We seek a collection $\{p_0, p_1, \ldots, p_N\}$ of piecewise linear functions such that, at any point $\mathbf{x} = (x, y) \in \overline{\Omega}$,

$$\hat{f}(\mathbf{x}) = \sum_{i=0}^{N} z_i p_i(\mathbf{x}).$$

Each function p_i will have the form

$$p_i(\mathbf{x}) = a_e x + b_e y + c_e, \qquad \text{for} \quad \mathbf{x} = (x, y) \in \overline{\Omega}_e,$$

with the coefficients a_e, b_e, c_e, $e = 1, 2, \ldots, E$, chosen to enforce the nodal constraints

$$p_i(\mathbf{x}_j) = \begin{cases} 1, & \text{if } i = j, \\ 0, & \text{if } i \neq j. \end{cases}$$

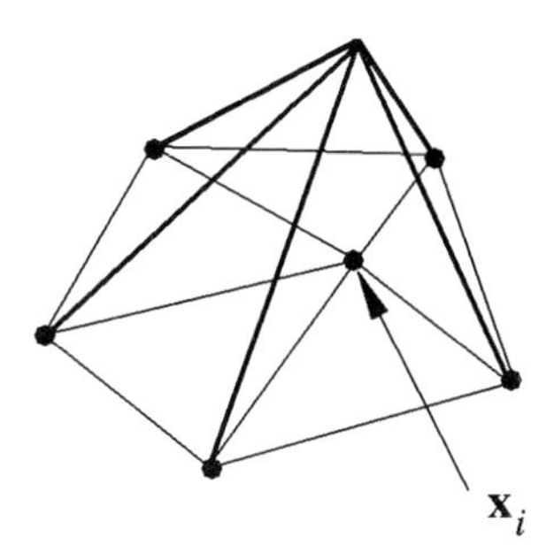

FIGURE 9. *Graph of a typical basis function p_i for piecewise linear interpolation on triangles.*

Figure 9 shows the graph of a typical basis function p_i. The function vanishes on any element $\overline{\Omega}_e$ for which the node $\mathbf{x}_i$ is not a vertex.

Using these constraints directly turns out not to be a desirable way to compute the basis functions. An equivalent but more geometrically motivated approach is to use the **areal coordinates** of points $\mathbf{x} = (x, y)$ with respect to the node $\mathbf{x}_i$. To compute $p_i(\mathbf{x})$, we first locate an element $\overline{\Omega}_e$ that contains $\mathbf{x}$. If $\mathbf{x}_i = (x_i, y_i)$ is not a vertex of $\overline{\Omega}_e$, then $p_i(\mathbf{x}) = 0$. Otherwise, denote the other vertices of $\overline{\Omega}_e$ as $\mathbf{x}_j = (x_j, y_j)$ and $\mathbf{x}_k = (x_k, y_k)$, as shown in Figure 10. The (signed) area of $\overline{\Omega}_e$ is

$$\begin{aligned} A_e &= \tfrac{1}{2}\det\begin{bmatrix} x_i & y_i & 1 \\ x_j & y_j & 1 \\ x_k & y_k & 1 \end{bmatrix} \\ &= \tfrac{1}{2}[x_i(y_j - y_k) - y_i(x_j - x_k) + x_j y_k + -x_k y_j]. \end{aligned}$$

Next, construct the triangle with vertices $\mathbf{x}$, $\mathbf{x}_j$, and $\mathbf{x}_k$. The (signed) area of this triangle is

$$A = \frac{1}{2}\det\begin{bmatrix} x & y & 1 \\ x_j & y_j & 1 \\ x_k & y_k & 1 \end{bmatrix}.$$

Finally, set $p_i(\mathbf{x}) = A/A_e$, if $\mathbf{x} \in \overline{\Omega}_e$. The function $p_i(\mathbf{x})$ vanishes at each of the nodes $\mathbf{x}_j, \mathbf{x}_k$ and takes the value 1 at $\mathbf{x}_i$. It is only slightly more difficult to check that p_i is linear in x and y inside $\overline{\Omega}_e$: For $\mathbf{x} \in \overline{\Omega}_e$, expand the determinant for the signed area A along its first row. Finally, $p_i(\mathbf{x})$ is well defined: If $\mathbf{x}$ lies on the boundary between two elements $\overline{\Omega}_e$ and $\overline{\Omega}_f$, then the value of $p_i(\mathbf{x})$ is the same whether one uses vertices from $\overline{\Omega}_e$ or $\overline{\Omega}_f$ to compute it.

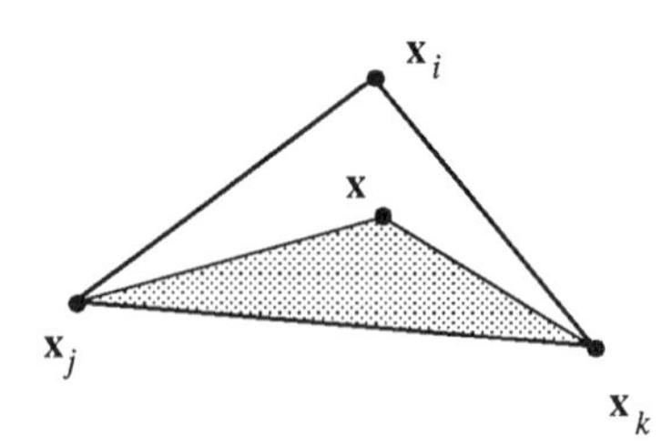

FIGURE 10. *An element* $\overline{\Omega}_e$ *with the triangular subset used to compute* $p_i(\mathbf{x})$.

Error Estimates for Interpolation on Triangles

Piecewise linear interpolation on triangles is a natural analog of piecewise linear interpolation in one dimension: It uses piecewise polynomials having the smallest degree needed to guarantee continuity. Heuristically, we might anticipate a comparable error estimate — namely, $\mathcal{O}(h^2)$. What is missing is a definition for the mesh size h of a triangulation.

DEFINITION. *The* **mesh size** *of a triangulation* $\overline{\Omega} = \bigcup_{e=1}^{E} \overline{\Omega}_e$ *is the length* h *of the longest edge occurring among the triangular sets* Ω_e.

To prove that the hunch is correct, we begin with the following:

LEMMA 1.14. *Let* $\mathsf{H} \in \mathbb{R}^{2\times 2}$ *have entries obeying the inequality* $|h_{i,j}| \leq M$ *for some* $M > 0$. *Then, for any* $\mathbf{u} \in \mathbb{R}^2$, $|\mathbf{u} \cdot \mathsf{H}\mathbf{u}| \leq 2M\|\mathbf{u}\|_2^2$.

PROOF: Let $\mathbf{u} = (u, v)$. We have

$$\begin{aligned} |\mathbf{u} \cdot \mathsf{H}\mathbf{u}| &\leq \begin{bmatrix} |u| \\ |v| \end{bmatrix} \cdot \begin{bmatrix} M & M \\ M & M \end{bmatrix} \begin{bmatrix} |u| \\ |v| \end{bmatrix} \\ &= M(u^2 + 2|u||v| + v^2) = M(|u| + |v|)^2. \end{aligned}$$

Since $(u-v)^2 \geq 0$, $2|u||v| \leq u^2+v^2$, and it follows from the triangle inequality that

$$|\mathbf{u} \cdot \mathsf{H}\mathbf{u}| \leq 2M(u^2 + v^2) = 2M\|\mathbf{u}\|_2^2. \quad \blacksquare$$

The proof of the main theorem also utilizes directional derivatives. If $\Psi: \mathbb{R}^2 \to \mathbb{R}$ is sufficiently smooth and $\mathbf{e} = (e_1, e_2)$ is a vector having unit length, then the **directional derivative** of Ψ in the direction of $\mathbf{e}$ at a point $\mathbf{x}$ is

$$\frac{\partial \Psi}{\partial e}(\mathbf{x}) := \nabla\Psi(\mathbf{x}) \cdot \mathbf{e}.$$

To justify this definition, consider the function $\eta(t) := \Psi(\mathbf{x} + t\mathbf{e})$. Here, the parameter t measures progress in the $\mathbf{e}$-direction, with $t = 0$ corresponding to the point $\mathbf{x}$. By the chain rule, the rate of change of η with respect to t — and hence of Ψ along the direction defined by $\mathbf{e}$ — is

$$\eta'(0) = \frac{\partial \Psi}{\partial x}(\mathbf{x})e_1 + \frac{\partial \Psi}{\partial y}(\mathbf{x})e_2 = \nabla\Psi(\mathbf{x}) \cdot \mathbf{e}.$$

Since

$$\eta''(0) = \frac{\partial^2 \Psi}{\partial x^2}(\mathbf{x})e_1^2 + 2\frac{\partial^2 \Psi}{\partial x \partial y}(\mathbf{x})e_1 e_2 + \frac{\partial^2 \Psi}{\partial y^2}(\mathbf{x})e_2^2,$$

the second derivative of Ψ in the direction of $\mathbf{e}$ is

$$\frac{\partial^2 \Psi}{\partial e^2}(\mathbf{x}) := \mathbf{e} \cdot \mathsf{H}_\Psi(\mathbf{x})\mathbf{e},$$

where $\mathsf{H}_\Psi(\mathbf{x})$ denotes the Hessian matrix of Ψ, defined in Section 0.4.

THEOREM 1.15. *Let $f \in C^2(\overline{\Omega})$, and suppose that $\hat{f} : \overline{\Omega} \to \mathbb{R}$ is the piecewise linear interpolant of f on a grid formed by a triangulation on Ω and having mesh size h. Then*

$$\|f - \hat{f}\|_\infty \leq 5Mh^2,$$

where

$$M = \max\left\{ \left\| \frac{\partial^2 f}{\partial x^2} \right\|_\infty, \left\| \frac{\partial^2 f}{\partial y^2} \right\|_\infty, \left\| \frac{\partial^2 f}{\partial x \partial y} \right\|_\infty \right\}.$$

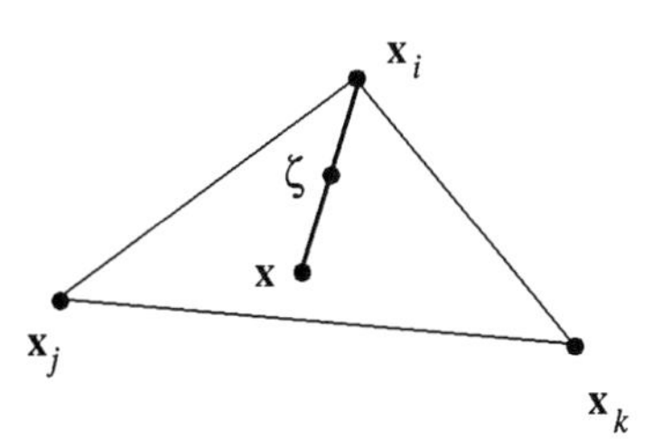

FIGURE 11. *Triangular element containing $\mathbf{x}$ and the point ζ guaranteed by the Taylor theorem.*

PROOF: Suppose that $\mathbf{x} = (x, y) \in \overline{\Omega}_e$, where $\overline{\Omega}_e$ is a triangular element having vertices $\mathbf{x}_i = (x_i, y_i)$, $\mathbf{x}_j = (x_j, y_j)$, and $\mathbf{x}_k = (x_k, y_k)$, as sketched in Figure 11. Since $\mathbf{x}$ is arbitrary, it suffices to establish that $|f(\mathbf{x}) - \hat{f}(\mathbf{x})| \leq$

$5Mh^2$. The theorem is trivially true if $\mathbf{x}$ is a vertex, so assume otherwise. Pick one of the vertices, say $\mathbf{x}_i$, and observe the following two facts:

$$(i) \qquad \|\mathbf{x} - \mathbf{x}_i\|_2 \leq h;$$

$$(ii) \qquad g := f - \hat{f} \in C^2(\overline{\Omega}_e).$$

The second fact allows us to apply the Taylor theorem with remainder (Theorem 0.17) to get

$$\begin{aligned} g(\mathbf{x}) = \quad & g(\mathbf{x}_i) + \nabla g(\mathbf{x}_i) \cdot (\mathbf{x} - \mathbf{x}_i) \\ & + \tfrac{1}{2}(\mathbf{x} - \mathbf{x}_i) \cdot \mathsf{H}_g(\boldsymbol{\zeta})(\mathbf{x} - \mathbf{x}_i). \end{aligned} \tag{1.5-2}$$

Here, $\boldsymbol{\zeta}$ is some point lying on the line segment connecting $\mathbf{x}_i$ and $\mathbf{x}$, as shown in Figure 11, and H_g is the Hessian matrix:

$$\mathsf{H}_g(\boldsymbol{\zeta}) = \begin{bmatrix} \dfrac{\partial^2 g}{\partial x^2}(\boldsymbol{\zeta}) & \dfrac{\partial^2 g}{\partial x \partial y}(\boldsymbol{\zeta}) \\ \\ \dfrac{\partial^2 g}{\partial y \partial x}(\boldsymbol{\zeta}) & \dfrac{\partial^2 g}{\partial y^2}(\boldsymbol{\zeta}) \end{bmatrix}.$$

The first term on the right side of Equation (1.5-2) vanishes, since f and $\hat{f}$ coincide at the node $\mathbf{x}_i$.

To analyze the second term on the right in Equation (1.5-2), rewrite it in the form

$$\nabla g(\mathbf{x}_i) \cdot \frac{\mathbf{x} - \mathbf{x}_i}{\|\mathbf{x} - \mathbf{x}_i\|_2} \|\mathbf{x} - \mathbf{x}_i\|_2. \tag{1.5-3}$$

This form is valid since $\mathbf{x}$ is not a vertex. The vector $\mathbf{e} := (\mathbf{x} - \mathbf{x}_i)/\|\mathbf{x} - \mathbf{x}_i\|_2$ has unit length, and we can resolve it into its components along the directions defined by the edges of $\overline{\Omega}_e$ that intersect at $\mathbf{x}_i$. Denoting the unit vectors in these directions by $\mathbf{e}_{i,j}$ and $\mathbf{e}_{i,k}$, as illustrated in Figure 12, we have $\mathbf{e} = \alpha \mathbf{e}_{i,j} + \beta \mathbf{e}_{i,k}$, where $0 < |\alpha|, |\beta| < 1$. Thus the quantity (1.5-3) is bounded above in magnitude by

$$\begin{aligned} & |\alpha| \; |\nabla g(\mathbf{x}_i) \cdot \mathbf{e}_{i,j}| \; \|\mathbf{x} - \mathbf{x}_i\|_2 \; + \; |\beta| \; |\nabla g(\mathbf{x}_i) \cdot \mathbf{e}_{i,k}| \; \|\mathbf{x} - \mathbf{x}_i\|_2 \\ & \qquad \leq \frac{\partial g}{\partial e_{i,j}}(\mathbf{x}_i) h + \frac{\partial g}{\partial e_{i,k}}(\mathbf{x}_i) h. \end{aligned}$$

But $\hat{f}$ interpolates f linearly along these edges, so each of the directional derivatives $\partial g / \partial e_{i,j}$ and $\partial g / \partial e_{i,k}$ obeys the bound (1.2-8) for derivatives of piecewise Lagrange linear interpolants. (If this assertion is not transparent, try mentally shifting the x-axis so that it lies along one of the edges in question, say, the one connecting $\mathbf{x}_i$ and $\mathbf{x}_j$. Then consider the x-derivative of

$f - \hat{f}$.) It follows that, on the edge joining $\mathbf{x}_i$ and $\mathbf{x}_j$,

$$\left\| \frac{\partial g}{\partial e_{i,j}} \right\|_\infty \leq \left\| \frac{\partial^2 g}{\partial e_{i,j}^2} \right\|_\infty h = \|\mathbf{e}_{i,j} \cdot \mathsf{H}_g \mathbf{e}_{i,j}\|_\infty h \leq 2Mh,$$

the last step being an application of Lemma 1.14. Similarly,

$$\left\| \frac{\partial g}{\partial e_{i,k}} \right\|_\infty \leq 2Mh,$$

and hence the quantity (1.5-3) has magnitude no greater than $4Mh^2$.

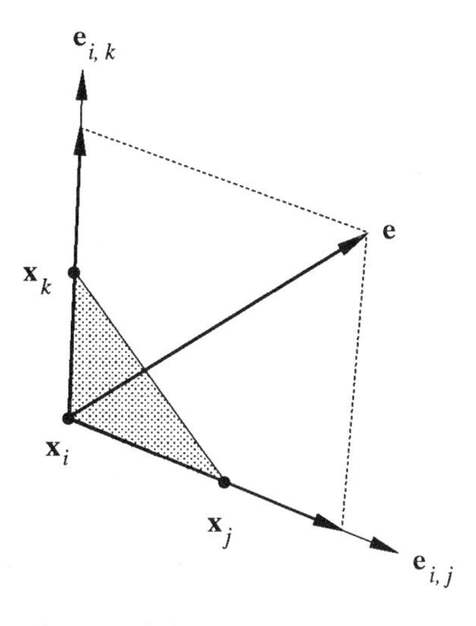

FIGURE 12. *Resolution of the unit vector* $\mathbf{e}$ *into directions defined by the edges of* Ω_e.

Finally, Lemma 1.14 implies that the third term on the right side of Equation (1.5-2) is no greater in magnitude than the quantity

$$\frac{1}{2}\begin{bmatrix} h \\ h \end{bmatrix} \cdot \begin{bmatrix} M & M \\ M & M \end{bmatrix} \begin{bmatrix} h \\ h \end{bmatrix} = Mh^2.$$

Applying the triangle inequality to Equation (1.5-2) therefore yields

$$|g(\mathbf{x})| \leq 0 + 4Mh^2 + Mh^2,$$

which proves the theorem. ∎

This error estimate is just the beginning of the theory of interpolation on triangles. In practice, there are several substantial issues that one must address in constructing triangulations, one of these issues being control over the shapes of the triangular elements used. Problem 12 scratches the surface of this problem by suggesting that triangles with edges meeting at very small angles can lead to poor approximations. Another aspect of interpolation on triangles that we have not considered is the use of interpolants having higher polynomial degree. Problem 13 provides an introduction to this question, and Lapidus and Pinder [9] give a more detailed discussion.

1.6 Splines

Motivation and Construction

Some applications call for interpolants that are smoother than piecewise Lagrange polynomials. However, using Hermite polynomials requires knowledge of derivative values at the nodes. This knowledge is not always available, nor is it necessary. Old-fashioned graphic artists used a thin, flexible strip, called a "spline," to draw smooth curves through plotted points. By analogy, it is numerically possible to pass a smooth interpolating curve through a set of points (x_i, y_i) without knowing slope values. The problem is this: Given a grid $\Delta = \{x_0, x_1, \ldots, x_N\}$ on $[a, b]$ and a set $\{y_0, y_1, \ldots, y_N\}$ of corresponding ordinates, determine a function $s \in C^1([a, b])$ such that $s(x_i) = y_i$ for $i = 0, \ldots, N$.

The following definition identifies a possible solution:

DEFINITION. *A function $s\colon [a, b] \to \mathbb{R}$ is a* **cubic spline** *interpolating the data $y_0, y_1, \ldots, y_N$ on the grid $\Delta = \{a = x_0, x_1, \ldots, x_N = b\}$ provided that*

$$\text{(i)} \qquad s(x_i) = y_i, \quad i = 0, 1, \ldots, N;$$

$$\text{(ii)} \qquad s \in \mathcal{M}_2^3(\Delta).$$

If s interpolates $\{f(x_0), f(x_1), \ldots, f(x_N)\}$ for some function $f\colon [a, b] \to \mathbb{R}$, then we say that s interpolates f on Δ.

This use of the word "spline" reflects an attempt to identify numerical analogs to the graphic artists' device. The definition requires the function s to interpolate the given data, to be piecewise cubic on the grid Δ, and to be twice continuously differentiable on $[a, b]$. The aim of this section is to show that such functions exist and to investigate some of their properties.

Functions satisfying requirements (i) and (ii) indeed exist. To define a cubic polynomial on each subinterval $[x_{i-1}, x_i]$ of the grid, we must determine four coefficients for each of N subintervals, giving a total of $4N$ parameters. The requirement that $s(x_i) = y_i$ for each index i yields $N + 1$ conditions. To see how many conditions the smoothness requirement imposes, notice that $s(x_{i-}) := \lim_{x \to x_i-} s(x)$ and $s(x_{i+}) := \lim_{x \to x_i+} s(x)$ must be equal at each of the interior nodes $x_1, x_2, \ldots, x_{N-1}$ if s is to be continuous. Similarly, since s' must be continuous, $s'(x_i-) = s'(x_i+)$ at each interior node. Finally, s'' must be continuous, so $s''(x_i-) = s''(x_i+)$ at each interior node, too. In all, the smoothness requirement imposes $3(N-1)$ constraints, so the definition of s provides $4N-2$ conditions with which to determine the $4N$ parameters defining the spline. This crude accounting shows that we have not *over*determined s.

However, we need two extra constraints to match the number of conditions with the number of parameters to be determined. To get them, we specify some aspect of the behavior of s at the boundary points $x_0 = a$ and $x_N = b$. Three common possibilities are as follows:

(*i*) $s''(a) = s''(b) = 0,$ **natural spline**;

(*ii*) $s^{(k)}(a) = s^{(k)}(b), \quad k = 1, 2,$ **periodic spline**;

(*iii*) $s'(a) = y_0', \quad s'(b) = y_N',$ **complete spline**.

Think of a natural spline as one whose graph extends along straight lines outside $[a, b]$. This choice may be adequate in the absence of better information at a and b. The periodic spline is often a reasonable choice when $y_0 = y_N$. In the complete-spline conditions, y_0' and y_N' denote known values of the slope of the interpolant at the endpoints of the interval. As we see below, the choice of end conditions can significantly affect the approximating power of s.

To compute splines, it helps to observe that s'' is a continuous, piecewise linear function on the grid Δ. Our strategy is as follows: Begin with a function in $\mathcal{M}_0^1(\Delta)$ having unknown coefficients, integrate it twice, and use the interpolation conditions, smoothness constraints, and end conditions to determine the coefficients in s'' and the constants of integration that arise.

We start with the expression

$$s''(x) = \sum_{i=0}^{N} m_i \ell_i(x), \tag{1.6-1}$$

where ℓ_i is the piecewise linear Lagrange basis function associated with the node x_i and m_i is an unknown coefficient, called a **moment** of s. If the subinterval lengths are $h_i = x_i - x_{i-1}$, then

$$s''(x) = m_i \frac{x_{i+1} - x}{h_{i+1}} + m_{i+1} \frac{x - x_i}{h_{i+1}}, \qquad x \in [x_i, x_{i+1}].$$

Integrating this equation twice yields

$$s(x) = m_i \frac{(x_{i+1} - x)^3}{6h_{i+1}} + m_{i+1} \frac{(x - x_i)^3}{6h_{i+1}} + \lambda_i (x - x_i) + \mu_i, \qquad x \in [x_i, x_{i+1}],$$

where λ_i and μ_i are constants of integration.

We determine the values of these constants by imposing the conditions $s(x_i+) = y_i$ and $s(x_{i+1}-) = y_{i+1}$. The first condition implies that

$$\frac{m_i h_{i+1}^2}{6} + \mu_i = y_i,$$

while the second reduces to the equation

$$\frac{m_{i+1} h_{i+1}^2}{6} + \lambda_i h_{i+1} + \mu_i = y_{i+1}.$$

Solving these equations, we find that

$$\lambda_i = \frac{y_{i+1} - y_i}{h_{i+1}} - \frac{h_{i+1}(m_{i+1} - m_i)}{6}, \qquad \mu_i = y_i - \frac{m_i h_{i+1}^2}{6}.$$

Following this procedure for each subinterval $[x_i, x_{i+1}]$, $i = 0, 1, \ldots, N-1$, yields the following piecewise cubic form for the spline s:

$$s(x) = \alpha_i + \beta_i(x - x_i) + \gamma_i(x - x_i)^2 + \delta_i(x - x_i)^3, \quad x \in [x_i, x_{i+1}]. \quad (1.6\text{-}2)$$

Here, the coefficients α_i, β_i, γ_i, and δ_i depend upon the interval $[x_i, x_{i+1}]$ in which x lies:

$$\alpha_i = y_i, \qquad \beta_i = \frac{y_{i+1} - y_i}{h_{i+1}} - h_{i+1}\frac{2m_i + m_{i+1}}{6}, \qquad \gamma_i = \frac{m_i}{2},$$

$$\delta_i = \frac{m_{i+1} - m_i}{6h_{i+1}}, \qquad i = 0, 1, \ldots, N-1. \quad (1.6\text{-}3)$$

We have reduced the problem to one of finding the moments m_i. We have already imposed the constraints $s''(x_i-) = s''(x_i+)$ by assuming the form (1.6-1) for s''. Also, our method of determining the constants λ_i and μ_i guarantees that s is continuous and passes through the given points (x_i, y_i). Among the conditions left to impose are $s'(x_i-) = s'(x_i+)$, $i = 1, 2, \ldots, N-1$. These each reduce to the form

$$a_i m_{i-1} + b_i m_i + c_i m_{i+1} = d_i, \qquad i = 1, 2, \ldots, N-1, \quad (1.6\text{-}4)$$

where

$$a_i = \frac{h_i}{h_i + h_{i+1}}, \qquad b_i = 2, \qquad c_i = 1 - a_i,$$

$$d_i = \frac{6}{h_i + h_{i+1}}\left(\frac{y_{i+1} - y_i}{h_{i+1}} - \frac{y_i - y_{i-1}}{h_i}\right).$$

Thus we have a system of $N-1$ linear equations for the $N+1$ moments $m_0, m_1, \ldots, m_N$.

The last two equations needed are the end conditions. For a *natural spline*, these conditions have the forms $s''(x_0) = m_0 = 0$ and $s''(x_N) = m_N = 0$. Thus the first and last equations in the set (1.6-4) collapse to

$$b_1 m_1 + c_1 m_2 = d_1; \qquad a_{N-1} m_{N-2} + b_{N-1} m_{N-1} = d_{N-1}.$$

For a *periodic spline*, the constraint $s''(x_0) = s''(x_N)$ means that $m_0 = m_N$, while the constraint $s'(x_0) = s'(x_N)$ implies that

$$a_N m_{N-1} + b_N m_N + c_N m_1 = d_N,$$

where

$$a_N = \frac{h_N}{h_N + h_1}, \qquad b_N = 2, \qquad c_N = 1 - a_N,$$

$$d_N = \frac{6}{h_N + h_1}\left(\frac{y_1 - y_N}{h_1} - \frac{y_N - y_{N-1}}{h_N}\right).$$

Finally, the end conditions $s'(x_0) = y_0'$ and $s'(x_N) = y_N'$ for a *complete spline* yield the following equations:

$$b_0 m_0 + c_0 m_1 = d_0, \qquad a_N m_{N-1} + b_N m_N = d_N,$$

where

$$c_0 = a_N = 1, \qquad b_0 = b_N = 2,$$

$$d_0 = \frac{6}{h_1}\left(\frac{y_1 - y_0}{h_1} - y_0'\right), \qquad d_N = \frac{6}{h_N}\left(y_N' - \frac{y_N - y_{N-1}}{h_N}\right).$$

It is illuminating to write the linear systems for the moments m_i in matrix form. For the natural spline, we get the $(N-1) \times (N-1)$ system

$$\begin{bmatrix} b_1 & c_1 & & & \\ a_2 & b_2 & c_2 & & \\ & & \ddots & & \\ & & a_{N-2} & b_{N-2} & c_{N-2} \\ & & & a_{N-1} & b_{N-1} \end{bmatrix} \begin{bmatrix} m_1 \\ m_2 \\ \vdots \\ m_{N-2} \\ m_{N-1} \end{bmatrix} = \begin{bmatrix} d_1 \\ d_2 \\ \vdots \\ d_{N-2} \\ d_{N-1} \end{bmatrix}. \tag{1.6-5}$$

For the periodic spline, we get the $N \times N$ system

$$\begin{bmatrix} b_1 & c_1 & & & a_1 \\ a_2 & b_2 & c_2 & & \\ & & \ddots & & \\ & & a_{N-1} & b_{N-1} & c_{N-1} \\ c_N & & & a_N & b_N \end{bmatrix} \begin{bmatrix} m_1 \\ m_2 \\ \vdots \\ m_{N-1} \\ m_N \end{bmatrix} = \begin{bmatrix} d_1 \\ d_2 \\ \vdots \\ d_{N-1} \\ d_N \end{bmatrix}. \tag{1.6-6}$$

Finally, for the complete spline, we get the $(N+1) \times (N+1)$ system

$$\begin{bmatrix} b_0 & c_0 & & & \\ a_1 & b_1 & c_1 & & \\ & & \ddots & & \\ & & a_{N-1} & b_{N-1} & c_{N-1} \\ & & & a_N & b_N \end{bmatrix} \begin{bmatrix} m_0 \\ m_1 \\ \vdots \\ m_{N-1} \\ m_N \end{bmatrix} = \begin{bmatrix} d_0 \\ d_1 \\ \vdots \\ d_{N-1} \\ d_N \end{bmatrix}. \tag{1.6-7}$$

Therefore, to determine s, we must solve a matrix equation for the moments m_i, then compute and store the coefficients $\alpha_i, \beta_i, \gamma_i, \delta_i$, $i = 0, \ldots, N-1$, as defined in Equations (1.6-3). To compute $s(x)$ for a particular $x \in [a, b]$, we must find the index i for which $x \in [x_i, x_{i+1}]$, look up the corresponding stored coefficients $\alpha_i, \beta_i, \gamma_i, \delta_i$, then apply Equation (1.6-2). Figure 1 shows the graph of a natural spline passing through a set of seven points.

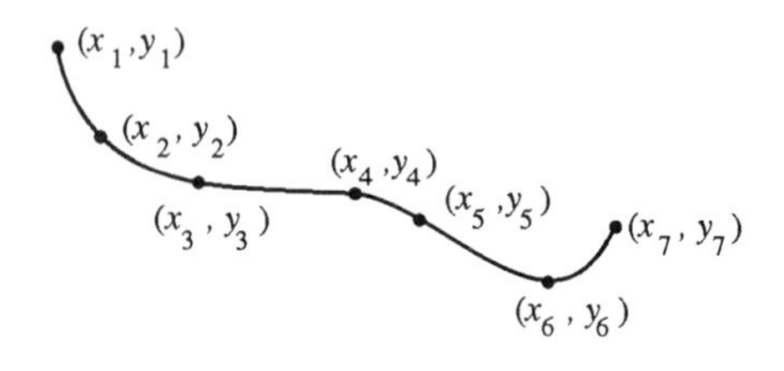

FIGURE 1. *Natural cubic spline passing through seven preassigned points* (x_i, y_i).

As we show later, the matrices in Equations (1.6-5), (1.6-6), and (1.6-7) are all nonsingular, so the matrix equations have unique solutions. Moreover, the matrices appearing in Equations (1.6-5) and (1.6-7) are **tridiagonal**, that is, they have nonzero entries only along the diagonal and in the positions immediately to the left and right of the diagonal. There is a very simple and efficient algorithm, called the **Thomas algorithm**, for solving tridiagonal matrix equations:

ALGORITHM 1.1 (THOMAS). *Consider the tridiagonal matrix equation* $\mathsf{T}\mathbf{m} = \mathbf{d}$, *where* $\mathbf{d}$ *has entries* $d_1, d_2, \ldots, d_N$ *and the* $N \times N$ *matrix* T *has diagonal entries* $b_1, b_2, \ldots, b_N$, *subdiagonal entries* $a_2, a_3, \ldots, a_N$, *and superdiagonal entries* $c_1, c_2, \ldots, c_{N-1}$. *Compute the entries* $m_1, m_2, \ldots, m_N$ *of the solution vector* $\mathbf{m}$ *as follows:*

1. $\beta_1 \leftarrow b_1$.
2. $\gamma_1 \leftarrow d_1/\beta_1$.
3. For $i = 2, 3, \ldots, N$:
4. $\quad\quad \beta_i \leftarrow b_i - (a_i c_{i-1}/\beta_{i-1})$.
5. $\quad\quad \gamma_i \leftarrow (d_i - a_i\gamma_{i-1})/\beta_i$.
6. Next i.
7. $m_N \leftarrow \gamma_N$.
8. For $j = 1, 2, \ldots, N-1$:

9. $m_{N-j} \leftarrow \gamma_{N-j} - (c_{N-j} m_{N-j+1})/\beta_{N-j}$.

10. Next j.

11. End.

This algorithm requires one division, followed by $5(N-1)$ arithmetic operations ($+$, $-$, $\times$, or $\div$) in the first loop, followed by $3(N-1)$ arithmetic operations in the second loop. The operation count for the Thomas algorithm is therefore $8N-7$, or roughly $8N$ when N is large, for a tridiagonal system of order N. We discuss further aspects of this algorithm in Section 2.4.

Practical Considerations

Users of cubic spline approximations should be aware of several theoretical results. One result of great practical significance is an error estimate for the moments m_i:

THEOREM 1.16. *Suppose that $f \in C^4([a,b])$, and let $\Delta = \{x_0, x_1, \ldots, x_N\}$ be a grid on $[a,b]$ having mesh size h. Denote by $\mathbf{m} \in \mathbb{R}^{N+1}$ the vector of moments m_i of the complete spline s interpolating f on Δ, and let $\mathbf{f} \in \mathbb{R}^{N+1}$ be the vector of true second-derivative values $f''(x_i)$ at the nodes of Δ. Then*

$$\|\mathbf{m} - \mathbf{f}\|_\infty = \max_{0 \le i \le N} |m_i - f''(x_i)| \le \frac{3}{2} \|f^{(4)}\|_\infty \, h^2.$$

(We use the symbol $\|\cdot\|_\infty$ to stand for both the maximum magnitude among the entries of a finite-dimensional vector and the supremum of a function over the interval $[a,b]$.) We prove this theorem later.

The importance of the theorem is twofold. First, it provides an essential ingredient in the proof of another theorem stating, in effect, that complete cubic splines approximate smooth functions with an error that is $\mathcal{O}(h^4)$. We investigate this line of reasoning later.

Second, and of more immediate interest, the fact that $\|\mathbf{m} - \mathbf{f}\|_\infty = \mathcal{O}(h^2)$ furnishes a useful device for checking computer programs. To do this, compute the moments m_i for a complete cubic spline approximating a known function f on several grids having a variety of mesh sizes h. Then, for each grid, compute the error norm $E = \|\mathbf{m} - \mathbf{f}\|_\infty$. Because $E \le Ch^2$ for a positive constant C, we have $\log E \le \log C + 2 \log h$. Consequently, a plot of $\log E$ versus $\log h$ for the various grids should yield points lying below a line of slope 2 with vertical intercept at $\log C$. We call such a diagram a **convergence plot**. In practice, the points on the convergence plot usually lie very close to a line of slope 2, as illustrated in Figure 2. Conscientious programmers take advantage of this idea to check computed results against theory.

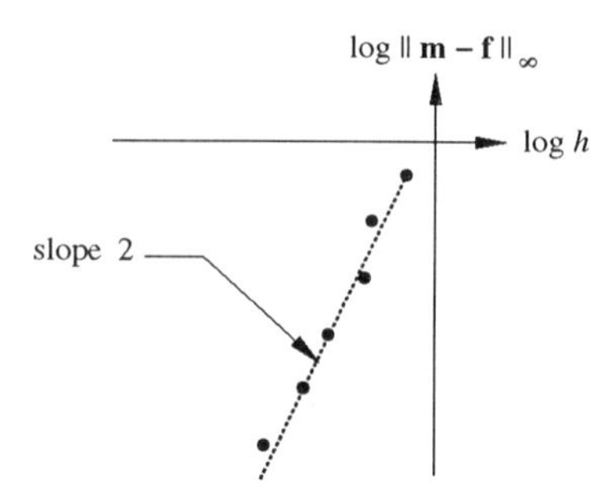

FIGURE 2. *A convergence plot for the vector* $\mathbf{m}$ *of moments for a complete spline approximation to a known function* f.

Cubic splines exhibit less of the "wiggly" behavior often associated with high-degree polynomial interpolation. Some people refer to this fact as the "minimum curvature" property of natural splines, although this term is misleading. The theorem, which we investigate rigorously below, says something like this: Among all reasonably smooth functions $\hat{f}$ that interpolate a given set of ordinates on a given grid, the natural spline minimizes the quantity $\int_a^b |\hat{f}''(x)|^2 \, dx$. This integral measures the average magnitude of $\hat{f}''$ over the interval $[a, b]$. In a rough sense, an interpolant $\hat{f}$ whose second derivative is small in magnitude will have a graph that does not "bend" very much in passing through the required points.

The trouble with interpreting this fact as a "minimum curvature" property is that $\hat{f}''$, strictly speaking, is not the curvature of $\hat{f}$. The curvature of $\hat{f}$ at a point x is $\hat{f}''(x)[1 + \hat{f}'(x)]^{-3/2}$, to which $\hat{f}''(x)$ is a good approximation only if the slope $\hat{f}'(x)$ has small magnitude. While cubic splines tend to be reasonably nonoscillatory, there are ways of tampering with the basic spline formulations to produce less "wiggly" interpolants; de Boor [4] gives a good introduction.

Mathematical Details

Let us first show that we can solve for the moments m_i.

PROPOSITION 1.17. *The matrices appearing in Equations (1.6-5), (1.6-6), and (1.6-7) are nonsingular.*

PROOF: In each equation, $a_i \geq 0$, $c_i \geq 0$, and $a_i + c_i = 1$. (Let us agree that $a_0 = c_N = 0$ in the cases where these entries are needed.) We use these facts to prove that the matrix in Equation (1.6-5) is nonsingular, the proofs for Equations (1.6-6) and (1.6-7) being similar. Abbreviate Equation (1.6-5) as $\mathsf{T}\mathbf{m} = \mathbf{d}$, where $\mathbf{m}, \mathbf{d} \in \mathbb{R}^{N-1}$ and T is the $(N-1) \times (N-1)$ tridiagonal matrix.

The argument rests on the following claim: For any $\mathbf{v} \in \mathbb{R}^{N-1}$,

$$\|\mathbf{v}\|_\infty \leq \|\mathsf{T}\mathbf{v}\|_\infty. \tag{1.6-8}$$

To see this, let j be the index for which $\|\mathbf{v}\|_\infty = |v_j|$, and formally set $v_0 = v_N = 0$. Since $b_1 = b_2 = \cdots = b_{N-1} = 2$,

$$\begin{aligned} \|\mathsf{T}\mathbf{v}\|_\infty \quad \geq |(\mathsf{T}\mathbf{v})_j| \quad &= |a_j v_{j-1} + 2v_j + c_j v_{j+1}| \\ &\geq 2|v_j| - a_j|v_{j-1}| - c_j|v_{j+1}| \\ &\geq \underbrace{(2 - a_j - c_j)}_{1}|v_j| = \|\mathbf{v}\|_\infty. \end{aligned}$$

This proves the claim.

We now show that T is nonsingular. If $\mathsf{T}\mathbf{v} = \mathbf{0}$, then $\|\mathbf{v}\|_\infty \leq \|\mathsf{T}\mathbf{v}\|_\infty = \|\mathbf{0}\|_\infty = 0$. It follows that $\mathbf{v} = \mathbf{0}$ and hence that T is nonsingular. ∎

Next we investigate the convergence properties of cubic splines. The reasoning given here shows that, when $f: [a, b] \to \mathbb{R}$ is smooth enough, complete spline approximations to f obey error estimates having the form $\|f - s\|_\infty = \mathcal{O}(h^4)$. We begin with Theorem 1.16:

THEOREM 1.16. *Suppose that* $f \in C^4([a, b])$, *and let* $\Delta = \{x_0, x_1, \ldots, x_N\}$ *be a grid on* $[a, b]$ *having mesh size* h. *Denote by* $\mathbf{m} \in \mathbb{R}^{N+1}$ *the vector of moments* m_i *of the complete spline interpolating* f *on* Δ, *and let* $\mathbf{f} \in \mathbb{R}^{N+1}$ *be the vector of true second-derivative values* $f''(x_i)$ *at the nodes of* Δ. *Then*

$$\|\mathbf{m} - \mathbf{f}\|_\infty \leq \tfrac{3}{2}\|f^{(4)}\|_\infty\, h^2.$$

PROOF: We have seen that $\mathbf{m}$ solves a tridiagonal matrix equation $\mathsf{T}\mathbf{m} = \mathbf{d}$. Denote by $\mathbf{r} \in \mathbb{R}^{N+1}$ the vector $\mathsf{T}(\mathbf{m}-\mathbf{f}) = \mathbf{d} - \mathsf{T}\mathbf{f}$. The strategy is to estimate $\|\mathbf{r}\|_\infty$, then to bound $\|\mathbf{m} - \mathbf{f}\|_\infty$ in terms of $\|\mathbf{r}\|_\infty$. For $i = 1, 2, \ldots, N-1$,

$$\begin{aligned} r_i &= d_i - [a_i f''(x_{i-1}) + b_i f''(x_i) + c_i f''(x_{i+1})] \\ &= \frac{6}{h_i + h_{i+1}}\left[\frac{f(x_{i+1}) - f(x_i)}{h_{i+1}} - \frac{f(x_i) - f(x_{i-1})}{h_i}\right] \\ &\quad - \frac{h_i}{h_i + h_{i+1}} f''(x_{i-1}) - 2f''(x_i) - \frac{h_{i+1}}{h_i + h_{i+1}} f''(x_{i+1}). \end{aligned} \tag{1.6-9}$$

To simplify this equation, we relate values of f and f'' at the nodes x_{i+1} and x_{i-1} to corresponding values at x_i using the Taylor theorem. Since

$f \in C^4([a,b])$, there are points $\zeta_1, \zeta_3 \in (x_i, x_{i+1})$ such that

$$\begin{aligned} f(x_{i+1}) &= f(x_i) + f'(x_i)h_{i+1} + \tfrac{1}{2}f''(x_i)h_{i+1}^2 + \tfrac{1}{6}f^{(3)}(x_i)h_{i+1}^3 \\ &\quad + \tfrac{1}{24}f^{(4)}(\zeta_1)h_{i+1}^4, \\ f''(x_{i+1}) &= f''(x_i) + f^{(3)}(x_i)h_{i+1} + \tfrac{1}{2}f^{(4)}(\zeta_3)h_{i+1}^2. \end{aligned}$$

Similarly, there exist $\zeta_2, \zeta_4 \in (x_{i-1}, x_i)$ such that

$$\begin{aligned} f(x_{i-1}) &= f(x_i) - f'(x_i)h_i + \tfrac{1}{2}f''(x_i)h_i^2 - \tfrac{1}{6}f^{(3)}(x_i)h_i^3 \\ &\quad + \tfrac{1}{24}f^{(4)}(\zeta_2)h_i^4, \\ f''(x_{i-1}) &= f''(x_i) - f^{(3)}(x_i)h_i + \tfrac{1}{2}f^{(4)}(\zeta_4)h_i^2. \end{aligned}$$

Substituting these expressions into Equation (1.6-9) and simplifying, we get

$$r_i = \frac{1}{h_i + h_{i+1}} \left[\frac{h_{i+1}^3}{4} f^{(4)}(\zeta_1) + \frac{h_i^3}{4} f^{(4)}(\zeta_2) - \frac{h_{i+1}^3}{2} f^{(4)}(\zeta_3) - \frac{h_i^3}{2} f^{(4)}(\zeta_4) \right].$$

Consequently,

$$|r_i| \le \frac{1}{h_i + h_{i+1}} \left(\frac{h_{i+1}^3}{4} + \frac{h_i^3}{4} + \frac{h_i^3}{2} + \frac{h_{i+1}^3}{2} \right) \|f^{(4)}\|_\infty.$$

The expression in parentheses reduces to $\frac{3}{4}(h_i^3 + h_{i+1}^3)$, and since $(h_i^3 + h_{i+1}^3)/(h_i + h_{i+1}) = h_i^2 - h_i h_{i+1} + h_{i+1}^2 < h_i^2 + h_{i+1}^2$, we obtain

$$|r_i| \le \frac{3}{4}(h_i^2 + h_{i+1}^2)\|f^{(4)}\|_\infty \le \frac{3}{2}h^2\|f^{(4)}\|_\infty.$$

Analogous arguments apply to r_0 and r_N, allowing us to deduce that $\|\mathbf{r}\|_\infty \le \frac{3}{2}h^2\|f^{(4)}\|_\infty$. Now it suffices to show that $\|\mathbf{m} - \mathbf{f}\|_\infty \le \|\mathbf{r}\|_\infty$. But this fact follows from the definition $\mathbf{r} := \mathsf{T}(\mathbf{m} - \mathbf{f})$ and the inequality (1.6-8). ∎

This theorem permits us to estimate the approximation errors associated with the complete spline and its first three derivatives. The estimates describe how fast the errors shrink as we refine the grid Δ, letting $h \to 0$. However, the estimates depend upon how nearly uniform the grid is, a property that we measure as follows.

DEFINITION. *The* **grid ratio** *of Δ is the maximum value of h/h_i, where $i = 1, 2, \ldots, N$.*

Thus uniform grids have grid ratio $\Gamma = 1$, and the larger Γ is, the greater the discrepancy is between the mesh size h and the length of the smallest subinterval $[x_{i-1}, x_i]$.

THEOREM 1.18. *Suppose that $f \in C^{(4)}([a,b])$, and let $\Delta = \{x_0, x_1, \ldots, x_N\}$ be a grid on $[a,b]$ having mesh size h and grid ratio Γ. If s is the complete spline interpolating f on Δ, then there are constants C_k such that*

$$\|f^{(k)} - s^{(k)}\|_\infty \leq C_k \Gamma \|f^{(4)}\|_\infty \, h^{4-k}, \qquad \text{for} \quad k = 0, 1, 2, 3.$$

The symbol C_k stands for a positive constant, independent of h, that may be different for different orders k of differentiation. The theorem guarantees that the interpolation error associated with complete cubic splines is $\mathcal{O}(h^4)$.

PROOF: We follow a "bootstrap" strategy, proving the estimate for $k = 3$, then using the result to argue for $k = 2$, and so forth. For $k = 3$, suppose that $x \in [x_{j-1}, x_j]$. By adding and subtracting in Equation (1.6-1), we get

$$\begin{aligned} |s^{(3)}(x) - f^{(3)}(x)| &= \left| \frac{m_j - m_{j-1}}{h_j} - f^{(3)}(x) \right| \\ &\leq \underbrace{\left| \frac{m_j - f''(x_j)}{h_j} - \frac{m_{j-1} - f''(x_{j-1})}{h_j} \right|}_{(\text{I})} \\ &\quad + \underbrace{\left| \frac{f''(x_j) - f''(x)}{h_j} - \frac{f''(x_{j-1}) - f''(x)}{h_j} - f^{(3)}(x) \right|}_{(\text{II})}. \end{aligned}$$

We estimate the terms labeled (I) and (II) separately. Theorem 1.16 and the definition of the grid ratio imply that

$$(\text{I}) \leq \frac{3\|f^{(4)}\|_\infty \, h^2}{2h_j} \leq \frac{3}{2}\Gamma \|f^{(4)}\|_\infty \, h.$$

To estimate (II), we use Taylor expansions about the point x, finding points $\eta_1, \eta_2 \in (x_{j-1}, x_j)$ for which

$$\begin{aligned} (\text{II}) = \Bigg| & \frac{1}{h_j}(x_j - x) f^{(3)}(x) + \frac{1}{2} \underbrace{\frac{(x_j - x)^2}{h_j}}_{(\text{III})} \underbrace{f^{(4)}(\eta_1)}_{(\text{IV})} \\ & - \frac{1}{h_j}(x_{j-1} - x) f^{(3)}(x) - \frac{1}{2} \underbrace{\frac{(x_{j-1} - x)^2}{h_j}}_{(\text{III})} \underbrace{f^{(4)}(\eta_2)}_{(\text{IV})} - f^{(3)}(x) \Bigg|. \end{aligned}$$

On the right side of this identity, the terms not marked with underbraces cancel, and the remaining terms obey the bounds

$$|(\text{III})| \leq \Gamma h, \qquad |(\text{IV})| \leq \|f^{(4)}\|_{\infty}.$$

Hence the triangle inequality yields (II) $\leq \Gamma \|f^{(4)}\|_{\infty}\, h$. Combining results gives

$$|s^{(3)}(x) - f^{(3)}(x)| \leq \frac{5}{2}\,\Gamma \|f^{(4)}\|_{\infty}\, h.$$

Since $x \in [a,b]$ is arbitrary, this proves the theorem for $k = 3$; in this case $C_3 = 5/2$.

For $k = 2$, let x_ℓ be the node in Δ closest to x. (If x is equidistant from two nodes, either will work.) Thus $|x_\ell - x| \leq h/2$. Having already analyzed $f^{(3)} - s^{(3)}$, we get to $f'' - s''$ by integrating. According to the fundamental theorem of calculus and the triangle inequality,

$$|f''(x) - s''(x)| \leq \underbrace{|f''(x_\ell) - s''(x_\ell)|}_{(\text{V})} + \underbrace{\left| \int_{x_\ell}^{x} \left[f^{(3)}(t) - s^{(3)}(t) \right] dt \right|}_{(\text{VI})}.$$

Again we estimate the underbraced terms separately. Using Theorem 1.16 and the fact that $\Gamma \geq 1$, we obtain

$$(\text{V}) \leq \frac{3}{2} \|f^{(4)}\|_{\infty}\, h^2 \leq \frac{3}{2} \Gamma \|f^{(4)}\|_{\infty}\, h^2.$$

Also, by the estimate for the case $k = 3$,

$$(\text{VI}) \leq \|f^{(3)} - s^{(3)}\|_{\infty} \left| \int_{x_\ell}^{x} dt \right| \leq \left(\frac{5}{2}\,\Gamma \|f^{(4)}\|_{\infty}\, h \right) \frac{h}{2}.$$

Therefore,

$$|f''(x) - s''(x)| \leq \frac{11}{4} \Gamma \|f^{(4)}\|_{\infty}\, h^2,$$

and the theorem holds for $k = 2$, with $C_2 = 11/4$.

For the case $k = 1$, we exploit the fact that $s(x_j) = f(x_j)$ at each node x_j, together with the fact that both s and f are in $C^1([a,b])$, to apply the Rolle theorem to $s - f$ on each subinterval of the grid. In particular, there exists a point ζ_j in each subinterval (x_{j-1}, x_j), $j = 1, 2, \ldots, N$, such that $s'(\zeta_j) = f'(\zeta_j)$. Moreover, since s is a complete spline, $s'(a) = f'(a)$ and $s'(b) = f'(b)$. For convenience, rename $a = \zeta_0$ and $b = \zeta_{N+1}$. Given any $x \in [a,b]$, let ζ_ℓ be the closest of the zeros ζ_j of $f' - s'$. Thus $|\zeta_\ell - x| \leq h$, as shown in Figure 3. Using the fundamental theorem of calculus and the triangle inequality as for $k = 2$, we find that

$$|f'(x) - s'(x)| \leq |f'(\zeta_\ell) - s'(\zeta_\ell)| + \left| \int_{\zeta_\ell}^{x} [f''(t) - s''(t)]\, dt \right|.$$

The first term on the right vanishes by our choice of ζ_ℓ, and the second term on the right is bounded above by $h(\frac{11}{4}\Gamma\|f^{(4)}\|_\infty h^2)$, according to the reasoning for the case $k = 2$. Therefore,

$$|f'(x) - s'(x)| \leq \frac{11}{4}\Gamma\|f^{(4)}\|_\infty h^3,$$

and we have settled the case $k = 1$ with $C_1 = 11/4$.

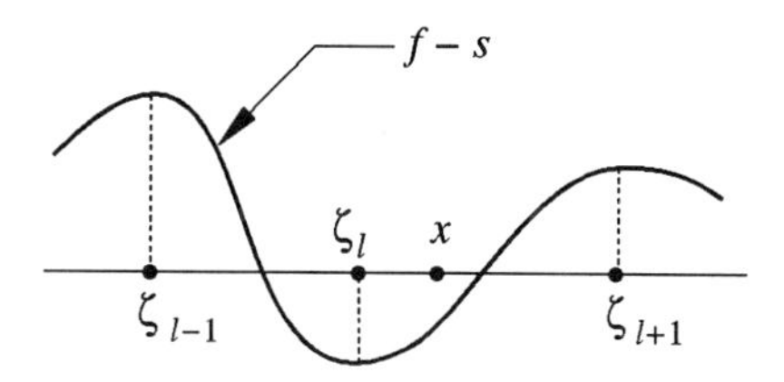

FIGURE 3. *Location of the zero ζ_ℓ of $f' - s'$ lying closest to x.*

Finally, we dispatch the case $k = 0$. If x_ℓ denotes the node of Δ closest to x, then tactics used in the previous two cases yield

$$|f(x) - s(x)| \leq |f(x_\ell) - s(x_\ell)| + \left| \int_{x_\ell}^{x} [f'(t) - s'(t)]\, dt \right|.$$

Again, the first term on the right vanishes, and the second is no larger than $(h/2)(\frac{11}{4}\Gamma\|f^{(4)}\|_\infty h^3)$ by virtue of the case $k = 1$. Therefore,

$$|f(x) - s(x)| \leq \frac{11}{8}\Gamma\|f^{(4)}\|_\infty h^4.$$

Hence the theorem holds for $k = 0$, with $C_0 = 11/8$. ∎

At the end of this section we discuss some ramifications of this error estimate.

Because the "minimum curvature" property of natural splines is prominent in the folklore of approximation theory, it is worthwhile to review the underlying logic. Throughout this review, we consider a fixed grid $\Delta = \{x_0, x_1, \ldots, x_N\}$ on an interval $[a, b]$. We begin by identifying a class of functions over which the minimization property holds.

DEFINITION. *$\mathcal{P}^2(\Delta)$ denotes the class of all functions $\phi: [a, b] \to \mathbb{R}$ such that*

(*i*) $\phi \in C^1([a, b])$;

(*ii*) $\phi\big|_{[x_{i-1}, x_i]} \in C^2([x_{i-1}, x_i])$, for $i = 1, 2, \ldots, N$.

The notation $\phi|_{[x_{i-1},x_i]}$ signifies the **restriction** of ϕ to the interval $[x_{i-1}, x_i]$. Think of $\mathcal{P}^2(\Delta)$ as the set of all "piecewise C^2" functions on $[a, b]$ for which jumps in the second derivative occur only at nodes of Δ. Any cubic spline $s \in \mathcal{M}_2^3(\Delta)$ clearly belongs to $\mathcal{P}^2(\Delta)$, since $s \in C^2([a, b])$.

The quantity to be minimized is the following:

DEFINITION. *If $\phi \in \mathcal{P}^2(\Delta)$, then*

$$|||\phi||| := \|\phi''\|_2 = \sqrt{\int_a^b |\phi''(t)|^2\, dt}.$$

(The integral exists and is finite, since ϕ'' is piecewise continuous on the closed interval $[a, b]$ and has, at worst, jump discontinuities at the nodes $x_1, x_2, \ldots, x_{N-1}$.) As mentioned earlier, $|||\phi|||$ does not measure the true curvature of ϕ, even though it is a related quantity. Also, the definition relates $|||\cdot|||$ to the standard norm $\|\cdot\|_2$ on the normed linear space $L^2[a, b]$, introduced in Chapter 0. However, $|||\cdot|||$ itself is *not* a norm, since it is possible to have $|||\phi||| = 0$ while $\phi \neq 0$. (Consider $\phi(x) = x$.) Since $|||\cdot|||$ possesses the other properties of norms — it scales as $|||c\phi||| = |c|\ |||\phi|||$ and obeys the triangle inequality — we call it a **seminorm**.

The following fact plays the role of a "Pythagorean theorem" for functions in $\mathcal{P}^2(\Delta)$.

THEOREM 1.19. *Let s be a cubic spline interpolating a function $\phi \in \mathcal{P}^2(\Delta)$ on the grid Δ. Suppose that s satisfies either of the following sets of end conditions:*

$$(i)\quad s''(a) = s''(b) = 0;$$

$$(ii)\quad s'(a) = \phi'(a), \quad s'(b) = \phi'(b).$$

Then

$$|||\phi - s|||^2 = |||\phi|||^2 - |||s|||^2.$$

Figure 4 suggests schematically how one might interpret this theorem, with $|||\phi|||$, $|||s|||$, and $|||\phi - s|||$ serving as analogs of the edges of a right triangle.

PROOF: The first task is to prove that, for *any* spline s on Δ and *any* function $\phi \in \mathcal{P}^2(\Delta)$,

$$\begin{aligned} |||\phi - s|||^2 = \;& |||\phi|||^2 - |||s|||^2 - 2\Big\{[\phi'(x) - s'(x)]\, s''(x)\Big\}\Big|_a^b \\ & + 2\sum_{i=1}^{N} [\phi(x) - s(x)]\, s^{(3)}(x)\Big|_{x_{i-1}+}^{x_i-}. \end{aligned} \tag{1.6-10}$$

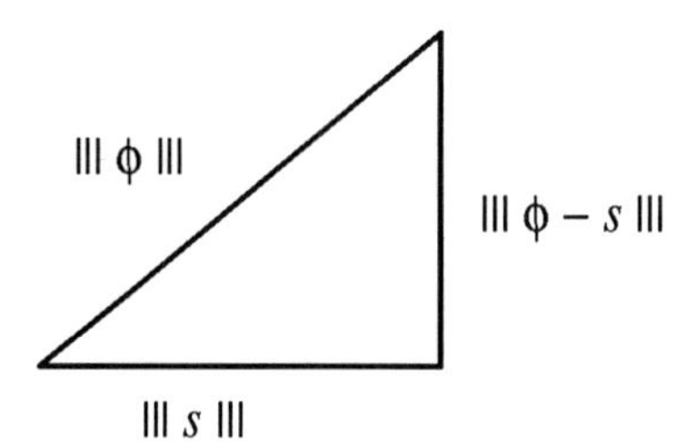

FIGURE 4. *Schematic diagram of the "Pythagorean theorem" for functions in* $\mathcal{P}^2(\Delta)$.

As we show below, Equation (1.6-10) yields the desired result when s interpolates ϕ and either of the end conditions (i) or (ii) holds.

We start with the identity

$$\begin{aligned} |||\phi - s|||^2 &= \int_a^b \left[(\phi'')^2 - 2\phi'' s'' + (s'')^2\right] dx \\ &= |||\phi|||^2 - 2 \underbrace{\int_a^b (\phi'' - s'')s''\, dx}_{(\mathrm{I})} - |||s|||^2, \end{aligned} \tag{1.6-11}$$

obtained by adding and subtracting $\int_a^b (s'')^2\, dx$ on the right side of the first equation. Integrating the expression (I) by parts twice on each subinterval $[x_{i-1}, x_i]$, we find that

$$\begin{aligned} \int_{x_{i-1}}^{x_i} (\phi'' - s'')s''\, dx = &\underbrace{\left[(\phi' - s')s''\right]}_{(\mathrm{II})}\Big|_{x_{i-1}}^{x_i} \\ &- \underbrace{\left[(\phi - s)s^{(3)}\right]\Big|_{x_{i-1}+}^{x_i-}}_{(\mathrm{III})} + \underbrace{\int_{x_{i-1}}^{x_i} (\phi - s)s^{(4)}\, dx}_{(\mathrm{IV})}. \end{aligned} \tag{1.6-12}$$

We pause for three observations. First, the quantity labeled (II) is continuous on $[a, b]$ by our assumptions about ϕ and s. Second, one-sided limits appear in the term labeled (III), owing to the fact that $s^{(3)}$ typically has jump discontinuities at nodes. Third, the term labeled (IV) vanishes, since s is cubic on each subinterval $[x_{i-1}, x_i]$. Summing Equation (1.6-12) over the indices $i = 1, 2, \ldots, N$ therefore produces the equation

$$\int_a^b (\phi'' - s'')s''\,dx = [(\phi' - s')s'']\Big|_a^b - \sum_{i=1}^{N} \left[(\phi - s)s^{(3)}\right]\Big|_{x_{i-1}+}^{x_i-}.$$

Substituting this result into Equation (1.6-11) yields the identity (1.6-10).

Now consider this identity under the assumptions that s interpolates ϕ on Δ and either of the end conditions (i) or (ii) holds. Since s and ϕ agree at the nodes, the sum on the right side of Equation (1.6-10) vanishes in both cases. If the condition (i) holds, then $s''(a) = s''(b) = 0$, and the proof is complete. On the other hand, if (ii) holds, then $\phi' - s'$ vanishes at the endpoints a and b, and again the proof is complete. ■

The "minimum curvature" property of cubic splines is an easy corollary:

COROLLARY 1.20. *Among all of the functions $\phi \in \mathcal{P}^2(\Delta)$ that interpolate the data $y_0, y_1, \ldots, y_N$ on Δ, the natural spline interpolant minimizes $|||\phi|||$.*

PROOF: The natural spline s interpolating $\{y_0, y_1, \ldots, y_N\}$ on Δ interpolates any other function $\phi \in \mathcal{P}^2(\Delta)$ that interpolates these data, so the theorem guarantees that

$$0 \leq |||\phi - s|||^2 = |||\phi|||^2 - |||s|||^2.$$

Therefore $|||s|||^2 \leq |||\phi|||^2$. ■

Another corollary is that the grid Δ and the data $y_0, y_1, \ldots, y_N$ uniquely determine the natural and complete cubic splines:

COROLLARY 1.21. *There is only one function $s \in \mathcal{M}_2^3(\Delta)$ that interpolates the data $y_0, y_1, \ldots, y_N$ and satisfies the end conditions $s''(x_0) = s''(x_N) = 0$. Similarly, for given values y_0' and y_N', there is only one function $s \in \mathcal{M}_2^3(\Delta)$ that interpolates $\{y_0, y_1, \ldots, y_N\}$ and satisfes $s'(x_0) = y_0'$ and $s'(x_N) = y_N'$.*

PROOF: This is Problem 14. ■

Further Remarks

Although it enjoys a charming name and an interesting minimization property, the natural spline is not always the best choice. Indeed, the proof of the $\mathcal{O}(h^4)$ interpolation error estimate works for complete splines but not for natural splines. Unless one approximates a function f for which f'' fortuitously vanishes at the endpoints, one might expect the natural-spline end conditions to yield poorer approximations near the endpoints. This in fact happens: Natural splines typically produce interpolation errors that are $\mathcal{O}(h^2)$ near the endpoints [1], even though the approximation in the middle of the interval is often much better. Problem 16 asks for a comparison of natural and complete spline interpolants.

The $\mathcal{O}(h^4)$ estimate that we prove for complete splines is not the best possible, even though the power of h is optimal. Hall and Meyer [5] show that the best possible estimate is

$$\|f - s\|_\infty \leq \frac{5}{384}\|f^{(4)}\|_\infty h^4.$$

They also derive best possible estimates, having the same powers of h as ours, for the first, second, and third derivatives of complete spline interpolants.

So far we have made no mention of how one might construct basis functions for cubic spline interpolation. The most common approach is to use **B-splines**. Consider a uniform grid $\Delta = \{x_0, x_1, \ldots, x_N\}$, extended via the addition of two new nodes, $x_{-1} := x_0 - h$ and $x_{N+1} := x_N + h$. On the new grid Δ^+, associate with each node $x_{-1}, x_0, \ldots, x_{N+1}$ a function $B_i \in \mathcal{M}_2^3(\Delta^+)$ satisfying the conditions

$$B_i(x_j) = \begin{cases} 1, & \text{if } \ j = i, \\ 0, & \text{if } \ |j - i| \geq 2. \end{cases}$$

To define the functions B_i uniquely, we impose the additional requirements $B_i'(x_j) = B_i''(x_j) = 0$ for $j = i \pm 2$. These functions are piecewise cubic and twice continuously differentiable, and they form an "almost nodal" basis for $\mathcal{M}_2^3(\Delta)$. Problem 15 asks for details.

Finally, Schumaker [12] generalizes the notion of splines on a grid Δ to mean any function in $\mathcal{M}_{n-1}^n(\Delta)$, for some polynomial degree n. The sets $\mathcal{M}_{-1}^0(\Delta)$ (piecewise constants), $\mathcal{M}_0^1(\Delta)$ (piecewise linears), and $\mathcal{M}_2^3(\Delta)$ (cubic splines) all constitute spline spaces that we have encountered so far. Is there a space $\mathcal{M}_1^2(\Delta)$? The construction of such functions, as Kammer et al. [7] propose, may be less than obvious. Problem 17 introduces this topic.

1.7 Least-Squares Methods

The approximation methods treated so far require the approximating function $\hat{f}$ to pass through known points (x_i, y_i) or their higher-dimensional analogs. For many applications such interpolation methods are inappropriate. This circumstance commonly occurs when the points (x_i, y_i) result from measurements, which may have errors or fluctuations attributable to incomplete control over the processes being measured. In such applications, the most appropriate approximating functions $\hat{f}$ typically have graphs lying "close to," but not precisely on, the measured points. In fact, scientific or statistical hypotheses often suggest simple forms for $\hat{f}$ — forms that require us to determine a small number of parameters using a multitude of measured data (x_i, y_i). The discrepancy between the number of measured data and the number of parameters to be determined generally makes it impossible to force $\hat{f}$ to pass through the points (x_i, y_i).

The most familiar example of such **overdetermined** systems arises when we wish to fit a line $y = c_1 + c_2 x$ through a set

$$\Big\{(x_0, y_0), (x_1, y_1), \ldots, (x_N, y_N)\Big\}$$

of points. Usually, $N > 1$, and consequently there is little hope that we can find constants c_1 and c_2 that force $y_i = c_1 + c_2 x_i$ for $i = 0, 1, \ldots, N$.

This section introduces the **method of least squares**, a common approach for finding approximating functions that "almost" agree with the known values (x_i, y_i). The method has many intricacies, including a variety of statistical properties, that we do not explore here.

Motivation and Construction

The first task is to identify an appropriate vector space in which to seek approximations. Suppose that we have some measured values $y_0, y_1, \ldots, y_N$, which we regard as approximate values of $f(x_0), f(x_1), \ldots, f(x_N)$ for some function f. Define

$$\mathbf{x} := \begin{bmatrix} x_0 \\ \vdots \\ x_N \end{bmatrix}, \qquad \mathbf{y} := \begin{bmatrix} y_0 \\ \vdots \\ y_N \end{bmatrix}.$$

We seek an approximating function $\hat{f}$ that is "close to" f in the (still imprecise) sense that $\mathbf{y} - \mathbf{\Phi} \simeq \mathbf{0}$, where

$$\mathbf{\Phi} := \begin{bmatrix} \Phi_0 \\ \vdots \\ \Phi_N \end{bmatrix} := \begin{bmatrix} \hat{f}(x_0) \\ \vdots \\ \hat{f}(x_N) \end{bmatrix}.$$

To make definite what we mean by $\mathbf{y} - \mathbf{\Phi} \simeq \mathbf{0}$, we introduce a norm in which to measure the distance between $\mathbf{y}$ and $\mathbf{\Phi}$. Recall from Chapter 0 that $\mathbb{R}^{N+1}$ is an inner-product space, with the inner product of two vectors $\mathbf{u}$ and $\mathbf{v}$ being given by the formula

$$\mathbf{u} \cdot \mathbf{v} = \begin{bmatrix} u_0 \\ \vdots \\ u_N \end{bmatrix} \cdot \begin{bmatrix} v_0 \\ \vdots \\ v_N \end{bmatrix} = \sum_{i=0}^{N} u_i v_i.$$

Associated with this inner product is the norm

$$\|\mathbf{v}\|_2 = \sqrt{\mathbf{v} \cdot \mathbf{v}} = \left(\sum_{i=0}^{N} u_i^2 \right)^{1/2}.$$

As the name "least squares" suggests, we want the vector $\mathbf{\Phi}$ to be close to $\mathbf{y}$ in the sense that $\|\mathbf{y} - \mathbf{\Phi}\|_2$ takes its minimum value.

How small this minimum is depends strongly on the choice of approximating function $\hat{f}$. One class of choices that compromise complete generality in favor of mathematical tractability are those that have the form

$$\hat{f}(x) = \sum_{j=1}^{n} c_j \varphi_j(x). \tag{1.7-1}$$

Here, $\hat{f}$ is a linear combination of **basis functions** $\varphi_1, \varphi_2, \ldots, \varphi_n$ that are chosen a priori, perhaps according to some hypothesis about the phenomena measured. The minimization process, which we discuss momentarily, determines the parameters $c_1, c_2, \ldots, c_n$.

The simple example $\hat{f}(x) = c_1 + c_2 x$, mentioned earlier, comes under this rubric, with $n = 2$. In this case, we can take as basis functions $\varphi_1(x) = 1$ and $\varphi_2(x) = x$, with the expectation that the data (x_i, y_i) reflect some straight-line relationship. However, the form (1.7-1) admits many other possibilities, including

$$\begin{aligned}
\hat{f}(x) &= c_1 + c_2 x + c_3 x^2 + c_4 x^3 && (n = 4), \\
\hat{f}(x) &= c_1 + c_2 \sin x + c_3 \cos x + c_4 \sin 2x + c_5 \cos 2x && (n = 5), \\
\hat{f}(x) &= c_1 + c_2 \ln x && (n = 2).
\end{aligned}$$

(The last example might be useful in fitting logarithms of data whose general functional form is ax^b.)

For any such choice, the vector $\boldsymbol{\Phi}$ has $N + 1$ entries

$$\Phi_i := \sum_{j=1}^{n} c_j \varphi_j(x_i)$$

and n undetermined parameters $c_1, c_2, \ldots, c_n$. We decompose these entries as follows:

$$\boldsymbol{\Phi} = \begin{bmatrix} \sum_{j=1}^{n} c_j \varphi_j(x_0) \\ \vdots \\ \sum_{j=1}^{n} c_j \varphi_j(x_N) \end{bmatrix} = c_1 \underbrace{\begin{bmatrix} \varphi_1(x_0) \\ \vdots \\ \varphi_1(x_N) \end{bmatrix}}_{\boldsymbol{\varphi}_1} + \quad \cdots \quad + c_n \underbrace{\begin{bmatrix} \varphi_n(x_0) \\ \vdots \\ \varphi_n(x_N) \end{bmatrix}}_{\boldsymbol{\varphi}_n}.$$

This decomposition casts $\boldsymbol{\Phi}$ as an undetermined vector belonging to $\mathcal{S} := \operatorname{span}\{\boldsymbol{\varphi}_1, \boldsymbol{\varphi}_2, \ldots, \boldsymbol{\varphi}_n\}$, where the vectors $\boldsymbol{\varphi}_1, \boldsymbol{\varphi}_2, \ldots, \boldsymbol{\varphi}_n$ are the elements of $\mathbb{R}^{N+1}$ identified by the underbraces.

Now we reformulate the approximation problem in the language of vector spaces: Find constants $c_1, c_2, \ldots, c_n$ such that, among all vectors in the subspace $\mathcal{S}$ of $\mathbb{R}^{N+1}$, the vector

$$\boldsymbol{\Phi} = \sum_{j=1}^{n} c_j \boldsymbol{\varphi}_j$$

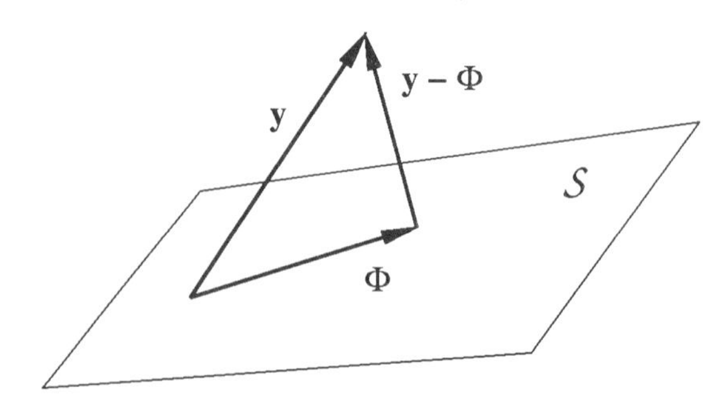

FIGURE 1. *A vector* $\mathbf{y}$ *and its best approximation* $\mathbf{\Phi}$ *in a subspace* $\mathcal{S}$*, represented by a plane in* $\mathbb{R}^3$.

lies closest to $\mathbf{y}$. Because $\mathbb{R}^{N+1}$ is an inner-product space, this problem has an easily identified solution: The vector $\mathbf{\Phi} \in \mathcal{S}$ lying closest to $\mathbf{y}$ in the norm $\|\cdot\|_2$ is the one for which the error $\mathbf{y} - \mathbf{\Phi}$ is orthogonal to every vector in $\mathcal{S}$. This basic geometric fact about inner-product spaces is the **projection principle**. Figure 1 illustrates the principle in $\mathbb{R}^3$ for the case when $\mathcal{S}$ is a plane.

We enforce the requirement that $\mathbf{y} - \mathbf{\Phi}$ be orthogonal to every vector in $\mathcal{S}$ by setting $\mathbf{y} - \mathbf{\Phi}$ orthogonal to each vector in a basis for $\mathcal{S}$. Since two vectors are orthogonal if their inner product vanishes, this approach yields the following **normal equations**:

$$\left(\mathbf{y} - \sum_{j=1}^{n} c_j \boldsymbol{\varphi}_j\right) \cdot \boldsymbol{\varphi}_k = 0, \qquad k = 1, 2, \ldots, n. \tag{1.7-2}$$

The matrix form of Equations (1.7-2) reveals the structure of the linear system that determines the least-squares coefficients c_j:

$$\underbrace{\begin{bmatrix} \boldsymbol{\varphi}_1 \cdot \boldsymbol{\varphi}_1 & \cdots & \boldsymbol{\varphi}_n \cdot \boldsymbol{\varphi}_1 \\ \vdots & & \vdots \\ \boldsymbol{\varphi}_1 \cdot \boldsymbol{\varphi}_n & \cdots & \boldsymbol{\varphi}_n \cdot \boldsymbol{\varphi}_n \end{bmatrix}}_{\mathsf{G}} \begin{bmatrix} c_1 \\ \vdots \\ c_n \end{bmatrix} = \begin{bmatrix} \boldsymbol{\varphi}_1 \cdot \mathbf{y} \\ \vdots \\ \boldsymbol{\varphi}_n \cdot \mathbf{y} \end{bmatrix} \tag{1.7-3}$$

The $n \times n$ matrix G whose (i, j)th entry is $\boldsymbol{\varphi}_j \cdot \boldsymbol{\varphi}_i$ is the **Gram matrix** for the basis $\{\boldsymbol{\varphi}_1, \boldsymbol{\varphi}_2, \ldots, \boldsymbol{\varphi}_n\}$. Later we examine conditions under which G is nonsingular and hence yields a system of normal equations (1.7-3) that has a unique solution. We examine methods for solving equations of the form (1.7-3) in Chapters 2 and 3.

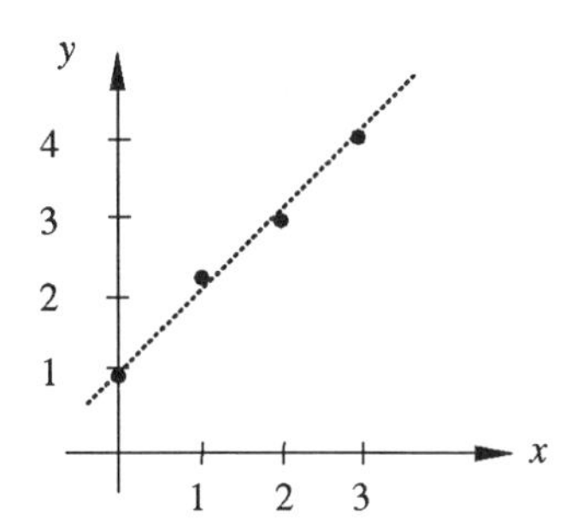

FIGURE 2. *The line $y = 0.98 + 1.03x$ giving a least-squares fit to data given in the text.*

For now, consider a simple example. Given the measured data

$$\mathbf{x} = \begin{bmatrix} x_0 \\ x_1 \\ x_2 \\ x_3 \end{bmatrix} = \begin{bmatrix} 0 \\ 1 \\ 2 \\ 3 \end{bmatrix}, \qquad \mathbf{y} = \begin{bmatrix} y_0 \\ y_1 \\ y_2 \\ y_3 \end{bmatrix} = \begin{bmatrix} 0.9 \\ 2.2 \\ 2.9 \\ 4.1 \end{bmatrix},$$

we find the line $y = \hat{f}(x) = c_1 + c_2 x$ that passes closest to the points (x_i, y_i) in the least-squares sense. Using the basis functions $\varphi_1(x) = 1$ and $\varphi_2(x) = x$, decompose the vector $\mathbf{\Phi}$ as follows:

$$\mathbf{\Phi} = c_1 \begin{bmatrix} \varphi_1(x_0) \\ \varphi_1(x_1) \\ \varphi_1(x_2) \\ \varphi_1(x_3) \end{bmatrix} + c_2 \begin{bmatrix} \varphi_2(x_0) \\ \varphi_2(x_1) \\ \varphi_2(x_2) \\ \varphi_2(x_3) \end{bmatrix} = c_0 \underbrace{\begin{bmatrix} 1 \\ 1 \\ 1 \\ 1 \end{bmatrix}}_{\boldsymbol{\varphi}_1} + c_1 \underbrace{\begin{bmatrix} 0 \\ 1 \\ 2 \\ 3 \end{bmatrix}}_{\boldsymbol{\varphi}_2}.$$

To form the normal equations, note that $\boldsymbol{\varphi}_1 \cdot \boldsymbol{\varphi}_1 = 4$, $\boldsymbol{\varphi}_1 \cdot \boldsymbol{\varphi}_2 = \boldsymbol{\varphi}_2 \cdot \boldsymbol{\varphi}_1 = 6$, $\boldsymbol{\varphi}_2 \cdot \boldsymbol{\varphi}_2 = 14$, $\boldsymbol{\varphi}_1 \cdot \mathbf{y} = 10.1$, and $\boldsymbol{\varphi}_2 \cdot \mathbf{y} = 20.3$. Thus the normal equations reduce to

$$\begin{bmatrix} 4 & 6 \\ 6 & 14 \end{bmatrix} \begin{bmatrix} c_1 \\ c_2 \end{bmatrix} = \begin{bmatrix} 10.1 \\ 20.3 \end{bmatrix}.$$

Solving this system yields $c_1 = 0.98$, $c_2 = 1.03$; therefore, the line passing closest to the given data in the least-squares sense is $y = \hat{f}(x) = 0.98 + 1.03\,x$, as Figure 2 depicts.

This framework also allows us to fit more general functions to the data given in this example. For the choice $\hat{f}(x) = c_1 + c_2 x + c_3 x^2$, for example, the basis functions $\varphi_1(x) = 1$, $\varphi_2(x) = x$, and $\varphi_3(x) = x^2$ correspond to the vectors φ_j defined as follows:

$$\varphi_1 = \begin{bmatrix} 1 \\ 1 \\ 1 \\ 1 \end{bmatrix}, \qquad \varphi_2 = \begin{bmatrix} 0 \\ 1 \\ 2 \\ 3 \end{bmatrix}, \qquad \varphi_3 = \begin{bmatrix} 0 \\ 1 \\ 4 \\ 9 \end{bmatrix}.$$

We leave the determination of the constants c_1, c_2, c_3 as an exercise.

Practical Considerations

Two questions of practical importance arise in the solution of least-squares problems. First, under what circumstances can we be sure that the linear systems of equations that arise possess unique solutions? Second, how can we solve the systems efficiently?

The answer to the first question has an appealing interpretation: Roughly speaking, the Gram matrix will be nonsingular provided each basis function describes some distinct aspect of the trends in the measured data. This somewhat vague prescription means that the vectors $\varphi_1, \varphi_2, \ldots, \varphi_n \in \mathbb{R}^{N+1}$ must form a linearly independent set. We prove later that the Gram matrix is nonsingular when the set $\{\varphi_1, \varphi_2, \ldots, \varphi_n\}$ is linearly independent.

A thorough answer to the second question hinges on the numerical linear algebra associated with the normal equations (1.7-3). We postpone much of this material to Chapters 2, 3, and 5. Of significance in this context is the structure of the Gram matrix G. In contrast to the tridiagonal matrices arising in cubic spline interpolation, G often has nonzero entries in every position. Therefore, we can rarely use algorithms based on "sparse" matrix structure for solving the system (1.7-3). This problem is hardly a cause for concern when n is small — say, five or less. However, the normal equations have two properties that are helpful when n is large: First, G is symmetric, in the sense that $\varphi_i \cdot \varphi_j = \varphi_j \cdot \varphi_i$. We investigate some of the numerical benefits of this property in Chapters 2 and 3.

The second property that G possesses also has important numerical implications:

DEFINITION. *An $n \times n$ real matrix M is* **positive definite** *if, for any nonzero vector $\mathbf{v} \in \mathbb{R}^n$, $\mathbf{v} \cdot \mathsf{M}\mathbf{v} > 0$.*

Chapter 2 reviews other characterizations of this property, and we prove that the Gram matrix is positive definite momentarily.

Mathematical Details

The fundamental result in least-squares theory is the projection principle, which is a consequence of the following lemma.

LEMMA 1.22 (PYTHAGOREAN THEOREM). *Let $\mathcal{V}$ be an inner-product space with inner product $\langle\cdot,\cdot\rangle$ and associated norm $\|\cdot\|_2$. If u and v are orthogonal vectors in $\mathcal{V}$, then $\|u+v\|_2^2 = \|u\|_2^2 + \|v\|_2^2$.*

PROOF: By definition, $\|u+v\|_2^2 = \langle u+v, u+v\rangle = \|u\|_2^2 + \langle u,v\rangle + \langle v,u\rangle + \|v\|_2^2$. The middle two terms in this expansion vanish, since u and v are orthogonal. ∎

(When $\mathcal{V} = \mathbb{R}^2$, this lemma reduces to the familiar Pythagorean theorem for right triangles in the plane.)

THEOREM 1.23 (PROJECTION PRINCIPLE). *Let $\mathcal{V}$ be an inner-product space with $v \in \mathcal{V}$, and let $\mathcal{S}$ be a subspace of $\mathcal{V}$. If $u \in \mathcal{S}$ is a vector for which $v-u$ is orthogonal to every vector in $\mathcal{S}$, then u minimizes the distance $\|v-w\|_2$ over all vectors $w \in \mathcal{S}$. Moreover, there is at most one such vector $u \in \mathcal{S}$.*

PROOF: Pick any vector $w \in \mathcal{S}$, and notice that $w-u \in \mathcal{S}$. Since $v-u$ is orthogonal to every vector in $\mathcal{S}$, it is orthogonal to $w-u$. Applying the Pythagorean theorem to the identity $w-v = (w-u)+(u-v)$, we find that

$$\|w-v\|_2^2 = \|w-u\|_2^2 + \|u-v\|_2^2 \geq \|u-v\|_2^2,$$

with equality holding if and only if $w = u$. This proves the theorem. ∎

The conclusion of this theorem holds when $v-u$ is orthogonal to each vector belonging to a basis for the subspace $\mathcal{S}$.

This section is concerned mainly with the inner-product space $\mathbb{R}^{N+1}$; however, the projection principle applies to any inner-product space.

While providing the theoretical basis for the least-squares method and the normal equations, the projection principle reveals little else about the linear algebra. We now demonstrate that the Gram matrix is symmetric and positive definite under reasonable hypotheses.

THEOREM 1.24. *Provided that the set $\{\varphi_1, \varphi_2, \ldots, \varphi_n\}$ is linearly independent, the Gram matrix G, as given in Equation (1.7-3), is symmetric and positive definite.*

The following proof, although prolix, further illuminates the structure of G.

PROOF: The original, overdetermined system of equations has the form

$$\begin{aligned} c_1\varphi_1(x_0) + c_2\varphi_2(x_0) + \cdots + c_n\varphi_n(x_0) &= y_0 \\ &\vdots \\ c_1\varphi_1(x_N) + c_2\varphi_2(x_N) + \cdots + c_n\varphi_n(x_N) &= y_N, \end{aligned}$$

which we rewrite in the matrix form

$$\left[\begin{pmatrix} \varphi_1(x_0) \\ \vdots \\ \varphi_1(x_N) \end{pmatrix} \cdots \begin{pmatrix} \varphi_n(x_0) \\ \vdots \\ \varphi_n(x_N) \end{pmatrix} \right] \begin{bmatrix} c_1 \\ \vdots \\ c_n \end{bmatrix} = \begin{bmatrix} y_0 \\ \vdots \\ y_N \end{bmatrix}.$$

Abbreviate this system as $\mathsf{A}\mathbf{c} = \mathbf{y}$, where $\mathsf{A} \in \mathbb{R}^{(N+1)\times n}$, $\mathbf{y} \in \mathbb{R}^{N+1}$, and $\mathbf{c}$ is the vector containing the n unknown coefficients c_i. It is a straightforward matter to check that $\mathsf{G} = \mathsf{A}^\top\mathsf{A} \in \mathbb{R}^{n\times n}$ and that the normal equations have the form

$$\mathsf{A}^\top\mathsf{A}\mathbf{c} = \mathsf{A}^\top\mathbf{y}.$$

Any matrix having the form $\mathsf{A}^\top\mathsf{A}$ is symmetric, since $(\mathsf{A}^\top\mathsf{A})^\top = \mathsf{A}^\top(\mathsf{A}^\top)^\top = \mathsf{A}^\top\mathsf{A}$.

Now we establish that G is positive definite. First notice that, whenever $\mathbf{v} \in \mathbb{R}^n$ is nonzero, $\mathsf{A}\mathbf{v} \neq \mathbf{0}$. (Otherwise, the columns of A, which are the basis vectors $\boldsymbol{\varphi}_1, \boldsymbol{\varphi}_2, \ldots, \boldsymbol{\varphi}_n$, could not form a linearly independent set.) Choose any nonzero vector $\mathbf{v} \in \mathbb{R}^n$. We must show that $\mathbf{v} \cdot \mathsf{G}\mathbf{v} = \mathbf{v} \cdot \mathsf{A}^\top\mathsf{A}\mathbf{v} > 0$. By using the associativity of matrix multiplication and manipulating the product of transposes, we can rewrite this expression as $\mathbf{v}^\top\mathsf{A}^\top\mathsf{A}\mathbf{v} = (\mathsf{A}\mathbf{v})^\top(\mathsf{A}\mathbf{v}) = \|\mathsf{A}\mathbf{v}\|_2^2$. But we have already argued that $\mathsf{A}\mathbf{v} \neq \mathbf{0}$, so $\mathbf{v} \cdot \mathsf{G}\mathbf{v} = \|\mathsf{A}\mathbf{v}\|_2^2 > 0$. ∎

The fact that the normal equations have a unique solution is an easy consequence of this theorem and the following general fact:

THEOREM 1.25. *Any positive definite matrix is nonsingular.*

PROOF: The proof is by contradiction: If M is a positive definite matrix that is singular, then there is some nonzero vector $\mathbf{v}$ for which $\mathsf{M}\mathbf{v} = \mathbf{0}$. In this case, $\mathbf{v}^\top\mathsf{M}\mathbf{v} = \mathbf{v}^\top\mathbf{0} = 0$, contradicting the fact that M is positive definite. ∎

Further Remarks

We finish with some brief remarks about orthogonal polynomials. It is heuristically desirable — though often computationally impractical — to choose the basis functions $\varphi_0, \varphi_1, \ldots, \varphi_n$ so that the vectors $\boldsymbol{\varphi}_1, \boldsymbol{\varphi}_2, \ldots, \boldsymbol{\varphi}_n$ are mutually orthogonal. For, with such a basis, the Gram matrix is diagonal and hence easy to invert. In some cases, it actually is possible to choose $\varphi_1, \varphi_2, \ldots, \varphi_n$ so that mutual orthogonality holds. The most common instances of this pleasant situation are problems in which one wishes to fit a polynomial or a trigonometric polynomial to data measured on a grid $\Delta = \{x_0, x_1, \ldots, x_N\}$ of uniformly spaced abscissae. In these cases, the inner product of two functions f and g defined on Δ is

$$\langle f, g\rangle := \sum_{i=0}^{N} f(x_i)g(x_i).$$

To produce a collection of mutually orthogonal basis vectors, we must abandon the monic polynomials $1, x, x^2, x^3, \ldots$. Instead, we define a new set $\{\phi_1, \phi_2, \phi_3, \ldots\}$ of basis functions using the following criteria: First, each $\phi_k : [x_0, x_N] \rightarrow \mathbb{R}$ is a polynomial having degree at most $k-1$. Second, $\phi_1 = \sqrt{N+1}$. Third,

$$\langle \phi_i, \phi_j \rangle = \begin{cases} 1, & \text{if} \quad i = j, \\ 0, & \text{if} \quad i > j \end{cases},$$

for $i = 2, 3, \ldots$. (The functions appearing in the inner product are, strictly speaking, restrictions of ϕ_i and ϕ_j to Δ.) These conditions suffice to define inductively a sequence $\{\phi_1, \phi_2, \phi_3, \ldots\}$ of basis vectors whose Gram matrix collapses to the identity matrix,

$$\mathsf{G} = \mathsf{I} = \begin{bmatrix} 1 & & \\ & 1 & \\ & & \ddots \end{bmatrix}.$$

The functions $\phi_1, \phi_2, \ldots$ are called the **Gram polynomials** for the grid Δ. Isaacson and Keller ([6], Section 3.5) furnish more details.

Since the theory used in this section applies to any inner-product space, it is possible to extend the least-squares method to continuous settings. Commonly, the vector space of concern is $L^2([a, b])$ instead of $\mathbb{R}^{N+1}$, and the appropriate inner product is

$$\langle f, g \rangle = \int_a^b f(x) g(x) \, dx.$$

Given a subspace $\mathcal{S} = \operatorname{span}\{\varphi_1, \varphi_2, \ldots, \varphi_n\}$ of $L^2([a, b])$, the idea is to approximate an arbitrary function $f \in L^2([a, b])$ using a function $\hat{f} \in \mathcal{S}$. According to the projection principle, we do this by forcing $\langle f - \hat{f}, \varphi_k \rangle = 0$ for $k = 1, 2, \ldots, n$. The normal equations in this setting therefore have the form

$$\sum_{j=1}^{n} c_j \int_a^b \varphi_j(x) \varphi_k(x) \, dx = \int_a^b \varphi_k(x) f(x) \, dx, \qquad k = 1, 2, \ldots, n, \qquad (1.7\text{-}4)$$

in which the coefficients c_j are unknown. The function

$$\hat{f}(x) := \sum_{j=1}^{n} c_j \varphi_j(x),$$

determined in this way, is called the L^2-**projection** of f onto $\mathcal{S}$.

In contrast to the sums required in the discrete least-squares method, the integrals in Equation (1.7-4) can be impossible to compute exactly. In these cases, it is necessary to use numerical approximations, a topic that we explore in Chapter 6. Aside from obstacles like this, though, many properties of the discrete least-squares method carry over to the continuous case, as Isaacson and Keller [6] explain. In particular, if one seeks a polynomial approximation to f, one can identify a collection of basis functions that are analogous to the Gram polynomials in that they reduce the Gram matrix to the identity matrix. These polynomials are the **Legendre polynomials**; we discuss them in more detail in Chapter 6.

1.8 Trigonometric Interpolation

While interpolation and the least-squares method seem to be fundamentally different approaches to approximating functions, there is a technique for which the two ideas are closely related. In this technique, called **trigonometric interpolation**, one develops linear approximations using trigonometric functions as basis functions. The theory underlying this method has strong and beautiful connections with classical Fourier analysis, some of which we exploit in this section.

Motivation and Construction

Consider a function $f\colon [0, 2\pi] \to \mathbb{R}$ that is periodic, meaning that $f(0) = f(2\pi)$. Let $\Delta = \{x_0, x_1, \ldots, x_{N+1}\}$, where $x_j = 2\pi j/(N+1)$, be a uniform grid on the interval $[0, 2\pi]$. Our goal is to find a periodic approximating function $\hat{f}$ that is a superposition of sines and cosines and that interpolates f in the sense that

$$\hat{f}(x_j) = f(x_j), \qquad j = 0, 1, \ldots, N. \tag{1.8-1}$$

(Periodicity then implies that $\hat{f}(x_{N+1}) = f(x_{N+1})$.)

If N is even, then $\hat{f}$ is to have the form

$$\hat{f}(x) = \frac{a_0}{2} + \sum_{n=1}^{M} (a_n \cos nx + b_n \sin nx), \tag{1.8-2a}$$

where $N = 2M$. If N is odd, say $N = 2M + 1$, then $\hat{f}$ must have the form

$$\hat{f}(x) = \frac{a_0}{2} + \sum_{n=1}^{M} (a_n \cos nx + b_n \sin nx) + \frac{a_{M+1}}{2} \cos(M+1)x. \tag{1.8-2b}$$

The idea behind these expansions is to approximate f using a finite superposition of sinusoidal components having various frequencies. Since f may have variations whose resolution requires infinitely many such components, $\hat{f}$ generally differs from f.

These finite trigonometric sums resemble the infinite sums that one encounters in classical Fourier analysis. Indeed, if $f \in C^1([0,\pi])$ and $f(0) = f(2\pi)$, then

$$f(x) = \frac{a_0}{2} + \sum_{n=1}^{\infty} (a_n \cos nx + b_n \sin nx),$$

where the **Fourier coefficients** are

$$a_n = \frac{1}{\pi}\int_{-\pi}^{\pi} f(x)\cos nx\, dx, \qquad b_n = \frac{1}{\pi}\int_{-\pi}^{\pi} f(x)\sin nx\, dx.$$

Moreover, the series converges uniformly in x. (See Kreider et al. [8, Section 10-4].) One can relate this classical theory heuristically to the problem of trigonometric interpolation by observing that classical Fourier analysis "samples" f — that is, it asks that the series agree with f — throughout the interval $[0,2\pi]$. In contrast, trigonometric interpolation "samples" f only at the finitely many points $x_0, x_1, \ldots, x_N \in [0,2\pi]$.

Instead of deriving the coefficients a_n and b_n in Equations (1.8-2) directly from the interpolation constraints (1.8-1), we first rewrite the trigonometric sums in a simpler form. When $N = 2M$, Equation (1.8-2a) reduces to the form

$$\hat{f}(x) = \sum_{n=-M}^{M} \phi(n)e^{inx},$$

where $i^2 = -1$, $a_n = \phi(n) + \phi(-n)$, and $b_n = i[\phi(n) - \phi(-n)]$. This fact follows from the identity $e^{inx} = \cos nx + i \sin nx$:

$$\begin{aligned}\sum_{n=-M}^{M} \phi(n)e^{inx} &= \phi(0) + \sum_{n=1}^{M}[\phi(n)(\cos nx + i\sin nx)\\ &\qquad +\phi(-n)(\cos nx - i\sin nx)]\\ &= \frac{a_0}{2} + \sum_{n=1}^{M}(a_n \cos nx + b_n \sin nx).\end{aligned}$$

Similar substitutions show that, when $N = 2M+1$, Equation (1.8-2b) simplifies to the complex exponential form

$$\hat{f}(x) = \sum_{n=-M}^{M+1} \phi(n)e^{inx}.$$

We subsume both of these cases in the notation

$$\hat{f}(x) = \sum_{n=-M}^{M+\theta} \phi(n)e^{inx}, \tag{1.8-3}$$

where $\theta = N - 2M$. This expression corresponds to the following representation, which is common in classical Fourier analysis:

$$\sum_{n=-\infty}^{\infty} f_n e^{inx} = \frac{a_0}{2} + \sum_{n=1}^{\infty} (a_n \cos nx + b_n \sin nx),$$

where

$$f_n := \frac{1}{2\pi} \int_{-\pi}^{\pi} f(x) e^{-inx}\, dx.$$

To determine the coefficients $\phi(n)$ in the expansion (1.8-3), it is profitable to cast the discussion in terms of inner products. Moreover, since we are working with complex exponentials, it is appropriate to use inner products for complex-valued functions $f\colon [0, 2\pi] \to \mathbb{C}$, sampled at the nodes $x_0, x_1, \ldots, x_N$ of the grid. For two such functions f and g, denote

$$\langle f, g \rangle := \sum_{j=0}^{N} f(x_j)\overline{g(x_j)},$$

where the overbar indicates complex conjugation. The function $\langle \cdot, \cdot \rangle$ is an inner product, provided we regard two complex-valued functions on $[0, 2\pi]$ as being equivalent if they have the same values at the nodes $x_0, x_1, \ldots, x_N$. Later, in discussing the phenomenon of aliasing, we examine peculiarities associated with this equivalence. This inner product gives rise to a norm, defined as follows:

$$\|f\|_\Delta := \sqrt{\langle f, f \rangle}.$$

We now establish an orthogonality relationship among the functions e^{inx}:

LEMMA 1.26. *For* $n = 0, \pm 1, \pm 2, \ldots$,

$$\langle e^{imx}, e^{inx} \rangle = \begin{cases} N+1, & \text{if } \ (m-n)/(N+1) \ \text{ is an integer,} \\ 0, & \text{otherwise.} \end{cases} \tag{1.8-4}$$

PROOF: We have

$$\begin{aligned} \langle e^{imx}, e^{inx} \rangle &= \sum_{j=0}^{N} \exp(imx_j) \exp(-inx_j) \\ &= \sum_{j=0}^{N} \exp\left[\frac{2\pi i j (m-n)}{N+1}\right]. \end{aligned}$$

This last sum has the form $1 + r + r^2 + \cdots + r^N$, where

$$r := \exp\left[\frac{2\pi i (m-n)}{N+1}\right].$$

If $(m-n)/(N+1)$ is an integer, then $r=1$, so the sum collapses to $N+1$, as claimed. Otherwise, $r \neq 1$, and we sum the geometric series to get $(r^{N+1} - 1)/(r-1)$. Since $r^{N+1} = \exp[2\pi i(m-n)] = 1$ in this case, the conclusion again follows. ∎

In the special case when $\hat{f} = f$ on the entire interval $[0, 2\pi]$, this orthogonality relationship makes it easy to determine the coefficients $\phi(n)$. We have

$$\langle f, e^{imx} \rangle = \langle \hat{f}, e^{imx} \rangle = \sum_{n=-M}^{M+\theta} \phi(n) \langle e^{inx}, e^{imx} \rangle = \phi(m) \langle e^{imx}, e^{imx} \rangle,$$

from which we conclude that

$$\phi(n) = \frac{\langle f, e^{inx} \rangle}{\langle e^{inx}, e^{inx} \rangle} = \frac{1}{N+1} \sum_{j=0}^{N} f(x_j) \exp(-inx_j). \tag{1.8-5}$$

By analogy with classical Fourier analysis, we call these coefficients the **discrete Fourier coefficients** of f. Another common name for the function $\phi(n)$ is the **discrete Fourier transform** of f.

More generally, f does not have the form (1.8-3). We must then regard the coefficients $\phi(n)$ determined by Equation (1.8-5) as furnishing an approximating function $\hat{f}$. Interestingly, the discrete Fourier coefficients $\phi(n)$ are precisely those that solve the trigonometric interpolation problem:

THEOREM 1.27. *The function $\hat{f}$ in Equation (1.8-3), with coefficients $\phi(n)$ given in Equation (1.8-5), satisfies the interpolation constraints $\hat{f}(x_j) = f(x_j)$, for $j = 0, 1, \ldots, N$.*

The proof requires a simple lemma:

LEMMA 1.28. *For j, k ranging over the indices $0, 1, \ldots, N$,*

$$\sum_{n=0}^{N} \exp[in(x_j - x_k)] = \begin{cases} N+1, & \text{if } j = k, \\ 0, & \text{if } j \neq k. \end{cases}$$

PROOF: The number $z := \exp[i(x_j - x_k)]$ is a zero of the polynomial

$$z^{N+1} - 1 = (z-1) \sum_{n=0}^{N} z^n.$$

One possibility is that $\exp[i(x_j - x_k)] = 1$, which occurs when $j = k$. The other, occurring when $j \neq k$, is that the sum vanishes. ∎

PROOF OF THEOREM 1.27: Consider the value of $\hat{f}$ at a node x_j:

$$\begin{aligned} \hat{f}(x_j) &= \sum_{n=-M}^{M+\theta} \phi(n)\exp(inx_j) \\ &= \sum_{n=1}^{M} \phi(-n)\exp(-inx_j) + \sum_{n=0}^{M+\theta} \phi(n)\exp(inx_j). \end{aligned} \tag{1.8-6}$$

We rewrite the first sum on the right by noting that

$$\begin{aligned} \exp(-inx_j) &= \exp\left(-inj\frac{2\pi}{N+1} + 2\pi ij\right) \\ &= \exp\left[ij(N+1-n)\frac{2\pi}{N+1}\right] \\ &= \exp\left[i(N+1-n)x_j\right]. \end{aligned}$$

A similar device, applied to the definition (1.8-5), shows that

$$\begin{aligned} \phi(-n) &= \frac{1}{N+1}\sum_{k=0}^{N} f(x_k)\exp(inx_j) \\ &= \sum_{k=0}^{N} f(x_k)\exp\left[-i(N+1-n)x_k\right] = \phi(N+1-n). \end{aligned}$$

As a consequence of these identities, we can reindex the first sum on the right in Equation (1.8-6) to get

$$\hat{f}(x_j) = \sum_{n=-M}^{M+\theta} \phi(n)\exp(inx_j) = \sum_{n=0}^{N} \phi(n)\exp(inx_j). \tag{1.8-7}$$

With this representation for $\hat{f}$ at the nodes of the grid, we find that

$$\begin{aligned} \hat{f}(x_j) &= \sum_{n=0}^{N}\left[\frac{1}{N+1}\sum_{k=0}^{N} f(x_k)\exp(-inx_k)\right]\exp(inx_j) \\ &= \frac{1}{N+1}\sum_{k=0}^{N} f(x_k)\sum_{n=0}^{N}\exp[in(x_j - x_k)] \\ &= \frac{1}{N+1}(N+1)f(x_j) = f(x_j), \end{aligned}$$

by Lemma 1.28. ∎

Practical Considerations: Fast Fourier Transform

Equation (1.8-5) furnishes an explicit formula for the discrete Fourier coefficients,

$$\phi(n) = \frac{1}{N+1}\sum_{j=0}^{N} f(x_j)\exp(-inx_j).$$

However, as a naive approach to computing $\phi(n)$, this formula is far from the best one. To see why, let us abbreviate the sum in Equation (1.8-5) by writing

$$\phi(n) = \sum_{j=0}^{N} \alpha_j \omega^{nj}, \tag{1.8-8}$$

where

$$\alpha_j := \frac{f(x_j)}{N+1}, \qquad \omega := \exp\left(-\frac{2\pi i}{N+1}\right).$$

One can compute the sum in Equation (1.8-8) from its coefficients using $N+1$ multiplications and N additions. Using this approach for each of the $N+1$ coefficients $\phi(n)$ requires $(N+1)(2N+1) = 2N^2+3N+1 = \mathcal{O}(N^2)$ arithmetic operations. This operation count inhibited the use of discrete Fourier analysis in signal processing and other applications for many years.

A class of algorithms called **fast Fourier transforms** (FFT), developed in the early 1960s, changed this picture radically. In general, if $N+1 = q_1q_2\cdots q_p$ is an integer factoring of $N+1$, then the FFT allows one to compute the $N+1$ discrete Fourier coefficients $\phi(n)$ in $\mathcal{O}(N(q_1+q_2+\cdots+q_p))$ operations — many fewer, typically, than $\mathcal{O}(N^2)$. The ideas are perhaps simplest in the case when $N+1 = 2^p$, which we use in the following exposition. In this special case, we show that the operation count is $\mathcal{O}(N\log_2 N)$. However, other cases are not only feasible but also, in many cases, quite efficient in comparison. We begin by sketching how the FFT accomplishes its task in so few operations. Then we outline the overall strategy for a common version of the algorithm. Finally, we discuss details of the algorithm's implementation.

To see why the FFT is so computationally efficient, consider the task of computing discrete Fourier coefficients $\phi(n)$, for the special case $N+1 = 2^p$. The following observation is central:

LEMMA 1.29 (DANIELSON-LANCZOS). *One can compute $\phi(n)$ by evaluating two sums of the type (1.8-8), each having $(N+1)/2 = 2^{p-1}$ terms.*

PROOF: Split the sum in Equation (1.8-8) into two sums, one over the even indices n and one over the odd indices. For $n = 0, 1, \ldots, N$, we obtain

$$\phi(n) = \sum_{j=0}^{(N-1)/2} \alpha_{2j}(\omega^2)^{nj} + \sum_{j=0}^{(N-1)/2} \alpha_{2j+1}(\omega^2)^{nj}\omega^n. \tag{1.8-9}$$

For $m = 0, 1, \ldots, (N-1)/2$, define

$$\psi_0(m) := \sum_{j=0}^{(N-1)/2} \alpha_{2j}(\omega^2)^{mj},$$

$$\psi_1(m) := \sum_{j=0}^{(N-1)/2} \alpha_{2j+1}(\omega^2)^{mj}.$$

The quantity $\psi_0(m)$ is obviously the first sum on the right side of Equation (1.8-9), when $n = m = 0, 1, \ldots, \frac{1}{2}(N-1)$. Also, for $n = \frac{1}{2}(N-1)+1, \frac{1}{2}(N-1)+2, \ldots, N$, we have $n = \frac{1}{2}(N+1)+m$, where $m = 0, 1, \ldots, \frac{1}{2}(N-1)$. For these values of n, the fact that $\omega^{N+1} = 1$ implies that

$$\begin{aligned}
\omega^n \psi_1(m) &= \sum_{j=0}^{(N-1)/2} \alpha_{2j+1}(\omega^2)^{mj}\omega^n \\
&= \sum_{j=0}^{(N-1)/2} \alpha_{2j+1}(\omega^2)^{mj+(N+1)j/2}\omega^n \\
&= \sum_{j=0}^{(N-1)/2} \alpha_{2j+1}(\omega^2)^{mj}\omega^n,
\end{aligned}$$

which is the second sum on the right in Equation (1.8-9). Therefore,

$$\phi(n) = \psi_0(m) + \omega^n \psi_1(m), \tag{1.8-10}$$

where m is the remainder on division of n by $(N-1)/2$. ∎

In other words, one can compute the discrete Fourier coefficients $\phi(n)$ for a grid having 2^p nodes by executing the operations needed to perform two such analyses on grids having 2^{p-1} nodes, then executing at most $2 \cdot 2^p$ operations to form the 2^p numbers $\phi(n)$ from Equation (1.8-10). The same reasoning applies to each of the analyses on the grids having 2^{p-1} nodes, and so forth. Thus, if $\Theta(p)$ denotes the number of arithmetic operations required for a grid having 2^p nodes, then recursion yields

$$\begin{aligned}
\Theta(p) &\leq 2\Theta(p-1) + 2 \cdot 2^p \\
&\leq 2[2\Theta(p-2) + 2 \cdot 2^{p-1}] + 2 \cdot 2^p = 2^2\Theta(p-2) + 4 \cdot 2^p \\
&\leq 2^3\Theta(p-3) + 6 \cdot 2^p \\
&\leq \cdots \leq 2^p\Theta(0) + 2^p \sum_{k=0}^{p} 2.
\end{aligned}$$

Since $\Theta(0) = 0$, we find that $\Theta(p) \leq 2p \cdot 2^p = 2(N+1)\log_2(N+1)$. This operation count is a tremendous improvement over the $\mathcal{O}(N^2)$ operations required in the naive approach. For example, for a problem involving $2^{10} = 1024$ nodes, the naive approach requires about 2.096×10^6 operations, while the FFT requires at most 2.048×10^4.

We turn now to the structure of the algorithm. Here it helps to look in more detail at the sums $\psi_0(m)$ and $\psi_1(m)$ used in Equation (1.8-10). Notice that

$$2 \sum_{j=0}^{(N-1)/2} \alpha_{2j}(\omega^2)^{mj}$$

is the mth discrete Fourier coefficient obtained by interpolating $f(x)$ only at the even-indexed nodes $x_0, x_2, \ldots, x_{N-1}$. Similarly, the quantity

$$2 \sum_{j=0}^{(N-1)/2} \alpha_{2j+1}(\omega^2)^{mj}$$

is the mth discrete Fourier coefficient for the interpolant of $f(x+2\pi/(N+1))$ that uses values of f at the odd-indexed nodes $x_1, x_3, \ldots, x_N$.

Therefore, one can compute the trigonometric interpolant on a grid having 2^p nodes by computing two trigonometric interpolants on grids having 2^{p-1} nodes. One of the latter interpolants interpolates f at the even-indexed nodes $x_0, x_2, \ldots, x_{N-1}$; the other interpolates f at the odd-indexed nodes $x_1, x_3, \ldots, x_N$. Thought of recursively, this observation suggests that we proceed in p stages: At stage r, we determine 2^{p-r} interpolants on grids having 2^r nodes. We calculate these interpolants using the 2^{p-r+1} interpolants, associated with grids having 2^{r-1} nodes, computed at stage $r-1$. Each interpolant computed at stage r has 2^r coefficients; we denote the coefficients for the qth interpolant at stage r as follows:

$$\phi^{(r)}(q,0), \phi^{(r)}(q,1), \ldots, \phi^{(r)}(q,2^r).$$

Using Equation (1.8-10), we compute these coefficients from those computed at stage $r-1$ as follows: For $q = 1, 2, \ldots, 2^{p-r}$,

$$2\phi^{(r)}(q,n) = \begin{cases} \phi^{(r-1)}(q,n) + \phi^{(r-1)}(q',n)\omega_r^n, & n = 0, 1, \ldots 2^{p-r} - 1, \\ \phi^{(r-1)}(q,n') - \phi^{(r-1)}(q',n')\omega_r^{n'}, & n = 2^{p-r}, \ldots, 2^{p-r+1}. \end{cases} \tag{1.8-11}$$

Here,

$$q' := 2^{p-r} + q, \qquad n' := n - 2^{r-1}, \qquad \omega_r := \exp(-2\pi i/2^r).$$

We start the recursion by setting $\phi^{(0)}(q,0)$ equal to one of the prescribed values $f(x_k)$, in a manner described shortly. After completing stage p, we set $\phi(n) := \phi^{(p)}(1,n)$, for $n = 0, 1, \ldots, N$. This strategy, based upon Equation

(1.8-11), lies at the heart of one version of the FFT, known as the **Cooley-Tukey algorithm**.

Finally, let us discuss the implementation of the FFT. In coding the algorithm, one must choose a data structure for the coefficients $\phi^{(r)}(q,n)$ computed at the various stages $r = 0, 1, \ldots, p$. Typically, one stores these quantities in an array $\Phi(k)$, where $k = k(r,q,n)$. The standard lexicographic ordering

$$k(r,q,n) := 2^r(q-1) + n$$

is straightforward, but it is inefficient: It requires two copies of $\Phi(k)$ to keep the right side of Equation (1.8-11) intact while computing the left side.

For a more economical indexing scheme, we can overwrite the quantities $\phi^{(r-1)}(q,n), \phi^{(r-1)}(q+2^{p-r},n)$ with the quantities $\phi^{(r)}(q,n), \phi^{(r)}(q,n+2^{r-1})$ computed from them. The price paid for this economy is the more intricate indexing scheme required. To see how this scheme works, consider Table 1.1 on page 103, which lists the coefficients $\phi^{(r)}(q,n)$ computed at the four stages of an FFT of length $N+1=8$. The rightmost column of the table is the output vector, containing the final discrete Fourier coefficients, $\phi(n)$, listed in their "natural" order. To the left of this column are the results of stage $r=3$, listing in each position the result $\phi^{(3)}(0,n)$ $(= \phi(n))$ alongside the node x_n associated with it in the expansion

$$\sum_{n=0}^{N} \phi^{(3)}(0,n) f(x_n).$$

We write the subscript n of x_n in the binary number system for illustrative purposes that soon become apparent.

To the left of the results for $r=3$ are those for $r=2$, listed in the format just described. For example, we use the coefficients $\phi^{(2)}(0,0)$ and $\phi^{(2)}(1,0)$ to compute $\phi^{(3)}(0,0)$ and $\phi^{(3)}(0,4)$, overwriting the former in the process. However, the fact that the coefficients

$$\phi^{(2)}(0,0), \phi^{(2)}(0,1), \phi^{(2)}(0,2), \phi^{(2)}(0,3)$$

are associated with the even-indexed nodes of the grid implies that the nodal indices undergo a permutation as we move from the columns associated with stage $r=3$ to those associated with $r=2$. Similar logic governs the transition from stage $r=2$ back to stage $r=1$ and the transition from $r=1$ to $r=0$, the stage at which we have the input data $f(x_0), f(x_1), \ldots, f(x_8)$.

As a consequence of the permutations on nodal indices, the vector of input values $f(x_n)$ is ordered differently than the vector of output values. Looking carefully, one can see that the order of the nodal indices in the input vector results from reversing the binary representations of the indices in the output vector. Under this **bit-reversal** mapping, $k(2^N-1,0,n) = n$, while $k(0,q,0)$ produces the bit-reverse of q. For example, writing q in binary, we have $k(0,101,0) = 101$, while $k(0,011,0) = 110$. The idea in this data structure is to load the input data in bit-reversed order, compute coefficients at successive stages "in place" by overwriting those at previous stages, and arrive at an output vector of coefficients listed in natural order.

The following pseudocode implements a bit-reversal algorithm:

ALGORITHM 1.2 (BIT-REVERSAL). *Given an array A of length $N+1 = 2^p$, the following algorithm permutes the entries of A using the bit-reversal mapping.*

1. $j \leftarrow 1$.
2. For $k = 1, 2, \ldots, N+1$:
3. If $j > k$ then:
4. $\alpha \leftarrow A(j)$.
5. $A(j) \leftarrow A(k)$.
6. $A(k) \leftarrow \alpha$.
7. End if.
8. $m \leftarrow (N+1)/2$.
9. If $m \geq 2$ and $j > m$ then:
10. $j \leftarrow j - m$.
11. $m \leftarrow m/2$.
12. Go to 9.
13. End if.
14. $j \leftarrow j + m$.
15. Next k.
16. End.

Given an array $A(j)$ ordered by bit-reversal, the following pseudocode implements the Cooley-Tukey algorithm:

ALGORITHM 1.3 (COOLEY-TUKEY FFT). *The following algorithm computes an FFT of length 2^p, given an array $A(j)$ of input data ordered by bit-reversal.*

1. For $r = 1, 2, \ldots, p$:
2. $m \leftarrow 2^r$.

3. $\omega \leftarrow 1$.
4. For $j = 1, 2, \ldots, 2^{r-1}$:
5. For $k = j, j + 2^r, \ldots, 2^p$:
6. $\alpha \leftarrow \omega A(k + 2^{r-1})$.
7. $A(k + 2^{r-1}) \leftarrow A(k) - \alpha$.
8. $A(k) \leftarrow A(k) + \alpha$.
9. Next k.
10. $\omega \leftarrow \omega \exp(2\pi i/2^r)$.
11. Next j.
12. Next r.
13. End.

Mathematical Details

Errors arise in any scheme for approximating general functions using finitely many degrees of freedom. In the case of trigonometric interpolation, one of the simplest ways to view the approximation error is to adopt the point of view of least-squares approximation, introduced in Section 1.7. If $\hat{f}$ is the trigonometric interpolant of f on a uniform grid $\{x_0, x_1, \ldots, x_{N+1}\}$ on $[0, 2\pi]$, then

$$\begin{aligned} \langle f - \hat{f}, e^{imx} \rangle &= \langle f, e^{imx} \rangle - \sum_{n=0}^{N} \phi(n) \langle e^{inx}, e^{imx} \rangle \\ &= \langle f, e^{imx} \rangle - \phi(m) \langle e^{imx}, e^{imx} \rangle = 0. \end{aligned}$$

In other words, if we regard the functions e^{inx} as basis functions, the discrete Fourier coefficients are precisely the coefficients that force $\hat{f}$ to satisfy the normal equations of the least-squares method. We conclude the following:

THEOREM 1.30. *The function $\hat{f}$, defined as in Equation (1.8-3) with coefficients $\phi(n)$ given in Equation (1.8-5), minimizes the least-squares error*

$$\left\| f - \sum_{n=-M}^{M+\theta} c_n e^{inx} \right\|_{\Delta}$$

over all possible choices of the coefficients c_n.

One should suspect that such an easily won theorem has limitations. In this case, the limitations arise from the fact that $\|\cdot\|_{\Delta}$ is a norm only if we regard functions $f: [0, 2\pi] \to \mathbb{R}$ as being equivalent when they agree at the nodes x_i of the grid. In particular, the theorem does not furnish any explicit estimate of the deviation of $\hat{f}$ from f at points *between* the nodes. As we briefly discuss below, such "global" error estimates are known but not easily accessible in a text at this level.

It is appropriate, however, to explore one consequence of the limitations associated with the norm $\|\cdot\|_\Delta$. Trigonometric interpolation is vulnerable to a characteristic type of error called **aliasing**. Consider a function f having an infinite Fourier-series representation:

$$f(x) = \sum_{n=-\infty}^{\infty} f_n e^{inx}, \qquad f_n := \frac{1}{2\pi} \int_{-\pi}^{\pi} f(x) e^{-inx}\, dx.$$

The trigonometric interpolant of f over a uniform grid $x_0, x_1, \ldots, x_N$ on $[0, 2\pi]$ has the form

$$\hat{f}(x) = \sum_{n=0}^{N} \phi(n) e^{inx}, \qquad \phi(n) := \frac{\langle f, e^{inx} \rangle}{N+1}.$$

Let us examine how the coefficients f_n and $\phi(n)$ are related. Observe that

$$\begin{aligned} \frac{\langle f, e^{imx} \rangle}{N+1} &= \frac{1}{N+1} \left\langle \left[\sum_{n=-\infty}^{\infty} \frac{1}{2\pi} \int_{-\pi}^{\pi} f(x) e^{-inx}\, dx \right] e^{inx}, e^{imx} \right\rangle \\ &= \sum_{n=-\infty}^{\infty} \frac{\langle e^{inx}, e^{imx} \rangle}{N+1} \frac{1}{2\pi} \int_{-\pi}^{\pi} f(x) e^{-inx}\, dx. \end{aligned}$$

By Lemma 1.26, $\langle e^{inx}, e^{imx} \rangle$ vanishes except when $n - m = j(N+1)$, for some integer j. It follows that

$$\frac{\langle f, e^{imx} \rangle}{N+1} = \sum_{j=-\infty}^{\infty} \frac{1}{2\pi} \int_{-\pi}^{\pi} f(x) \exp\Big\{-i[m + j(N+1)]x\Big\}\, dx. \qquad (1.8\text{-}12)$$

Equation (1.8-12) reveals that the discrete Fourier coefficient $\phi(m)$ contains information not only from the classical Fourier coefficient f_m having the same frequency $m/(4\pi)$ but also from Fourier coefficients $f_{m+j(N+1)}$ associated with higher frequencies. In a sense, trigonometric interpolation "conserves" the oscillatory information in the function f, subject to the impossibility of representing oscillations having frequency greater than or equal to $F_{\text{crit}} := (N+1)/(4\pi)$ on the grid $\{x_0, x_1, \ldots, x_N\}$. This critical frequency F_{crit} is called the **Nyquist frequency** for the grid. The interpolant $\hat{f}$ assigns information associated with frequencies greater than or equal to the Nyquist frequency to low-frequency discrete Fourier coefficients, thereby distorting the representation of low-frequency modes.

Aliasing is precisely this distortion. Figure 1 illustrates the phenomenon, showing that the functions $\sin x$ and $\sin 5x$ have the same nodal values, and hence the same trigonometric interpolants, on a grid with $N+1=4$. On this grid, any function f having nonzero classical Fourier coefficients f_1 and f_5 will have a trigonometric interpolant in which the oscillations associated with f_5 are aliased as contributions to the discrete Fourier coefficient $\phi(1)$.

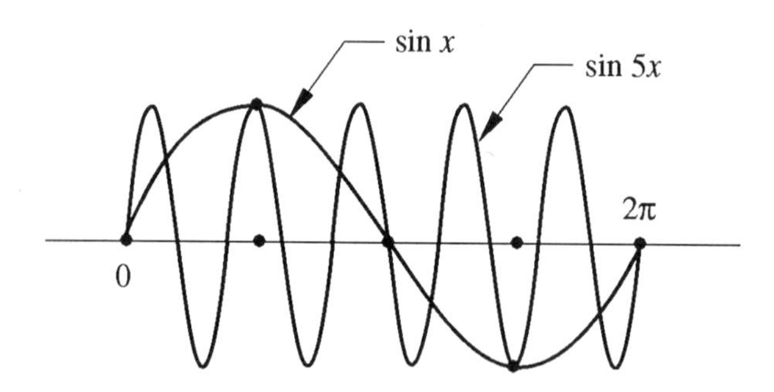

FIGURE 1. *The functions* $\sin x$ *and* $\sin 5x$*, sampled on a four-interval grid on* $[0, 2\pi]$ *to illustrate aliasing.*

Further Remarks

This section presents the barest rudiments of the FFT, which one might best regard as a huge class of algorithms. The literature on the FFT is extensive, dating from before the widely cited work of Cooley and Tukey [3] to the present. While we restrict our discussion to FFTs of length 2^p, in applications FFTs of more general lengths arise quite commonly. In fact, the computations in these cases, based upon factorings of the lengths of the FFTs, can be quite efficient. Moreover, reasonable algorithms exist even for FFTs having prime length. Curious readers should consult Rader [10].

Noticeably absent from this section is an estimate of the error associated with trigonometric interpolation. Suppose that $f \in C^{(m)}([0, 2\pi])$, and let $\hat{f}$ denote the trigonometric interpolant of f on a uniform grid with nodes $x_k = 2\pi k/(N+1)$. One can show that there is a positive constant C such that

$$\|f - \hat{f}\|_{L^2} \leq \frac{C}{(N+1)^m} \|f^{(m)}\|_{L^2}.$$

Here, $\|\cdot\|_{L^2}$ signifies the usual L^2 norm on $[0, 2\pi]$, and $f^{(m)}$ is the mth derivative of f. The proof of this estimate is too involved to lie within our scope. Canuto et al. ([1], Section 9.1) supply details and draw connections with the analysis of aliasing errors.

Table 1.1: Progress of an FFT of length eight, illustrating bit reversal.

input	$m = 0$		$m = 1$		$m = 2$		$m = 3$		output
value	coeff.	node	coeff.	node	coeff.	node	coeff.	node	value
$f(x_0)$	$\phi^{(0)}(0,0)$	x_{000}	$\phi^{(1)}(0,0)$	x_{000}	$\phi^{(2)}(0,0)$	x_{000}	$\phi^{(3)}(0,0)$	x_{000}	$\phi(0)$
$f(x_4)$	$\phi^{(0)}(1,0)$	x_{100}	$\phi^{(1)}(0,1)$	x_{100}	$\phi^{(2)}(0,1)$	x_{010}	$\phi^{(3)}(0,1)$	x_{001}	$\phi(1)$
$f(x_2)$	$\phi^{(0)}(2,0)$	x_{010}	$\phi^{(1)}(1,0)$	x_{010}	$\phi^{(2)}(0,2)$	x_{100}	$\phi^{(3)}(0,2)$	x_{010}	$\phi(2)$
$f(x_6)$	$\phi^{(0)}(3,0)$	x_{110}	$\phi^{(1)}(1,1)$	x_{110}	$\phi^{(2)}(0,3)$	x_{110}	$\phi^{(3)}(0,3)$	x_{011}	$\phi(3)$
$f(x_1)$	$\phi^{(0)}(4,0)$	x_{001}	$\phi^{(1)}(2,0)$	x_{001}	$\phi^{(2)}(1,0)$	x_{001}	$\phi^{(3)}(0,4)$	x_{100}	$\phi(4)$
$f(x_5)$	$\phi^{(0)}(5,0)$	x_{101}	$\phi^{(1)}(2,1)$	x_{101}	$\phi^{(2)}(1,1)$	x_{011}	$\phi^{(3)}(0,5)$	x_{101}	$\phi(5)$
$f(x_3)$	$\phi^{(0)}(6,0)$	x_{011}	$\phi^{(1)}(3,0)$	x_{011}	$\phi^{(2)}(1,2)$	x_{101}	$\phi^{(3)}(0,6)$	x_{110}	$\phi(6)$
$f(x_7)$	$\phi^{(0)}(7,0)$	x_{111}	$\phi^{(1)}(3,1)$	x_{111}	$\phi^{(2)}(1,3)$	x_{111}	$\phi^{(3)}(0,7)$	x_{111}	$\phi(7)$

1.9 Problems

PROBLEM 1. Examine the Runge phenomenon for Lagrange polynomial interpolation of $f(x) = (1 + 25x^2)^{-1}$ for $-1 \leq x \leq 1$. Let $\Delta = \{-1 = x_0, x_1, \ldots, x_9, x_{10} = 1\}$, where $x_k = -1 + 0.2k$, and compare the graph of the tenth-degree Lagrange interpolant for f on Δ with the graph of f. Investigate what happens when $\Delta = \{z_0, z_1, \ldots, z_{10}\}$, where the z_k are the **Chebyshev abscissae**,

$$z_k = \cos\left(\frac{2k+1}{N+1}\frac{\pi}{2}\right), \qquad k = 0, 1, \ldots, N = 10.$$

PROBLEM 2. Prove the inequality (1.2-7).

PROBLEM 3. If $L_0, L_1, \ldots L_N$ are the basis functions for Lagrange interpolation of degree N on a grid $\{x_0, \ldots x_N\}$, show that $\sum_{k=0}^{N} L_k(x) = 1$.

PROBLEM 4. Show that the functions defined in Equations (1.4-3) and (1.4-4) satisfy the conditions (1.4-5) and (1.4-6).

PROBLEM 5. Let $f \in C^1([a, b])$. Call

$$M = \sup_{x \in [a,b]} f(x), \qquad m = \inf_{x \in [a,b]} f(x),$$

and consider the constant function $\Phi(x) = (M + m)/2$, for $x \in [a, b]$, as an approximation to f. Show that $|f(x) - \Phi(x)|$ is bounded by a quantity proportional to the interval length $b - a$. This observation lies behind an error estimate for piecewise constant approximations that has the form $\|f - \Phi\|_\infty \leq \text{const}\|f'\|_\infty h$. (The fundamental theorem of calculus is a useful weapon here.)

PROBLEM 6. Prove Proposition 1.4.

PROBLEM 7. Prove Theorem 1.10.

PROBLEM 8. To six significant places, $\sin(0.40) = 0.389418$, $\cos(0.40) = 0.921061$, $\sin(0.45) = 0.434966$, and $\cos(0.45) = 0.900447$. Use these data, together with Hermite cubic interpolation, to estimate $\sin(0.43)$. What is the error? How does it compare with the error estimate for Hermite cubic interpolation? Comment.

PROBLEM 9. Let Δ_x and Δ_y be grids on the intervals $[a, b]$ and $[c, d]$, respectively, and suppose that $\mathcal{M}(\Delta_x)$ and $\mathcal{N}(\Delta_y)$ are piecewise polynomial interpolation spaces associated with these grids. Show that $\mathcal{M}(\Delta_x) \otimes \mathcal{N}(\Delta_y)$ is independent of the choice of bases for $\mathcal{M}(\Delta_x)$ and $\mathcal{N}(\Delta_y)$.

PROBLEM 10. If $\mathcal{V}$ is a vector space, then a function $p : \mathcal{V} \to \mathcal{V}$ is a **projection** provided (*i*) p is linear, that is, $p(\alpha u + \beta v) = \alpha p(u) + \beta p(v)$ for all $u, v \in \mathcal{V}$ and all scalars α and β;

(*ii*) p is idempotent, that is, $p(p(v)) = p(v)$ for all $v \in \mathcal{V}$.

Consider the interpolation operator $\pi : C^0([a,b]) \to \mathcal{M}_0^1(\Delta)$, where Δ is a grid on $[a,b]$. In this case, $\pi(f)$ is the piecewise Lagrange linear interpolant of f on Δ. Show that π is a projection.

PROBLEM 11. Prove Theorem 1.13. Do the interpolatory projections π_x and π_y commute?

PROBLEM 12. For piecewise planar interpolation on triangles, Theorem 1.14 gives the error estimate $\|f - \Phi\|_\infty \leq 4Mh_{\text{max}}^2$. This estimate fails to show how interpolation on "long, skinny" triangles might differ from interpolation on more "regular" triangles. Denote by h_{max} and h_{min} the lengths of the longest and shortest sides, respectively. From the law of sines for arbitrary triangles,

$$\frac{h_{\text{max}}}{\sin\theta_{\text{max}}} = \frac{h_{\text{min}}}{\sin\theta_{\text{min}}},$$

where θ_{max} is the angle opposite the longest side and θ_{min} is the angle opposite the shortest side. Derive an error estimate of the form $\|f - \Phi\|_\infty \leq C/(\sin\theta_{\text{min}})^2$. What implications does this estimate have for the use of "long, skinny" triangles?

PROBLEM 13. The general quadratic polynomial in x and y has the form $\Phi(x,y) = a + bx + cy + dx^2 + exy + fy^2$, where a, b, c, d, e, f are constants. Develop a scheme for quadratic interpolation over a single triangle. Where are the nodes? What are the basis functions? If you use your scheme over each triangle Ω_e of a triangulated region Ω, for what nonnegative integer values of p will your piecewise polynomial interpolants lie in $C^p(\Omega)$?

PROBLEM 14. Prove Corollary 1.21.

PROBLEM 15. The purpose of this exercise is to construct bases for spaces of natural cubic splines on $[a,b]$ such that each function in the basis has nonzero values only over a small subset of $[a,b]$.

(A) Let $\Delta = \{x_0, \ldots, x_N\}$ be a uniform grid on $[a,b]$, so $x_i = x_0 + ih$. Extend Δ by adding new nodes $x_{-1} = x_0 - h$ and $x_{N+1} = x_0 + (N+1)h$, and call the larger grid Δ'. For each $i = -1, 0, \ldots, N, N+1$, determine a function $B_i \in \mathcal{M}_2^3(\Delta')$ that satisfies the conditions

$$B_i(x_j) = \begin{cases} 0, & \text{if } |j-i| \geq 2, \\ 1, & \text{if } j = i; \end{cases}$$

and $B_i'(x_j) = B_i''(x_j) = 0, \quad j = i \pm 2.$

(B) Observe that $B_i(x) = 0$ whenever $|x - x_i| \geq 2h$. Plot $B_i(x)$.

(C) Show that every $s \in \mathcal{M}_2^3(\Delta)$ can be written as a linear combination of the functions B_i, that is, there are constants $c_{-1}, c_0, \ldots, c_{N+1}$ such that $s(x) = \sum_{i=-1}^{N+1} c_i B_i(x)$ for $x \in [a, b]$. (Remember, Δ does not include x_{-1} or x_{N+1}.) The functions B_i are called **B-splines**; "B" stands for "bell-shaped."

PROBLEM 16. Using some reasonable choice of f, computationally verify the error estimate $\|\mathbf{f} - \mathbf{m}\|_\infty \leq \mathcal{O}(h^2)$ of Theorem 1.16 by means of a convergence plot.

PROBLEM 17. Schumaker [11] generalizes the notion of a spline on a grid $\Delta = \{a = x_0, \ldots, x_N = b\}$ to mean a function in $\mathcal{M}_{n-1}^n(\Delta)$, for some degree n. Familiar cases include $\mathcal{M}_2^3(\Delta)$ (cubic splines), $\mathcal{M}_0^1(\Delta)$ (piecewise Lagrange linears), and $\mathcal{M}_{-1}^0(\Delta)$ (piecewise constants). What sense can we make out of $\mathcal{M}_1^2(\Delta)$? Show that the "obvious" conditions,

(*i*) S is quadratic on each $[x_{i-1}, x_i]$;

(*ii*) $S \in C^1([a, b])$;

(*iii*) $S(x_i) = f(x_i)$ for $i = 0, \ldots, N$,

lead to fewer equations than unknown parameters. Another approach, suggested by Kammerer et al. [6], is the following: Set

$$S(x_0) = f(x_0); \qquad S(x_N) = f(x_N);$$

$$S\left(\frac{x_{i-1} + x_i}{2}\right) = f\left(\frac{x_{i-1} + x_i}{2}\right), \qquad i = 1, \ldots, N;$$

and require $S \in C^1([a, b])$. Show that these conditions lead to a match between the number of equations and the number of undetermined parameters in S. Describe the resulting interpolants.

1.10 References

1. C. Canuto, M.Y. Hussaini, A. Quarteroni, and T.A. Zang, *Spectral Methods in Fluid Dynamics*, Springer-Verlag, Berlin, 1988.

2. E.W. Cheney, *Introduction to Approximation Theory*, Chelsea, New York, 1966.

3. J.W. Cooley and J.W. Tukey, "An algorithm for the machine computation of complex Fourier series," *Math. Comp. 19* (1965), 297–301.

4. C. de Boor, *A Practical Guide to Splines* (2nd ed.), Academic Press, New York, 1984.

5. C.A. Hall and W.W. Meyer, "Optimal error bounds for cubic spline interpolation," *Jour. Approx. Theory 16* (1976), 105–122.

6. E. Isaacson and H.B. Keller, *Analysis of Numerical Methods*, Wiley, New York, 1966.

7. W.J. Kammerer, G.W. Reddien, and R.S. Varga, "Quadratic splines," *Numer. Math. 22* (1974), 241–259.

8. D.L. Kreider, R.G. Kuller, D.R. Ostberg, and F.W. Perkins, *An Introduction to Linear Analysis*, Addison-Wesley, Reading, MA, 1966.

9. L. Lapidus and G.F. Pinder, *Numerical Solution of Partial Differential Equations in Science and Engineering*, Wiley, New York, 1982.

10. C.M. Rader, "Discrete Fourier transforms when the number of data samples is prime," *Proc. IEEE 56* (1968), 1107–1108.

11. C. Runge, "Uber die Darstellung willkürlicher Funktionen und die Interpolation zwischen aquidistantent Ordinaten," *Z. Math. Phys. 46* (1901), 224–243.

12. L. Schumaker, *Spline Functions: Basic Theory*, Wiley, New York, 1981.

13. G. Strang and G. Fix, *An Analysis of the Finite Element Method*, Prentice-Hall, Englewood Cliffs, NJ, 1973.

Chapter 2

Direct Methods for Linear Systems

2.1 Introduction

In Chapter 1 we encounter several sets of simultaneous algebraic equations. These have the form

$$\begin{aligned} a_{1,1}x_1 + a_{1,2}x_2 + \cdots + a_{1,n}x_n &= b_1 \\ a_{2,1}x_1 + a_{2,2}x_2 + \cdots + a_{2,n}x_n &= b_2 \\ &\vdots \\ a_{n,1}x_1 + a_{n,2}x_2 + \cdots + a_{n,n}x_n &= b_n, \end{aligned} \tag{2.1-1}$$

where the coefficients $a_{i,j}, b_i$ are known real numbers and the variables x_i are unknown.[1] Any such set of equations is a **linear system**. This chapter discusses numerical methods for determining $x_1, x_2, \ldots, x_n$.

Of special interest are cases in which n is large. A key consideration for any method in this setting is the **operation count**, which is a tally of the number of additions, subtractions, multiplications, and divisions required to produce answers for $x_1, x_2, \ldots, x_n$. Algorithms for solving Equations (2.1-1) can have operation counts that grow rapidly with the order n of the system, so applications for which n is large require careful choices of numerical methods.

In the language of matrices and vectors, the linear system (2.1-1) takes the form

$$\begin{bmatrix} a_{1,1} & \cdots & a_{1,n} \\ \vdots & & \vdots \\ a_{n,1} & \cdots & a_{n,n} \end{bmatrix} \begin{bmatrix} x_1 \\ \vdots \\ x_n \end{bmatrix} = \begin{bmatrix} b_1 \\ \vdots \\ b_n \end{bmatrix},$$

[1]Many ideas presented in this chapter extend naturally to complex linear systems.

or, more briefly, $\mathsf{A}\mathbf{x} = \mathbf{b}$, where $\mathbf{x}, \mathbf{b} \in \mathbb{R}^n$ and A belongs to the vector space $\mathbb{R}^{n\times n}$ of all $n \times n$ real matrices. We make frequent reference to the transposes of matrices as well as to the notions of symmetry and positive definiteness introduced in Section 1.7.

We assume that A is nonsingular, so that the system (2.1-1) has a unique solution vector $\mathbf{x} = \mathsf{A}^{-1}\mathbf{b}$ for any right-side vector $\mathbf{b} \in \mathbb{R}^n$. This formal statement hides an important practical caveat: Computing the inverse A^{-1} from A usually requires much more arithmetic than is necessary to solve for $\mathbf{x}$. One might broadly characterize this chapter as a study of methods for finding $\mathbf{x}$ without computing A^{-1} explicitly.

For now, the methods of interest are **direct**. That is, they produce a definite n-tuple $\hat{\mathbf{x}}$ of entries of $\mathbf{x}$ after a predetermined, finite sequence of arithmetic operations. Were we to execute the arithmetic without error, $\hat{\mathbf{x}}$ would satisfy the equation set exactly. However, in practice the computed values generally do not solve the linear system (2.1-1) exactly, since errors associated with machine arithmetic contaminate the calculations. One of our goals is to assess these errors. Intuitively speaking, a "reasonable" direct method applied to a "reasonable" linear system yields an approximate solution $\hat{\mathbf{x}}$ such that some norm of the error $\mathbf{x} - \hat{\mathbf{x}}$ is "small." We make these notions more precise below.

In contrast to direct methods are **indirect** or **iterative** methods, which produce sequences $\{\mathbf{x}^{(m)}\}$ of approximate solutions. One hopes that these sequences converge to the true solutions in the sense that $\|\mathbf{x} - \mathbf{x}^{(m)}\| \to 0$ as $m \to \infty$. Chapter 4 discusses iterative methods for linear systems.

A general strategy called **factorization** often guides the development of efficient numerical algorithms. The idea is to construct a decomposition $\mathsf{A} = \mathsf{BC}$, then to solve $\mathsf{BC}\mathbf{x} = \mathbf{b}$ in two stages. First we define an intermediate unknown $\mathbf{z} := \mathsf{C}\mathbf{x}$ and solve $\mathsf{B}\mathbf{z} = \mathbf{b}$ for $\mathbf{z}$. Then we recover the original unknown $\mathbf{x}$ by solving $\mathsf{C}\mathbf{x} = \mathbf{z}$. This strategy is worth pursuing if we can define B and C so that linear systems involving them are easy to solve. The next section introduces one such factorization. The overall strategy generalizes to more factors: One can imagine solving systems decomposed as $\mathsf{BCD}\mathbf{x} = \mathsf{b}$, and so forth, using essentially the same ideas.

A warning is in order: One commonly taught direct method, known as **Cramer's rule**, is extraordinarily poor as a numerical algorithm. In Cramer's rule, one forms the n matrices $\mathsf{A}_{(j)}$, $j = 1, 2, \ldots, n$, by replacing the jth column of A with the column vector $\mathbf{b}$. One then calculates the $n+1$ determinants $\det \mathsf{A}$, $\det \mathsf{A}_{(1)}, \det \mathsf{A}_{(2)}, \ldots, \det \mathsf{A}_{(n)}$, whereupon $x_j = \det \mathsf{A}_{(j)} / \det \mathsf{A}$. Since $\det \mathsf{A} = 0$ if and only if A is singular (see [4], Section 4.2), our assumption that A is nonsingular guarantees that this expression for x_j is well defined.

Cramer's rule possesses siren-like appeal, since it furnishes an apparently closed-form solution to the linear system and also shows plainly that something goes wrong when A is singular. However, computing the $n+1$ determinants via standard expansion by minors requires an astronomical $\mathcal{O}((n+1)n!)$

arithmetic operations as $n \to \infty$. In contrast, the most operation-intensive direct method that we discuss below — Gauss elimination — requires only $\mathcal{O}(n^3)$ operations. As we show later, one can reduce the operation count of Cramer's rule by using some of the steps of Gauss elimination to compute each of the $n+1$ determinants required. This approach still requires $\mathcal{O}((n+1)\cdot n^3) = \mathcal{O}(n^4)$ operations! Cramer's rule serves as a useful theoretical tool in some circumstances, but even from a charitable point of view it is inappropriate as a numerical method when $n > 2$.

2.2 Gauss Elimination

Motivation and Construction

A common strategy for solving the linear system (2.1-1) is to convert it to an equivalent **upper triangular** system, that is, one having the form

$$\begin{bmatrix} u_{1,1} & u_{1,2} & \cdots & u_{1,n} \\ & u_{2,2} & \cdots & u_{2,n} \\ & & \ddots & \vdots \\ & & & u_{n,n} \end{bmatrix} \begin{bmatrix} x_1 \\ x_2 \\ \vdots \\ x_n \end{bmatrix} = \begin{bmatrix} c_1 \\ c_2 \\ \vdots \\ c_n \end{bmatrix}. \tag{2.2-1}$$

Afterward, we can conveniently solve the individual equations in (2.2-1) in reverse order, from last to first, to get

$$\begin{aligned} x_n &= c_n / u_{n,n} \\ x_{n-1} &= \left(c_{n-1} - u_{n-1,n}x_n\right) / u_{n-1,n-1} \\ &\vdots \\ x_1 &= \left(c_1 - u_{1,2}x_2 - \cdots - u_{1,n}x_n\right) / u_{1,1}. \end{aligned}$$

We call this procedure **backward substitution**.

There is a straightforward algorithm for converting the system (2.1-1) to one of the form (2.2-1). This process, called **row reduction**, has $(n-1)$ steps. In the first step, we eliminate the variable x_1 from each of the equations corresponding to rows $2, 3, \ldots, n$. We do this by forming the factors $l_{i,1} = a_{i,1}/a_{1,1}$, then subtracting $l_{i,1} \times$ (row 1) from row i, for $i = 2, 3, \ldots, n$. (For now, assume that division by 0 does not occur.) The resulting system, which has the same solution vector $\mathbf{x} = (x_1, x_2, \ldots, x_n)^\top$ as the system (2.1-1), has the structure

$$\underbrace{\begin{bmatrix} a_{1,1} & a_{1,2} & \cdots & a_{1,n} \\ 0 & a_{2,2}^{[1]} & \cdots & a_{2,n}^{[1]} \\ \vdots & \vdots & & \vdots \\ 0 & a_{n,2}^{[1]} & \cdots & a_{n,n}^{[1]} \end{bmatrix}}_{\mathsf{A}^{[1]}} \begin{bmatrix} x_1 \\ x_2 \\ \vdots \\ x_n \end{bmatrix} = \underbrace{\begin{bmatrix} b_1 \\ b_2^{[1]} \\ \vdots \\ b_n^{[1]} \end{bmatrix}}_{\mathbf{b}^{[1]}},$$

where $a_{i,j}^{[1]} = a_{i,j} - l_{i,1}a_{1,j}$ and $b_i^{[1]} = b_i - l_{i,1}b_1$.

In the second step, we repeat this procedure for the remaining $(n-1)\times(n-1)$ submatrix below row 1 of the partially reduced system $\mathsf{A}^{[1]}\mathbf{x} = \mathbf{b}^{[1]}$. Thus we form the factors $l_{i,2} = a_{i,2}^{[1]}/a_{2,2}^{[1]}$, $i = 3, 4, \ldots, n$, and subtract $l_{i,2} \times$ (row 2) from row i, for $i = 3, 4, \ldots, n$, to get a system of the form

$$\underbrace{\begin{bmatrix} a_{1,1} & a_{1,2} & a_{1,3} & \cdots & a_{1,n} \\ 0 & a_{2,2}^{[1]} & a_{2,3}^{[1]} & \cdots & a_{2,n}^{[1]} \\ 0 & 0 & a_{3,3}^{[2]} & \cdots & a_{3,n}^{[2]} \\ \vdots & \vdots & \vdots & & \vdots \\ 0 & 0 & a_{n,3}^{[2]} & \cdots & a_{n,n}^{[2]} \end{bmatrix}}_{\mathsf{A}^{[2]}} \begin{bmatrix} x_1 \\ x_2 \\ x_3 \\ \vdots \\ x_n \end{bmatrix} = \underbrace{\begin{bmatrix} b_1 \\ b_2^{[1]} \\ b_3^{[2]} \\ \vdots \\ b_n^{[2]} \end{bmatrix}}_{\mathbf{b}^{[2]}}.$$

Again, this system has the same solution vector $\mathbf{x}$ as the system (2.1-1).

After $n-1$ steps like this, we are left with an upper triangular system,

$$\underbrace{\begin{bmatrix} a_{1,1} & \cdots & a_{1,n} \\ & \ddots & \vdots \\ & & a_{n,n}^{[n-1]} \end{bmatrix}}_{\mathsf{A}^{[n-1]} = \mathsf{U}} \begin{bmatrix} x_1 \\ \vdots \\ x_n \end{bmatrix} = \underbrace{\begin{bmatrix} b_1 \\ \vdots \\ b_n^{[n-1]} \end{bmatrix}}_{\mathbf{b}^{[n-1]} = \mathbf{c}},$$

having the same solution vector $\mathbf{x}$ as the original system (2.1-1). For brevity, let us denote $u_{i,j} = a_{i,j}^{[i-1]}$ and $c_i = b_i^{[i-1]}$ and write this upper triangular system as $\mathsf{U}\mathbf{x} = \mathbf{c}$. We can now apply backward substitution to solve for the unknowns $x_1, x_2, \ldots, x_n$.

The entire procedure, using row reduction followed by backward substitution to solve the system (2.1-1), is called **Gauss elimination**. In addition to being a numerical method in its own right, it serves as a conceptual foundation for other techniques discussed in this chapter.

Practical Considerations

Row reduction is the most expensive part of Gauss elimination. At step j, the process requires $n-j$ divisions to compute the factors $l_{i,j}$, $(n-j)(n-j+1)$

multiplications to form the products of these factors with entries in row j, including the jth entry of $\mathbf{b}$, and $(n-j)(n-j+1)$ subtractions of these products from corresponding entries in rows $j+1, j+2, \ldots, n$. The total number of operations for row reduction is therefore

$$\sum_{j=1}^{n-1} [(n-j) + 2(n-j)(n-j+1)] = \frac{2n^3}{3} + \frac{n^2}{2} - \frac{7n}{6}.$$

In simplifying this sum, we have used the identities

$$\sum_{j=1}^{m} j = \frac{m(m+1)}{2}, \qquad \sum_{j=1}^{m} j^2 = \frac{m(m+1)(2m+1)}{6}.$$

Backward substitution requires substantially less arithmetic. Solving for the unknown x_j requires $n-j$ multiplications, $n-j$ subtractions, and one division. The total number of operations is therefore

$$\sum_{j=1}^{n} [2(n-j)+1] = n^2.$$

Hence Gauss elimination requires $\frac{2}{3}n^3 + \frac{3}{2}n^2 - \frac{7}{6}n$ arithmetic operations. When the number n of unknowns is large, the term $\frac{2}{3}n^3$ dominates those having lower degree. Since Gauss elimination plays such a fundamental role as a direct solution method, $\mathcal{O}(n^3)$ serves as a benchmark, against which to measure operation counts for all related algorithms.

(Our operation count for Gauss elimination differs from that presented by many authors. People traditionally ignore additions and subtractions in operation counts, since multiplications and divisions require substantially more time on standard, single-instruction, single-dataset computers. How the operations compare on more advanced computers depends on both the machine's architecture and the programmer's skill at exploiting it. Not wanting to devote detailed attention to this issue, we count all arithmetic operations.)

Some applications call for the solution of several linear systems having the same matrix but different right-side vectors, as in

$$\mathsf{A}\mathbf{x}_1 = \mathbf{b}_1, \quad \mathsf{A}\mathbf{x}_2 = \mathbf{b}_2, \quad \ldots. \tag{2.2-2}$$

In such problems the factorization strategy mentioned in Section 2.1 is useful. In the context of Gauss elimination, we save computational effort in systems like (2.2-2) if we avoid repeating the arithmetic required for row reduction. These savings are possible, provided that we systematically store the factors $l_{i,j} = a_{i,j}^{[j-1]}/a_{j,j}^{[j-1]}$ calculated at each step j of the row reduction. Define the lower triangular matrix $\mathsf{L} \in \mathbb{R}^{n\times n}$ as follows:

$$\mathsf{L} := \begin{bmatrix} 1 & & & \\ l_{2,1} & 1 & & \\ \vdots & \ddots & \ddots & \\ l_{n,1} & \cdots & l_{n,n-1} & 1 \end{bmatrix}.$$

One can confirm that $\mathsf{A} = \mathsf{LU}$. We call this decomposition of A an **LU factorization**. When A is nonsingular, it has a unique LU factorization in which L has only unit diagonal entries; the proof of this fact is an exercise.

Now we solve each of the linear systems $\mathsf{A}\mathbf{x}_k = \mathbf{b}_k$ in (2.2-2) by using the factorization strategy: First solve $\mathsf{L}\mathbf{z}_k = \mathbf{b}_k$ for the intermediate vector $\mathbf{z}_k \in \mathbb{R}^n$. This step proceeds by a **forward substitution** algorithm analogous to backward substitution, and it requires $\mathcal{O}(n^2)$ arithmetic operations. Then solve the upper triangular system $\mathsf{U}\mathbf{x}_k = \mathbf{z}_k$ using ordinary backward substitution. Therefore, once we have performed the $\mathcal{O}(n^3)$ operations needed to obtain the LU factorization of A, solving any linear system involving A requires only $\mathcal{O}(n^2)$ additional operations.

The LU factorization enjoys two other properties. First, storing both L and U requires no more space than storing A, since there is no need to store the entries that are 0 or 1 a priori. Second, the LU factorization provides a cheap way to compute the determinant of A. Since $\mathsf{A} = \mathsf{LU}$, $\det \mathsf{A} = \det \mathsf{L} \det \mathsf{U}$. But the determinant of any upper or lower triangular matrix is simply the product of its diagonal entries, and hence $\det \mathsf{L} = 1$. It follows that $\det \mathsf{A} = \det \mathsf{U} = u_{1,1}u_{2,2}\cdots u_{n,n}$. Consequently, by using row reduction, one can compute $\det \mathsf{A}$ in $\mathcal{O}(n^3)$ operations, as opposed to the $\mathcal{O}(n!)$ operations required in ordinary expansion by minors.

The success of Gauss elimination hinges on the assumption that none of the **pivots** $a_{j,j}^{[j-1]}$ vanishes. From a computational point of view, if any of the pivots is extremely small in magnitude, then the factors $l_{i,j} = a_{i,j}^{[j-1]}/a_{j,j}^{[j-1]}$ become large in magnitude, and the products

$$l_{i,j}a_{j,j+1}^{[j-1]},\ l_{i,j}a_{j,j+2}^{[j-1]},\ \ldots,\ l_{i,j}a_{j,n}^{[j-1]}$$

formed with these factors during step j of row reduction tend to magnify any existing errors in the entries of row j. To avoid the difficulties associated with small or vanishing pivots, one can adopt a **pivoting** strategy, of which several exist.

Perhaps the simplest and most common pivoting strategy is **partial pivoting**. To see how it works, assume that we have row-reduced A through step $j-1$. Rows $1, 2, \ldots, j-1$ are then the only ones in the partially reduced matrix $\mathsf{A}^{[j-1]}$ that contain nonzero entries in columns $1, \ldots, j-1$. Scan the entries

$$a_{j,j}^{[j-1]},\ a_{j+1,j}^{[j-1]},\ \ldots,\ a_{n,j}^{[j-1]}$$

in column j below the $(j-1)$st row to find the one having largest magnitude, say $a_{k,j}^{[j-1]}$. Problem 8, which uses ideas developed later in this section, asks for proof that *some* nonzero candidate $a_{k,j}^{[j-1]}$ exists among the scanned entries, unless the original matrix A is singular. (If more than one of the entries scanned has this same magnitude, let k be the smallest of their row indices.) Then interchange rows j and k in the matrix $\mathsf{A}^{[j-1]}$ and in the right-side vector $\mathbf{b}^{[j-1]}$, as illustrated schematically in Figure 1, and proceed to eliminate column j from rows $j+1, j+2, \ldots, n$ via row reduction. Since interchanging rows simply changes the order of the equations being solved, the resulting linear system still has the same solution vector as the original system (2.1-1).

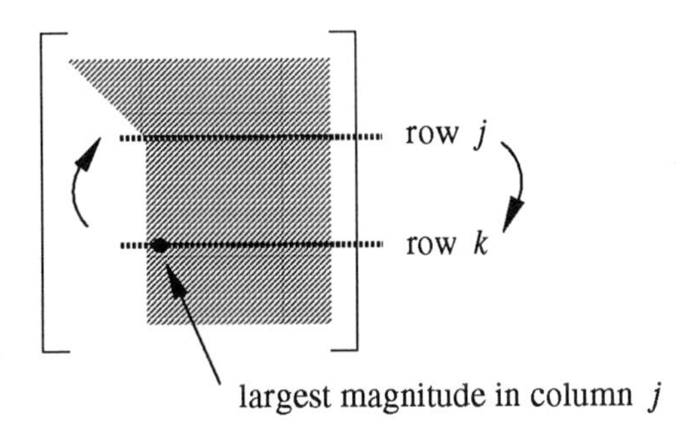

FIGURE 1. *Schematic illustration of a row interchange in partial pivoting.*

As an example, consider the linear system

$$\begin{bmatrix} 4.0000 & 2.0000 & 10.0000 & 7.0000 \\ -3.0000 & 1.0000 & 8.0000 & 2.0000 \\ 0.0000 & 5.0000 & -1.0000 & 4.0000 \\ 1.0000 & 6.0000 & -5.0000 & 2.0000 \end{bmatrix} \begin{bmatrix} x_1 \\ x_2 \\ x_3 \\ x_4 \end{bmatrix} = \begin{bmatrix} 1.0000 \\ -2.0000 \\ 0.0000 \\ -6.0000 \end{bmatrix}. \tag{2.2-3}$$

The entry $a_{1,1} = 4.0000$ already has the largest magnitude of any entry in column 1, so we proceed without any initial row interchange. The first step of row reduction, using the factors $l_{2,1} = -0.7500$, $l_{3,1} = 0.0000$, and $l_{4,1} = 0.2500$, yields

$$\begin{bmatrix} 4.0000 & 2.0000 & 10.0000 & 7.0000 \\ 0 & 2.5000 & 15.5000 & 7.2500 \\ 0 & 5.0000 & -1.0000 & 4.0000 \\ 0 & 5.5000 & -7.5000 & 0.2500 \end{bmatrix} \begin{bmatrix} x_1 \\ x_2 \\ x_3 \\ x_4 \end{bmatrix} = \begin{bmatrix} 1.0000 \\ -1.2500 \\ 0.0000 \\ -6.2500 \end{bmatrix}.$$

Scanning column 2 below the first row, we see that the entry having largest magnitude, namely 5.5000, occurs in row 4. Therefore we interchange rows 2 and 4 to get

$$\begin{bmatrix} 4.0000 & 2.0000 & 10.0000 & 7.0000 \\ 0 & 5.5000 & -7.5000 & 0.2500 \\ 0 & 5.0000 & -1.0000 & 4.0000 \\ 0 & 2.5000 & 15.5000 & 7.2500 \end{bmatrix} \begin{bmatrix} x_1 \\ x_2 \\ x_3 \\ x_4 \end{bmatrix} = \begin{bmatrix} 1.0000 \\ -6.2500 \\ 0.0000 \\ -1.2500 \end{bmatrix}.$$

Now we apply the second step of row reduction, using the factors $l_{3,2} = 0.9091$ and $l_{4,2} = 0.4545$ (to four decimal places) to obtain

$$\begin{bmatrix} 4.0000 & 2.0000 & 10.0000 & 7.0000 \\ 0 & 5.5000 & -7.5000 & 0.2500 \\ 0 & 0 & 5.8182 & 3.7727 \\ 0 & 0 & 18.9091 & 7.1364 \end{bmatrix} \begin{bmatrix} x_1 \\ x_2 \\ x_3 \\ x_4 \end{bmatrix} = \begin{bmatrix} 1.0000 \\ -6.2500 \\ -5.6819 \\ -1.5906 \end{bmatrix}.$$

Scanning column 3 below row 2 reveals that the entry 18.9091 having largest magnitude lies in row 4, so we interchange rows 3 and 4 to get

$$\begin{bmatrix} 4.0000 & 2.0000 & 10.0000 & 7.0000 \\ 0 & 5.5000 & -7.5000 & 0.2500 \\ 0 & 0 & 18.9091 & 7.1364 \\ 0 & 0 & 5.8182 & 3.7727 \end{bmatrix} \begin{bmatrix} x_1 \\ x_2 \\ x_3 \\ x_4 \end{bmatrix} = \begin{bmatrix} 1.0000 \\ -6.2500 \\ 1.5906 \\ 5.6819 \end{bmatrix}.$$

The final step of row reduction, using the factor $l_{4,3} = 0.3077$, yields the upper triangular system

$$\begin{bmatrix} 4.0000 & 2.0000 & 10.0000 & 7.0000 \\ 0 & 5.5000 & -7.5000 & 0.2500 \\ 0 & 0 & 18.9091 & 7.1364 \\ 0 & 0 & 0 & 1.5769 \end{bmatrix} \begin{bmatrix} x_1 \\ x_2 \\ x_3 \\ x_4 \end{bmatrix} = \begin{bmatrix} 1.0000 \\ -6.2500 \\ 1.5906 \\ 5.1925 \end{bmatrix}.$$

By storing the factors $l_{i,j}$ in the order indicated by their indices, we get the lower triangular matrix

$$\begin{bmatrix} & 1 & 0 & 0 & 0 \\ - & 0.7500 & 1 & 0 & 0 \\ & 0.0000 & 0.9091 & 1 & 0 \\ & 0.2500 & 0.4545 & 0.3077 & 1 \end{bmatrix}.$$

However, because of the row interchanges introduced during row reduction, this matrix no longer gives a valid LU factorization of A. It is possible, nevertheless, to salvage an LU factorization from this procedure. We discuss the details shortly, but the upshot is this: So long as A is nonsingular, it is possible to permute its rows so that the resulting matrix has an LU factorization. The lower triangular matrix L in this factorization contains the factors $l_{i,j}$ used in the row reduction, but, owing to the row interchanges, these factors may appear in positions different from those suggested by their indices.

Some pathologic linear systems are so intractable that partial pivoting still leaves the row reduction vulnerable to unacceptable roundoff errors, even though A is nonsingular. In such cases, one can resort to **total pivoting**. Here, after eliminating column $j-1$ from rows $j, j+1, \ldots, n$, we scan the entire $(n-j+1) \times (n-j+1)$ submatrix below row $j-1$ and to the right of column $j-1$ to find the entry $a_{k,m}^{[j-1]} \equiv a_{\max}$ having largest magnitude. Then we interchange rows j and k and columns j and m, so that $a_{\max}$ occupies the pivotal position, and proceed with row reduction. Figure 2 illustrates this scheme schematically. Such a column interchange calls for extra bookkeeping to keep track of the corresponding switch in the order of the unknowns $x_1, x_2, \ldots, x_n$.

Compared with partial pivoting, total pivoting requires a much more extensive search at each step of row reduction. Before performing stage j of row reduction, we must determine the largest-magnitude entry of an $(n-j+1) \times (n-j+1)$ submatrix. This task requires $(n-j+1)^2 - 1$

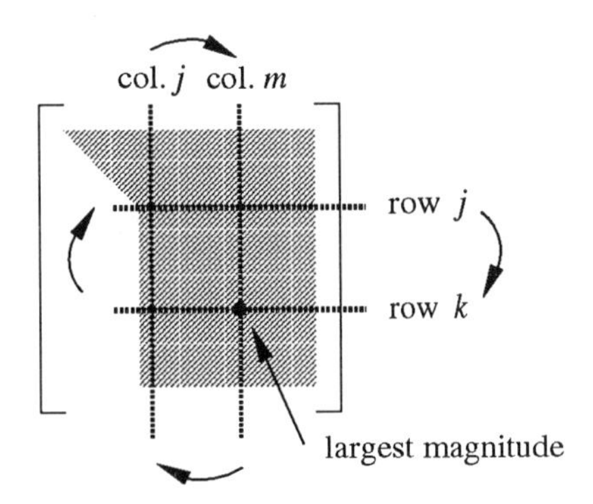

FIGURE 2. *Schematic illustration of row and column interchanges in total pivoting.*

arithmetic comparisons. Therefore, total pivoting requires

$$\sum_{j=1}^{n-1} [(n-j+1)^2 - 1] = \mathcal{O}(n^3)$$

comparisons, together with the extra storage required to keep track of row and column interchanges. Since a comparison typically requires at least as much execution time as an arithmetic operation such as addition, it seems prudent to use total pivoting only for the most stubborn of linear systems.

Mathematical Details

We devote the rest of this section to the theoretical details behind pivoting. The discussion emphasizes a matrix-theoretic approach, which lends an algebraic flavor to the arguments.

As hinted earlier, it is possible to incorporate the row interchanges associated with partial pivoting into the LU factorization. To do this, we employ a matrix representation for row reduction. Suppose that we have completed step $k-1$ of row reduction to get a partially reduced matrix $\mathsf{A}^{[k-1]} \in \mathbb{R}^{n\times n}$. Given the factors $l_{k+1,k}, l_{k+2,k}, \ldots, l_{n,k}$, the process of subtracting $l_{i,k}\times$(row k) from row i for $i = k+1, k+2, \ldots, n$ is equivalent to multiplying $\mathsf{A}^{[k-1]}$ (as well as $\mathbf{b}^{[k-1]}$) on the left by the $n \times n$ matrix

$$\mathsf{G}_k = \begin{bmatrix} 1 & & & & & \\ & \ddots & & & & \\ & & 1 & & & \\ & & -l_{k+1,k} & 1 & & \\ & & \vdots & & \ddots & \\ & & -l_{n,k} & & & 1 \end{bmatrix}. \qquad (2.2\text{-}4)$$

That is,

$$\mathsf{A}^{[k]} = \mathsf{G}_k \mathsf{A}^{[k-1]} = \mathsf{G}_k \mathsf{G}_{k-1} \cdots \mathsf{G}_1 \mathsf{A},$$
$$\mathbf{b}^{[k]} = \mathsf{G}_k \mathbf{b}^{[k-1]} = \mathsf{G}_k \mathsf{G}_{k-1} \cdots \mathsf{G}_1 \mathbf{b}.$$

We call any matrix having the structure (2.2-4) a **column-k Gauss transformation**. The inverse of such a matrix corresponds to adding the row multiples instead of subtracting them. Thus,

$$\mathsf{G}_k^{-1} = \begin{bmatrix} 1 & & & & & \\ & \ddots & & & & \\ & & 1 & & & \\ & & l_{k+1,k} & 1 & & \\ & & \vdots & & \ddots & \\ & & l_{n,k} & & & 1 \end{bmatrix},$$

which is also a column-k Gauss transformation.

PROPOSITION 2.1. *If $\mathsf{G}_k \in \mathbb{R}^{n \times n}$ is a column-k Gauss transformation for each index k, $k = 1, 2, \ldots n-1$, then the product $\mathsf{G}_1^{-1}\mathsf{G}_2^{-1} \cdots \mathsf{G}_{n-1}^{-1}$ is a lower triangular matrix with unit diagonal entries.*

PROOF: This is an exercise. ∎

We can represent row reduction *without* pivoting as a succcession of multiplications on the left by appropriate Gauss transformations:

$$\mathsf{G}_{n-1}\mathsf{G}_{n-2} \cdots \mathsf{G}_1 \mathsf{A} = \mathsf{U}, \quad \mathsf{G}_{n-1}\mathsf{G}_{n-2} \cdots \mathsf{G}_1 \mathbf{b} = \mathbf{c}.$$

By applying the inverses of these Gauss transformations in the opposite order, we can solve for A to find

$$\mathsf{A} = \underbrace{\mathsf{G}_1^{-1}\mathsf{G}_2^{-1} \cdots \mathsf{G}_{n-1}^{-1}}_{\mathsf{L}} \mathsf{U}.$$

The product $\mathsf{G}_1^{-1}\mathsf{G}_2^{-1} \cdots \mathsf{G}_{n-1}^{-1}$ is precisely the lower triangular matrix L in the LU factorization of A, assuming that no zero pivots arise.

To incorporate partial pivoting into this formalism, we need a matrix representation for the interchange of rows. We begin with the following:

DEFINITION. *The matrix $\mathsf{P} \in \mathbb{R}^{n \times n}$ is a* **permutation matrix** *if (i) P has exactly one entry whose value is 1 in each row and in each column and (ii) all other entries of P are 0.*

To justify this terminology, we note that, if $\mathsf{P} \in \mathbb{R}^{n \times n}$ is a permutation matrix and $\mathbf{x} \in \mathbb{R}^n$, then the entries in $\mathsf{P}\mathbf{x} \in \mathbb{R}^n$ are obtained by permuting the entries of $\mathbf{x}$. A trivial example of a permutation matrix is the $n \times n$ identity matrix I, whose action on any vector $\mathbf{x} \in \mathbb{R}^n$ leaves the entries unchanged.

For us, a useful class of permutation matrices is the set of **elementary permutation matrices** $\mathsf{P}_{k,m}$, obtained from I by interchanging two columns k and m. For example, in $\mathbb{R}^{4\times 4}$,

$$\mathsf{P}_{2,3} = \begin{bmatrix} 1 & 0 & 0 & 0 \\ 0 & 0 & 1 & 0 \\ 0 & 1 & 0 & 0 \\ 0 & 0 & 0 & 1 \end{bmatrix}$$

is an elementary permutation matrix. (With this notation, $\mathsf{P}_{k,k}$ denotes the identity matrix I for any index k.) Any elementary permutation matrix $\mathsf{P}_{k,m}$ has the properties $\mathsf{P}_{k,m}^{-1} = \mathsf{P}_{k,m} = \mathsf{P}_{k,m}^{\mathsf{T}}$.

A key property of elementary permutation matrices is the following:

PROPOSITION 2.2. *If* $\mathsf{A} \in \mathbb{R}^{n\times n}$ *and* $\mathsf{P}_{k,m} \in \mathbb{R}^{n\times n}$ *is an elementary permutation matrix, then* $\mathsf{P}_{k,m}\mathsf{A}$ *differs from* A *by the interchange of rows* k *and* m. *Similarly,* $\mathsf{A}\mathsf{P}_{k,m}$ *differs from* A *by the interchange of columns* k *and* m.

PROOF: This is an exercise. ∎

The main consequence is that elementary permutation matrices serve as matrix representations of the row interchanges in partial pivoting. Another consequence is that any permutation matrix is the product of elementary permutation matrices, and the set of all $n \times n$ permutation matrices is closed under matrix multiplication. For a group-theoretic view of these observations, see Herstein ([2], Section 2.10).

In row reduction with partial pivoting, we apply to the partially reduced matrix $\mathsf{A}^{[k-1]}$ an elementary permutation matrix $\mathsf{P}_{k,m}$, where $m \geq k$, before applying a Gauss transformation G_k. If no row interchange is necessary, the appropriate permutation matrix is the identity matrix, $\mathsf{I} = \mathsf{P}_{k,k}$. Therefore, we can represent the entire procedure as a sequence of matrix multiplications:

$$\mathsf{G}_{n-1}\mathsf{P}_{n-1}\cdots\mathsf{G}_2\mathsf{P}_2\mathsf{G}_1\mathsf{P}_1\mathsf{A} = \mathsf{U},$$

$$\mathsf{G}_{n-1}\mathsf{P}_{n-1}\cdots\mathsf{G}_2\mathsf{P}_2\mathsf{G}_1\mathsf{P}_1\mathbf{b} = \mathbf{c}.$$

(To fight indicial clutter, we use the notation P_k as shorthand for the elementary permutation $\mathsf{P}_{k,m}$ applied at step k of row reduction.) Solving for A and using the fact that $\mathsf{P}_k^{-1} = \mathsf{P}_k$, we find that

$$\mathsf{A} = \mathsf{P}_1\mathsf{G}_1^{-1}\mathsf{P}_2\mathsf{G}_2^{-1}\cdots\mathsf{P}_{n-1}\mathsf{G}_{n-1}^{-1}\mathsf{U}. \tag{2.2-5}$$

Unfortunately, the matrix $\mathsf{P}_1\mathsf{G}_1^{-1}\mathsf{P}_2\mathsf{G}_2^{-1}\cdots\mathsf{P}_{n-1}\mathsf{G}_{n-1}^{-1}$ is not lower triangular unless each P_k is the identity matrix — that is, unless no row interchanges occur. For this reason, Equation (2.2-5) does not qualify as an LU factorization in general. However, the following theorem indicates how to produce an LU factorization of a permutation of the rows of A.

THEOREM 2.3. *Suppose that* A *has the row reduction (2.2-5). Then* $\mathsf{PA} = \mathsf{LU}$, *where* $\mathsf{P} := \mathsf{P}_{n-1}\mathsf{P}_{n-2}\cdots\mathsf{P}_1$ *is a permutation matrix and*

$$\mathsf{L} := \mathsf{P}(\mathsf{P}_1\mathsf{G}_1^{-1}\mathsf{P}_2\mathsf{G}_2^{-1}\cdots\mathsf{P}_{n-1}\mathsf{G}_{n-1}^{-1})$$

is lower triangular with unit diagonal.

The proof uses a lemma:

LEMMA 2.4. *Let* $\mathsf{G}_k^{-1} \in \mathbb{R}^{n\times n}$ *be a column-k Gauss transformation for* $k < n-1$, *and suppose that* $\mathsf{P}_j \in \mathbb{R}^{n\times n}$ *interchanges rows j and i, where* $j, i > k$. *Then the matrix* $\mathsf{P}_j\mathsf{G}_k^{-1}\mathsf{P}_j$ *is a column-k Gauss transformation as well. Moreover,* $\mathsf{P}_j\mathsf{G}_k^{-1}\mathsf{P}_j$ *differs from* G_k^{-1} *only in the interchange of the entries in rows j and i of column k.*

PROOF: Multiplying G_k^{-1} on the left by P_j accomplishes the following row interchange:

$$\begin{bmatrix} \ddots & & & & \\ & 1 & & & \\ & \vdots & \ddots & & \\ & l_{j,k} & 1 & 0 & \\ & \vdots & & \ddots & \\ & l_{i,k} & 0 & 1 & \\ & \vdots & & & \ddots \end{bmatrix} \mapsto \begin{bmatrix} \ddots & & & & \\ & 1 & & & \\ & \vdots & \ddots & & \\ & l_{i,k} & 0 & 1 & \\ & \vdots & & & \\ & l_{j,k} & 1 & 0 & \\ & \vdots & & & \ddots \end{bmatrix}.$$

Multiplying the result on the right by P_j then effects a column interchange to the right of column k:

$$\begin{bmatrix} \ddots & & & & \\ & 1 & & & \\ & \vdots & \ddots & & \\ & l_{i,k} & 0 & 1 & \\ & \vdots & & & \\ & l_{j,k} & 1 & 0 & \\ & \vdots & & & \ddots \end{bmatrix} \mapsto \begin{bmatrix} \ddots & & & & \\ & 1 & & & \\ & \vdots & \ddots & & \\ & l_{i,k} & 1 & 0 & \\ & \vdots & & \ddots & \\ & l_{j,k} & 0 & 1 & \\ & \vdots & & & \ddots \end{bmatrix}.$$

This proves the lemma. ∎

PROOF OF THEOREM 2.3: Multiplying both sides of Equation (2.2-5) by P clearly shows that $\mathsf{PA} = \mathsf{LU}$. Hence, we need only prove that L, as defined, is lower triangular with unit diagonal. The overall idea is to write L as a product of appropriate Gauss transformations. Toward this end, Lemma 2.4 has a helpful consequence: Any matrix product $\mathsf{G}_k^\bullet$ having the form

$$\mathsf{G}_k^\bullet := \mathsf{P}_{n-1}\mathsf{P}_{n-2}\cdots\mathsf{P}_{k+1}\mathsf{G}_k^{-1}\mathsf{P}_{k+1}\mathsf{P}_{k+2}\cdots\mathsf{P}_{n-1}$$

is a column-k Gauss transformation that differs from G_k^{-1} only by a permutation of the entries below the diagonal in column k.

To unravel the matrix product that defines L, observe that

$$\begin{aligned}\mathsf{P}\left(\mathsf{P}_1\mathsf{G}_1^{-1}\cdots\mathsf{P}_{n-1}\mathsf{G}_{n-1}^{-1}\right)\mathsf{U} &= \mathsf{P}_{n-1}\cdots\underbrace{\mathsf{P}_1\mathsf{P}_1}_{\mathsf{I}}\mathsf{G}_1^{-1}\cdots\mathsf{P}_{n-1}\mathsf{G}_{n-1}^{-1}\mathsf{U}\\ &= \mathsf{P}_{n-1}\cdots\mathsf{P}_2\mathsf{G}_1^{-1}\underbrace{\mathsf{P}_2\mathsf{G}_2^{-1}}_{\mathsf{P}_2\,\mathsf{I}\,\mathsf{G}_2^{-1}}\cdots\mathsf{P}_{n-1}\mathsf{G}_{n-1}^{-1}\mathsf{U}.\end{aligned}$$

If we replace the identity matrix I in $\mathsf{P}_2\,\mathsf{I}\,\mathsf{G}_2^{-1}$ by the equivalent expression $\mathsf{P}_3\cdots\mathsf{P}_{n-1}\mathsf{P}_{n-1}\cdots\mathsf{P}_3$, then there appears a factor

$$\mathsf{P}_{n-1}\cdots\mathsf{P}_2\mathsf{G}_1^{-1}\mathsf{P}_2\cdots\mathsf{P}_{n-1},$$

which collapses to a column-1 Gauss transformation $\mathsf{G}_1^\bullet$. Therefore,

$$\mathsf{P}\left(\mathsf{P}_1\mathsf{G}_1^{-1}\cdots\mathsf{P}_{n-1}\mathsf{G}_{n-1}^{-1}\right)\mathsf{U} = \mathsf{G}_1^\bullet\mathsf{P}_{n-1}\cdots\mathsf{P}_3\mathsf{G}_2^{-1}\underbrace{\mathsf{P}_3\mathsf{G}_3^{-1}}_{\mathsf{P}_3\,\mathsf{I}\,\mathsf{G}_3^{-1}}\cdots\mathsf{P}_{n-1}\mathsf{G}_{n-1}^{-1}\mathsf{U}.$$

By replacing I again, now with the equivalent expression

$$\mathsf{P}_4\cdots\mathsf{P}_{n-1}\mathsf{P}_{n-1}\cdots\mathsf{P}_4,$$

we can reason as before, discovering a column-2 Gauss transformation $\mathsf{G}_2^\bullet$ such that

$$\mathsf{P}\left(\mathsf{P}_1\mathsf{G}_1^{-1}\cdots\mathsf{P}_{n-1}\mathsf{G}_{n-1}^{-1}\right)\mathsf{U} = \mathsf{G}_1^\bullet\mathsf{G}_2^\bullet\mathsf{P}_{n-1}\cdots\mathsf{P}_4\mathsf{G}_3^{-1}\underbrace{\mathsf{P}_4\mathsf{G}_4^{-1}}_{\mathsf{P}_4\,\mathsf{I}\,\mathsf{G}_4^{-1}}\cdots\mathsf{P}_{n-1}\mathsf{G}_{n-1}^{-1}\mathsf{U}.$$

Continuing in this way yields the identity

$$\mathsf{P}\left(\mathsf{P}_1\mathsf{G}_1^{-1}\cdots\mathsf{P}_{n-1}\mathsf{G}_{n-1}^{-1}\right)\mathsf{U} = \underbrace{\mathsf{G}_1^\bullet\cdots\mathsf{G}_{n-2}^\bullet\mathsf{G}_{n-1}^{-1}}_{\mathsf{L}}\,\mathsf{U}.$$

But the matrix $\mathsf{L} = \mathsf{G}_1^\bullet\cdots\mathsf{G}_{n-2}^\bullet\mathsf{G}_{n-1}^{-1}$ is lower triangular, with unit diagonal, and each column k contains, below the diagonal, some permutation of the factors $l_{k,j}$ used in the row reduction. ∎

The linear system (2.2-3) furnishes an example. In this case, the permutation matrix P is

$$\underbrace{\begin{bmatrix} 1 & 0 & 0 & 0 \\ 0 & 1 & 0 & 0 \\ 0 & 0 & 0 & 1 \\ 0 & 0 & 1 & 0 \end{bmatrix}}_{\mathsf{P}_{3,4}} \underbrace{\begin{bmatrix} 1 & 0 & 0 & 0 \\ 0 & 0 & 0 & 1 \\ 0 & 0 & 1 & 0 \\ 0 & 1 & 0 & 0 \end{bmatrix}}_{\mathsf{P}_{2,4}} \underbrace{\begin{bmatrix} 1 & 0 & 0 & 0 \\ 0 & 1 & 0 & 0 \\ 0 & 0 & 1 & 0 \\ 0 & 0 & 0 & 1 \end{bmatrix}}_{\mathsf{P}_{1,1}} = \underbrace{\begin{bmatrix} 1 & 0 & 0 & 0 \\ 0 & 0 & 0 & 1 \\ 0 & 1 & 0 & 0 \\ 0 & 0 & 1 & 0 \end{bmatrix}}_{\mathsf{P}}.$$

The correct lower triangular factor for PA is therefore

$$\mathsf{L} = \begin{bmatrix} 1 & 0 & 0 & 0 \\ 0.2500 & 1 & 0 & 0 \\ 0.7500 & 0.4545 & 1 & 0 \\ 0.0000 & 0.9091 & 0.3077 & 1 \end{bmatrix},$$

to four decimal places.

One can still compute $\det \mathsf{A}$ from $\det \mathsf{U}$; however, partial pivoting affects the calculation. In particular, the relationship $\det \mathsf{P} \det \mathsf{A} = \det \mathsf{L} \det \mathsf{U}$ shows that $\det \mathsf{A} = \det \mathsf{U} / \det \mathsf{P}$. The factor $\det \mathsf{U}$ is the product of diagonal entries, as before, but $\det \mathsf{P} = (-1)^p$, where p is the number of steps in row reduction requiring an actual row interchange. For the example system (2.2-3), we performed row interchanges in steps 2 and 3, so $\det \mathsf{P} = (-1)^2 = 1$, and $\det \mathsf{A} = \det \mathsf{U} \simeq 655.9907$, to four decimal places. (The exact value of $\det \mathsf{A}$ is 656; the discrepancy results from accumulated roundoff errors in the row reduction used to compute U.)

Strictly speaking, one can construct an LU factorization with pivoting for *any* matrix $\mathsf{A} \in \mathbb{R}^{n \times n}$. However, in cases where no nonzero pivot exists at some stage of the row reduction, the resulting upper triangular matrix U is singular. To visualize this concept, imagine that the first two steps of row reduction on $\mathsf{A} \in \mathbb{R}^{5 \times 5}$ yield

$$\mathsf{A}^{[2]} = \begin{bmatrix} \star & \star & \star & \star & \star \\ 0 & \star & \star & \star & \star \\ 0 & 0 & 0 & \star & \star \\ 0 & 0 & 0 & \star & \star \\ 0 & 0 & 0 & \star & \star \end{bmatrix}.$$

Here the symbol $\star$ stands for an arbitrary nonzero entry. $\mathsf{A}^{[2]}$ has no nonzero pivot below row 2 in column 3, so standard row reduction with partial pivoting cannot proceed. However, we can "skip" the third step of row reduction, formally taking $\mathsf{P}_3 = \mathsf{G}_3 = \mathsf{I}$ to get $\mathsf{A}^{[3]} = \mathsf{A}^{[2]}$. We then move on to the fourth step. Assuming that a nonzero pivot is available at that stage, we expect the final result to have the upper triangular structure

$$\mathsf{U} = \begin{bmatrix} \star & \star & \star & \star & \star \\ 0 & \star & \star & \star & \star \\ 0 & 0 & 0 & \star & \star \\ 0 & 0 & 0 & \star & \star \\ 0 & 0 & 0 & 0 & \star \end{bmatrix}.$$

Notice that $\det \mathsf{U} = 0$.

This observation helps in establishing the following result.

THEOREM 2.5. *$\mathsf{A} \in \mathbb{R}^{n\times n}$ is nonsingular if and only if, at each step k of row reduction, it is possible to find a nonzero pivot in column k below row $k-1$.*

PROOF: See Problem 8. ∎

There is some utility in characterizing matrices that are amenable to row reduction without partial pivoting. For this task, some additional terminology is helpful. Given a matrix $\mathsf{A} \in \mathbb{R}^{n\times n}$, we speak of a **block partitioning**

$$\mathsf{A} = \begin{pmatrix} \mathsf{A}_{1,1} & \mathsf{A}_{1,2} \\ \mathsf{A}_{2,1} & \mathsf{A}_{2,2} \end{pmatrix},$$

where the matrices $\mathsf{A}_{1,1} \in \mathbb{R}^{k\times k}$, $\mathsf{A}_{1,2} \in \mathbb{R}^{k\times(n-k)}$, $\mathsf{A}_{2,1} \in \mathbb{R}^{(n-k)\times k}$, and $\mathsf{A}_{2,2} \in \mathbb{R}^{(n-k)\times(n-k)}$ are the **blocks**. We often indicate the block dimensions associated with such a partitioning as follows:

$$\mathsf{A} \quad = \quad \begin{matrix} & k & n-k \\ k & \mathsf{A}_{1,1} & \mathsf{A}_{1,2} \\ n-k & \mathsf{A}_{2,1} & \mathsf{A}_{2,2} \end{matrix}, \qquad (2.2\text{-}6)$$

and we agree that all block partitionings appearing in a single equation have the same block dimensions unless otherwise stated. One can easily verify that the familiar rules of matrix multiplication hold "blockwise:"

$$\begin{pmatrix} \mathsf{A}_{1,1} & \mathsf{A}_{1,2} \\ \mathsf{A}_{2,1} & \mathsf{A}_{2,2} \end{pmatrix} \begin{pmatrix} \mathsf{B}_{1,1} & \mathsf{B}_{1,2} \\ \mathsf{B}_{2,1} & \mathsf{B}_{2,2} \end{pmatrix}$$

$$= \begin{pmatrix} \mathsf{A}_{1,1}\mathsf{B}_{1,1} + \mathsf{A}_{1,2}\mathsf{B}_{2,1} & \mathsf{A}_{1,1}\mathsf{B}_{1,2} + \mathsf{A}_{1,2}\mathsf{B}_{2,2} \\ \mathsf{A}_{2,1}\mathsf{B}_{1,1} + \mathsf{A}_{2,2}\mathsf{B}_{2,1} & \mathsf{A}_{2,1}\mathsf{B}_{1,2} + \mathsf{A}_{2,2}\mathsf{B}_{2,2} \end{pmatrix},$$

provided that we respect the noncommutativity of matrix multiplication in forming the block products.

DEFINITION. *If $\mathsf{A} \in \mathbb{R}^{n\times n}$, then the k***th leading principal submatrix** *A_k of A, for $k = 1, 2, \ldots, n$, is the block*

$$\mathsf{A}_k = \mathsf{A}_{1,1} = \begin{bmatrix} a_{1,1} & \cdots & a_{1,k} \\ \vdots & & \vdots \\ a_{k,1} & \cdots & a_{k,k} \end{bmatrix}$$

in the block partitioning (2.2-6).

The leading principal submatrices determine whether A is amenable to row reduction without pivoting:

THEOREM 2.6. *Let $\mathsf{A} \in \mathbb{R}^{n\times n}$ be nonsingular. Row reduction of A without partial pivoting is possible if and only if all leading principal submatrices of A are nonsingular.*

PROOF: First assume that row reduction without partial pivoting is possible. Thus we can compute an LU factorization $\mathsf{A} = \mathsf{LU}$. Since $0 \neq \det \mathsf{A} = \det \mathsf{L} \det \mathsf{U}$, neither L nor U is singular. To show that the leading principal submatrix A_k is nonsingular, we examine the following block partitionings:

$$\begin{matrix} & k & n-k \\ k & \mathsf{A}_{1,1} & \mathsf{A}_{1,2} \\ n-k & \mathsf{A}_{2,1} & \mathsf{A}_{2,2} \end{matrix} = \begin{pmatrix} \mathsf{L}_{1,1} & 0 \\ \mathsf{L}_{2,1} & \mathsf{L}_{2,2} \end{pmatrix} \begin{pmatrix} \mathsf{U}_{1,1} & \mathsf{U}_{1,2} \\ 0 & \mathsf{U}_{2,2} \end{pmatrix}.$$

Since $\mathsf{A}_k = \mathsf{A}_{1,1} = \mathsf{L}_{1,1}\mathsf{U}_{1,1}$ and $\det \mathsf{L}_{1,1} = 1$, A_k is nonsingular provided that $\det \mathsf{U}_{1,1} \neq 0$. But the fact that U is upper triangular implies that

$$\det \mathsf{U}_{1,1} \det \mathsf{U}_{2,2} = \det \mathsf{U} = \det \mathsf{A} \neq 0,$$

from which it follows that $\det \mathsf{U}_{1,1} \neq 0$.

Now assume that each of the leading principal submatrices $\mathsf{A}_1, \mathsf{A}_2, \ldots, \mathsf{A}_n$ is nonsingular. We demonstrate by induction on k that each pivot $a_{k,k}^{[k-1]}$ encountered during row reduction is nonzero. When $k = 1$, the pivot is $a_{1,1}^{[0]} = \mathsf{A}_1$, which is nonzero by the hypothesis of nonsingularity. If none of the pivots $a_{1,1}^{[0]}, a_{2,2}^{[1]}, \ldots, a_{k-1,k-1}^{[k-2]}$ vanishes, then we prove that $a_{k,k}^{[k-1]} \neq 0$ by examining the row reduction through step $k-1$. At that point, the application of Gauss transformations $\mathsf{G}_1, \mathsf{G}_2, \ldots, \mathsf{G}_{k-1}$ to A has generated a partially reduced matrix $\mathsf{A}^{[k-1]}$. We block partition the process as follows:

$$\mathsf{A}^{[k-1]} = \begin{matrix} & k & n-k \\ k & \mathsf{A}_{1,1}^{[k-1]} & \mathsf{A}_{1,2}^{[k-1]} \\ n-k & \mathsf{A}_{2,1}^{[k-1]} & \mathsf{A}_{2,2}^{[k-1]} \end{matrix}$$

$$= \begin{pmatrix} (\mathsf{G}_{k-1})_{1,1} & 0 \\ (\mathsf{G}_{k-1})_{2,1} & (\mathsf{G}_{k-1})_{2,2} \end{pmatrix} \cdots \begin{pmatrix} (\mathsf{G}_1)_{1,1} & 0 \\ (\mathsf{G}_1)_{2,1} & (\mathsf{G}_1)_{2,2} \end{pmatrix} \begin{pmatrix} \mathsf{A}_{1,1}^{[0]} & \mathsf{A}_{1,2}^{[0]} \\ \mathsf{A}_{1,1}^{[0]} & \mathsf{A}_{2,2}^{[0]} \end{pmatrix}.$$

The $k \times k$ block $\mathsf{A}_{1,1}^{[k-1]}$, which has $a_{k,k}^{[k-1]}$ in its lower right corner, is upper triangular. Moreover,

$$\mathsf{A}_{1,1}^{[k-1]} = (\mathsf{G}_{k-1})_{1,1}(\mathsf{G}_{k-2})_{1,1} \cdots (\mathsf{G}_1)_{1,1}\mathsf{A}_{1,1}^{[0]}.$$

But by hypothesis $\mathsf{A}_{1,1}^{[0]} = \mathsf{A}_k$ is nonsingular and hence has nonzero determinant, and $\det(\mathsf{G}_i)_{1,1} = 1$ for $i = 1, 2, \ldots, k-1$. Therefore,

$$\det \mathsf{A}_{1,1}^{[k-1]} = a_{1,1}^{[0]} a_{2,2}^{[1]} \cdots a_{k,k}^{[k-1]} \neq 0,$$

and hence $a_{k,k}^{[k-1]} \neq 0$. ∎

Using this theorem, one can identify several common types of matrices that are amenable to row reduction without pivoting.

DEFINITION. *The matrix* $\mathsf{A} \in \mathbb{R}^{n \times n}$ *is* **strictly row diagonally dominant** *if*

$$|a_{i,i}| > \sum_{j \neq i} |a_{i,j}|$$

for every row index $i = 1, 2, \ldots, n$.

In common usage, we omit the word "row," so that "strictly diagonally dominant" means "strictly row diagonally dominant."

COROLLARY 2.7. *Any strictly diagonally dominant matrix is amenable to row reduction without pivoting.*

PROOF: See Problem 2. ∎

As another example, recall that $\mathsf{A} \in \mathbb{R}^{n \times n}$ is symmetric if $\mathsf{A} = \mathsf{A}^\top$ and positive definite if, for any nonzero vector $\mathbf{x} \in \mathbb{R}^n$, $\mathbf{x}^\top \mathbf{A} \mathbf{x} > 0$. Theorem 1.25 guarantees that positive definite matrices are nonsingular.

THEOREM 2.8. *For a symmetric matrix* $\mathsf{A} \in \mathbb{R}^{n \times n}$, *the following statements are equivalent:*

1. A *is positive definite.*
2. *Every eigenvalue of* A *is a positive real number.*
3. *Every leading principal submatrix of* A *is positive definite.*

PROOF: See Problem 7. ∎

Statement 3 has an immediate consequence:

COROLLARY 2.9. *Symmetric, positive definite matrices are amenable to row reduction without pivoting.*

In the next section we examine an efficient scheme for solving linear systems involving symmetric, positive definite systems.

2.3 Variants of Gauss Elimination

Motivation

LU factorization offers a powerful approach to solving linear systems, and Gauss elimination provides a systematic and easily motivated alogorithm for computing L and U. Still, Gauss elimination is quite complicated from a computational viewpoint. Row reduction alone requires multiplication of machine-rounded numbers by factors determined during the execution of the algorithm, subtraction of the results from other rounded numbers, and repeated storage and retrieval of intermediate results from registers in the computer's memory. Even though pivoting strategies allow some control over the amplification of roundoff errors, detailed analysis of the machine errors incurred during Gauss elimination is an intricate task.

One crude but common technique for controlling the effects of roundoff errors is to compute intermediate quantities in double-precision arithmetic, trusting that the accumulated machine error will stay small enough not to affect digits in the single-precision answer. However, in ordinary Gauss elimination, storing intermediate results in double precision can significantly increase the memory needed to execute the algorithm. Methods that avoid intermediate storage therefore have some appeal.

In this section, we examine several algorithms that produce triangular factorizations while avoiding some of the complexity associated with Gauss elimination. In particular, we present the Doolittle and Crout methods, which obviate storage and retrieval of intermediate results in computing L and U. We also examine the Cholesky decomposition, which produces a space-saving triangular factorization for symmetric, positive definite matrices.

The Doolittle and Crout Methods

Knowing that an LU factorization of $\mathsf{A} \in \mathbb{R}^{n\times n}$ exists, we can use the n^2 relationships

$$a_{i,j} = \sum_{p=1}^{\min\{i,j\}} l_{i,p} u_{p,j} \tag{2.3-1}$$

to solve for the factors $l_{i,j}$ and $u_{i,j}$. Since L and U are lower and upper triangular, together they comprise only $n^2 + n$ nonzero entries. Moreover, if we adopt the convention that $l_{i,i} = 1$ for $i = 1, 2, \ldots, n$, then we need to determine only n^2 quantities $l_{i,j}$ and $u_{i,j}$, and the equations (2.3-1) suffice. In particular, we have

$$u_{i,j} = a_{i,j} - \sum_{p=1}^{i-1} l_{i,p} u_{p,j}, \qquad \text{for} \quad j = i, i+1, \ldots, n, \tag{2.3-2}$$

and

$$l_{i,j} = \frac{1}{u_{j,j}} \left(\sum_{p=1}^{j-1} l_{i,p} u_{p,j} \right), \qquad \text{for} \quad i = j+1, j+2, \ldots, n. \qquad (2.3\text{-}3)$$

While Equations (2.3-2) and (2.3-3) seem unremarkable, a clever solution strategy turns them into a useful algorithm, called the **Doolittle method.**

ALGORITHM 2.1 (DOOLITTLE). *Given a nonsingular matrix* $\mathsf{A} \in \mathbb{R}^{n \times n}$, *compute the matrices* $\mathsf{L}, \mathsf{U} \in \mathbb{R}^{n \times n}$ *in the* LU *factorization of* A *as follows:*

1. For $k = 1, 2, \ldots, n$:
2. For $j = k, k+1, \ldots, n$:
3. $u_{k,j} \leftarrow a_{k,j} - \sum_{p=1}^{k-1} l_{k,p} u_{p,j}$.
4. Next j.
5. For $i = k+1, k+2, \ldots, n$:
6. $l_{i,k} \leftarrow \left(a_{i,k} - \sum_{p=1}^{k-1} l_{i,p} u_{p,k} \right) / u_{k,k}$.
7. Next i.
8. Next k.
9. End.

(When $k = 1$, we assign the value 0 to the sums in this algorithm.) Figure 1 schematically shows how the computations progress through the entries in an $n \times n$ matrix.

None of the calculations in the kth pass through the outermost loop requires us to know values for variables $l_{i,j}$ or $u_{i,j}$ to be determined in subsequent steps $k+1, k+2, \ldots$. Furthermore, if we solve for the unknowns in this step *in the order listed*, then all of the quantities on the right sides of Equations (2.3-2) and (2.3-3) are known by the time we need them. We can therefore accumulate each of the sums appearing in these equations in a single register — or in double precision, if desired — without retrieving and re-storing intermediate results.

An example with n small illustrates the scheme. Consider the LU factorization

$$\mathsf{A} = \begin{bmatrix} 2 & 1 & 2 \\ 1 & 2 & 3 \\ 4 & 1 & 2 \end{bmatrix} = \underbrace{\begin{bmatrix} 1 & 0 & 0 \\ l_{2,1} & 1 & 0 \\ l_{3,1} & l_{3,2} & 1 \end{bmatrix}}_{\mathsf{L}} \underbrace{\begin{bmatrix} u_{1,1} & u_{1,2} & u_{1,3} \\ 0 & u_{2,2} & u_{2,3} \\ 0 & 0 & u_{3,3} \end{bmatrix}}_{\mathsf{U}}.$$

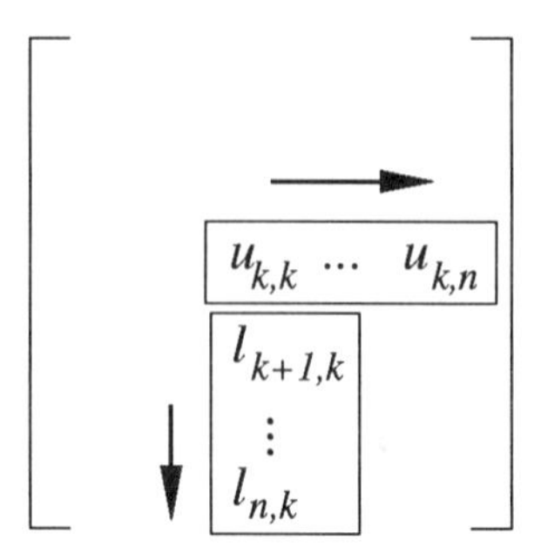

FIGURE 1. *Entries of the* LU *factorization determined during the kth pass through the outer loop in the Doolittle method.*

In the first pass ($k = 1$) through the outer loop, we solve for the factors $u_{1,1}, u_{1,2}, u_{1,3}, l_{2,1}, l_{3,1}$, in that order, getting

$$u_{1,1} = a_{1,1} = 2, \qquad u_{1,2} = a_{1,2} = 1, \qquad u_{1,3} = a_{1,3} = 2,$$

$$l_{2,1} = a_{2,1}/u_{1,1} = \tfrac{1}{2}, \qquad l_{3,1} = a_{3,1}/u_{1,1} = 2.$$

In the second pass ($k = 2$), we solve for $u_{2,2}, u_{2,3}, l_{3,2}$ in order, getting

$$u_{2,2} = a_{2,2} - l_{2,1}u_{2,1} = \tfrac{3}{2}, \qquad u_{2,3} = a_{2,3} - l_{2,1}u_{1,3} = 2,$$

$$l_{3,2} = (a_{3,2} - l_{3,1}u_{1,2})/u_{2,2} = -\tfrac{2}{3}.$$

Finally, in the third pass, we solve for $u_{3,3}$:

$$u_{3,3} = a_{3,3} - l_{3,1}u_{1,3} - l_{3,2}u_{2,3} = -\tfrac{2}{3}.$$

Therefore,

$$\mathsf{L} = \begin{bmatrix} 1 & 0 & 0 \\ \frac{1}{2} & 1 & 0 \\ 2 & -\frac{2}{3} & 1 \end{bmatrix}, \qquad \mathsf{U} = \begin{bmatrix} 2 & 1 & 2 \\ 0 & \frac{3}{2} & 2 \\ 0 & 0 & -\frac{2}{3} \end{bmatrix}.$$

An alternative to the Doolittle method, the **Crout method**, uses a different orchestration in solving for the entries of L and U.

ALGORITHM 2.2 (CROUT). *Given a nonsingular matrix $\mathsf{A} \in \mathbb{R}^{n\times n}$, compute the matrices $\mathsf{L}, \mathsf{U} \in \mathbb{R}^{n\times n}$ in the* LU *factorization of* A *as follows:*

1. For $j = 1, 2, \ldots, n$:
2. For $i = 1, 2, \ldots, j$:

3. $u_{i,j} \leftarrow a_{i,j} - \sum_{p=1}^{i-1} l_{i,p} u_{p,j}$.
4. Next i.
5. For $i = j+1, j+2, \ldots, n$:
6. $l_{i,j} \leftarrow \left(a_{i,j} - \sum_{p=1}^{j-1} l_{i,p} u_{p,j}\right) / u_{j,j}$.
7. Next i.
8. Next j.
9. End.

Figure 2 schematically illustrates the order of the computations.

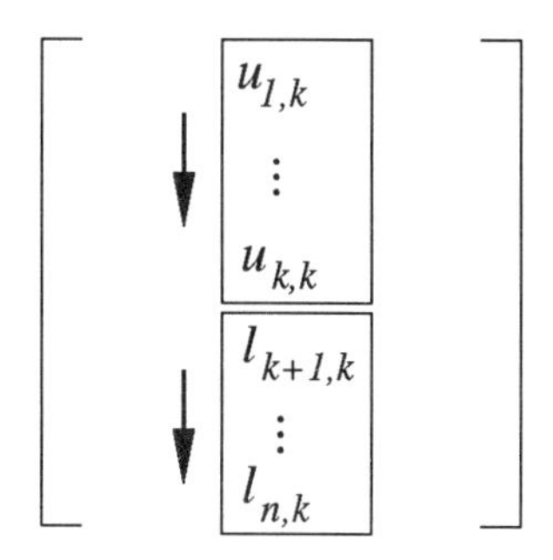

FIGURE 2. *Entries determined during the kth pass through the outer loop in the Crout method.*

The sleekness of these methods scarcely reduces the value of pivoting. The aim, as before, is to avoid the magnification of roundoff errors by choosing divisors that are as large in magnitude as possible. For the Doolittle method, the numbers $u_{k,k}$ appear as divisors in step 7 of Algorithm 2.2. Actually performing the computations in the order listed, however, leaves little leeway for choosing other divisors. The trick is to notice that, for a given value of k, one can compute all of the sums

$$S_{i,k} = a_{i,k} - \sum_{p=1}^{k-1} l_{i,p} u_{p,k}, \qquad i = k, k+1, \ldots, n,$$

at the start of the kth pass through the outer loop. Once these values are available, scan them for the one with the largest magnitude, say $S_{m,k} = S_{\max}$. Then interchange rows k and m in A and interchange the row vectors

$(l_{k,1}, l_{k,2}, \ldots, l_{k,k-1})$ and $(l_{m,1}, l_{m,2}, \ldots, l_{m,k-1})$ of previously determined entries in L. The number $S_{\max}$ now serves as the divisor $u_{k,k}$ in the new, row-permuted matrix. Finally, compute the remaining entries $u_{k+1,k}, \ldots, u_{n,k}$ in the new row k of U according to Equation (2.3-2), then compute the entries $l_{k+1,k}, \ldots, l_{n,k}$ in column k of L by dividing the remaining sums $S_{i,k}$ by $u_{k,k}$.

Partial pivoting for the Crout method follows a similar idea. We leave details for Problem 3.

Cholesky Decomposition

When $\mathsf{A} \in \mathbb{R}^{n\times n}$ is symmetric and positive definite, an efficient triangular factorization known as **Cholesky decomposition** is possible. This scheme factors A as a product of the form $\mathsf{CC}^\top$, where $\mathsf{C} \in \mathbb{R}^{n\times n}$ is a lower triangular matrix called the **Cholesky triangle** for A. Just as the LU factorization of an arbitrary nonsingular matrix in $\mathbb{R}^{n\times n}$ requires no more storage than does the original matrix, the factorization $\mathsf{CC}^\top$ requires $(n^2+n)/2$ storage locations — the same number needed to store the distinct entries of the original, symmetric matrix A.

The following theorem justifies the Cholesky decomposition.

THEOREM 2.10. *If $\mathsf{A} \in \mathbb{R}^{n\times n}$ is symmetric and positive definite, then there exists a lower triangular matrix $\mathsf{C} \in \mathbb{R}^{n\times n}$ such that $\mathsf{A} = \mathsf{CC}^\top$. Moreover, the diagonal entries of* C *are all positive.*

PROOF: We use induction on n. When $n = 1$, the matrix A is a positive real number $a_{1,1}$. In this case, the matrix C is $c_{1,1} = \sqrt{a_{1,1}}$. Now assume that the theorem holds for matrices in $\mathbb{R}^{n\times n}$, and let $\mathsf{A} \in \mathbb{R}^{(n+1)\times(n+1)}$ be symmetric and positive definite. To construct the Cholesky triangle C for A, start with a block partitioning of A:

$$\mathsf{A} = \begin{matrix} & n & 1 \\ n & \mathsf{A}_n & \mathbf{a} \\ 1 & \mathbf{a}^\top & a_{n+1,n+1} \end{matrix}, \qquad \mathsf{A} = \begin{pmatrix} \mathsf{A}_n & \mathbf{a} \\ \mathbf{a}^\top & a_{n+1,n+1} \end{pmatrix}.$$

Here, $\mathsf{A}_n \in \mathbb{R}^{n\times n}$ is a leading principal submatrix of A and is therefore itself symmetric and positive definite, and $\mathbf{a} \in \mathbb{R}^n$. The entry $a_{n+1,n+1}$ must be positive; otherwise, we could make $\mathbf{x}^\top \mathsf{A}\mathbf{x} \leq 0$ by choosing $\mathbf{x} = (0, \ldots, 0, 1)^\top$. By the inductive hypothesis, there is a lower triangular matrix $\mathsf{C}_n \in \mathbb{R}^{n\times n}$, having positive diagonal entries, such that $\mathsf{C}_n\mathsf{C}_n^\top = \mathsf{A}_n$. To complete the proof, it suffices to find a vector $\boldsymbol{\chi} \in \mathbb{R}^{n+1}$ and a number $c > 0$ such that

$$\mathsf{CC}^\top = \begin{pmatrix} \mathsf{C}_n & \mathbf{0} \\ \boldsymbol{\chi}^\top & c \end{pmatrix} \begin{pmatrix} \mathsf{C}_n^\top & \boldsymbol{\chi} \\ \mathbf{0}^\top & c \end{pmatrix} = \begin{pmatrix} \mathsf{A}_n & \mathbf{a} \\ \mathbf{a}^\top & a_{n+1,n+1} \end{pmatrix}.$$

(Block sizes: rows n, 1; columns n, 1.)

To do this, we solve the equations

$$\begin{aligned} \mathsf{C}_n \boldsymbol{\chi} &= \mathbf{a}, \\ \boldsymbol{\chi}^\top \boldsymbol{\chi} + c^2 &= a_{n+1,n+1}. \end{aligned}$$

The first of these equations has a solution $\boldsymbol{\chi} = \mathsf{C}_n^{-1}\mathbf{a}$, the nonsingularity of C_n following from that of A_n and the identity $\det \mathsf{A}_n = (\det \mathsf{C}_n)^2$. The second equation clearly has a solution

$$c = \sqrt{a_{n+1,n+1} - \boldsymbol{\chi}^\top \boldsymbol{\chi}},$$

the only possible issue being whether $c^2 > 0$. We settle this issue by appealing to the fact that A is symmetric and positive definite. Observe that

$$\boldsymbol{\chi}^\top \boldsymbol{\chi} = \left(\mathsf{C}_n^{-1}\mathbf{a}\right)^\top \mathsf{C}_n^{-1}\mathbf{a} = \mathbf{a}^\top \left(\mathsf{C}_n^{-1}\right)^\top \mathsf{C}_n^{-1}\mathbf{a} = \mathbf{a}^\top \mathsf{A}_n^{-1}\mathbf{a}.$$

Therefore,

$$a_{n+1,n+1} - \boldsymbol{\chi}^\top \boldsymbol{\chi} = a_{n+1,n+1} - \mathbf{a}^\top \mathsf{A}_n^{-1}\mathbf{a}.$$

Now define $\mathbf{x} \in \mathbb{R}^{n+1}$ as follows:

$$\mathbf{x} := \left((\mathsf{A}_n^{-1}\mathbf{a})^\top, -1\right)^\top \neq \mathbf{0}.$$

A straightforward calculation shows that

$$\mathbf{x}^\top \mathsf{A}\mathbf{x} = a_{n+1,n+1} - \mathbf{a}^\top \mathsf{A}_n^{-1}\mathbf{a} = a_{n+1,n+1} - \boldsymbol{\chi}^\top \boldsymbol{\chi},$$

which must be positive since A is positive definite. ■

To derive a practical algorithm for computing the Cholesky triangle C, one can proceed as for the Doolittle and Crout schemes, solving the equations $a_{i,j} = \sum_{k=1}^{n} c_{i,k} c_{k,j}$ for the entries $c_{i,j}$:

ALGORITHM 2.3 (CHOLESKY) *Let $\mathsf{A} \in \mathbb{R}^{n \times n}$ be positive definite. The following steps compute the entries $c_{i,j}$ of the Cholesky triangle for A:*

1. $c_{1,1} \leftarrow \sqrt{a_{1,1}}$.
2. For $i = 2, 3, \ldots, n$:
3. $\quad c_{i,1} \leftarrow a_{i,1}/c_{1,1}$.
4. Next i.
5. For $j = 2, 3, \ldots, n-1$:
6. $\quad c_{j,j} \leftarrow \left(a_{j,j} - \sum_{k=1}^{j-1} c_{j,k}^2\right)^{1/2}$.

7. For $i = j+1, j+2, \ldots, n$:
8. $c_{i,j} \leftarrow \left(a_{i,j} - \sum_{k=1}^{j-1} c_{i,k}c_{j,k}\right)/c_{j,j}$.
9. Next i.
10. Next j.
11. $c_{n,n} \leftarrow \left(a_{n,n} - \sum_{k=1}^{n-1} c_{n,k}^2\right)^{1/2}$.
12. End.

Theorem 2.10 ensures that none of the entries $c_{j,j}$ in the denominator of step 8 vanishes. Moreover, the relationship $a_{i,i} = \sum_{k=1}^{n} c_{i,k}^2$ implies that $|c_{i,k}| \leq \sqrt{a_{i,i}}$. Therefore the entries in A control the magnitudes of those in C, even without a pivoting strategy.

One possible objection to Algorithm 2.3 is that it calls for the extraction of n square roots. The Cholesky algorithm requires $\mathcal{O}(n^3)$ arithmetic operations (see Problem 4), so for large n the computational burden associated with these square roots is small compared with the arithmetic required in the rest of the algorithm.

It is possible, however, to avoid taking square roots in factoring symmetric, positive definite matrices. The idea, *in theory*, is to construct a further factorization, $\mathsf{C} = \mathsf{LS}$. (For the moment, we neglect issues of computational efficiency.) Here S is a **diagonal** matrix, that is, one whose off-diagonal entries $s_{i,j}$, $i \neq j$, are all 0. The diagonal entries are $s_{i,i} = c_{i,i}$. With this definition of S, we can easily construct the entries of L from those of C. This construction reveals that L is lower triangular with unit diagonal entries. We now have the **LDL^{T} factorization**,

$$\mathsf{CC}^\top = \mathsf{LS}(\mathsf{LS})^\top = \mathsf{L}\underbrace{\mathsf{SS}^\top}_{\mathsf{D}}\mathsf{L}^\top = \mathsf{LDL}^\top,$$

where D is a diagonal matrix whose diagonal entries $d_{i,i} = c_{i,i}^2$ are positive and can be computed without extracting square roots.

As one might expect, computing the LDL^{T} factorization by first computing the Cholesky triangle C is inefficient. The following algorithm accomplishes the task more economically.

ALGORITHM 2.4 (LDL^{T}). *Let $\mathsf{A} \in \mathbb{R}^{n\times n}$ be positive definite. The following steps compute the entries $d_{i,i}$ and $l_{i,j}$, $i > j$, of the LDL^T factorization for A:*

1. For $i = 1, 2, \ldots, n$:
2. For $j = 1, 2, \ldots, i-1$:

3. $\qquad p_j \leftarrow a_{i,j} d_j$.
4. $\qquad$ Next j.
5. $\qquad d_{i,i} \leftarrow a_{i,i} - \sum_{k=1}^{i-1} a_{i,k} p_k$.
6. $\qquad$ For $j = i+1, i+2, \ldots, n$:
7. $\qquad l_{j,i} \leftarrow \left(a_{j,i} - \sum_{k=1}^{i-1} a_{j,k} p_k \right) / d_{i,i}$.
8. $\qquad$ Next j.
9. Next i.
10. End.

The loop initiated in step 2 of this algorithm is vacuous if $i = 1$, so $d_{1,1} = a_{1,1}$. Problem 4 asks for an operation count for this algorithm.

2.4 Band Matrices

Motivation and Construction

Many applications give rise to matrices that have a small number of nonzero entries. A simple example appears in Section 1.6 in the discussion of cubic splines. There we encounter tridiagonal systems, in which the matrices have the structure

$$\mathsf{A} = \begin{bmatrix} b_1 & c_1 & & & & \\ a_2 & b_2 & c_2 & & & \\ & a_3 & b_3 & c_3 & & \\ & & & \ddots & & \\ & & & a_{n-1} & b_{n-1} & c_{n-1} \\ & & & & a_n & b_n \end{bmatrix}. \tag{2.4-1}$$

Tridiagonal systems also arise from discrete approximations to second-order partial differential equations, as discussed in Chapter 8.

Difference approximations to higher-order differential equations typically lead to matrices having more general patterns of nonzero entries, such as the following 5-diagonal structure:

$$\mathsf{A} = \begin{bmatrix} \star & \star & \star & & & & & & \\ \star & \star & \star & \star & & & & & \\ \star & \star & \star & \star & \star & & & & \\ & \star & \star & \star & \star & \star & & & \\ & & & & \ddots & & & & \\ & & & \star & \star & \star & \star & \star & \\ & & & & \star & \star & \star & \star & \star \\ & & & & & \star & \star & \star & \star \\ & & & & & & \star & \star & \star \end{bmatrix}. \tag{2.4-2}$$

Here, the entries not marked by the symbol "$\star$" are zero. This notation is often quite useful, since it shows where the only possible nonzero entries occur without requiring us to write them explicitly. We call (2.4-2) the **zero structure** of the matrix.

A third example arises in Chapter 8, when we discuss finite-difference methods for the Laplace equation. There we approximate certain boundary-value problems by linear systems in which the matrices are **block tridiagonal**:

$$\mathsf{A} = \begin{array}{c} m \\ m \\ \vdots \\ \\ \\ \end{array} \overset{\begin{array}{ccc} m & m & \cdots \end{array}}{\begin{pmatrix} \mathsf{B}_1 & \mathsf{C}_1 & & & \\ \mathsf{A}_2 & \mathsf{B}_2 & \mathsf{C}_2 & & \\ & & \ddots & & \\ & & \mathsf{A}_{n-1} & \mathsf{B}_{n-1} & \mathsf{C}_{n-1} \\ & & & \mathsf{A}_n & \mathsf{B}_n \end{pmatrix}}. \tag{2.4-3}$$

Each of the blocks in this block partitioning is a matrix in $\mathbb{R}^{m\times m}$, and the overall matrix A lies in $\mathbb{R}^{mn\times mn}$.

All of these structures have in common the fact that, for appreciable values of n, most of the entries are zero. In other words, the matrices are **sparse**. Sparse matrices present clear opportunities for computational savings if one can design algorithms that avoid trivial calculations involving the zero entries. The examples listed belong to a special class of sparse matrices called **band matrices**, since their entries vanish outside a relatively narrow band about the diagonal. In this section we examine numerical methods for solving systems involving such matrices.

We begin with tridiagonal matrices. As shown in Section 1.6, linear systems involving these matrices admit an especially attractive solution scheme, namely, Algorithm 1.1. Recall that this algorithm produces the solution to a tridiagonal system of the form (2.4-1) in $\mathcal{O}(n)$ operations — a noteworthy improvement over the $\mathcal{O}(n^3)$ estimate for straightforward Gauss elimination.

A close look at Algorithm 1.1 reveals that it is simply a version of row reduction (steps 1 through 6) followed by backward substitution (steps 7 through 10). In fact, the algorithm implicitly constructs an LU factorization of the matrix A in Equation (2.4-1) having the form

$$\mathsf{A} = \begin{bmatrix} \beta_1 & & & & \\ a_2 & \beta_2 & & & \\ & \ddots & \ddots & & \\ & & a_{n-1} & \beta_{n-1} & \\ & & & a_n & \beta_n \end{bmatrix} \begin{bmatrix} 1 & \eta_1 & & & \\ & 1 & \eta_2 & & \\ & & \ddots & \ddots & \\ & & & 1 & \eta_{n-1} \\ & & & & 1 \end{bmatrix}, \tag{2.4-4}$$

where $\eta_i = c_i/\beta_i$. In this instance, we adopt the convention that the upper triangular factor has unit diagonal, while the lower triangular factor contains the pivots β_i.

Since Algorithm 1.1 implements no pivoting strategy, it fails if any of the pivots vanishes. The proof of Proposition 1.16 shows that tridiagonal matrices arising in the cubic spline applications considered in Chapter 1 are strictly diagonally dominant, so Corollary 1.7 guarantees that zero pivots will never occur for these problems. Later in this section we prove that strict diagonal dominance also prevents the growth of roundoff errors.

For more general band matrices, operation counts depend upon the structure of the band. In these cases it is frequently useful to compute LU factorizations. Before discussing details, let us introduce some terminology:

DEFINITION. *A matrix* $\mathsf{A} \in \mathbb{R}^{n \times n}$ *has* **lower bandwidth** p *if* p *is the smallest nonnegative integer for which* $a_{i,j} = 0$ *whenever* $j < i - p$. A *has* **upper bandwidth** q *if* q *is the smallest nonnegative integer for which* $a_{i,j} = 0$ *whenever* $j > i + q$. *The* **bandwidth** *of a matrix with lower and upper bandwidths* p *and* q, *respectively, is* $b = p + q + 1$.

For example, a matrix with lower bandwidth $p = 1$ and upper bandwidth $q = 2$ has bandwidth 4 and the following structure:

$$\begin{bmatrix} a_{1,1} & a_{1,2} & a_{1,3} & & & \\ a_{2,1} & a_{2,2} & a_{2,3} & a_{3,3} & & \\ & & \ddots & & & \\ & & a_{n-2,n-3} & a_{n-2,n-2} & a_{n-2,n-1} & a_{n-2,n} \\ & & & a_{n-1,n-2} & a_{n-1,n-1} & a_{n-1,n} \\ & & & & a_{n,n-1} & a_{n,n} \end{bmatrix}.$$

Actually storing the matrix in this form is wasteful, since the numbers p and q suffice to specify all of the entries outside the central band of width $b = p + q + 1$. The **compact storage mode**

$$\begin{bmatrix} 0 & a_{1,1} & a_{1,2} & a_{1,3} \\ a_{2,1} & a_{2,2} & a_{2,3} & a_{3,3} \\ & \vdots & & \\ a_{n-2,n-3} & a_{n-2,n-2} & a_{n-2,n-1} & a_{n-2,n} \\ a_{n-1,n-2} & a_{n-1,n-1} & a_{n-1,n} & 0 \\ a_{n,n-1} & a_{n,n} & 0 & 0 \end{bmatrix},$$

which stores the diagonals in columns, offers a more efficient alternative.

In addition to saving on memory, one can usually rewrite LU factorization algorithms to operate only on those entries of A that appear in the compact storage mode. Most of the tedium associated with this effort involves the bookkeeping needed to shift column indices in converting from the standard storage mode to the compact storage mode. As we demonstrate later in this section, there is a payoff: If one computes the LU factorization without pivoting, then L has lower bandwidth p and U has upper bandwidth q. This fact enables us to store the LU factorization in the same compact storage mode used for the original matrix A.

The following algorithm, designed for band matrices with $p = q$ (and hence bandwidth $b = 2p + 1$), goes one step further, overwriting the storage locations assigned to A with the entries of its LU factorization. The algorithm utilizes Equations (2.3-2) and (2.3-3).

ALGORITHM 2.5. *Given a nonsingular band matrix* $\mathsf{A} \in \mathbb{R}^{n\times n}$ *having lower and upper bandwidth p, the following steps compute the factors* L *and* U *in the* LU *factorization, without pivoting. The algorithm stores the entries of* L *and* U *in the array allocated for* A, *destroying the original matrix.*

1. For $i = 1, 2, \ldots, n-1$:
2. $\quad j \leftarrow p + 1$.
3. $\quad$ For $k = 1, 2, \ldots, n-i$:
4. $\quad\quad j \leftarrow j - 1$.
5. $\quad\quad$ If $j \geq 0$ then:
6. $\quad\quad\quad a_{i+k,j} \leftarrow a_{i+k,j}/a_{i,p+1}$.
7. $\quad\quad\quad$ For $m = 1, 2, \ldots, p$:
8. $\quad\quad\quad\quad a_{i+k,j+m} \leftarrow a_{i+k,j+m} - a_{i+k,j}a_{i,j}$.
9. $\quad\quad\quad$ Next m.
10. $\quad\quad$ End if.
11. $\quad$ Next k.
12. Next i.
13. End.

This algorithm requires $\mathcal{O}(np^2)$ operations. When p is much smaller than n, this operation count is greatly preferable to the $\mathcal{O}(n^3)$ operations needed for an arbitrary nonsingular matrix in $\mathbb{R}^{n\times n}$.

Having computed the LU factorization, we can use it to solve a linear system $\mathsf{A}\mathbf{x} = \mathbf{b}$, given any vector $\mathbf{b} \in \mathbb{R}^n$. The idea is to use the factors L and U, stored in the compact storage mode, to solve the systems $\mathsf{L}\mathbf{z} = \mathbf{b}$, $\mathsf{U}\mathbf{x} = \mathbf{z}$. Recall from Section 2.2 that these two systems require simple forward and backward substitution. The next algorithm performs these tasks.

ALGORITHM 2.6. *Given the* LU *factorization produced by Algorithm 2.5 and stored in* A *in compact storage mode, the following steps compute the solution* $\mathbf{x}$ *to* $\mathsf{A}\mathbf{x} = \mathsf{b}$ *for any* $\mathbf{b} \in \mathbb{R}^n$. *The algorithm overwrites* $\mathbf{b}$ *with* $\mathbf{x}$.

1. For $i = 2, 3, \ldots n$:
2. $\quad k \leftarrow p + 2 - i$.
3. $\quad l \leftarrow 1$.
4. $\quad$ If $k \leq 0$ then:
5. $\quad\quad k \leftarrow 1$.
6. $\quad\quad l \leftarrow i - p$.
7. $\quad$ End if.
8. $\quad b_i \leftarrow \sum_{j=k}^{p} a_{i,j} b_{j-k+l}$.
9. Next i.
10. $b_n \leftarrow b_n / a_{n,p+1}$.
11. For $k = 2, 3, \ldots, n$:
12. $\quad i \leftarrow n + 1 - k$.
13. $\quad J \leftarrow \min\{b, p + k\}$.
14. $\quad b_i \leftarrow \left(b_i - \sum_{j=p+2}^{J} a_{i,j} b_{j-p} \right) / a_{i,p+1}$.
15. Next k.
16. End.

Steps 1 through 9 execute the forward substitution, while steps 10 through 15 complete the backward substitution.

For systems involving symmetric, positive definite matrices with banded structure, even more compact algorithms are possible. We leave the design of a scheme for Cholesky decomposition of such matrices for Problem 5.

Finally, consider block tridiagonal matrices. Even though these are special cases of band matrices, they occur frequently enough to have inspired a variety of specialized direct solution techniques. We examine a scheme that generalizes Algorithm 1.1. Specifically, we seek a "block LU factorization" of Equation (2.4-3) having a format similar to that given in Equation (2.4-4):

$$\mathsf{A} = \begin{pmatrix} \mathsf{E}_1 & & & & \\ \mathsf{A}_2 & \mathsf{E}_2 & & & \\ & \ddots & \ddots & & \\ & & \mathsf{A}_{n-1} & \mathsf{E}_{n-1} & \\ & & & \mathsf{A}_n & \mathsf{E}_n \end{pmatrix} \begin{pmatrix} \mathsf{I} & \mathsf{F}_1 & & & \\ & \mathsf{I} & \mathsf{F}_2 & & \\ & & \ddots & \ddots & \\ & & & \mathsf{I} & \mathsf{F}_n \\ & & & & \mathsf{I} \end{pmatrix}, \quad (2.4\text{-}5)$$

where $\mathsf{I} \in \mathbb{R}^{m\times m}$ is the identity matrix and each of the blocks $\mathsf{E}_i, \mathsf{F}_i$ is an $m \times m$ real matrix.

Formally solving for the blocks in this factorization, we obtain

$$\begin{aligned} \mathsf{E}_1 &= \mathsf{B}_1, \\ \mathsf{F}_i &= \mathsf{E}_i^{-1}\mathsf{C}_i, \qquad i = 1, 2, \ldots, n-1. \\ \mathsf{E}_i &= \mathsf{B}_i - \mathsf{A}_i\mathsf{F}_{i-1}, \qquad i = 2, 3, \ldots, n. \end{aligned}$$

To compute the blocks F_i, it is not necessary to form explicit inverses for the matrices E_i. A more efficient strategy is to solve systems of the form $\mathsf{E}_i\mathbf{f}_{i,j} = \mathbf{c}_{i,j}$, where $\mathbf{c}_{i,j} \in \mathbb{R}^m$ denotes the jth column of the matrix C_i, which is known, and $\mathbf{f}_{i,j} \in \mathbb{R}^m$ signifies the unknown jth column of F_i. This approach allows us to compute an LU factorization for each matrix E_i just once, using forward and backward substitution to determine each of the m columns $\mathbf{f}_{i,j}$.

Once we have the block factorization (2.4-5), we can use blockwise versions of forward and backward substitution to determine the solution to any linear system having the form

$$\begin{pmatrix} \mathsf{B}_1 & \mathsf{C}_1 & & \\ \mathsf{A}_2 & \mathsf{B}_2 & \mathsf{C}_2 & \\ & & \ddots & \\ & & \mathsf{A}_n & \mathsf{B}_n \end{pmatrix} \begin{pmatrix} \mathbf{x}_1 \\ \mathbf{x}_2 \\ \vdots \\ \mathbf{x}_n \end{pmatrix} = \begin{pmatrix} \mathbf{b}_1 \\ \mathbf{b}_2 \\ \vdots \\ \mathbf{b}_n \end{pmatrix},$$

where each of the blocks $\mathbf{x}_i, \mathbf{b}_i$ is a vector in $\mathbb{R}^m$. For the forward substitution stage, we solve the systems

$$\begin{aligned} \mathsf{E}_1\mathbf{z}_1 &= \mathbf{b}_1, \\ \mathsf{E}_i\mathbf{z}_i &= \mathbf{b}_i - \mathsf{A}_i\mathbf{z}_{i-1}, \qquad i = 1, 2, \ldots, n, \end{aligned}$$

for the intermediate vectors $\mathbf{z}_1, \mathbf{z}_2, \ldots, \mathbf{z}_n \in \mathbb{R}^m$. For each of these systems, we can exploit the fact that we have already computed LU factorizations for $\mathsf{E}_1, \mathsf{E}_2, \ldots, \mathsf{E}_n$. Backward substitution then calls for simple matrix multiplication:

$$\begin{aligned} \mathbf{x}_n &= \mathbf{z}_n, \\ \mathbf{x}_i &= \mathbf{z}_i - \mathsf{F}_i\mathbf{x}_{i+1}, \qquad i = n-1, n-2, \ldots, 1. \end{aligned}$$

Practical Considerations

The compactness that we associate with LU factorizations of band matrices unfortunately applies only to the most favorable of settings. If pivoting strategies are necessary, then row interchanges typically disrupt the band structures of L and U. One can gain some appreciation for what happens by considering changes in zero structure that occur when we apply row reduction with partial pivoting to a simple example. We begin with a band matrix $\mathsf{A} \in \mathbb{R}^{5\times 5}$ having lower bandwidth $p = 1$ and upper bandwidth $q = 2$. We expect the zero structure of the LU factorization without pivoting to look like this:

$$\underbrace{\begin{bmatrix} \star & \star & \star & & \\ \star & \star & \star & \star & \\ & \star & \star & \star & \star \\ & & \star & \star & \star \\ & & & \star & \star \end{bmatrix}}_{\mathsf{A}} = \underbrace{\begin{bmatrix} \star & & & & \\ \star & \star & & & \\ & \star & \star & & \\ & & \star & \star & \\ & & & \star & \star \end{bmatrix}}_{\mathsf{L}} \underbrace{\begin{bmatrix} \star & \star & \star & & \\ & \star & \star & \star & \\ & & \star & \star & \star \\ & & & \star & \star \\ & & & & \star \end{bmatrix}}_{\mathsf{U}}.$$

Suppose, however, that we interchange rows k and $k+1$ before each step k of row reduction. We can implement this process by alternately applying row interchanges, represented by permutation matrices $\mathsf{P}_{k,k+1}$, and row reductions, represented by Gauss transformations G_k. The following schematic shows the zero structures that occur during the process:

$$\mathsf{A} \underset{\mathsf{P}_{1,2}}{\longmapsto} \begin{bmatrix} \star & \star & \star & \star & \\ \star & \star & \star & & \\ & \star & \star & \star & \star \\ & & \star & \star & \star \\ & & & \star & \star \end{bmatrix} \underset{\mathsf{G}_1}{\longmapsto} \begin{bmatrix} \star & \star & \star & \star & \\ & \star & \star & \star & \\ & \star & \star & \star & \star \\ & & \star & \star & \star \\ & & & \star & \star \end{bmatrix}$$

$$\underset{\mathsf{P}_{2,3}}{\longmapsto} \begin{bmatrix} \star & \star & \star & \star & \\ & \star & \star & \star & \star \\ & \star & \star & \star & \\ & & \star & \star & \star \\ & & & \star & \star \end{bmatrix} \underset{\mathsf{G}_2}{\longmapsto} \begin{bmatrix} \star & \star & \star & \star & \\ & \star & \star & \star & \star \\ & & \star & \star & \star \\ & & \star & \star & \star \\ & & & \star & \star \end{bmatrix}$$

$$\underset{\mathsf{P}_{3,4}}{\longmapsto} \begin{bmatrix} \star & \star & \star & \star & \\ & \star & \star & \star & \star \\ & & \star & \star & \star \\ & & \star & \star & \star \\ & & & \star & \star \end{bmatrix} \underset{\mathsf{G}_3}{\longmapsto} \begin{bmatrix} \star & \star & \star & \star & \\ & \star & \star & \star & \star \\ & & \star & \star & \star \\ & & & \star & \star \\ & & & \star & \star \end{bmatrix}$$

$$\underset{\mathsf{P}_{4,5}}{\longmapsto} \begin{bmatrix} \star & \star & \star & \star & \\ & \star & \star & \star & \star \\ & & \star & \star & \star \\ & & & \star & \star \\ & & & \star & \star \end{bmatrix} \underset{\mathsf{G}_4}{\longmapsto} \begin{bmatrix} \star & \star & \star & \star & \\ & \star & \star & \star & \star \\ & & \star & \star & \star \\ & & & \star & \star \\ & & & & \star \end{bmatrix}.$$

It instructive to derive, step by step, the zero structure of the matrix L associated with this row reduction, as given by Theorem 2.3. The final LU factorization for this example has the zero structure

$$\mathsf{PA} = \begin{bmatrix} \star & & & & \\ & \star & & & \\ & & \star & & \\ & & & \star & \\ \star & \star & \star & \star & \star \end{bmatrix} \begin{bmatrix} \star & \star & \star & \star & \\ & \star & \star & \star & \star \\ & & \star & \star & \star \\ & & & \star & \star \\ & & & & \star \end{bmatrix}. \tag{2.4-6}$$

This example suggests two facts about LU factorization of band matrices with partial pivoting. First, the upper triangular matrix U is still a band matrix. However, its upper bandwidth may no longer be the same as that of the original matrix A, as it would be with no pivoting strategy. In fact, the upper bandwidth of U in our example is $p+q$. This fact holds generally, as we prove below.

Second, the lower triangular matrix L no longer is banded. A priori, any position below the diagonal in L may be occupied by some nonzero factor $l_{i,j}$ by the time the row reduction is complete. However, no column of L can contain more than p such factors below the diagonal. Clearly, when it is feasible, LU factorization of band matrices without partial pivoting enjoys advantages that partial pivoting destroys.

The practical aspects of block tridiagonal matrices bring unpleasant news of a similar nature. This sparse structure often arises in connection with the numerical approximation of partial differential equations, and in this context the individual blocks $\mathsf{A}_i, \mathsf{B}_i, \mathsf{C}_i \in \mathbb{R}^{m\times m}$ are themselves typically sparse. For example, finite-difference approximations of the type mentioned at the beginning of this section lead to block structures of the form

$$\mathsf{A} = \begin{array}{c} \\ m \\ m \\ \vdots \\ \\ \\ \end{array}\!\!\begin{array}{c} \begin{array}{ccccc} m & m & \cdots & & \end{array} \\ \left(\begin{array}{ccccc} \mathsf{T}_1 & \mathsf{D}_1 & & & \\ \mathsf{D}_2 & \mathsf{T}_2 & \mathsf{D}_2 & & \\ & & \ddots & & \\ & & \mathsf{D}_{n-1} & \mathsf{T}_{n-1} & \mathsf{D}_{n-1} \\ & & & \mathsf{D}_n & \mathsf{T}_n \end{array}\right) \end{array}.$$

Here, each $\mathsf{T}_i \in \mathbb{R}^{m\times m}$ is tridiagonal, while each $\mathsf{D}_i \in \mathbb{R}^{m\times m}$ is diagonal. This structure requires the storage of $\mathcal{O}(nm)$ nonzero entries.

What is disappointing is that the blocks E_i and F_i appearing in the block LU factorization (2.4-5) are typically full matrices in $\mathbb{R}^{m\times m}$, despite the sparseness of the original blocks. Thus the block LU factorization generally requires the storage of $\mathcal{O}(nm^2)$ nonzero entries. In cases where m is large, converting a block tridiagonal matrix to its block LU factorization can therefore entail tremendous increases in the memory requirements of a computer program. Chapter 4 discusses alternative methods that avoid this difficulty.

Mathematical Details

Among the business left unfinished is an analysis of the Thomas algorithm for tridiagonal matrices. At issue is whether the entries β_i and η_i in the LU factorization (2.4-4) obey bounds that limit the growth of roundoff errors in Algorithm 1.1. Since the algorithm uses the parameters β_i as divisors, we seek lower bounds on $|\beta_i|$; similarly, since the parameters η_i serve as multipliers, we would like some guarantee that the magnitudes $|\eta_i|$ remain small. The following theorem establishes the desired bounds.

THEOREM 2.11. *If the tridiagonal matrix* $\mathsf{A} \in \mathbb{R}^{n\times n}$ *in Equation (2.4-1) is strictly diagonally dominant, then the following inequalities hold:*

$$|\eta_i| \quad < \quad 1, \qquad\qquad i = 1, 2, \ldots, n-1,$$

$$|b_i| - |a_i| \quad \leq \quad |\beta_i|, \qquad\qquad i = 1, 2, \ldots, n.$$

In this case, strict diagonal dominance means that $|b_i| > |a_i| + |c_i|$ for each row $i = 1, 2, \ldots, n$, where we agree that $a_1 = c_n = 0$.

PROOF: We prove the first inequality by induction on i. Since $\eta_1 = c_1/b_1$, the fact that $|\eta_1| < 1$ follows directly from strict diagonal dominance. Assume that $|\eta_{i-1}| < 1$. Then, according to Algorithm 1.1,

$$|\eta_i| = \left|\frac{c_i}{\beta_i}\right| = \frac{|c_i|}{|b_i - a_i\eta_{i-1}|}.$$

By using the triangle inequality in the form $|b - a| \geq \big|\,|b| - |a|\,\big|$, strict diagonal dominance, and the inductive hypothesis, we deduce that

$$|\eta_i| \leq \frac{|c_i|}{|b_i| - |a_i|}.$$

But strict diagonal dominance also implies that $|b_i| - |a_i| > |c_i|$, so we conclude that $|\eta_i| < 1$, completing the induction. The second inequality follows for the case $i = 1$, since $\beta_1 = b_1$. For $i = 2, 3, \ldots, n$ we have

$$|\beta_i| = |b_i - a_i\eta_{i-1}| \geq |b_i| - |a_i\eta_{i-1}| \geq |b_i| - |a_i|. \qquad \blacksquare$$

We turn now to more general band matrices. In devising Algorithms 2.5 and 2.6, we exploit the fact that, for any nonsingular band matrix, the LU factorization remains just as compact as the original matrix, so long as row reduction without pivoting is possible.

THEOREM 2.12. *Let* $\mathsf{A} \in \mathbb{R}^{n \times n}$ *be a nonsingular band matrix with lower bandwidth* p *and upper bandwidth* q. *If* A *has an* LU *factorization* $\mathsf{A} = \mathsf{LU}$, *where* L *has unit diagonal entries, then* L *has lower bandwidth* p *and* U *has upper bandwidth* q.

PROOF: We argue by induction on n. When $n = 1$ there is nothing to prove, since $\mathsf{A} = a_{1,1}$. Assume that the theorem is true for band matrices in $\mathbb{R}^{n \times n}$, and let the nonsingular matrix $\mathsf{A} \in \mathbb{R}^{(n+1)\times(n+1)}$ have lower and upper bandwidths p and q, respectively. Let L and U be the upper and lower triangular factors of A, with L having unit diagonal entries. Consider the following block partitionings:

$$\mathsf{A} = \underbrace{\begin{matrix} & n & 1 \\ n & \mathsf{L}_n & \mathbf{0} \\ 1 & \boldsymbol{\lambda}^\top & 1 \end{matrix}}_{\mathsf{L}} \underbrace{\begin{pmatrix} \mathsf{U}_n & \mathbf{u} \\ \mathbf{0}^\top & u \end{pmatrix}}_{\mathsf{U}} = \begin{pmatrix} \mathsf{A}_n & \mathsf{L}_n\mathbf{u} \\ \boldsymbol{\lambda}^\top\mathsf{U}_n & \boldsymbol{\lambda}^\top\mathbf{u}+u \end{pmatrix},$$

where $\mathsf{A}_n := \mathsf{L}_n\mathsf{U}_n$. The matrices L and U have lower bandwidth p and upper bandwidth q, respectively, if and only if each of the following conditions holds:

(*i*) L_n has lower bandwidth p;

(*ii*) U_n has upper bandwidth q;

(*iii*) the first $n-p$ entries of $\boldsymbol{\lambda}$ vanish;

(*iv*) the first $n-q$ entries of $\mathbf{u}$ vanish.

The first two of these statements follow from the induction hypothesis: The matrix $\mathsf{A}_n \in \mathbb{R}^{n\times n}$ is the nth leading principal submatrix of A, so it also has lower and upper bandwidths p and q, respectively. Since A_n has LU factorization $\mathsf{L}_n\mathsf{U}_n$, L_n and U_n have the desired lower and upper bandwidths.

To establish the third fact, we note that the band structure of A forces the first $n-p$ entries of the row vector $\boldsymbol{\lambda}^\top\mathsf{U}_n$ to vanish. The jth entry of $\boldsymbol{\lambda}^\top\mathsf{U}_n$ is

$$\sum_{i=1}^{j} \lambda_i u_{i,j} = \lambda_j u_{j,j} + \sum_{i=1}^{j-1} \lambda_i u_{i,j}.$$

The fact that $0 \neq \det\mathsf{A} = u \det\mathsf{U}_n$ implies that neither u nor any of the diagonal entries $u_{i,i}$ of U_n vanishes. Since U_n is upper triangular with nonzero diagonal entries, the first entry of $\boldsymbol{\lambda}^\top\mathsf{U}_n$ is $\lambda_1 u_{1,1} = 0$, and therefore $\lambda_1 = 0$. The second entry is $\lambda_1 u_{1,2} + \lambda_2 u_{2,2} = 0$, which implies that $\lambda_2 = 0$ since $\lambda_1 = 0$. We continue reasoning in this way, until we come to the $(n-p)$th entry, which is

$$\lambda_1 u_{1,n-p} + \cdots + \lambda_{n-p-1} u_{n-p-1,n-p} + \lambda_{n-p} u_{n-p,n-p} = 0.$$

Knowing at this point that $\lambda_1 = \cdots = \lambda_{n-p-1} = 0$, we deduce that $\lambda_{n-p} = 0$. Therefore the first $n-p$ entries of $\boldsymbol{\lambda}$ vanish. A similar argument, using the observation that the first $n-q$ entries of $\mathsf{L}_n\mathbf{u}$ vanish, proves the fourth fact, concluding the proof. ∎

The next theorem analyzes the effects of partial pivoting on the sparseness of U.

THEOREM 2.13. *Let* $\mathsf{A} \in \mathbb{R}^{n\times n}$ *be a nonsingular matrix having lower and upper bandwidths* p *and* q*, respectively. Suppose that* $\mathsf{P} \in \mathbb{R}^{n\times n}$ *is a permutation matrix that allows the* LU *factorization* $\mathsf{PA} = \mathsf{LU}$. *Then* U *has upper bandwidth* $p + q$.

PROOF: The matrix PA has the form

$$\mathsf{PA} = \begin{bmatrix} a_{\pi(1),1} & \cdot & \cdot & \cdot & a_{\pi(1),n} \\ \vdots & & & & \\ a_{\pi(i),1} & \cdot & \cdot & \cdot & a_{\pi(i),n} \\ \vdots & & & & \\ a_{\pi(n),1} & \cdot & \cdot & \cdot & a_{\pi(n),n} \end{bmatrix}, \quad \longleftarrow \quad \text{row } i$$

where $(\pi(1), \pi(2), \ldots, \pi(n))$ is a permutation on $(1, 2, \ldots, n)$. By hypothesis, this matrix has an LU factorization that does not require partial pivoting. The band structure of A implies that $a_{i,j} = 0$ whenever $i+p < j$ or $i+q > j$. It follows that

$$a_{\pi(i),j} = 0 \quad \begin{cases} \text{for} \quad j = \pi(i) + p + 1, \ldots, n-1, n \quad \text{and} \\ \\ \text{for} \quad j = 1, 2, \ldots, \pi(i) + q - 1. \end{cases} \tag{2.4-7}$$

Since Theorem 2.12 guarantees that U has the same upper bandwidth as PA, we can finish the proof by showing that PA has upper bandwidth $p+q$, that is, that $a_{\pi(i),j} = 0$ whenever $i + q + p < j$. For this, it suffices to prove that $\pi(i) \le i + q$ for $i = 1, 2, \ldots, n$. The proof is by contradiction: If $\pi(i) > i + q$, then Equation (2.4-6) implies that

$$\left(a_{\pi(i),1}\; a_{\pi(i),2} \; \cdot \quad \cdot \quad \cdot \; a_{\pi(i),i}\right) = (0\;0 \; \cdot \quad \cdot \quad \cdot \; 0)\,.$$

But this i-tuple is row i of the ith leading principal submatrix A_i of A, and if its entries are all zero then A_i is singular. By Theorem 2.6, this conclusion contradicts the existence of the LU factorization $\mathsf{PA} = \mathsf{LU}$. Therefore $\pi(i) \le i + q$ for $i = 1, 2, \ldots, n$, and the proof is complete. ∎

As mentioned earlier, the lower triangular factor L can have nonzero entries anywhere below its diagonal, depending on the row interchanges performed.

Further Remarks

To derive a scheme for solving block tridiagonal systems, we exploit an analogy between the entries of tridiagonal matrices and the blocks in block tridiagonal matrices. It is therefore natural to ask whether one can find conditions, analogous to Theorem 2.11, that prevent the growth of roundoff errors in such block LU factorizations. A condition of this type exists, but its statement makes reference to matrix norms, which we discuss in the next section. Golub and Van Loan ([1], Section 5.5) give details.

2.5 Matrix Norms

Motivation and Construction

We now analyze the difference between the computed solution $\hat{\mathbf{x}}$ to $\mathsf{A}\mathbf{x} = \mathbf{b}$ and the exact solution $\mathbf{x}$. At issue is the size of the **error** $\boldsymbol{\varepsilon} := \mathbf{x} - \hat{\mathbf{x}}$, measured using some norm $\|\cdot\|$ as introduced in Section 0.3. Since $\mathbf{x}$ is unknown, we hope to estimate $\|\boldsymbol{\varepsilon}\|$ in terms of quantities that are computable from $\hat{\mathbf{x}}$, A, and $\mathbf{b}$. Section 2.6 pursues this idea. In preparation for that discussion, we devote this section to the establishment of matrix norms. These functions measure relationships between the sizes of vectors and the sizes of their images under matrix multiplication.

We begin with some preliminary remarks about the eigenvalues of a matrix $\mathsf{A} \in \mathbb{R}^{n \times n}$.

DEFINITION. *A number $\lambda \in \mathbb{C}$ is an* **eigenvalue** *of $\mathsf{A} \in \mathbb{R}^{n \times n}$ if there is a nonzero vector $\mathbf{x} \in \mathbb{C}^n$ for which $\mathsf{A}\mathbf{x} = \lambda\mathbf{x}$. Any such vector $\mathbf{x}$ is an* **eigenvector** *of A associated with λ. The collection σ_A of all eigenvalues of A is the* **spectrum** *of A, and the number*

$$\varrho(\mathsf{A}) = \max_{\lambda \in \sigma_A} |\lambda|$$

is the **spectral radius** *of A.*

Eigenvalues are the (possibly complex-valued) factors by which A stretches its eigenvectors. The identity $\mathsf{A}\mathbf{x} = \lambda\mathbf{x}$ with $\mathbf{x} \neq \mathbf{0}$ implies that the matrix $\lambda\mathsf{I} - \mathsf{A}$ is singular, and hence any eigenvalue λ of A is a zero of the **characteristic polynomial** $\det(\lambda\mathsf{I} - \mathsf{A})$, which has degree n in λ. Chapter 5 discusses numerical methods for determining eigenvalues and eigenvectors of matrices.

The following theorem summarizes important properties of eigenvalues and eigenvectors. For proofs, we refer to Strang [4].

THEOREM 2.14. *Let $\mathsf{A} \in \mathbb{R}^{n \times n}$. Then*

(i) *A is singular if and only if 0 is an eigenvalue of A.*

(ii) *If A is upper or lower triangular, then its eigenvalues are its diagonal entries.*

(iii) *If A is symmetric, then all of its eigenvalues are real numbers.*

(iv) *If A is symmetric and* **nonnegative** *— that is, $\mathbf{x}^\top \mathsf{A}\mathbf{x} \geq 0$ for every $\mathbf{x} \in \mathbb{R}^n$ — then all eigenvalues of A are nonnegative.*

(v) *If A is symmetric and positive definite, then all of its eigenvalues are positive.*

(vi) *If* A *is symmetric, then there exists an orthonormal basis for* $\mathbb{R}^n$*, each of whose elements is an eigenvector of* A.

The sixth assertion means that we can find a set $\{\boldsymbol{v}_1, \boldsymbol{v}_2, \ldots, \boldsymbol{v}_n\}$ of eigenvectors of A such that (1) each eigenvector $\boldsymbol{v}_i$ has unit Euclidean length $\|\boldsymbol{v}_i\|_2 = (\boldsymbol{v}_i^\top \boldsymbol{v}_i)^{1/2}$; (2) distinct eigenvectors $\boldsymbol{v}_i, \boldsymbol{v}_j$ in the set are orthogonal, that is, $\boldsymbol{v}_i^\top \boldsymbol{v}_j = 0$; and (3) any vector $\mathbf{y} \in \mathbb{R}^n$ has an expansion

$$\mathbf{y} = \sum_{i=1}^{n} c_i \boldsymbol{v}_i, \tag{2.5-1}$$

for some real coefficients $c_1, c_2, \ldots, c_n$. Straightforward calculation shows that the coefficients in such an expansion are $c_i = \boldsymbol{v}_i^\top \mathbf{y}$. Moreover,

$$\|\mathbf{y}\|_2^2 = \mathbf{y}^\top \mathbf{y} = c_1^2 + c_2^2 + \cdots + c_n^2. \tag{2.5-2}$$

The last identity generalizes the Pythagorean theorem.

In some cases, the spectrum of a matrix A yields scanty information about the relationship between the size of $\mathbf{x}$ and the size of $\mathsf{A}\mathbf{x}$. For example, consider the matrix

$$\mathsf{A}_1 = \begin{bmatrix} 0 & 2 \\ 0 & 0 \end{bmatrix}.$$

This matrix has characteristic polynomial $\det(\lambda \mathsf{I} - \mathsf{A}_1) = \lambda^2$, which has a double root $\lambda = 0$. Therefore the spectrum of A_1 is $\{0\}$, and $\varrho(\mathsf{A}_1) = 0$. However, for any vector $\mathbf{x} = (0, x_2)^\top \in \mathbb{R}^2$, the image vector $\mathsf{A}_1\mathbf{x} = (2x_2, 0)^\top$ has Euclidean length twice that of $\mathbf{x}$. In this case, eigenvalues reveal very little about how multiplication by the matrix changes the size of an arbitrary vector.

Recall from Chapter 0 that *norms* are the natural devices for measuring the sizes of vectors. For a mapping $\|\cdot\|: \mathbb{R}^n \to \mathbb{R}$ to be a norm, it must satisfy three conditions:

(i) For any vector $\mathbf{x} \in \mathbb{R}^n$, $\|\mathbf{x}\| \geq 0$, and $\|\mathbf{x}\| = 0$ if and only if $\mathbf{x} = \mathbf{0}$.

(ii) Whenever $\mathbf{x} \in \mathbb{R}^n$ and $c \in \mathbb{R}$, $\|c\,\mathbf{x}\| = |c|\|\mathbf{x}\|$.

(iii) Whenever $\mathbf{x}, \mathbf{y} \in \mathbb{R}^n$, $\|\mathbf{x} + \mathbf{y}\| \leq \|\mathbf{x}\| + \|\mathbf{y}\|$.

An extension of this concept allows us to gauge the size of $\mathsf{A}\mathbf{x} \in \mathbb{R}^n$ in terms of the size of $\mathbf{x} \in \mathbb{R}^n$, for any matrix $\mathsf{A} \in \mathbb{R}^{n \times n}$. The following definition captures the idea.

DEFINITION. *If* $\mathsf{A} \in \mathbb{R}^{n \times n}$ *and* $\|\cdot\|: \mathbb{R}^n \to \mathbb{R}$ *is a norm, then the* **subordinate matrix norm** $\|\cdot\|: \mathbb{R}^{n \times n} \to \mathbb{R}$ *is defined as follows:*

$$\|\mathsf{A}\| := \sup_{\mathbf{x} \neq \mathbf{0}} \frac{\|\mathsf{A}\mathbf{x}\|}{\|\mathbf{x}\|}. \tag{2.5-3}$$

As an immediate consequence of this definition,

$$\|\mathsf{A}\mathbf{x}\| \leq \|\mathsf{A}\| \, \|\mathbf{x}\|, \tag{2.5-4}$$

for any vector $\mathbf{x} \in \mathbb{R}^n$. It is an easy exercise to deduce that, for any matrices $\mathsf{A}, \mathsf{B} \in \mathbb{R}^{n \times n}$,

$$\|\mathsf{AB}\| \leq \|\mathsf{A}\| \, \|\mathsf{B}\|. \tag{2.5-5}$$

Also, the following three formulas for $\|\mathsf{A}\|$ are equivalent to Equation (2.5-3):

$$\begin{aligned} \|\mathsf{A}\| &= \sup_{\|\mathbf{x}\|=1} \|\mathsf{A}\mathbf{x}\|, \\ \|\mathsf{A}\| &= \inf\left\{M \geq 0 : \|\mathsf{A}\mathbf{x}\| \leq M\|\mathbf{x}\| \text{ for all } \mathbf{x} \in \mathbb{R}^n\right\}, \\ \|\mathsf{A}\| &= \inf\left\{M \geq 0 : \|\mathsf{A}\mathbf{x}\| \leq M \text{ for all } \mathbf{x} \in \mathbb{R}^n \text{ with } \|\mathbf{x}\| = 1\right\}. \end{aligned} \tag{2.5-6}$$

Therefore, if $\|\mathsf{A}\mathbf{x}\| \leq M\|\mathbf{x}\|$ for all $\mathbf{x}$, then $\|\mathsf{A}\| \leq M$. On the other hand, if $\|\mathsf{A}\mathbf{x}\| \geq M\|\mathbf{x}\|$ for some $\mathbf{x} \neq \mathbf{0}$, then $\|\mathsf{A}\| \geq M$.

Problem 10 asks for verification that subordinate matrix norms satisfy the three conditions required to be a norm on the vector space $\mathbb{R}^{n \times n}$. (However, not every norm on $\mathbb{R}^{n \times n}$ is subordinate to a vector norm. Problem 14 examines this fact.) In particular, any subordinate matrix norm obeys the triangle inequality. Here lies a crucial defect in the spectral radius as a measure of size: If $n > 1$, it is possible to find matrices $\mathsf{A}, \mathsf{B} \in \mathbb{R}^{n \times n}$ for which $\varrho(\mathsf{A} + \mathsf{B}) > \varrho(\mathsf{A}) + \varrho(\mathsf{B})$, and consequently the triangle inequality fails. Problem 9 asks for details.

While matrix norms typically give better characterizations than the spectrum of the "stretching" power of a matrix, one can derive a simple lower estimate for $\|\mathsf{A}\|$ if one knows an eigenvalue λ of A. Since $\mathsf{A}\mathbf{x} = \lambda\mathbf{x}$, we have $\|\mathsf{A}\mathbf{x}\| = |\lambda| \, \|\mathsf{x}\|$. From the inequality (2.5-3) it follows that $\|\mathsf{A}\| \geq |\lambda|$ and hence that

$$\|\mathsf{A}\| \geq \varrho(\mathsf{A}). \tag{2.5-7}$$

Each of the vector norms $\|\cdot\|_1$, $\|\cdot\|_2$, and $\|\cdot\|_\infty$ gives rise to a useful subordinate matrix norm. Later in this section we prove the following characterizations:

$$\|\mathsf{A}\|_\infty = \max_{1 \leq i \leq n} \sum_{j=1}^{n} |a_{i,j}|, \text{ the "maximum row sum" of } \mathsf{A}.$$

$$\|\mathsf{A}\|_1 = \max_{1 \leq j \leq n} \sum_{i=1}^{n} |a_{i,j}|, \text{ the "maximum column sum" of } \mathsf{A}.$$

$$\|\mathsf{A}\|_2 = \sqrt{\varrho(\mathsf{A}^\top \mathsf{A})}.$$

When A is symmetric, one can calculate $\|\mathsf{A}\|_2$ more simply. Symmetry implies that $\mathsf{A}^\top\mathsf{A} = \mathsf{A}^2$. But the eigenvalues of A^2 are simply the squares of the eigenvalues of A (see Problem 15). Therefore, when A is symmetric,

$$\|\mathsf{A}\|_2 = \sqrt{\varrho(\mathsf{A}^2)} = \varrho(\mathsf{A}).$$

Some simple examples illustrate these norms. Consider once more the matrix

$$\mathsf{A}_1 = \begin{bmatrix} 0 & 2 \\ 0 & 0 \end{bmatrix}.$$

Geometric reasoning suggests that the largest value of $\|\mathsf{A}_1\mathbf{x}\|_2$, where $\mathbf{x}$ ranges over the vectors having unit Euclidean length, occurs when $\mathbf{x} = \mathbf{e}_2 = (0,1)^\top$. But $\|\mathsf{A}_1\mathbf{e}_2\|_2 = \|(2,0)^\top\|_2 = 2\|\mathbf{e}_2\|_2$, and therefore simple geometry suggests that $\|\mathsf{A}_1\|_2 = 2$. The eigenvalues of $\mathsf{A}_1^\top\mathsf{A}_1$ are 0 and 4, so indeed $\|\mathsf{A}_1\|_2 = \sqrt{4} = 2$. Checking column and row sums, we find that $\|\mathsf{A}_1\|_1 = \|\mathsf{A}_1\|_\infty = 2$. However, both eigenvalues of A_1 are 0, so $\|\mathsf{A}_1\|_2 \neq \varrho(\mathsf{A})$.

Next consider

$$\mathsf{A}_2 = \begin{bmatrix} 0 & 1 \\ -1 & 0 \end{bmatrix}.$$

In this case, $\mathsf{A}_2^\top\mathsf{A}_2$ is the identity matrix I, both of whose eigenvalues are 1. Therefore

$$\|\mathsf{A}_2\|_2 = \sqrt{\varrho(\mathsf{A}_2^\top\mathsf{A}_2)} = 1 = \|\mathsf{A}_2\|_1 = \|\mathsf{A}_2\|_\infty.$$

In the geometric view, multiplying $\mathbf{x}$ on the left by A_2 rotates $\mathbf{x}$ about the origin by $-\pi/2$ radians without changing its Euclidean length, as shown in Figure 1.

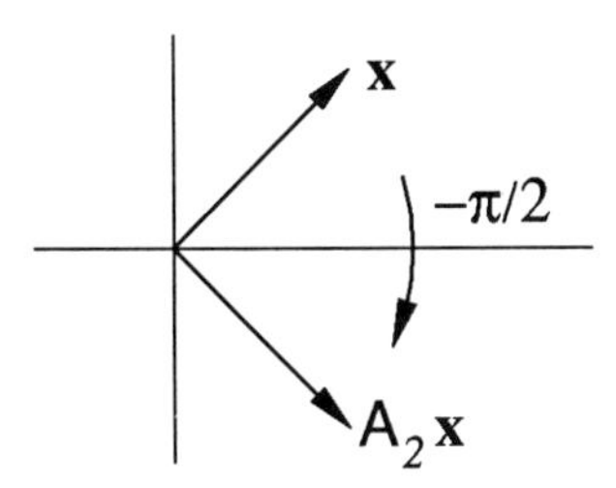

FIGURE 1. *Geometric action of the matrix* A_2.

Finally, consider the 3×3 matrix

$$\mathsf{A}_3 = \begin{bmatrix} 1 & -1 & 0 \\ -1 & 2 & -1 \\ 0 & -1 & 1 \end{bmatrix}.$$

In this case A_3 is symmetric, so $||A_3||_2$ is just the largest value of $|\lambda|$, where λ ranges over the eigenvalues of A_3. Solving the cubic equation $\det(\lambda I - A_3) = 0$, we find that the eigenvalues of A_3 are 0, 1, and 3. Therefore, $||A_3||_2 = 3$. However, in this case $||A_3||_1 = ||A_3||_\infty = 4$.

Mathematical Details

We now prove the characterizations of the matrix norms $||\cdot||_1$, $||\cdot||_2$, and $||\cdot||_\infty$ stated earlier.

THEOREM 2.15. *Let* $A \in \mathbb{R}^{n\times n}$. *Then*

$$(i) \quad ||A||_\infty = \max_{1\le i\le n} \sum_{j=1}^{n} |a_{i,j}|.$$

$$(ii) \quad ||A||_1 = \max_{1\le j\le n} \sum_{i=1}^{n} |a_{i,j}|.$$

$$(iii) \quad ||A||_2 = \sqrt{\varrho(A^\top A)}.$$

PROOF: To prove (i), let $\mathbf{x} = (x_1, x_2, \ldots, x_n) \in \mathbb{R}^n$, and call

$$N := \max_{1\le i\le n} \sum_{j=1}^{n} |a_{i,j}|.$$

Using the definition of the vector norm $||\cdot||_\infty$ and the triangle inequality, we find that

$$\begin{aligned} ||A\mathbf{x}||_\infty &= \max_{1\le i\le n} \left| \sum_{j=1}^{n} a_{i,j} x_j \right| \le \max_{1\le i\le n} \sum_{j=1}^{n} |a_{i,j}| \max_{1\le j\le n} |x_j| \\ &= N||\mathbf{x}||_\infty. \end{aligned}$$

Therefore $||A||_\infty \le N$. It now suffices to show that $||A|| \ge N$, which we do by showing that $||A\mathbf{x}||_\infty$ actually attains the value N for some unit vector $\mathbf{x}$. If $A = 0$, the result is clear, so assume that $A \neq 0$. Choose i so that $\sum_{j=1}^{n} |a_{i,j}| = N$, and define $\mathbf{x}$ by

$$x_j := \begin{cases} a_{i,j}/|a_{i,j}|, & \text{if } a_{i,j} \neq 0, \\ 0, & \text{if } a_{i,j} = 0. \end{cases}$$

It is now straightforward to check that $||\mathbf{x}||_\infty = 1$ and $||A\mathbf{x}||_\infty = N$.

The characterization (*ii*) is an exercise, the argument being similar in spirit to the one just given.

To prove (*iii*), choose $\mathbf{y} \in \mathbb{R}^n$ such that $\|\mathbf{y}\|_2 = 1$ and $\|\mathsf{A}\mathbf{y}\|_2 = \|\mathsf{A}\|_2$. (This is possible because $\|\cdot\|_2$ is a continuous function on the compact set S_2 defined in Equation (0.3-2); hence $\|\cdot\|_2$ attains a maximum value at some point $\mathbf{x} \in S_2$.) Since $(\mathsf{A}\mathbf{y})^\top = \mathbf{y}^\top\mathsf{A}^\top$, we have

$$\|\mathsf{A}\|_2^2 = \|\mathsf{A}\mathbf{y}\|_2^2 = (\mathsf{A}\mathbf{y})^\top(\mathsf{A}\mathbf{y}) = \mathbf{y}^\top\mathsf{A}^\top\mathsf{A}\mathbf{y}. \tag{2.5-8}$$

But $\mathsf{A}^\top\mathsf{A} \in \mathbb{R}^{n\times n}$ is symmetric, so Theorem 2.14 guarantees that there exists an orthonormal basis $\{\boldsymbol{v}_1, \boldsymbol{v}_2, \ldots, \boldsymbol{v}_n\}$ for $\mathbb{R}^n$ consisting entirely of eigenvectors of $\mathsf{A}^\top\mathsf{A}$. Moreover, all of the corresponding eigenvalues are nonnegative, since $\mathsf{A}^\top\mathsf{A}\mathbf{x} = \lambda\mathbf{x}$ implies that

$$0 \le \|\mathsf{A}\mathbf{x}\|_2^2 = (\mathsf{A}\mathbf{x})^\top(\mathsf{A}\mathbf{x}) = \mathbf{x}^\top\mathsf{A}^\top\mathsf{A}\mathbf{x} = \mathbf{x}^\top\lambda\mathbf{x} = \lambda\|\mathbf{x}\|_2^2.$$

Denote by λ_i the eigenvalue of $\mathsf{A}^\top\mathsf{A}$ associated with $\boldsymbol{v}_i$. If we substitute an expansion of the form (2.5-1) into Equation (2.5-8), then we obtain

$$\begin{aligned}
\|\mathsf{A}\|_2^2 &= \sum_{i=1}^{n} c_i\boldsymbol{v}_i^\top\mathsf{A}^\top\mathsf{A}\sum_{j=1}^{n} c_j\boldsymbol{v}_j \\
&= \sum_{i=1}^{n} c_i\boldsymbol{v}_i^\top\sum_{j=1}^{n} c_j\lambda_j\boldsymbol{v}_j \\
&= \sum_{j=1}^{n} \lambda_j c_j^2 \le \varrho(\mathsf{A}^\top\mathsf{A})\sum_{j=1}^{n} c_j^2 = \varrho(\mathsf{A}^\top\mathsf{A}),
\end{aligned}$$

the last step following from the fact that $\|\mathbf{y}\|_2^2 = 1$. Hence $\|\mathsf{A}\|_2^2 \le \varrho(\mathsf{A}^\top\mathsf{A})$.

To finish the proof, we show that $\|\mathsf{A}\|_2^2 \ge \varrho(\mathsf{A}^\top\mathsf{A})$. Suppose that $\boldsymbol{v}_k$ is an eigenvector of $\mathsf{A}^\top\mathsf{A}$, chosen from the orthonormal basis, and that its associated eigenvalue $\lambda_k = \varrho(\mathsf{A}^\top\mathsf{A})$. The inequality (2.5-4) and the fact that $\boldsymbol{v}_k$ has unit length imply that

$$\begin{aligned}
\|\mathsf{A}\|_2^2 = \|\mathsf{A}\|_2^2\,\|\boldsymbol{v}_k\|_2^2 \;\ge\; \|\mathsf{A}\boldsymbol{v}_k\|_2^2 &= \boldsymbol{v}_k^\top\mathsf{A}^\top\mathsf{A}\boldsymbol{v}_k \\
= \lambda_k\,\boldsymbol{v}_k^\top\boldsymbol{v}_k &= \varrho(\mathsf{A}^\top\mathsf{A}),
\end{aligned}$$

as claimed. ∎

Further Remarks

Our focus on linear systems having the same number of equations as unknowns may obscure the fact that one can define norms for the more general vector spaces $\mathbb{R}^{m\times n}$. Let $\mathsf{A} \in \mathbb{R}^{m\times n}$, and suppose that $\|\cdot\|_{\mathrm{I}}$ is a norm on $\mathbb{R}^m$

and $||\cdot||_{\mathrm{II}}$ is a norm on $\mathbb{R}^n$. Since the mapping $\mathbf{x} \mapsto \mathsf{A}\mathbf{x}$ sends vectors $\mathbf{x} \in \mathbb{R}^n$ to images $\mathsf{A}\mathbf{x} \in \mathbb{R}^m$, the natural extension of our definition of subordinate matrix norms is the following:

$$||\mathsf{A}||_{\mathrm{I,II}} := \sup_{\mathbf{x}\neq\mathbf{0}} \frac{||\mathsf{A}\mathbf{x}||_{\mathrm{I}}}{||\mathbf{x}||_{\mathrm{II}}}.$$

Much of the theory developed in this section translates in a straightforward manner to this more general setting. In particular, we have

$$||\mathsf{A}\mathbf{x}||_{\mathrm{II}} \leq ||\mathsf{A}||_{\mathrm{I,II}}\, ||\mathbf{x}||_{\mathrm{I}}.$$

Also, if $\mathsf{B} \in \mathbb{R}^{p\times m}$ and $||\cdot||_{\mathrm{III}}$ is a norm on $\mathbb{R}^p$, then $\mathsf{BA} \in \mathbb{R}^{p\times n}$, and

$$||\mathsf{BA}||_{\mathrm{III,II}} \leq ||\mathsf{B}||_{\mathrm{III,I}}\, ||\mathsf{A}||_{\mathrm{I,II}}.$$

The proofs are exercises.

2.6 Errors and Iterative Improvement

Think of direct methods for $\mathsf{A}\mathbf{x} = \mathbf{b}$ as "black-box solvers": Once encoded, the algorithms take as inputs the data defining $\mathsf{A} \in \mathbb{R}^{n\times n}$ and $\mathbf{b} \in \mathbb{R}^n$ and produce certain ouputs, including a numerical approximation $\hat{\mathbf{x}}$ to the exact solution $\mathbf{x}$. With exact arithmetic, we could safely assume that $\hat{\mathbf{x}} = \mathbf{x}$. However, any direct method implemented on a finite-precision machine suffers from accumulated errors, beginning with the initial machine representations of the input data A and $\mathbf{b}$. Furthermore, in applications A and $\mathbf{b}$ may have measurement errors. How these problems affect the output $\hat{\mathbf{x}}$ depends, to a great extent, upon properties of the matrix A. This section examines ways to estimate the errors $\mathbf{x} - \hat{\mathbf{x}}$ for a given matrix A and to calculate improved approximations to $\mathbf{x}$ when the computed output $\hat{\mathbf{x}}$ is suspect.

Error Estimates for Linear Systems

The first task is a bit delicate, since we usually have no way of knowing the actual error $\boldsymbol{\varepsilon} := \mathbf{x} - \hat{\mathbf{x}} = \mathsf{A}^{-1}\mathbf{b} - \hat{\mathbf{x}}$. It *is* possible, at least within the constraints imposed by machine arithmetic and measurement errors, to compute the **residual** $\mathbf{r} = \mathbf{b} - \mathsf{A}\hat{\mathbf{x}}$, which vanishes if and only if $\boldsymbol{\varepsilon} = \mathbf{0}$. Does a small residual guarantee that the error is small?

The system

$$\mathsf{A}\mathbf{x} = \begin{bmatrix} 1 & 0 & 0 \\ 0 & 2 & 0 \\ 0 & 0 & 10^{-9} \end{bmatrix} \begin{bmatrix} x_1 \\ x_2 \\ x_3 \end{bmatrix} = \begin{bmatrix} 2 \\ 4 \\ 10^{-9} \end{bmatrix} = \mathbf{b} \qquad (2.6\text{-}1)$$

furnishes a counterexample. The exact solution is $\mathbf{x} = (2, 2, 1)^\top$. For argument's sake, consider the erroneous solution $\hat{\mathbf{x}} = (2, 2, -1)^\top$. In this case the error is $\varepsilon = (0,0,2)^{\mathsf{T}}$, with norm $\|\varepsilon\|_\infty = 2 = \|\mathbf{x}\|_\infty$. Surely an error having the same magnitude as $\mathbf{x}$ is large! However, the residual in this example is $\mathbf{r} = (0,0,2\times 10^{-9})^{\mathsf{T}}$, whose norm $\|\mathbf{r}\|_\infty = 2\times 10^{-9}$ seems small, especially in comparison with $\|\mathbf{b}\|_\infty = 4$. In larger systems the symptoms of pathology may be much more subtle.

To analyze the relationship between ε and $\mathbf{r}$ more generally, select a norm $\|\cdot\|$ in which to measure $\mathbf{r}$ and ε. Observe that $\mathbf{r} = \mathsf{A}\mathbf{x} - \mathsf{A}\hat{\mathbf{x}} = \mathsf{A}\varepsilon$, that is, $\varepsilon = \mathsf{A}^{-1}\mathbf{r}$. The inequality (2.5-4) therefore yields

$$\frac{\|\varepsilon\|}{\|\mathsf{A}\mathbf{x}\|} = \frac{\|\mathsf{A}^{-1}\mathbf{r}\|}{\|\mathbf{b}\|} \le \frac{\|\mathsf{A}^{-1}\|\,\|\mathbf{r}\|}{\|\mathbf{b}\|}.$$

But $\|\mathsf{A}\mathbf{x}\| \le \|\mathsf{A}\|\,\|\mathbf{x}\|$, so we can replace the denominator on the left and multiply through by $\|\mathsf{A}\|$ to obtain the estimate

$$R_x := \frac{\|\varepsilon\|}{\|\mathbf{x}\|} \le \|\mathsf{A}\|\,\|\mathsf{A}^{-1}\|\frac{\|\mathbf{r}\|}{\|\mathbf{b}\|}. \tag{2.6-2}$$

This inequality bears interpretation. The quantity $R_x := \|\varepsilon\|/\|\mathbf{x}\|$ appearing on the left is the norm of the error ε, scaled by the "size" of the exact solution $\mathbf{x}$. We call this ratio the **relative error**. The ratio $\|\mathbf{r}\|/\|\mathbf{b}\|$ also represents a relative quantity, namely the magnitude of the residual relative to that of the right-side vector $\mathbf{b}$. The inequality (2.6-2) asserts that the relative error is small when the relative magnitude of the residual is small, *provided that* $\|\mathsf{A}\|\,\|\mathsf{A}^{-1}\|$ *is not too large.*

DEFINITION. *If* $\mathsf{A} \in \mathbb{R}^{n\times n}$ *is nonsingular, then its* **condition number** *with respect to the norm* $\|\cdot\|$ *is* $\operatorname{cond}(\mathsf{A}) = \|\mathsf{A}\|\,\|\mathsf{A}^{-1}\|$.

When $\operatorname{cond}(\mathsf{A})$ is large, the inequality (2.6-2) allows large errors despite small residuals, and we say that the matrix A is **poorly conditioned**.

The actual value of $\operatorname{cond}(\mathsf{A})$ may depend upon the norm used to compute it. The results of Problem 13 suggest, though, that when $\operatorname{cond}(\mathsf{A})$ is large with respect to one norm, it is typically large with respect to others. When we need to specify which norm we are using, we do so with subscripts. Thus $\operatorname{cond}_\infty(\mathsf{A}) := \|\mathsf{A}\|_\infty\|\mathsf{A}^{-1}\|_\infty$. In the example (2.6-1), $\operatorname{cond}_\infty(\mathsf{A}) = 2\times 10^9$.

In no case can $\operatorname{cond}(\mathsf{A})$ be smaller than 1. To see this, let $\mathbf{x} \in \mathbb{R}^n$ be any nonzero vector, and observe that

$$\|\mathsf{A}\|\,\|\mathsf{A}^{-1}\|\,\|\mathbf{x}\| \ge \|\mathsf{A}\,\mathsf{A}^{-1}\|\,\|\mathbf{x}\| \ge \|\mathsf{A}\mathsf{A}^{-1}\mathbf{x}\| = \|\mathbf{x}\|.$$

Dividing through by $\|\mathbf{x}\|$ shows that $\operatorname{cond}(\mathsf{A}) \ge 1$. The identity matrix I actually attains this "ideal" value: $\operatorname{cond}(\mathsf{I}) = 1$, with respect to any subordinate matrix norm (see Problem 11).

Another approach to estimating the error in direct solution methods is to ask how the inevitable errors in input data compare with the resulting errors in the computed solution $\hat{\mathbf{x}}$. For example, when do small perturbations in $\mathbf{b}$ lead to small errors $\boldsymbol{\varepsilon}$? To answer this question, assume for a moment that the perturbation in $\mathbf{b}$, which we denote as $\boldsymbol{\delta}$, is the only source of error in $\hat{\mathbf{x}}$.

Under this assumption, we have $\mathsf{A}\mathbf{x} = \mathbf{b}$, by definition of $\mathbf{x}$, and $\mathsf{A}\hat{\mathbf{x}} = \mathbf{b}+\boldsymbol{\delta}$. Therefore $\mathsf{A}(\mathbf{x}-\hat{\mathbf{x}}) = \mathbf{b}-(\mathbf{b}+\boldsymbol{\delta})$, which implies that $\boldsymbol{\varepsilon} = -\mathsf{A}^{-1}\boldsymbol{\delta}$. Taking norms yields $\|\boldsymbol{\varepsilon}\| \leq \|\mathsf{A}^{-1}\|\,\|\boldsymbol{\delta}\|$, and now we use the fact that $\|\mathsf{A}\|\,\|\mathbf{x}\| \geq \|\mathbf{b}\|$ to deduce that

$$\frac{\|\boldsymbol{\varepsilon}\|}{\|\mathsf{A}\|\,\|\mathbf{x}\|} \leq \|\mathsf{A}^{-1}\|\frac{\|\boldsymbol{\delta}\|}{\|\mathbf{b}\|}.$$

Consequently,

$$\underbrace{\|\boldsymbol{\varepsilon}\|/\|\mathbf{x}\|}_{R_x} \leq \text{cond}\,(\mathsf{A})\ \underbrace{\|\boldsymbol{\delta}\|/\|\mathbf{b}\|}_{R_b}. \tag{2.6-3}$$

When $\text{cond}\,(\mathsf{A})$ is large, small relative errors $R_b := \|\boldsymbol{\delta}\|/\|\mathbf{b}\|$ in the right-side vector can lead to large relative errors R_x in the computed solution $\hat{\mathbf{x}}$. More precisely, if $\text{cond}\,(\mathsf{A}) \simeq 10^s$, then we expect to lose s significant digits in computing an approximation to $\mathbf{x}$. In this view, $\text{cond}\,(\mathsf{A})$ serves as an indicator of the sensitivity of a linear system to errors in the input data. On a machine whose single-precision representations of real numbers are accurate to six decimal digits, for example, condition numbers larger than 10^4 can pose serious practical difficulties.

To complete this picture, we should also ask when small perturbations in the matrix A lead to small errors $\boldsymbol{\varepsilon}$. If $\mathsf{D} \in \mathbb{R}^{n\times n}$ denotes the perturbation in A, then the relative error in A is

$$R_A := \|\mathsf{D}\|/\|\mathsf{A}\|.$$

The following theorem incorporates perturbations in both A and $\mathbf{b}$.

THEOREM 2.16. *Suppose that* $\mathsf{A} \in \mathbb{R}^{n\times n}$ *is nonsingular,* $\mathbf{b} \in \mathbb{R}^n$ *is nonzero, and* $\mathbf{x} \in \mathbb{R}^n$ *satisfies the linear system* $\mathsf{A}\mathbf{x} = \mathbf{b}$. *Let* $\mathsf{D} \in \mathbb{R}^{n\times n}$, $\boldsymbol{\delta} \in \mathbb{R}^n$, *and* $\boldsymbol{\varepsilon} \in \mathbb{R}^n$ *satisfy the perturbed equation*

$$(\mathsf{A}+\mathsf{D})(\mathbf{x}+\boldsymbol{\varepsilon}) = \mathbf{b}+\boldsymbol{\delta}.$$

If $R_A\,\text{cond}\,(\mathsf{A}) < 1$, *then*

$$R_x \leq \frac{\text{cond}\,(\mathsf{A})}{1 - R_A\,\text{cond}\,(\mathsf{A})}(R_b + R_A). \tag{2.6-4}$$

(Inequality (2.6-3) is the special case in which $R_A = 0$.)

PROOF: The assumptions imply that $\mathsf{A}\boldsymbol{\varepsilon} = \boldsymbol{\delta} - \mathsf{D}\mathbf{x} - \mathsf{D}\boldsymbol{\varepsilon}$, which we multiply by A^{-1} to obtain $\boldsymbol{\varepsilon} = \mathsf{A}^{-1}\boldsymbol{\delta} - \mathsf{A}^{-1}\mathsf{D}\mathbf{x} - \mathsf{A}^{-1}\mathsf{D}\boldsymbol{\varepsilon}$. Taking norms yields

$$\|\boldsymbol{\varepsilon}\| \leq \|\mathsf{A}^{-1}\|\,\|\boldsymbol{\delta}\| + \|\mathsf{A}^{-1}\|\,\|\mathsf{D}\|\,\|\mathbf{x}\| + \|\mathsf{A}^{-1}\|\,\|\mathsf{D}\|\,\|\boldsymbol{\varepsilon}\|,$$

or

$$(1 - \|\mathsf{A}^{-1}\| \, \|\mathsf{D}\|) \, \|\varepsilon\| \leq \|\mathsf{A}^{-1}\| \, \|\delta\| + \|\mathsf{A}^{-1}\| \, \|\mathsf{D}\| \, \|\mathbf{x}\|.$$

Now divide by $\|\mathbf{x}\|$ and note that $\|\mathsf{A}^{-1}\| \, \|\mathsf{D}\| = R_A \operatorname{cond}(\mathsf{A})$ to get

$$[1 - R_A \operatorname{cond}(\mathsf{A})] \, R_x \leq \frac{\|\mathsf{A}^{-1}\| \, \|\delta\|}{\|\mathbf{x}\|} + R_A \operatorname{cond}(\mathsf{A}). \tag{2.6-5}$$

But $\mathsf{A}\mathbf{x} = \mathbf{b}$, so $1/\|\mathbf{x}\| \leq \|\mathsf{A}\|/\|\mathbf{b}\|$. Substituting this inequality into (2.6-5) yields the estimate (2.6-4). ∎

It is possible in some cases to draw a connection between $\operatorname{cond}(\mathsf{A})$ and the spectrum of A. When A is symmetric and positive definite, $\|\mathsf{A}\|_2 = \lambda_{\text{max}}$, the largest eigenvalue of A, and $\|\mathsf{A}^{-1}\|_2 = 1/\lambda_{\text{min}}$, where λ_{min} is the smallest eigenvalue of A (see Problem 9). Therefore, for symmetric, positive definite matrices, $\operatorname{cond}_2(\mathsf{A}) = \lambda_{\text{max}}/\lambda_{\text{min}}$.

A classic example illustrates poor conditioning in symmetric, positive definite matrices. The **Hilbert matrices** $\mathsf{H}_n \in \mathbb{R}^{n \times n}$ are defined by $h_{i,j} = 1/(i+j-1)$. Such matrices arise in the context of certain least-squares problems, as discussed in Section 1.7: If we choose the polynomials $1, x, x^2, \ldots, x^{n-1}$ as basis functions and use the inner product $\langle f, g \rangle = \int_0^1 f(x)g(x)\,dx$, then the Gram matrix is

$$\begin{bmatrix} \langle 1,1 \rangle & \cdots & \langle x^{n-1},1 \rangle \\ \vdots & & \vdots \\ \langle 1,x^{n-1} \rangle & \cdots & \langle x^{n-1},x^{n-1} \rangle \end{bmatrix} = \begin{bmatrix} 1 & \cdots & 1/n \\ \vdots & & \vdots \\ 1/n & \cdots & 1/(2n-1) \end{bmatrix} = \mathsf{H}_n.$$

In contrast to the matrix in Equation (2.6-1), H_n shows few overt signs of poor conditioning. However, $\operatorname{cond}(\mathsf{H}_n)$ grows rapidly with n. For example, $\operatorname{cond}_2(\mathsf{H}_3) \simeq 5 \times 10^2$, while $\operatorname{cond}_2(\mathsf{H}_8) \simeq 1.5 \times 10^{10}$ (Ortega [3], p. 35).

The characterization of $\operatorname{cond}_2(\mathsf{A})$ in terms of eigenvalues has intuitive appeal, but it does not apply universally. A famous example, which we owe to Wilkinson ([5], p. 195), thwarts any misconception that $\lambda_{\text{max}}/\lambda_{\text{min}}$ is a reliable indicator of poor conditioning. Consider the linear system

$$\underbrace{\begin{bmatrix} 0.501 & -1 & & & \\ & 0.502 & -1 & & \\ & & \ddots & \ddots & \\ & & & 0.599 & -1 \\ & & & & 0.600 \end{bmatrix}}_{\mathsf{A}} \underbrace{\begin{bmatrix} x_1 \\ x_2 \\ \vdots \\ x_{99} \\ x_{100} \end{bmatrix}}_{\mathbf{x}} = \underbrace{\begin{bmatrix} 1 \\ 0 \\ \vdots \\ 0 \\ 0 \end{bmatrix}}_{\mathbf{b}} + \underbrace{\begin{bmatrix} 0 \\ 0 \\ \vdots \\ 0 \\ \delta \end{bmatrix}}_{\delta}. \tag{2.6-6}$$

When $\delta = 0$, this system has a solution $\mathbf{x}$ in which $x_1 = 1/0.501 \simeq 2$. For $\delta \neq 0$, backward substitution shows that

$$x_1 > \frac{\delta}{0.600^{100}} > 10^{22}\delta.$$

In this case, an extremely small perturbation δ can destroy the accuracy of the computed solution. Notice that $\lambda_{\max}/\lambda_{\min} = 0.600/0.501 \simeq 1.2$, giving no hint of poor conditioning. Problem 12 asks for a proof that $\text{cond}_\infty(\mathsf{A}) > 10^{21}$ for this example.

Iterative Improvement of Computed Solutions

Short of computing A^{-1} — a costly chore — we have so far given no general method for computing or even estimating $\text{cond}(\mathsf{A})$. Furthermore, even if we knew that $\text{cond}(\mathsf{A})$ were large, it is not yet clear what we could do about the situation. We close this section by discussing an approach for computing an improved approximation to $\mathbf{x} = \mathsf{A}^{-1}\mathbf{b}$, given an erroneous approximation $\hat{\mathbf{x}}$. The scheme automatically produces an estimate of $\text{cond}(\mathsf{A})$.

Once we know how to generate an improved approximation from a given approximation $\hat{\mathbf{x}}$, we can repeat the procedure as part of an *iterative* method. The idea is as follows: Given an approximate solution $\hat{\mathbf{x}}^{(0)}$ to $\mathsf{A}\mathbf{x} = \mathbf{b}$, with $\mathbf{r}^{(0)} = \mathbf{b} - \mathsf{A}\mathbf{x}^{(0)} \neq \mathbf{0}$, use computable information to generate a sequence

$$\left\{\hat{\mathbf{x}}^{(1)}, \hat{\mathbf{x}}^{(2)}, \ldots\right\}$$

of iterates that give successively better approximations to $\mathbf{x}$. Associated with each iterate is a residual, $\mathbf{r}^{(k)} = \mathbf{b} - \mathsf{A}\hat{\mathbf{x}}^{(k)}$, and we expect that $\|\mathbf{r}^{(k)}\| \to \mathbf{0}$ as $k \to \infty$. Therefore, we can keep producing iterates until we decide that $\|\mathbf{r}^{(k)}\|$ is small enough to make $\hat{\mathbf{x}}^{(k)}$ an acceptable approximation to the exact solution $\mathbf{x}$.

The estimate κ of $\text{cond}(\mathsf{A})$, whose calculation we discuss below, affords a quantifiable way to make this decision. Equation (2.6-2) suggests that the relative error associated with any iterate $\hat{\mathbf{x}}^{(k)}$ is

$$\frac{\|\boldsymbol{\varepsilon}^{(k)}\|}{\|\mathbf{x}\|} \simeq \kappa \frac{\|\mathbf{r}^{(k)}\|}{\|\mathbf{b}\|}, \tag{2.6-7}$$

where $\boldsymbol{\varepsilon}^{(k)} = \mathbf{x} - \hat{\mathbf{x}}^{(k)}$. We can therefore stop iterating when the right side of this estimate is smaller than some prescribed tolerance.

It remains to specify how to generate the iterates and to estimate $\text{cond}(\mathsf{A})$. The method relies on a simple heuristic. If we had an LU factorization of A and access to exact arithmetic, then we could compute the residual $\mathbf{r} = \mathbf{b} - \mathsf{A}\hat{\mathbf{x}}$ exactly. Then we could solve

$$\mathsf{A}\boldsymbol{\varepsilon} = \mathbf{r} \tag{2.6-8}$$

very cheaply to find the exact error $\boldsymbol{\varepsilon}$. Thus $\mathbf{x} = \hat{\mathbf{x}} + \boldsymbol{\varepsilon}$ would be the exact solution. This scenario is clearly fictional, since $\mathbf{r}$ typically contains errors that contaminate the correction computed from Equation (2.6-8).

Still, the heuristic is salvageable: We can compute the residual using high-precision arithmetic — double-precision, if the main algorithm employs single-precision arithmetic — and then use Equation (2.6-8) in an iterative sense. The following algorithm results.

ALGORITHM 2.7. *Given an* LU *factorization* $\mathsf{LU} = \mathsf{A} \in \mathbb{R}^{n\times n}$ *and an approximate solution* $\hat{\mathbf{x}}^{(0)}$ *to the linear system* $\mathsf{A}\mathbf{x} = \mathbf{b}$, *all computed using single-precision arithmetic, the following steps generate a sequence* $\{\hat{\mathbf{x}}^{(k)}\}$ *of improved approximations to* $\mathbf{x}$. *The words* SINGLE *and* DOUBLE *indicate the precision to be used in each step, and* $\tau > 0$ *is a prescribed tolerance.*

1. $k \leftarrow 0$.
2. $\mathbf{r}^{(k)} \leftarrow \mathbf{b} - \mathsf{A}\hat{\mathbf{x}}^{(k)}$ (DOUBLE).
3. If $\|\mathbf{r}^{(k)}\| \geq \tau$, then:
4. Solve $\mathsf{L}\mathbf{z} = \mathbf{r}^{(k)}$ for $\mathbf{z}$ (SINGLE).
5. Solve $\mathsf{U}\boldsymbol{\varepsilon}^{(k)} = \mathbf{z}$ for $\boldsymbol{\varepsilon}^{(k)}$ (SINGLE).
6. $\hat{\mathbf{x}}^{(k+1)} \leftarrow \hat{\mathbf{x}}^{(k)} + \boldsymbol{\varepsilon}^{(k)}$ (SINGLE).
7. $k \leftarrow k + 1$
8. Go to 2.
9. End if.
10. End.

One can modify this algorithm to accommodate partial pivoting.

As Equation (2.6-7) indicates, choosing a reasonable tolerance τ requires an estimate κ of cond (A). To help derive such an estimate, we establish the following fact about perturbations of linear systems.

PROPOSITION 2.17. *Let* $\mathsf{A} \in \mathbb{R}^{n\times n}$, $\mathbf{b} \in \mathbb{R}^n$, *and* $\mathbf{x} \in \mathbb{R}^n$ *satisfy the linear system* $\mathsf{A}\mathbf{x} = \mathbf{b}$, *and suppose that the vector* $\hat{\mathbf{x}} := \mathbf{x} + \boldsymbol{\varepsilon}$ *is nonzero and satisfies the perturbed system* $(\mathsf{A} + \mathsf{D})\hat{\mathbf{x}} = \mathbf{b}$, *where* $\mathsf{D} \in \mathbb{R}^{n\times n}$. *Then*

$$\frac{\|\boldsymbol{\varepsilon}\|}{\|\hat{\mathbf{x}}\|} \leq \operatorname{cond}(\mathsf{A}) \frac{\|\mathsf{D}\|}{\|\mathsf{A}\|}.$$

PROOF: The hypotheses imply that $\mathsf{D}\hat{\mathbf{x}} = \mathbf{b} - \mathsf{A}\hat{\mathbf{x}} = \mathsf{A}(\mathbf{x} - \hat{\mathbf{x}})$, that is, $\boldsymbol{\varepsilon} = \mathsf{A}^{-1}\mathsf{D}\mathbf{x}$. Therefore, $\|\boldsymbol{\varepsilon}\| \leq \|\mathsf{A}^{-1}\| \, \|\mathsf{D}\| \, \|\hat{\mathbf{x}}\|$. Since $\hat{\mathbf{x}} \neq \mathbf{0}$, we can divide through by $\|\hat{\mathbf{x}}\|$ to complete the proof. ∎

For a machine on which single-precision arithmetic is accurate to s decimal digits, the perturbation to A arising from roundoff errors has relative magnitude roughly 10^{-s}. Hence,

$$\frac{\|\boldsymbol{\varepsilon}\|}{\|\hat{\mathbf{x}}\|} \leq \operatorname{cond}(\mathsf{A}) \frac{\|\mathsf{D}\|}{\|\mathsf{A}\|} \simeq 10^{-s} \operatorname{cond}(\mathsf{A}).$$

Therefore, during the first step of Algorithm 2.7, we can set

$$\operatorname{cond}(\mathsf{A}) \simeq \kappa := 10^s \frac{\|\boldsymbol{\varepsilon}^{(0)}\|}{\|\hat{\mathbf{x}}^{(0)}\|}.$$

This estimate of cond (A) is plainly a crude one. For more sophisticated estimates that do not rely on explicit knowledge of a machine's precision, see Golub and Van Loan ([1], Section 4.5).

2.7 Problems

PROBLEM 1. Compute LU factorizations of the matrix

$$\mathsf{A} = \begin{bmatrix} 4 & 0 & 2 & 4 \\ 0 & 1 & 1 & 1 \\ 2 & 1 & 11 & 9 \\ 4 & 1 & 9 & 25 \end{bmatrix},$$

using the Doolittle and Crout methods.

PROBLEM 2. Prove Corollary 2.7.

PROBLEM 3. Devise a partial pivoting scheme for the Crout method.

PROBLEM 4.

(A) Program Algorithm 2.3, and use it to compute the Cholesky decomposition of the following matrix:

$$\mathsf{A} = \begin{bmatrix} 4 & 0 & 2 & 4 \\ 0 & 1 & 1 & 1 \\ 2 & 1 & 11 & 9 \\ 4 & 1 & 9 & 25 \end{bmatrix} = \begin{bmatrix} 2 & 0 & 0 & 0 \\ 0 & 1 & 0 & 0 \\ 1 & 1 & 3 & 0 \\ 2 & 1 & 2 & 4 \end{bmatrix} \begin{bmatrix} 2 & 0 & 1 & 2 \\ 0 & 1 & 1 & 1 \\ 0 & 0 & 3 & 2 \\ 0 & 0 & 0 & 4 \end{bmatrix}.$$

(B) Give operation counts for Algorithms 2.3 and 2.4.

PROBLEM 5. Devise a compact storage mode suitable for symmetric, positive definite band matrices. Write an algorithm to compute the Cholesky decomposition using this storage scheme.

PROBLEM 6. Derive an operation count for the block-tridiagonal algorithm applied to a matrix of the following block-partitioned form:

$$\mathsf{A} = \begin{pmatrix} \mathsf{T} & \mathsf{D} & & & \\ \mathsf{D} & \mathsf{T} & \mathsf{D} & & \\ & & \ddots & & \\ & & \mathsf{D} & \mathsf{T} & \mathsf{D} \\ & & & \mathsf{D} & \mathsf{T} \end{pmatrix}.$$

Here, $\mathsf{T} \in \mathbb{R}^{m\times m}$ is tridiagonal, $\mathsf{D} \in \mathbb{R}^{m\times m}$ is diagonal, and $\mathsf{A} \in \mathbb{R}^{mn\times mn}$.

PROBLEM 7. Prove Theorem 2.8.

PROBLEM 8. Prove Theorem 2.5.

PROBLEM 9.

(A) Show that $\varrho(\mathsf{A}) \leq ||\mathsf{A}||$ for any subordinate matrix norm $||\cdot||$. (Hint: Consider eigenvectors having unit length.)

(B) Show that the spectral radius $\varrho: \mathbb{R}^{n\times n} \to \mathbb{R}$ is not a matrix norm by finding matrices A and B such that $\varrho(\mathsf{A}+\mathsf{B}) > \varrho(\mathsf{A}) + \varrho(\mathsf{B})$.

(C) Show that $1/||\mathsf{A}^{-1}|| = \inf_{||\mathbf{x}||=1} ||\mathsf{A}\mathbf{x}||$.

(D) Let A be symmetric and positive definite with smallest eigenvalue λ_{min}. Show that $||\mathsf{A}^{-1}||_2 = 1/\lambda_{\text{min}}$.

PROBLEM 10. Matrix norms $||\cdot||: \mathbb{R}^{n\times n} \to \mathbb{R}$ inherit nice properties of the vector norms that define them:

(A) Prove that any subordinate matrix norm satisfies the requirements to be a norm.

(B) Prove that any matrix norm $||\mathsf{A}||$ is a continuous function of the n^2 entries of A.

(C) Prove that all matrix norms on $\mathbb{R}^{n\times n}$ are equivalent.

(Propositions (B) and (C) do not require the norm to be subordinate to a vector norm.)

PROBLEM 11.

(A) Prove part 2 of Theorem 2.15.

(B) Prove that $||\mathsf{I}|| = 1$ in any subordinate matrix norm.

PROBLEM 12.

(A) For each of the following matrices, sketch the image $\{\mathsf{A}\mathbf{x} : \mathbf{x} \in S_2\}$ of the unit sphere in $\mathbb{R}^2$. Give a geometric interpretation of $||\mathsf{A}||_2$ in each case.

$$\mathsf{A} = \begin{bmatrix} 0 & 2 \\ 0 & 0 \end{bmatrix}, \qquad \mathsf{A} = \begin{bmatrix} 0 & 1 \\ -1 & 0 \end{bmatrix}, \qquad \mathsf{A} = \begin{bmatrix} 2 & 1 \\ 1 & 3 \end{bmatrix}.$$

(B) For the matrix A in Equation (2.6-6), show that $\text{cond}_\infty(\mathsf{A}) > 10^{21}$.

PROBLEM 13.

(A) Find constants m_1, M_1, m_2, M_2 such that

$$m_1 \operatorname{cond}_2(\mathsf{A}) \leq \operatorname{cond}_1(\mathsf{A}) \leq M_1 \operatorname{cond}_2(\mathsf{A})$$

and

$$m_2 \operatorname{cond}_\infty(\mathsf{A}) \leq \operatorname{cond}_2(\mathsf{A}) \leq M_2 \operatorname{cond}_\infty(\mathsf{A}).$$

(B) Let $||\cdot||$ be a subordinate matrix norm, and let $\mathsf{A}, \mathsf{B} \in \mathbb{R}^{n\times n}$. Prove that $\operatorname{cond}(\mathsf{AB}) \leq \operatorname{cond}(\mathsf{A}) \operatorname{cond}(\mathsf{B})$.

PROBLEM 14.

(A) Show that the **Frobenius norm**

$$||\mathsf{A}||_F := \left(\sum_{i=1}^{n}\sum_{j=1}^{n} |a_{i,j}|^2\right)^{1/2}$$

satisfies the three conditions required of norms, as does the function

$$||\mathsf{A}||_{\max} := \max |a_{i,j}|.$$

(B) Neither of the norms in (A) is subordinate to a vector norm when $n > 1$. Therefore, we have no guarantee that the inequality (2.5-5) holds. Show that it fails for the norm $||\cdot||_{\max}$.

(C) Show that $||\cdot||_F$ is not subordinate to any vector norm for $n > 1$. (Hint: Consider $||\mathsf{I}||_F$.)

PROBLEM 15. Suppose that $\mathsf{A} \in \mathbb{R}^{n\times n}$ and that p is a polynomial. Show the following:

(A) If λ is an eigenvalue of A, then $p(\lambda)$ is an eigenvalue of $p(\mathsf{A})$.

(B) If $\mathbb{R}^n$ has a basis consisting of eigenvectors of A and μ is an eigenvalue of $p(\mathsf{A})$, then there is an eigenvalue λ of A for which $\mu = p(\lambda)$. (Actually, the assumption that eigenvectors of A form a basis is not necessary.)

2.8 References

1. G.H. Golub and C.F. Van Loan, *Matrix Computations*, Johns Hopkins University Press, Baltimore, MD, 1983.

2. I.N. Herstein, *Topics in Algebra*, Xerox Publishing Co., Lexington, MA, 1964.

3. J. Ortega, *Numerical Analysis: A Second Course*, Academic Press, New York, 1972.

4. G. Strang, *Linear Algebra and its Applications* (2nd ed.), Academic Press, New York, 1980.

5. J. Wilkinson, *The Algebraic Eigenvalue Problem*, Oxford University Press, New York, 1965.

Chapter 3

Solution of Nonlinear Equations

3.1 Introduction

In this chapter we consider numerical methods for finding real solutions to equations of the form

$$f(x) = 0. \tag{3.1-1}$$

Here, f is some nonlinear function of the unknown variable x. Any number $x^* \in \mathbb{R}$ that satisfies this equation is a **real zero** of f. Examples of such equations include

$$\begin{aligned}
\sin(x^3 + 2x^2 - 1) &= 0, \\
\exp\left[\frac{x^3 - \tan^{-1} x}{\exp(x^2)}\right] &= x, \\
\cosh(x^2) + \pi &= 0.
\end{aligned}$$

(We can convert the second example to the form (3.1-1) by subtracting x from both sides.)

Numerical methods for solving Equation (3.1-1) share two general features. First, they are **iterative** methods. That is, given an initial guess $x^{(0)}$, they produce a sequence $\{x^{(m)}\} = \{x^{(0)}, x^{(1)}, x^{(2)}, \ldots\}$ of real numbers, called **iterates**. If the equation, the method, and the initial guess are all "reasonable," then we expect that $|x^* - x^{(m)}| \to 0$ as $m \to \infty$. In this case, the method **converges** to x^*. Otherwise, the sequence $\{x^{(m)}\}$ may converge to a different point in $\mathbb{R}$, or it may not converge to any number. In the latter case the method **diverges**. There are two key questions concerning a numerical method for solving Equation (3.1-1). First, for what initial guesses $x^{(0)}$ does the method converge to x^*? Second, for efficiency's sake, how fast does the sequence of iterates converge?

The second general feature of numerical methods for nonlinear equations is that they require informed users. Except in special cases, one cannot hope to solve Equation (3.1-1) numerically without first analyzing f. Look again at the three examples listed above. The third example has no real solutions, so solving for them numerically is futile. Even in the first two examples, some analysis is needed to determine how many real solutions each equation possesses. Moreover, many numerical methods do not converge to a sought zero x^* unless the initial guess $x^{(0)}$ is close to x^*. To solve Equation (3.1-1) numerically, one should know something about the number and locations of the zeros of f.

One can easily imagine more complicated examples involving less familiar transcendental functions, functions having discontinuities, or functions defined by algorithms that themselves may be quite involved. In this chapter we restrict attention to functions that are at least continuous in some neighborhood of the sought solution.

Even under restrictive smoothness assumptions, difficulties can arise in numerical work. Problems often occur when the zeros of f are extremely sensitive to small numerical errors. Polynomials are notorious in this regard. Consider the following example, introduced by Wilkinson [5]:

$$f(x) = (x-1)(x-2)\cdots(x-20) = \sum_{n=0}^{20} a_n x^n,$$

where $a_{20} = 1$, $a_{19} = -(1+2+\cdots+20) = -210$, ..., $a_0 = 20!$. The zeros $1, 2, \ldots, 20$ of this polynomial are all real. Also, like every polynomial, this one has derivatives of all orders at each $x \in \mathbb{R}$, so smoothness of f as a function of x is not an issue. Now define $\hat{f}(x)$ to be the polynomial whose coefficients $\hat{a}_n$ are identical to those of f, except that $\hat{a}_{19} = a_{19} + 2^{-23} \simeq -(210 - 10^{-7})$. Even though f and $\hat{f}$ have coefficients that are "close," the zeros of $\hat{f}$ differ significantly from those of f. In particular, $\hat{f}$ has a conjugate pair of complex zeros $16.7307 \pm 2.8126\,i$, correct to four decimal places.

To accommodate phenomena like this, several specialized methods exist for finding polynomial zeros. We do not investigate these methods here; rather, we refer to Press et al. [4, Section 9.5] for an introduction. For our purposes, the instability of some polynomial zeros plays a cautionary role: Nonlinear functions, even when smooth, can have zeros that are difficult to approximate numerically.

A much milder source of difficulty arises at zeros where the graph of f is tangent to the x-axis, as shown in Figure 1. These zeros may be difficult to detect, since the graph of f may not cross the x-axis as it passes through $(x^*, f(x^*))$. Also, one of the most powerful zero-finding schemes that we discuss — Newton's method — loses some of its power at such zeros. In preparation for later discussion, we digress briefly to characterize this type of zero.

DEFINITION. *A zero x^* of a function $f\colon [a,b] \to \mathbb{R}$ has* **multiplicity** q *if there is some continuous function $g\colon [a,b] \to \mathbb{R}$ such that*

(i) $g(x^*) \neq 0$,

(ii) *for every $x \in [a,b]$, $f(x) = (x - x^*)^q g(x)$.*

If x^* is a zero of f having multiplicity 1, then x^* is a **simple zero** *of f.*

The following proposition connects the multiplicity of zeros with the nature of the tangency of the graph of f to the x-axis:

PROPOSITION 3.1. *A function $f \in C^q([a,b])$ has a zero of multiplicity q at $x^* \in [a,b]$ if and only if*

$$0 = f(x^*) = f'(x^*) = \cdots = f^{(q-1)}(x^*) \quad \text{and} \quad f^{(q)}(x^*) \neq 0. \tag{3.1-2}$$

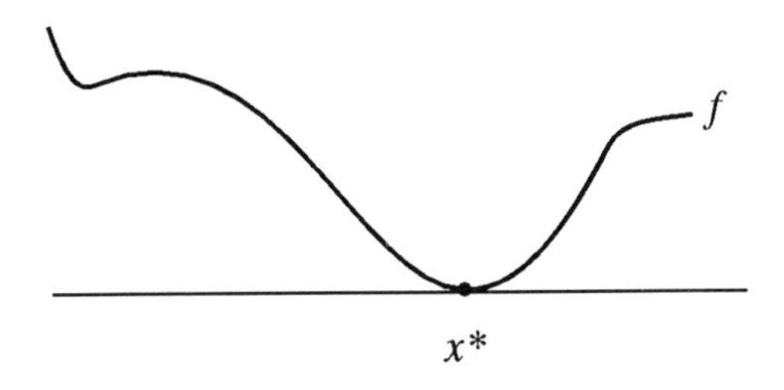

FIGURE 1. *A point where $f(x^*) = f'(x^*) = 0$.*

PROOF: If f has a zero of multiplicity q at x^*, then $f(x) = (x - x^*)^q g(x)$, where g is continuous at x^* and $g(x^*) \neq 0$. Since $f \in C^q([a,b])$, $g \in C^q([a,b] \setminus \{x^*\})$. It is straightforward to check by induction that, for $x \in [a,b] \setminus \{x^*\}$ and $0 \leq k \leq q$,

$$f^{(k)}(x) = c_k(x - x^*)^{q-k} g(x) + (x - x^*)^{q-k+1} g_k(x),$$

where c_k is a nonzero constant and g_k is some function that is continuous on $[a,b]$ with $g_k \in C^{q-k}([a,b] \setminus \{x^*\})$. Consequently, $f^{(k)}(x^*) = 0$ for $0 \leq k < q$, and $f^{(q)}(x^*) = c_q g(x^*) \neq 0$.

Conversely, if Equations (3.1-2) hold, then the Taylor expansion (Theorem 0.13) for f about x^* has the form

$$\begin{aligned} f(x) &= f(x^*) + f'(x^*)(x - x^*) + \cdots + \frac{f^{(q)}(\zeta)}{q!}(x - x^*)^q \\ &= \frac{f^{(q)}(\zeta)}{q!}(x - x^*)^q, \end{aligned}$$

for some point ζ lying between x^* and x. Since ζ depends on x, we may as well write $\zeta = \zeta(x)$ and note that $\zeta(x)$ is continuous at x^*, since $\lim_{x \to x^*} \zeta(x) = x^* = \zeta(x^*)$. Thus $f(x) = (x - x^*)^q g(x)$, where $g(x) := f^{(q)}(\zeta(x))/q!$. ∎

The remainder of this chapter has the following format: Sections 3.2 through 3.5 discuss one-dimensional methods, designed to find real zeros of functions $f: [a,b] \to \mathbb{R}$. The analysis of these methods builds intuition for problems involving several equations in several unknowns. In this more complicated setting we seek solutions $\mathbf{x}^* := (x_1^*, x_2^*, \ldots, x_n^*)^\top \in \mathbb{R}^n$ to systems of equations having the form

$$\mathbf{f}(\mathbf{x}) := \begin{bmatrix} f_1(x_1, x_2, \ldots, x_n) \\ f_2(x_1, x_2, \ldots, x_n) \\ \vdots \\ f_n(x_1, x_2, \ldots, x_n) \end{bmatrix} = \mathbf{0}. \tag{3.1-3}$$

Here, $\mathbf{f}: \Omega \to \mathbb{R}^n$, where the domain Ω is a suitable subset of $\mathbb{R}^n$. We often stipulate that Ω be **convex**. This means that, whenever $\mathbf{x}, \mathbf{y} \in \Omega$, the line segment

$$\left\{\mathbf{x} + t(\mathbf{y} - \mathbf{x}) \in \mathbb{R}^n : 0 \le t \le 1\right\},$$

which connects $\mathbf{x}$ and $\mathbf{y}$, lies entirely in Ω. Figure 2 illustrates convex and nonconvex sets in $\mathbb{R}^2$. Sections 3.6 and 3.7 treat various generalizations of one-dimensional methods to the numerical solution of Equation (3.1-3).

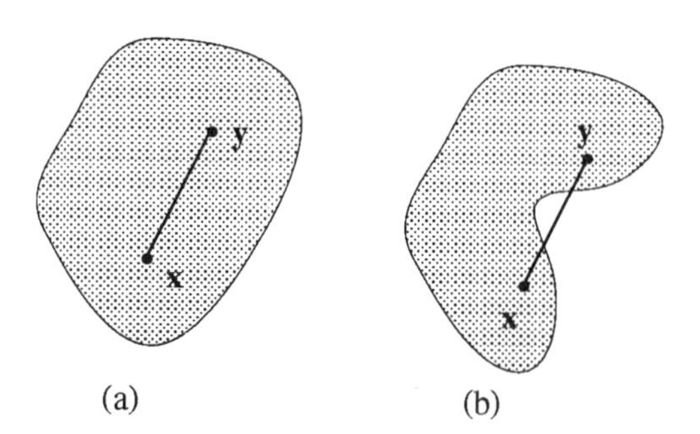

FIGURE 2. *A convex set (a) and a nonconvex set (b) in* $\mathbb{R}^2$.

3.2 Bisection

Motivation and Construction

Suppose that $f: [a,b] \to \mathbb{R}$ is continuous and that $f(a)f(b) < 0$. Thus f changes sign on the closed interval $[a,b]$. The intermediate value theorem

(Theorem 0.8) guarantees that f has a zero x^* somewhere in the open interval (a,b). Figure 1 depicts this idea, showing that the graph of f may actually cross the x-axis more than once in (a,b). To approximate x^*, we might use the midpoint $\overline{x} := (a+b)/2$ of the interval. This approximation is admittedly crude, but its error is easy to estimate: $|x^* - \overline{x}| < (b-a)/2$. As we detail below, this reasoning gives rise to the **bisection method**.

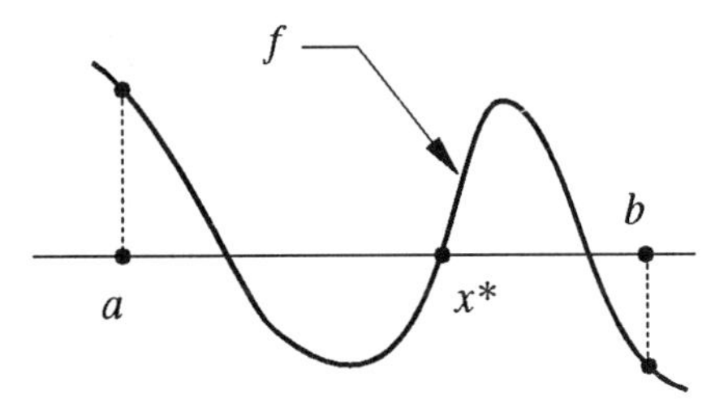

FIGURE 1. *The existence of at least one zero in $[a,b]$ for a continuous function f satisfying $f(a)f(b) < 0$.*

Let us call the zero x^* **bracketed** if we can identify a **bracketing interval** $[a,b]$ such that $f(a)f(b) < 0$ and x^* is the only zero of f lying in $[a,b]$. Starting with an initial bracketing interval $[a_0,b_0]$, the bisection method generates a sequence

$$\Big\{[a_0,b_0],[a_1,b_1],[a_2,b_2],\ldots\Big\}$$

of successively smaller bracketing intervals for x^*.

The idea is this: Having computed the bracketing interval $[a_m,b_m]$, we regard the midpoint $x^{(m)} := (a_m+b_m)/2$ as our next estimate for x^*. If $f(a_m)f(x^{(m)}) < 0$, then $x^* \in (a_m, x^{(m)})$, and hence $[a_{m+1},b_{m+1}] := [a_m, x^{(m)}]$ becomes the next bracketing interval. (In the unlikely event that $f(x^{(m)}) = 0$, the method would stop.) On the other hand, if $f(x^{(m)})f(b_m) < 0$, then $[a_{m+1},b_{m+1}] := [x^{(m)},b_m]$ is the next bracketing interval. Once we have determined $[a_{m+1},b_{m+1}]$, the midpoint $x^{(m+1)} := (a_{m+1}+b_{m+1})/2$ becomes the next approximation to x^*, and we repeat the process. The intervals generated in this fashion "trap" the zero x^*, in the sense that x^* lies in each interval and

$$\begin{aligned} b_{m+1} - a_{m+1} = 2^{-1}(b_m - a_m) = \cdots &= 2^{-(m+1)}(b_0 - a_0) \\ &\to 0 \quad \text{as} \quad m \to \infty. \end{aligned} \tag{3.2-1}$$

As Figure 2 illustrates, this procedure generates an iterative sequence $\{x^{(m)}\}$ of interval midpoints that approximate the exact zero x^*. In practice we stop generating new iterates as soon as $|x^* - x^{(m)}| < \tau$, where $\tau > 0$ is some prescribed tolerance. Since $|x^* - x^{(m)}| < (b_m - a_m)/2$, this stopping criterion is easy to check: It holds whenever $(b_m - a_m)/2 = 2^{-(m+1)}(b_0 - a_0) < \tau$, that is, when $m > -\log_2[\tau/(b_0 - a_0)]$. The following algorithm incorporates these ideas.

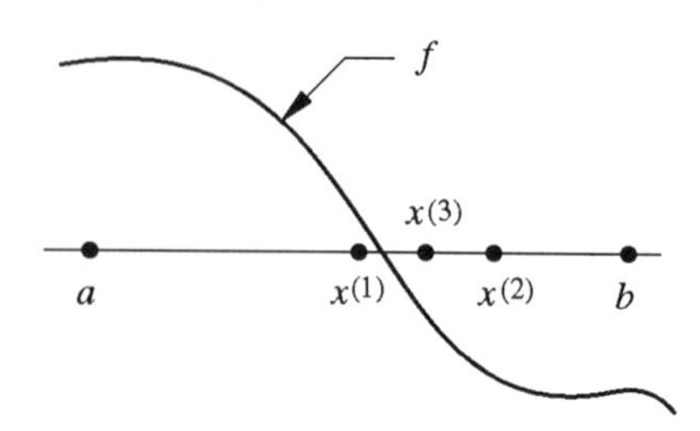

FIGURE 2. *Example of a sequence of iterates $x^{(m)}$ generated by the bisection method.*

ALGORITHM 3.1 (BISECTION). *Given a continuous function $f\colon [a_0, b_0] \to \mathbb{R}$, where $[a_0, b_0]$ brackets the zero x^*, and a tolerance $\tau > 0$, this algorithm computes an approximate value to x^*.*

1. $m \leftarrow 0$.
2. $x^{(m)} \leftarrow (a_m + b_m)/2$.
3. If $(b_m - a_m)/2 \geq \tau$, then:
4. If $f(a_m)f(x^{(m)}) \leq 0$ then:
5. $a_{m+1} \leftarrow a_m$ and $b_{m+1} \leftarrow x^{(m)}$.
6. Else:
7. $a_{m+1} \leftarrow x^{(m)}$ and $b_{m+1} \leftarrow b_m$.
8. End if.
9. $m \leftarrow m + 1$.
10. Go to 2.
11. End if.
12. End.

It is logically possible to have $f(x^{(m)}) = 0$ at some stage of the iteration. In this case $f(a_m)f(x^{(m)}) = f(x^{(m)})f(b_m) = 0$ and $x^{(m)} = x^*$, and theoretically we should stop iterating. However, this case occurs so rarely in numerical practice that it is not worth the extra computing time to test whether $f(x^{(m)}) = 0$.

Practical Considerations

The bisection method has an attractive property: It always converges. More precisely:

PROPOSITION 3.2. *Suppose that* $f: [a_0, b_0] \to \mathbb{R}$ *is continuous and that* $[a_0, b_0]$ *brackets exactly one zero* x^* *of* f. *Then bisection generates a sequence* $\{x^{(m)}\}$ *that converges to* x^*.

PROOF: The relationship (3.2-1) implies that $|x^{(m)} - x^*| \leq 2^{-m}(b_0 - a_0)$, which tends to 0 as $m \to \infty$. ∎

The hypotheses that the function f have exactly one zero in (a_0, b_0) and that $f(a_0)f(b_0) < 0$ are essential in this theorem. They also indicate how much one should know about f before embarking on a search via bisection. If f has more than one zero in (a_0, b_0) and $f(a_0)f(b_0) < 0$, then the bisection method converges to one of the zeros. However, we may have no way to determine in advance which one. Also, a continuous function f can have zeros in an interval (a_0, b_0) without having $f(a_0)f(b_0) < 0$. Figure 3 shows the graph of such a function. It is doubtful that one can formulate a bisection algorithm that obviates all prior analysis of f; the hypothesis of continuity by itself simply leaves too much latitude for constructing counterexamples.

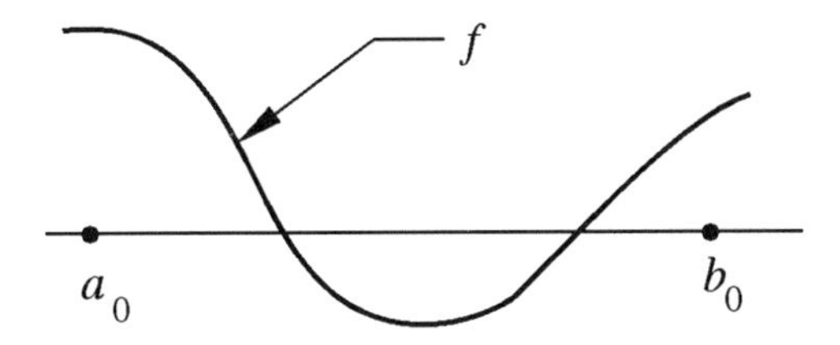

FIGURE 3. *The graph of a continuous function* f *for which* $f(a_0)f(b_0) > 0$, *even though* f *has zeros in* (a_0, b_0).

The argument used to prove Propositon 3.2 indicates how fast we can expect the bisection iterates to converge to x^*. Roughly speaking, each iteration reduces the error $|x^* - x^{(m)}|$ by the factor $\frac{1}{2}$. Therefore, to reduce this error by the factor 10^{-1}, we expect to require M iterations, where $2^{-M} = 10^{-1}$, that is, $M = 1/\log_{10} 2 \simeq 3.3$. In other words, bisection takes about 3.3 iterations to gain one decimal digit of accuracy in the approximation to x^*. Compared with schemes presented in subsequent sections, bisection converges slowly. One might regard this slowness as the price paid for guaranteed convergence.

The dichotomy between this scheme, which is slow but sure, and the faster, more temperamental schemes discussed later in this chapter suggests an important practical view of bisection: Since it converges reliably, even from poor starting guesses, it makes a reasonable initializing algorithm. In this view, one can often use a few iterations of the bisection method to generate an iterate $x^{(m)}$ lying close enough to x^* to guarantee that a faster "polishing" method, using $x^{(m)}$ as an initial guess, converges to x^*.

One can implement such a hybrid strategy at several levels of sophistication. The following naive approach uses the polishing scheme whenever possible, but it never accepts iterates $x^{(m)}$ that lie outside the smallest known bracketing interval. In the following pseudocode, the notation $x^{(m+1)} \leftarrow \text{POLISH}\,(x^{(m)})$ indicates that we are to use the polishing scheme to compute the new iterate x. Sections 3.3 through 3.5 discuss candidates for such a scheme.

ALGORITHM 3.2. *Given a continuous function f defined on an interval $[a_0, b_0]$ that brackets exactly one zero x^* of f, the following algorithm uses bisection and a faster, more sensitive polishing algorithm to compute a sequence $\{x^{(m)}\}$ of approximations to x^*.*

1. $m \leftarrow 0$.
2. $a \leftarrow a_0$, $b \leftarrow b_0$
3. $x^{(0)} \leftarrow (a+b)/2$.
4. $x^{(m+1)} \leftarrow \text{POLISH}\,(x^{(m)})$.
5. If $x^{(m+1)} \leq a$ or $x^{(m+1)} \geq b$ then:
6. If $f(a)f(x^{(m)}) \leq 0$ then:
7. $b \leftarrow x^{(m)}$.
8. Else:
9. $a \leftarrow x^{(m)}$.
10. End if.
11. $x^{(m+1)} \leftarrow (a+b)/2$.
12. End if.
13. $m \leftarrow m+1$.
14. If convergence test fails, go to 4.
15. End.

(If the polishing algorithm is not sophisticated, then this algorithm can fail to locate a zero.) We discuss appropriate convergence tests in the next few sections.

3.3 Successive Substitution in One Variable

Motivation and Construction

An intuitively appealing method for solving $f(x) = 0$ arises if we recast the equation in the form $x = \Phi(x)$ for some **iteration function** Φ. We call any solution x^* to this latter equation a **fixed point** of Φ. Given an initial guess $x^{(0)}$, we compute new approximations to x^* simply by setting

$$x^{(m+1)} \leftarrow \Phi\left(x^{(m)}\right). \tag{3.3-1}$$

If all goes well, the sequence $\{x^{(m)}\}$ of iterates generated in this way converges to x^*. We call this scheme **successive substitution**. Implicit lies the hope that Φ somehow "shoves" the iterates $x^{(m)}$ toward a fixed point.

The form $x = \Phi(x)$ is not as special as it may appear at first. Defining $\Phi(x) := x - f(x)$ automatically converts the equation $f(x) = 0$ to $x = \Phi(x)$ for any function f. Problem 2 provides opportunities for making this conversion in other ways. Problem 15 illustrates that fixed points are far from rare.

To illustrate successive substitution, consider the function $f(x) := x - \frac{1}{2}\cos x$. This function has one zero x^* in the interval $[0, \pi/2]$, as Figure 1 illustrates. To find an approximation to this zero, we set $\Phi(x) := \frac{1}{2}\cos x$ and use successive substitution. Table 3.1 lists, to four decimal places, iterates $x^{(m)}$ that result when we use the initial guess $x^{(0)} = \frac{1}{2}$. After six iterations, the two most recent iterates $x^{(5)}$ and $x^{(6)}$ agree to four decimal places.

Table 3.1: Successive substitution iterates for $f(x) = x - \frac{1}{2}\cos x$.

m	$x^{(m)}$	$\lvert f(x^{(m)}\rvert$	$\lvert x^{(m)} - x^*\rvert$
0	0.5000	6.121×10^{-2}	4.982×10^{-2}
1	0.4388	1.384×10^{-2}	1.139×10^{-2}
2	0.4526	2.984×10^{-3}	2.449×10^{-3}
3	0.4496	6.504×10^{-4}	5.342×10^{-4}
4	0.4503	1.414×10^{-4}	1.162×10^{-4}
5	0.4502	3.078×10^{-5}	2.798×10^{-5}
6	0.4502	6.695×10^{-6}	5.499×10^{-6}

While it may be tempting to stop iterating when changes in successive iterates become small, this halting criterion can conceivably be a poor one. Small changes in successive iterates may indicate simply that the scheme is converging very slowly. What we really want is some assurance that the

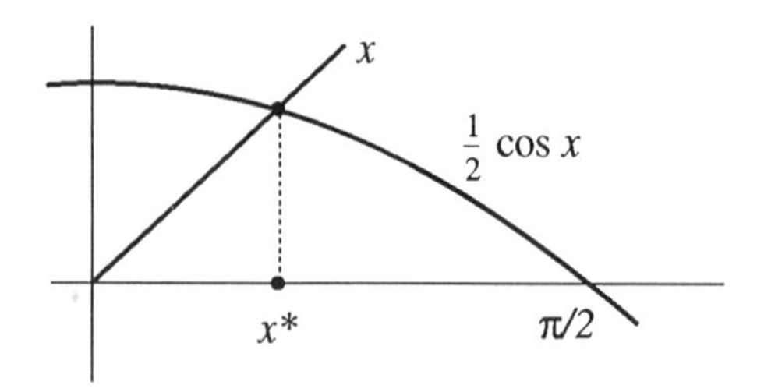

FIGURE 1. *The graphs of x and $\frac{1}{2}\cos x$, showing that $x - \frac{1}{2}\cos x$ has a zero in the interval $[0, \pi/2]$.*

error $\varepsilon^{(m)} := x^* - x^{(m)}$ is small in magnitude before we accept $x^{(m)}$ as a reasonable approximation. Later in this section we discuss halting criteria more rigorously.

Practical Considerations

Successive substitution does not always converge. A geometric view of the difficulty lends insight not only into what is needed to guarantee convergence but also into the rate at which $x^{(m)} \to x^*$ when convergence occurs.

Figure 2 shows graphs of two smooth functions $y = \Phi(x)$. Each graph intersects that of the identity function $y = x$ at the point $(x^*, \Phi(x^*))$ corresponding to a fixed point. For each function Φ, the figure also shows the iterates generated using successive substitution with initial guesses $x^{(0)}$. In constructing the figure, we reflect the ordinates $x^{(m+1)} := \Phi(x^{(m)})$ across the lines $y = x$ to locate them as arguments of Φ for the next iteration. This graphic evidence suggests that successive substitution converges to the fixed point x^* when the graph of Φ is not too steep. The scheme diverges, however, when the graph of Φ is steep.

Figure 3 shows similar plots for two functions Φ whose slopes are negative. Here again successive substitution converges when the graph of Φ is not too steep, but the scheme diverges when the graph of Φ is steep.

Geometrically, the steepness of the graph of Φ indicates the "stretching power" of Φ. Thus Φ has a steep graph in a region if two nearby points x and y in the region have images $\Phi(x)$ and $\Phi(y)$ that are far apart, as Figure 4 illustrates. The following definition quantifies this notion.

DEFINITION. *Let $S \subset \mathbb{R}$. A function $\Phi: S \to \mathbb{R}$ satisfies a* **Lipschitz condition** *on S if there exists a constant $L > 0$ such that, for any two points $x, y \in S$,*

$$|\Phi(x) - \Phi(y)| \leq L|x - y|. \tag{3.3-2}$$

The greatest lower bound for such constants is the **Lipschitz constant** *for*

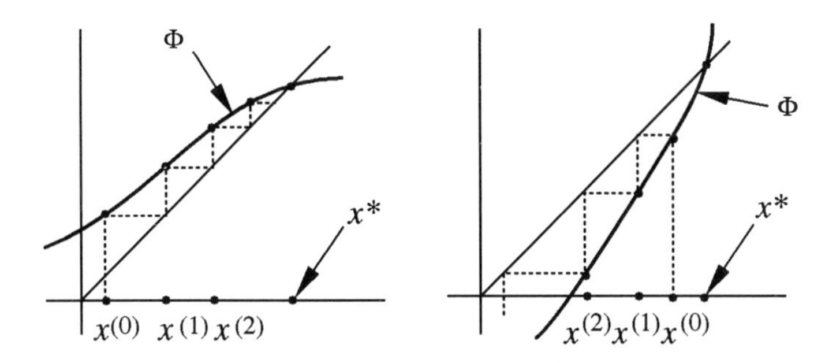

FIGURE 2. *Schematic illustration of successive substitution for two smooth functions having positive slopes.*

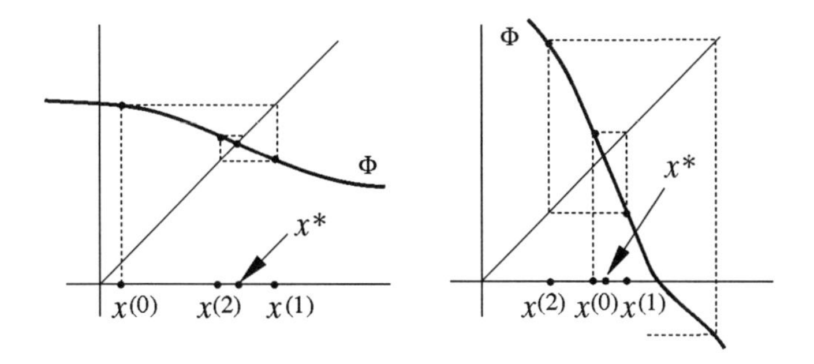

FIGURE 3. *Schematic illustration of successive substitution for two smooth functions having negative slopes.*

Φ on S. If Φ has Lipschitz constant $L < 1$ on S, then Φ is a **contraction** *on S.*

An easy argument shows that any function that satisfies a Lipschitz condition is continuous. But the Lipschitz condition also tells us something about how fast the function values $\Phi(x)$ can change as the argument x changes. We regard Φ as having a steep graph when it has a Lipschitz constant $L \geq 1$; thus, contractions are functions whose graphs are not steep.

This definition has connections with a more familiar measure of steepness, namely the derivative of Φ. If $\Phi \in C^1([a,b])$, then Φ satisfies the Lipschitz condition (3.3-2) on $[a,b]$, with $\sup_{x\in[a,b]} |\Phi'(x)| = L$. To see this, recall the mean value theorem (Theorem 0.14), which guarantees the existence of a point $\zeta \in (a,b)$ such that $\Phi(x) - \Phi(y) = \Phi'(\zeta)(x-y)$. Since $|\Phi'(\zeta)| \leq L$,

$$|\Phi(x) - \Phi(y)| = |\Phi'(\zeta)|\,|x-y| \leq L|x-y|.$$

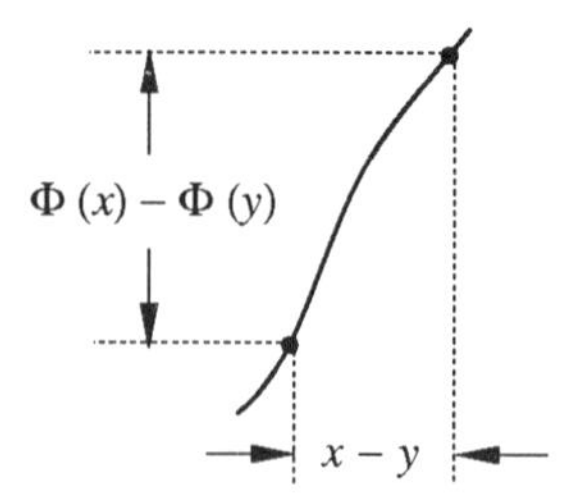

FIGURE 4. *Correspondence between the steepness of the graph of* Φ *and the increase* $|\Phi(x) - \Phi(y)|$ *over the distance* $|x - y|$.

In particular, a continuously differentiable function Φ is a contraction on $[a, b]$ if $|\Phi'| < 1$ everywhere on the interval $[a, b]$.

Now we can state the principle governing convergence of the iterative scheme (3.3-1): *If* Φ *is a contraction in some neighborhood* $(x^* - \delta, x^* + \delta)$ *of the fixed point* x^*, *then successive substitution, starting with* $x^{(0)} \in (x^* - \delta, x^* + \delta)$, *shrinks the distance between the iterates and the fixed point. Moreover, the iterates stay inside the interval* $(x^* - \delta, x^* + \delta)$. We make a rigorous statement to this effect below.

For now, assume that Φ is a contraction and let us examine the *rate* at which $x^{(m)} \to x^*$. The inequality (3.3-2) and the fact that $x^* = \Phi(x^*)$ allow us to estimate the error $\varepsilon^{(m+1)} := x^* - x^{(m+1)}$ in terms of previous values:

$$\begin{aligned} \left|\varepsilon^{(m+1)}\right| = \left|x^* - x^{(m+1)}\right| &= \left|\Phi(x^*) - \Phi(x^{(m)})\right| \\ &\leq L\left|x^* - x^{(m)}\right| = L\left|\varepsilon^{(m)}\right|. \end{aligned} \tag{3.3-3}$$

Therefore, with each iteration, successive substitution reduces the magnitude of the error at least by a factor L.

Measuring the error reduction associated with one iteration furnishes a standard way to gauge the convergence rates of iterative schemes:

DEFINITION. *Let* $p \geq 1$. *An iterative scheme that produces sequences* $\{x^{(m)}\}$ *of approximations to* x^* **converges with order** p *if there exists a constant* $C > 0$ *and an integer* $M \geq 0$ *such that*

$$\left|x^* - x^{(m+1)}\right| \leq C\left|x^* - x^{(m)}\right|^p \tag{3.3-4}$$

whenever $m \geq M$. *If* $p = 1$, *convergence occurs when* $0 < C < 1$, *and we say that the scheme converges* **linearly**. *If* $p = 2$, *the scheme converges* **quadratically**. *If*

$$\lim_{m\to\infty} \frac{|x^* - x^{(m+1)}|}{|x^* - x^{(m)}|^p} = C,$$

then we call C the **asymptotic error constant**.

One can interpret the order of convergence p in terms of decimal places of accuracy. If $|x^* - x^{(m)}| = 10^{-q}$, then $|x^* - x^{(m+1)}| \leq C \cdot 10^{-pq}$. So, if C is not too large, each iterate has roughly p times as many decimal places of accuracy as the previous iterate. For example, if $p = 2$, once the iterates become close to x^*, each iterate is accurate to roughly twice as many decimal places as the previous one.

When programming any iterative scheme, one should check the order of convergence using a **convergence plot**. By applying the algorithm to a problem for which the zero x^* is known, one can compute a sequence of errors $\varepsilon^{(m)} := x^* - x^{(m)}$. According to the definition (3.3-4),

$$\log\left|\varepsilon^{(m+1)}\right| \leq p \log\left|\varepsilon^{(m)}\right| + \log C.$$

Consequently, a plot of $\log|\varepsilon^{(m+1)}|$ versus $\log|\varepsilon^{(m)}|$, such as the one drawn in Figure 5, typically yields points lying roughly on a line having slope p.

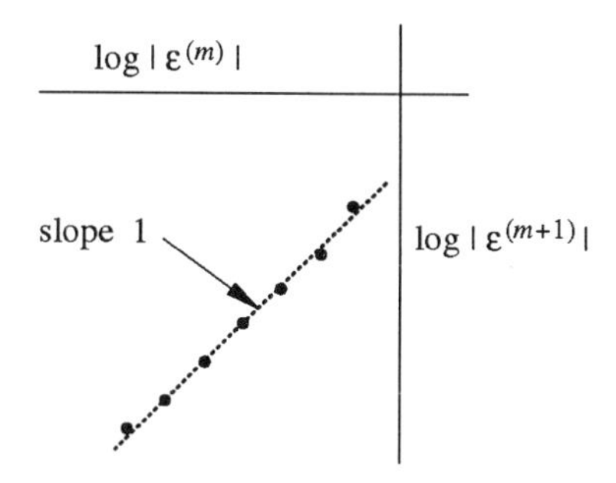

FIGURE 5. *A convergence plot for the problem illustrated in Figure 1, confirming that successive substitution converges linearly.*

For some schemes it is difficult to identify a power $p > 1$ for which the condition (3.3-4) holds, yet to say that the scheme converges linearly understates its actual performance. In these cases, the following definition may apply:

DEFINITION. *An iterative scheme producing sequences $\{x^{(m)}\}$ of approximations to x^** **converges superlinearly** *if there exists a sequence $\{C_m\}$ such that $C_m \to 0$ and*

$$\left|x^* - x^{(m+1)}\right| \leq C_m \left|x^* - x^{(m)}\right|. \tag{3.3-5}$$

By establishing the inequality (3.3-3), we have proved the following theorem:

THEOREM 3.3. *If the iteration function Φ is a contraction on some interval containing the iterates $x^{(0)}, x^{(1)}, x^{(2)}, \ldots$, then the successive substitution scheme (3.3-1) converges at least linearly.*

Figure 5 confirms this result graphically.

Under some circumstances, one can construct iteration functions Φ for which successive substitution converges with order 2 or greater. Problem 5 investigates this possibility.

The Lipschitz condition (3.3-2) also leads to two practical halting criteria. To derive them, we establish an inequality that has applications in several subsequent arguments.

LEMMA 3.4. *Let m, n be positive integers, and let $\{x^{(m)}\}$ be a sequence of iterates generated by the successive substitution scheme (3.3-1), where the iteration function Φ has Lipschitz constant $L < 1$ on some interval containing every iterate $x^{(m)}$. For $j = 0, 1, \ldots, m$,*

$$\left|x^{(m+n)} - x^{(m)}\right| \leq L^{m-j}\frac{1-L^n}{1-L}\left|x^{(j+1)} - x^{(j)}\right|. \tag{3.3-6}$$

This inequality relates the difference between any two iterates to the difference between an arbitrary pair of successive iterates that occur earlier in the sequence. The proof illustrates several standard techniques for working with contractions.

PROOF: We begin by writing $|x^{(m+n)} - x^{(m)}|$ as a telescoping sum and applying the triangle inequality:

$$\left|x^{(m+n)} - x^{(m)}\right| = \left|\sum_{i=m}^{m+n-1}\left(x^{(i+1)} - x^{(i)}\right)\right| \leq \sum_{i=m}^{m+n-1}\left|x^{(i+1)} - x^{(i)}\right|. \tag{3.3-7}$$

By reasoning as we did for the inequality (3.3-3), we estimate each term in this last sum as follows: For $j = 0, 1, \ldots, i$,

$$\begin{aligned}\left|x^{(i+1)} - x^{(i)}\right| = \left|\Phi(x^{(i)}) - \Phi(x^{(i-1)})\right| &\leq L\left|x^{(i)} - x^{(i-1)}\right| \\ \leq \cdots &\leq L^{i-j}\left|x^{(j+1)} - x^{(j)}\right|.\end{aligned} \tag{3.3-8}$$

Substituting this estimate into the relationship (3.3-7) and using the identity

$$(1-L)\sum_{i=0}^{n-1} L^i = 1 - L^n$$

shows that

$$\begin{aligned} \left|x^{(m+n)} - x^{(m)}\right| &\leq L^{m-j} \sum_{i=0}^{n-1} L^i \left|x^{(j+1)} - x^{(j)}\right| \\ &= L^{m-j} \frac{1-L^n}{1-L} \left|x^{(j+1)} - x^{(j)}\right|. \end{aligned}$$

This completes the proof. ∎

As an immediate consequence, we can estimate the error $\varepsilon^{(m)}$ in terms of the computed quantities $x^{(j+1)} - x^{(j)}$ for any j, $j = 0, 1, \ldots, m-1$. Since the iterative sequence converges to x^*, letting $n \to \infty$ in the inequality (3.3-7) yields

$$\lim_{n\to\infty} \left|x^{(m+n)} - x^{(m)}\right| = \left|x^* - x^{(m)}\right| = \left|\varepsilon^{(m)}\right| \leq \frac{L^{m-j}}{1-L} \left|x^{(j+1)} - x^{(j)}\right|.$$

The two special cases $j = 0$ and $j = m-1$ yield useful estimates:

COROLLARY 3.5 *Under the hypotheses of Lemma 3.4, the iterates $x^{(m)}$ generated using successive substitution obey the error estimates*

(*i*) $\left|x^* - x^{(m)}\right| \leq \dfrac{L^m}{1-L} \left|x^{(1)} - x^{(0)}\right|,$

(*ii*) $\left|x^* - x^{(m)}\right| \leq \dfrac{L}{1-L} \left|x^{(m)} - x^{(m-1)}\right|.$

The estimate (*ii*) vindicates the practice of halting successive substitution when the difference between successive iterates becomes small. We call this inequality an **a posteriori** error estimate, since it gauges the magnitude of the error in terms of the most recent information generated by the iterative scheme. In contrast, the inequality (*i*) is an **a priori** error estimate. It allows us to determine the number of iterations needed to satisfy a prescribed error tolerance as soon as we have taken one iteration. Both estimates require knowledge of L.

Mathematical Details

The following theorem guarantees that successive substitution converges if the initial guess is close to a fixed point.

THEOREM 3.6. *Suppose that the iteration function Φ has a fixed point x^* and that Φ is a contraction on some neighborhood $(x^* - \delta, x^* + \delta)$ of x^*, where $\delta > 0$. Let $x^{(0)}$ be any initial guess lying in $(x^* - \delta, x^* + \delta)$, and denote by*

$\{x^{(m)}\}$ the sequence of iterates generated by the successive substitution scheme (3.3-1). Then each $x^{(m)} \in (x^* - \delta, x^* + \delta)$, and $x^{(m)} \to x^*$ as $m \to \infty$.

PROOF: We show that $|x^* - x^{(m)}| < \delta$ by induction on m. The hypotheses ensure that $x^{(0)} \in (x^* - \delta, x^* + \delta)$. Assume that $x^{(i)} \in (x^* - \delta, x^* + \delta)$ for $i = 0, 1, \ldots, m$. To prove that $x^{(m+1)} \in (x^* - \delta, x^* + \delta)$, we repeatedly apply the argument used to establish the inequality (3.3-3), deducing that

$$\left|x^* - x^{(m+1)}\right| \leq L\left|x^* - x^{(m)}\right| \leq \cdots \leq L^{m+1}\left|x^* - x^{(0)}\right|.$$

But $|x^* - x^{(0)}| < \delta$, so $|x^* - x^{(m+1)}| < L^{m+1}\delta < \delta$, completing the induction. Since $L^{m+1} \to 0$ as $m \to \infty$, we have also shown that $|x^* - x^{(m+1)}| \to 0$ as $m \to \infty$. ∎

With a few alterations in the hypotheses, we can prove a more powerful theorem.

THEOREM 3.7. *Let Φ be a contraction with Lipschitz constant L on some closed interval $[x^{(0)} - \delta, x^{(0)} + \delta]$ about an initial guess $x^{(0)}$. Suppose that the successive substitution scheme (3.3-1) satisfies the condition*

$$\left|x^{(1)} - x^{(0)}\right| = \left|\Phi(x^{(0)}) - x^{(0)}\right| \leq (1 - L)\delta. \tag{3.3-9}$$

Then

(A) *Each iterate $x^{(m)} \in (x^{(0)} - \delta, x^{(0)} + \delta)$.*

(B) *The sequence $\{x^{(m)}\}$ converges to a point $x^* \in [x^{(0)} - \delta, x^{(0)} + \delta]$.*

(C) *The limit x^* is the unique fixed point of Φ in $[x^{(0)} - \delta, x^{(0)} + \delta]$.*

There are two reasons for the greater utility of this theorem. First, its hypotheses do not require us to know in advance that a fixed point x^* exists for Φ. Second, the parameter δ in this theorem is the radius of a ball centered at a *known* point $x^{(0)}$, and consequently we may be able to determine δ more realistically here than in Theorem 3.6.

PROOF: We prove (A) by induction on m. By hypothesis $x^{(1)} \in (x^{(0)} - \delta, x^{(0)} + \delta)$. Assume that $x^{(i)} \in (x^{(0)} - \delta, x^{(0)} + \delta)$ for $i \leq m$. To show that $x^{(m+1)} \in (x^{(0)} - \delta, x^{(0)} + \delta)$, we apply Lemma 3.4, observing that

$$\left|x^{(m+1)} - x^{(0)}\right| \leq \frac{1 - L^{m+1}}{1 - L}\left|x^{(1)} - x^{(0)}\right|.$$

But $|x^{(1)} - x^{(0)}| \leq (1 - L)\delta$, so

$$\left|x^{(m+1)} - x^{(0)}\right| \leq \left(1 - L^{m+1}\right)\delta < \delta,$$

which completes the induction.

To establish (B), it suffices to show that $\{x^{(m)}\}$ is a Cauchy sequence. Given $\epsilon > 0$, we find an integer M so large that $|x^{(m+n)} - x^{(m)}| < \epsilon$ whenever $n > 0$ and $m \geq M$. Lemma 3.4 with $j = 0$ yields

$$\left|x^{(m+n)} - x^{(m)}\right| \leq L^m \frac{1 - L^n}{1 - L} \left|x^{(1)} - x^{(0)}\right| \leq L^m (1 - L^n)\,\delta < L^m \delta.$$

It follows that $|x^{(m+n)} - x^{(m)}| \leq L^m\delta$, so $|x^{(m+n)} - x^{(m)}| < \epsilon$ whenever m is large enough to make $L^m\delta < \epsilon$. We therefore choose M to be any integer such that

$$M > \frac{\log(\epsilon/\delta)}{\log L}.$$

To prove (C), call $\lim_{m\to\infty} x^{(m)} = x^*$, and use the continuity of Φ:

$$x^* = \lim_{m\to\infty} x^{(m+1)} = \lim_{m\to\infty} \Phi(x^{(m)}) = \Phi(x^*).$$

Thus x^* is a fixed point of Φ. Uniqueness follows from the Lipschitz condition: If x^{**} is a fixed point of Φ in $(x^{(0)} - \delta, x^{(0} + \delta)$, then

$$|x^* - x^{**}| = |\Phi(x^*) - \Phi(x^{**})| \leq L\,|x^* - x^{**}|.$$

When $L = 0$, we have $|x^* - x^{**}| \leq 0$, so $x^* = x^{**}$. When $L \in (0, 1)$, the inequality yields the absurd conclusion $|x^* - x^{**}| < |x^* - x^{**}|$ unless $x^* = x^{**}$. ∎

3.4 Newton's Method in One Variable

Motivation and Construction

Figure 1 shows the graph of a function f defining a straight line. For such functions there is a simple scheme for solving the equation $f(x) = 0$: Given a point $(x^{(0)}, y^{(0)}) = (x^{(0)}, f(x^{(0)}))$ lying on the graph of f, set

$$x^{(1)} \leftarrow x^{(0)} - \left(\frac{1}{\text{slope}}\right) y^{(0)} = x^{(0)} - \frac{f(x^{(0)})}{f'(x^{(0)})}.$$

This scheme converges in one iteration to the point x^* where the graph of f crosses the x-axis. Newton's method exploits this idea: At each iterative level m we approximate the graph of f near $x^{(m)}$ by a straight line passing through the point $(x^{(m)}, f(x^{(m)}))$ and having slope $f'(x^{(m)})$. The zero $x^{(m+1)} := x^{(m)} - f(x^{(m)})/f'(x^{(m)})$ of this approximating function then becomes the next iterate. Figure 2 shows how this scheme works, at least ideally.

A more abstract picture of Newton's method facilitates analysis and produces an algorithm that extends more readily to several dimensions. Assume

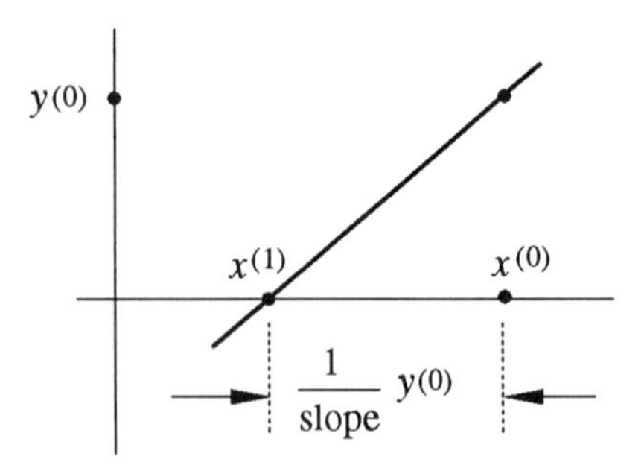

FIGURE 1. *How to find the zero of a straight line.*

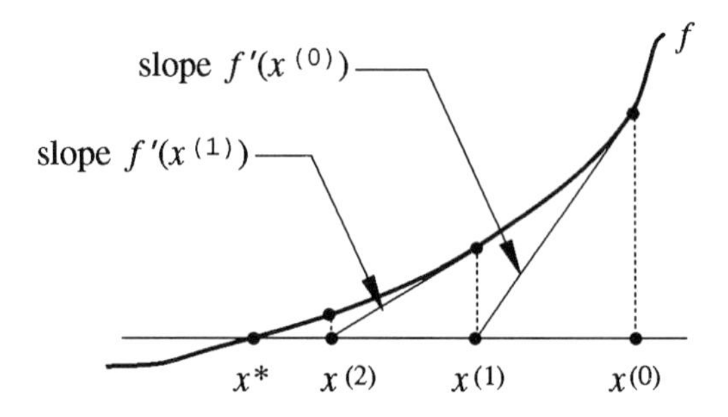

FIGURE 2. *Progress of the iterations in Newton's method.*

that $f \in C^1([a,b])$ and that the zero x^* and the iterate $x^{(m)}$ lie in (a,b). The fundamental theorem of calculus (Theorem 0.11) asserts that

$$0 = f(x^*) = f(x^{(m)}) + \int_{x^{(m)}}^{x^*} f'(t)\,dt.$$

Now approximate the integral by the expression $f'(x^{(m)})(x^* - x^{(m)})$:

$$0 \simeq f(x^{(m)}) + f'(x^{(m)})\left(x^* - x^{(m)}\right). \tag{3.4-1}$$

If we change the sign $\simeq$ to $=$, then x^* typically does not satisfy this equation exactly, so we must regard the equation as defining a new approximation $x^{(m+1)}$ to x^*. We solve for this approximation via the following steps:

$$\begin{aligned} &(i) \quad \delta^{(m+1)} &:=& \quad x^{(m+1)} - x^{(m)} \leftarrow -f(x^{(m)})/f'(x^{(m)}), \\ &(ii) \quad x^{(m+1)} &\leftarrow& \quad x^{(m)} + \delta^{(m+1)}. \end{aligned} \tag{3.4-2}$$

Practical Considerations

There are two topics of practical concern in the application of Newton's method: the convergence rate of the method and its sensitivity to initial guesses.

Under many circumstances Newton's method converges quadratically. To appreciate how fast this rate can be, consider a hypothetical sequence of errors $\varepsilon^{(m)} := x^* - x^{(m)}$ satisfying the quadratic relationship $|\varepsilon^{(m+1)}| \leq 1 \cdot |\varepsilon^{(m)}|^2$. If $\varepsilon^{(0)} = 0.1$, we have

$$\begin{aligned}
|\varepsilon^{(1)}| &\leq 0.01 \\
|\varepsilon^{(2)}| &\leq 0.0001 \\
|\varepsilon^{(3)}| &\leq 0.00000001 \\
|\varepsilon^{(4)}| &\leq 0.0000000000000001 \\
&\vdots
\end{aligned}$$

Thus each iteration gains roughly twice as many significant digits as the one before it. This convergence is much faster than the linear convergence discussed in the previous section, where each iteration reduces the errror roughly by a constant factor.

One way to understand the convergence rate of Newton's method is to write the method as a successive substitution scheme, using the iteration function

$$\Phi(x) := x - \frac{f(x)}{f'(x)}.$$

For the moment, let us proceed formally, assuming that Φ satisfies all of the conditions necessary to justify our manipulations. Expanding Φ using the Taylor theorem (Theorem 0.12) about a fixed point x^* yields

$$\begin{aligned}
x^{(m+1)} - x^* &= \Phi(x^{(m)}) - \Phi(x^*) \\
&= \Phi(x^*) + \Phi'(x^*)\left(x^{(m)} - x^*\right) \\
&\quad + \tfrac{1}{2}\Phi''(\zeta_m)\left(x^{(m)} - x^*\right)^2 - \Phi(x^*) \qquad (3.4\text{-}3)\\
&= \Phi'(x^*)\left(x^{(m)} - x^*\right) + \tfrac{1}{2}\Phi''(\zeta_m)\left(x^{(m)} - x^*\right)^2,
\end{aligned}$$

where ζ_m denotes a point lying strictly between $x^{(m)}$ and x^*. Since $f(x^*) = 0$,

$$\Phi'(x^*) = \frac{f(x^*)f''(x^*)}{[f'(x^*)]^2} = 0.$$

(We discuss below what happens when $f'(x^*) = 0$.)

Here lies the core of one possible convergence proof: Since $\Phi'(x^*) = 0$, smoothness of Φ implies that there must be a neighborhood of x^* in which Φ' is "close" to 0. Specifically, there must be a region about x^* in which $|\Phi'| < 1$, and in this region Φ is a contraction. It follows that any initial guess chosen inside this region yields an iterative sequence $\{x^{(m)}\}$ that converges to x^*.

The fact that $x^{(m)} \to x^*$ as $m \to \infty$ forces $\zeta_m \to x^*$ as $m \to \infty$. Therefore, unless we are lucky enough to have $x^{(m)} = x^*$, we can divide through by $(x^{(m)} - x^*)^2$ in Equation (3.4-3) to get

$$\begin{aligned} \frac{x^{(m+1)} - x^*}{\left(x^{(m)} - x^*\right)^2} &= \tfrac{1}{2}\Phi''(\zeta_m) \\ &\to \tfrac{1}{2}\Phi''(x^*) = \frac{f''(x^*)}{2f'(x^*)}, \quad \text{as} \quad m \to \infty. \end{aligned} \tag{3.4-4}$$

In particular, if $|\Phi''|$ is bounded above by a constant $M > 0$, then

$$\left|x^* - x^{(m+1)}\right| \leq \frac{M}{2}\left|x^* - x^{(m)}\right|^2.$$

This inequality establishes quadratic convergence for Newton's method, at least in the "generic" case.

One can make this argument rigorous by adding appropriate hypotheses about the behavior of f and its derivatives. Problem 5 asks for details. The problem also suggests a generalization of the Taylor expansion approach that produces iterative schemes converging with order $p = 3, 4, \ldots$. However, arguments along these lines require f to possess greater smoothness than that needed to derive the scheme via the fundamental theorem of calculus. Later in this section we analyze the convergence of Newton's method without relying on this extra smoothness.

As with any iterative method, it is worthwhile to construct a convergence plot for coded versions of Newton's method. Table 3.2 lists the results, to five decimal places, of Newton's method applied to the model equation $x^2 - 1 = 0$. The initial guess is $x^{(0)} = 9$, and the iterates converge to the exact solution $x^* = 1$. Figure 3 shows the convergence plot of $\ln|x^* - x^{(m+1)}|$ versus $\ln|x^* - x^{(m)}|$, illustrating that the points lie close to a line having slope 2.

The iterates generated for this model equation raise the issue of halting criteria. One simple criterion uses the mean value theorem: When f is continuously differentiable near x^*, $f(x^{(m)}) - f(x^*) = f'(\zeta)(x^{(m)} - x^*)$ for some ζ lying between $x^{(m)}$ and x^*. Therefore,

$$\left|x^* - x^{(m)}\right| \leq \frac{|f(x^{(m)})|}{\mu}, \tag{3.4-5}$$

where μ is a lower bound for $|f'(x)|$ on an interval containing x^* and the iterate $x^{(m)}$. (Problem 7 considers the case when $f'(x^*) = 0$.) This a posteriori estimate, which makes no use of any special properties of Newton's method, bounds the magnitude of the error $x^* - x^{(m)}$ in terms of the **residual** $f(x^{(m)})$, which is a computable quantity.

Table 3.2: Iterates generated by Newton's method for $x^2 - 1 = 0$.

m	$x^{(m)}$	$\ln\|x^* - x^{(m)}\|$	$f(x^{(m)})$
0	9.00000	2.07944	8.00000×10^1
1	4.55556	1.26851	1.97531×10^1
2	2.38753	0.32753	4.70032×10^0
3	1.40319	-0.90835	9.68937×10^{-1}
4	1.05793	-2.84860	1.19206×10^{-1}
5	1.00159	-6.44665	3.17415×10^{-3}
6	1.00000	-13.5880	2.51084×10^{-6}

Pretend, for example, that we do not know that $x^* = 1$ in the table above. By the sixth iteration, we have some confidence that x^* lies in the interval $(0.75, 1.25)$, and over this interval 1.50 is a lower bound for $f'(x) = 2x$. Since $f(x^{(6)}) \simeq 2.51084 \times 10^{-6}$, we conclude that $x^{(6)}$ differs from x^* by at most $2.51084 \times 10^{-6}/1.5 \simeq 1.67 \times 10^{-6}$.

The following algorithm incorporates this error estimate.

Algorithm 3.3 (Newton's method). *Let f be a differentiable function defined on an open interval (a, b) containing a zero x^* of f, and let $\mu := \inf_{x \in (a,b)} |f'(x)|$. Given an initial guess $x^{(0)} \in (a, b)$ and an error tolerance $\tau > 0$, the following algorithm generates a sequence $\{x^{(m)}\}$ of approximations to x^*.*

1. $m \leftarrow 0$.
2. If $|f(x^{(m)})| \geq \mu\tau$, then:
3. $\quad \delta^{(m+1)} \leftarrow -f(x^{(m)})/f'(x^{(m)})$.
4. $\quad x^{(m+1)} \leftarrow x^{(m)} + \delta^{(m+1)}$.
5. $\quad m \leftarrow m + 1$.
6. $\quad$ Go to 2.
7. End if.
8. End.

As the informality of the Taylor expansion argument may suggest, Newton's method does not always converge quadratically. Equation (3.4-4) clearly

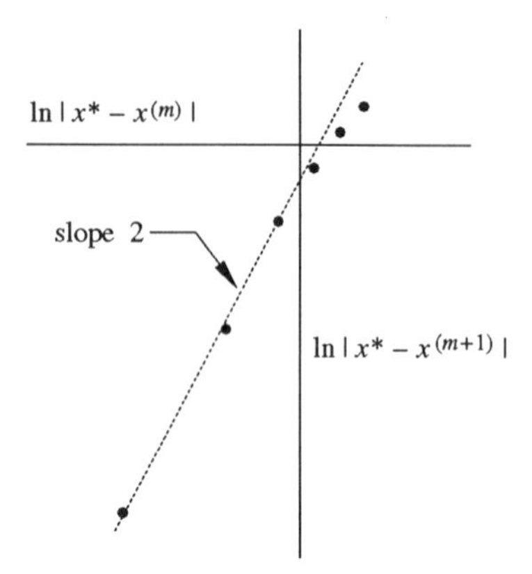

FIGURE 3. *Convergence plot for Newton's method applied to* $f(x) = x^2 - 1$, *with initial guess* $x^{(0)} = 9$.

shows that the argument itself runs into trouble when f' vanishes at x^*. This situation occurs when x^* is a zero having multiplicity 2 or greater, as discussed in Section 3.1. The Taylor expansion argument indicates that Newton's method converges quadratically near *simple* zeros of f. One can use a slightly more delicate analysis to show that *Newton's method converges linearly near zeros having multiplicity greater than 1.* Problem 5 asks for the details of this approach.

This reduction in the convergence rate pales in comparison with a more serious difficulty, namely, the sensitivity of the method to choices of initial guess $x^{(0)}$. Unfortunately, Newton's method may fail to converge to *any* zero of f if one chooses $x^{(0)}$ carelessly. Figure 4 illustrates, for example, how the existence of local extrema between $x^{(0)}$ and the exact zero x^* can lead to the calculation of iterates that diverge wildly. Such difficulties argue strongly for the hybrid strategy mentioned in Section 3.2: Use a slow but sure scheme like bisection to draw the iterates close to the sought zero, then use a fast scheme like Newton's method as a polisher to produce highly accurate approximations to x^*.

To implement such a hybrid scheme, it helps to know a set of conditions under which Newton's method is sure to converge. Figure 5 suggests one test of this sort. The figure shows the graph of a functions $f(x)$ over the interval $[a, b]$. The interval contains exactly one zero x^* of f. Graphically, at least, the following crucial observation seems clear: Once any iterate $x^{(m)}$ lands inside $[a, b]$, the behavior of f forces subsequent iterates to converge to x^*.

The function f in Figure 5 has three key properties. First, f has exactly one zero x^* in (a, b), and the zero is not a point where the graph of f is tangent to the x-axis. Second, f is either concave from above or concave from below on $[a, b]$. Third, Newton's method using either a or b as an initial guess

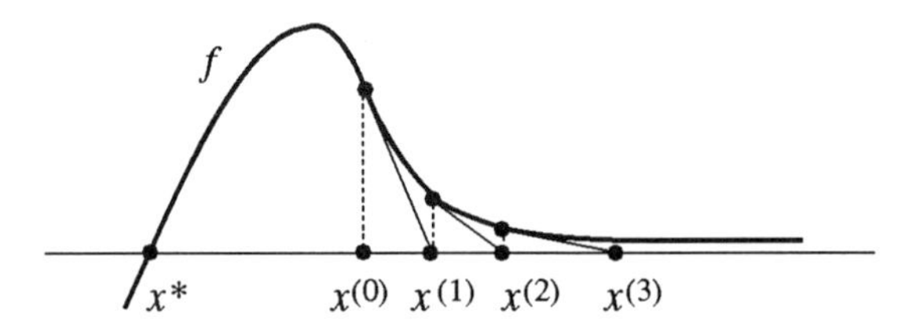

FIGURE 4. *Divergent sequence of iterates generated by Newton's method when a local extremum lies between the exact zero x^* and the initial guess $x^{(0)}$.*

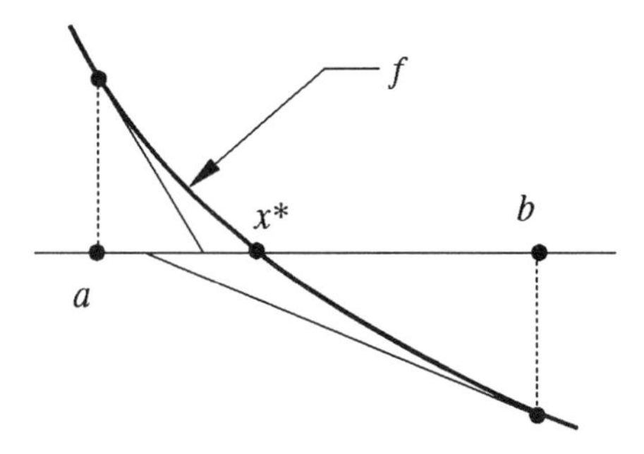

FIGURE 5. *Graph of a function f on an interval $[a,b]$ in which iterates generated by Newton's method are sure to converge to x^*.*

$x^{(0)}$ produces a subsequent iterate $x^{(1)}$ that lies inside (a,b). The following conditions capture these properties:

(i) $f(a)f(b) < 0$, and $f'(x) \neq 0$ for all $x \in [a,b]$.

(ii) Either $f''(x) \geq 0$ for all $x \in [a,b]$ or else $f''(x) \leq 0$ for all $x \in [a,b]$.

(iii) $|f(a)/f'(a)| < b-a$ and $|f(b)/f'(b)| < b-a$.

We prove shortly that Newton's method converges to x^* for any initial guess $x^{(0)}$ chosen in an interval $[a,b]$ satisfying these three conditions.

Mathematical Details

We now turn to a formal proof that Newton's method converges at least quadratically near simple zeros. At issue is how well the **affine model** $f(x^{(m)}) + f'(x^{(m)})(x - x^{(m)})$ approximates $f(x)$ near $x^{(m)}$. Instead of employing the more common argument based on the Taylor theorem, we use the following estimate.

LEMMA 3.8. *Let $f\colon (a,b) \to \mathbb{R}$ be differentiable, and suppose that f' satisfies a Lipschitz condition with Lipschitz constant $L > 0$ on (a,b). Then*

$$|f(y) - f(x) - f'(x)(y-x)| \leq \frac{L(y-x)^2}{2}. \tag{3.4-6}$$

This lemma readily extends to higher-dimensional settings, as we see in Section 3.7.

PROOF: Assume without loss of generality that $y > x$. Differentiability implies continuity, so by the fundamental theorem of calculus (Theorem 0.11),

$$f(y) - f(x) = \int_x^y f'(t)\,dt.$$

Rearranging this identity and taking absolute values yields

$$\begin{aligned}
|f(y) - f(x) - f'(x)(y-x)| &= \left| \int_x^y [f'(t) - f'(x)]\,dt \right| \\
&\leq \int_x^y |f'(t) - f'(x)|\,dt \\
&\leq L \int_x^y |t - x|\,dt = \frac{L(y-x)^2}{2}.
\end{aligned}$$ ∎

The main theorem asserts that, under certain conditions, Newton's method (3.4-2) yields a sequence $\{x^{(m)}\}$ that converges to x^*, so long as the initial guess $x^{(0)}$ lies close enough to x^*.

THEOREM 3.9. *Let $f\colon (a,b) \to \mathbb{R}$ be a differentiable function obeying the following conditions:*

(i) *f' has Lipschitz constant $L > 0$ on (a,b).*

(ii) *$|f'|$ is bounded away from 0 on (a,b), and $\mu := \inf_{x\in(a,b)} |f'(x)| > 0$.*

(iii) *There exists a point $x^* \in (a,b)$ such that $f(x^*) = 0$.*

Then there is a radius $\delta > 0$ such that, whenever the initial guess $x^{(0)} \in (x^* - \delta, x^* + \delta)$, the sequence $\{x^{(m)}\}$ of iterates produced by Newton's method lies in $(x^* - \delta, x^* + \delta)$, and $x^{(m)} \to x^*$ as $m \to \infty$.

PROOF: We begin with an observation. If Newton's method generates an iterate $x^{(m)}$ that lies in (a,b), then the fact that $f(x^*)=0$ implies that

$$\begin{aligned} \left|x^* - x^{(m+1)}\right| &= \left|x^* - \left[x^{(m)} - \frac{f(x^{(m)})}{f'(x^{(m)})}\right]\right| \\ &= \frac{1}{|f'(x^{(m)})|}\left|-f(x^*) + f(x^{(m)}) + f'(x^{(m)})\left(x^* - x^{(m)}\right)\right| \\ &\le \frac{L}{2\mu}\left|x^* - x^{(m)}\right|^2. \end{aligned} \tag{3.4-7}$$

The last line follows from Lemma 3.8 and the fact that $\mu \le |f'(x^{(m)})|$.

Now we identify δ. Let $\delta_0 := \min\{|x^*-a|, |x^*-b|\}$, so that δ_0 is the largest radius for which $(x^*-\delta_0, x^*+\delta_0) \subset (a,b)$. Then choose any $\theta \in (0,1)$ and set $\delta := \min\{\delta_0, \theta(2\mu/L)\}$. Let $\{x^{(m)}\}$ be a sequence of iterates generated by Newton's method with initial guess $x^{(0)} \in (x^*-\delta, x^*+\delta)$. We prove by induction that each $x^{(m)} \in (x^*-\delta, x^*+\delta)$. When $m=0$, the claim is true by hypothesis. If $x^{(m)} \in (x^*-\delta, x^*+\delta)$, then the inequality (3.4-7) holds. But the inductive hypothesis implies that

$$\left|x^* - x^{(m)}\right| < \delta \le \theta\left(\frac{2\mu}{L}\right),$$

so

$$\left|x^* - x^{(m+1)}\right| \le \frac{L}{2\mu}|x - x^*|^2 < \theta|x - x^{(m)}| < \theta\delta, \tag{3.4-8}$$

which completes the induction.

The estimate (3.4-8) also reveals that

$$\left|x^* - x^{(m)}\right| < \theta\left|x^* - x^{(m-1)}\right| < \cdots < \theta^m\left|x^* - x^{(0)}\right|.$$

Since $\theta \in (0,1)$, $x^{(m)} \to x^*$ as $m \to \infty$. ∎

The inequality (3.4-7) has the following immediate consequence:

COROLLARY 3.10. *Under the hypotheses of Theorem 3.9, Newton's method with initial guess $x^{(0)} \in (x^*-\delta, x^*+\delta)$ converges at least quadratically.*

Theorem 3.9 is a **local convergence theorem**, since it establishes the *existence* of an interval $(x^*-\delta, x^*+\delta)$ in which Newton's method converges but gives very little information about that interval. Indeed, we typically do not know x^*, and the proof itself gives only sketchy information about δ. In applications one often wants a **global convergence theorem**, that is, one that permits the positive identification of intervals on which the iterative scheme converges with certainty. The next theorem, confirming the conditions characterizing Figure 5, furnishes such a result.

THEOREM 3.11. *Let $f \in C^2([a,b])$ satisfy the following conditions:*

(i) *$f(a)f(b) < 0$, and $f'(x) \neq 0$ for all $x \in [a,b]$.*

(ii) *Either $f''(x) \geq 0$ for all $x \in [a,b]$ or else $f''(x) \leq 0$ for all $x \in [a,b]$.*

(iii) *$|f(a)/f'(a)| < b-a$ and $|f(b)/f'(b)| < b-a$.*

Then f has a unique zero $x^* \in (a,b)$, and Newton's method converges to x^* for any initial guess $x^{(0)} \in [a,b]$.

Our proof follows Henrici ([2], Section 4.8). The argument rests on a result from the theory of real variables: Every bounded sequence of real numbers that is monotonic (that is, nonincreasing or nondecreasing) has a limit. Problem 6 reviews a proof of this result. The idea is to establish that Newton's method generates a monotonic sequence of iterates bounded by x^*, then to show that the limit of this sequence must be x^*.

PROOF: The fact that f has a exactly one zero $x^* \in (a,b)$ follows from condition (i). We leave the proof as an exercise. The conditions (i) – (iii) comprise the following four cases:

(A) $f(a) < 0$, $f(b) > 0$, and $f'' \leq 0$;

(B) $f(a) < 0$, $f(b) > 0$, and $f'' \geq 0$;

(C) $f(a) > 0$, $f(b) < 0$, and $f'' \geq 0$;

(D) $f(a) > 0$, $f(b) < 0$, and $f'' \leq 0$.

The truth of the theorem for case (C) follows from the proof for case (A) if we consider the function $-f$ instead of f. Case (D) follows similarly from case (B). Moreover, by the change of variables $x \mapsto -x$, we can prove the theorem for case (B) by appealing to case (C) (and hence (A)), the only changes being that Newton's method generates a sequence corresponding to $\{-x^{(m)}\}$ and the zero in $[-b,-a]$ is now $-x^*$. Therefore it suffices to establish the theorem for case (A).

We use two properties of f'. First, $f' > 0$. To justify this assertion, notice that the mean value theorem guarantees the existence of some point $\zeta \in (a,b)$ for which

$$f'(\zeta) = \frac{f(b) - f(a)}{b-a} > 0.$$

Since f' is continuous and never passes through 0 on $[a,b]$, it therefore must be positive throughout the interval. Second, f' is nondecreasing on $[a,b]$, that is, $f'(y) \leq f'(x)$ whenever $y > x$ on $[a, b]$. This fact follows from the hypothesis that $f'' \leq 0$ on $[a, b]$.

For any initial guess $x^{(0)} \in [a, b]$, either $x^{(0)} \in [a, x^*]$ or $x^{(0)} \in [x^*, b]$. We give a detailed proof for the former case, the argument in the latter case being similar.

We use induction on m to prove that $\{x^{(m)}\}$ is bounded above by x^* and nondecreasing. To start the argument, observe that $x^{(0)} \leq x^*$ by hypothesis. Also, $f(x^{(0)}) \leq 0$, and $f'(x^{(0)}) > 0$, so

$$x^{(1)} := x^{(0)} - \frac{f(x^{(0)})}{f'(x^{(0)})} \geq x^{(0)}.$$

Now assume that $x^{(m-1)} \leq x^*$. We must show that $x^{(m)} \leq x^*$ and $x^{(m)} \leq x^{(m+1)}$. If $x^{(m-1)} = x^*$, then $x^{(m)} = x^{(m+1)} = x^*$ since x^* is a fixed point of the iteration, and in this trivial case the induction is complete. Otherwise, the mean value theorem implies the existence of a point $\zeta \in (x^{(m-1)}, x^*)$ such that

$$\begin{aligned} -f(x^{(m-1)}) = f(x^*) - f(x^{(m-1)}) &= f'(\zeta)\left(x^* - x^{(m-1)}\right) \\ &\leq f'(x^{(m-1)})\left(x^* - x^{(m-1)}\right). \end{aligned}$$

This inequality follows from the fact that f' is nonincreasing. Since $f' > 0$,

$$-\frac{f(x^{(m-1)})}{f'(x^{(m-1)})} \leq x^* - x^{(m-1)}.$$

As a consequence,

$$x^{(m)} = x^{(m-1)} - \frac{f(x^{(m-1)})}{f'(x^{(m-1)})} \leq x^{(m-1)} + \left(x^* - x^{(m-1)}\right) = x^*.$$

In addition to establishing x^* as an upper bound for the iterative sequence, this inequality shows that $f(x^{(m)}) \leq 0$, from which it follows that

$$x^{(m+1)} = x^{(m)} - \frac{f(x^{(m)})}{f'(x^{(m)})} \geq x^{(m)}.$$

Thus the sequence $\{x^{(m)}\}$ is nondecreasing, and we have finished the induction.

Being monotonic and bounded in $[a, b]$, the sequence $\{x^{(m)}\}$ converges to some point $\overline{x} \in [a, b]$. To prove that $\overline{x} = x^*$, it suffices to show that $f(\overline{x}) = 0$, since x^* is the only zero of f in $[a, b]$. But the continuity of f and f' and the fact that f' never vanishes on $[a, b]$ imply that the iteration map $\Phi(x) := x - f(x)/f'(x)$ is continuous on $[a, b]$. Therefore,

$$\overline{x} = \lim_{m \to \infty} x^{(m+1)} = \Phi\left(\lim_{m \to \infty} x^{(m)}\right) = \Phi(\overline{x}),$$

which is possible only if $f(\overline{x}) = 0$. ∎

3.5 The Secant Method

Motivation and Construction

Despite the power of Newton's method in solving $f(x) = 0$, its use of the derivative f' can be troublesome. For one thing, f' often requires substantially more effort to evaluate than f. The function

$$f(x) = \frac{\exp\left[\sin^2\left(x^3 + x\right)\right]}{\ln\left[\tan(1/x)\right]}$$

is a case in point. More significantly, many applications involve functions whose evaluation requires extensive use of subroutines, and for such functions there may be no readily available, closed-form expressions for derivatives. Such functions abound in chemical engineering, for example, where the evaluation of thermodynamic properties of various mixtures often involves complicated sequences of table look-ups, interpolation, and the numerical approximation of definite integrals. We devote this section to a discussion of modifications to Newton's method that avoid the exact calculation of derivatives.

The main idea is to replace the Newton scheme

$$x^{(m+1)} \leftarrow x^{(m)} - \frac{f(x^{(m)})}{f'(x^{(m)})}$$

with an analog having the form

$$x^{(m+1)} \leftarrow x^{(m)} - \frac{f(x^{(m)})}{D_m}. \tag{3.5-1}$$

Here,

$$D_m := \frac{f(x^{(m)} + h_m) - f(x^{(m)})}{h_m}$$

is a difference quotient approximating $f'(x^{(m)})$. We call Equation (3.5-1) a **finite-difference Newton method**. The choice of the **offsets** h_m is clearly crucial to the definition of such a scheme. Having chosen a particular sequence $\{h_m\}$ of offsets, we say that $\{h_m\}$ **generates** the finite-difference Newton method (3.5-1).

Since D_m is supposed to approximate $f'(x^{(m)})$, we expect to get viable substitutes for Newton's method by using sequences $\{h_m\}$ of "small" numbers. One crude idea is to use the same small offset $h_m = h$ at every iteration. The following finite-difference Newton method results:

$$x^{(m+1)} \leftarrow x^{(m)} - \frac{hf(x^{(m)})}{f(x^{(m)} + h) - f(x^{(m)})}. \tag{3.5-2}$$

We show later that, if f has reasonable properties and h is small enough, then this scheme converges.

The key to *rapid* convergence, however, is to choose the generating sequence $\{h_m\}$ so that $h_m \to 0$ as $m \to \infty$. This way, the finite-difference analogs D_m presumably give better approximations to $f'(x^{(m)})$ — forcing the scheme to behave more like Newton's method proper — as the iterations progress.

An elegant way to implement this idea is to take $h_m = x^{(m-1)} - x^{(m)}$ for $m = 1, 2, 3, \ldots$. This choice yields the **secant method**,

$$x^{(m+1)} \leftarrow x^{(m)} - f(x^{(m)}) \frac{x^{(m-1)} - x^{(m)}}{f(x^{(m-1)}) - f(x^{(m)})}. \tag{3.5-3}$$

Figure 1 illustrates the scheme. While early values of the difference $x^{(m-1)} - x^{(m)}$ may not be very small, we expect that $x^{(m-1)} - x^{(m)} \to 0$ as $m \to \infty$. An analysis given later in this section shows that the secant method merits serious consideration as an alternative to Newton's method.

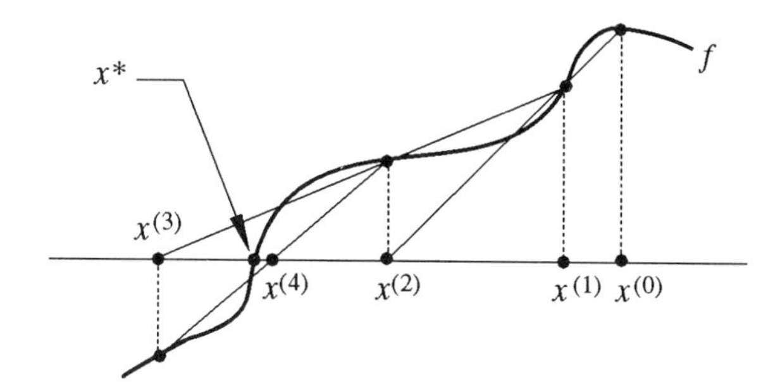

FIGURE 1. *Progress of iterates generated by the secant method.*

Other choices for h_m are also possible. The analysis given below offers several guidelines for constructing such methods, but for a detailed treatment we recommend Ortega and Rheinboldt [3].

Practical Considerations

The crude finite-difference Newton scheme (3.5-2) has three flaws. First, it demands that we decide what size offset h will be "small." This issue can be a thorny one. If h is too large, then the difference quotients D_m may not yield good approximations to $f'(x^{(m)})$. On the other hand, if h is too small, then $f(x^{(m)} + h)$ may be close in value to $f(x^{(m)})$. As a consequence, we may lose most of the significant digits in their machine representations, and thus in the value of D_m, by computing the difference $f(x^{(m)} + h) - f(x^{(m)})$. The resolution of this issue depends strongly on the behavior of f, and we do not pursue it further.

Second, the scheme (3.5-2) calls for two separate evaluations of f at each iteration. This feature can be unattractive if f itself requires extensive computation. Third, as we prove, the scheme converges only linearly and is therefore slow in comparison with Newton's method. One might regard linear convergence as a "fallback position:" Being comparable in convergence rate to bisection and successive substitution, the crude scheme (3.5-2) hardly warrants special attention.

The secant method answers these criticisms. By specifying the values of h_m for $m \geq 1$, the scheme eliminates the need to decide how small these offsets should be. It is necessary, however, to choose two values $x^{(0)}$ and $x^{(1)}$ to initialize the scheme. In practice, this task fits naturally into the hybrid strategy advocated in Section 3.2. For example, one can use bisection on a bracketing interval $[a, b]$ to produce the first few iterates $a = x^{(0)}, b = x^{(1)}, x^{(2)}, \ldots, x^{(m)}$, switching to the secant method as soon as Equation (3.5-3) produces an iterate $x^{(m+1)}$ that lies strictly between $x^{(m-1)}$ and $x^{(m)}$.

Also, the method requires only one new evaluation of f at each iteration. In this respect, the secant method has an advantage over Newton's method: Not only does it obviate evaluations of f', but it also demands less computation per iteration than Newton's method.

There remains the issue of convergence rate. While the secant method does not match the quadratic convergence of Newton's method, it does converge superlinearly. We show below that, under certain conditions on f, the convergence rate of the secant method is the **golden ratio**, $(1 + \sqrt{5})/2 \simeq 1.618$. This reasonably rapid convergence, together with the fact that the secant method requires just one evaluation of f per iteration, provokes the following observation: If f' is expensive to evaluate, then Newton's method exacts a high cost per iteration. For such functions, it may be more efficient to take a somewhat larger number of much cheaper iterations via the secant method.

It is difficult to quantify this idea rigorously, but a rough calculation suggests an interesting rule of thumb. Denote by $\varepsilon_N^{(m)}$ and $\varepsilon_S^{(m)}$ the errors $x^* - x^{(m)}$ in the mth iterates of the Newton and secant methods, respectively, and call $\gamma := (1 + \sqrt{5})/2$. Since the asymptotic error constants for the two methods depend on the function f, let us assume for argument's sake that both constants are 1. The convergence rates of the two schemes then yield

$$\left|\varepsilon_N^{(m)}\right| \simeq \left|\varepsilon_N^{(0)}\right|^{2^m}, \qquad \left|\varepsilon_S^{(m)}\right| \simeq \left|\varepsilon_S^{(0)}\right|^{\gamma^m}.$$

Therefore, to satisfy a halting criterion of the form $|\varepsilon_N^{(m)}| < \tau$ for $0 < \tau < 1$, we must iterate on Newton's method until

$$2^m \simeq R := \frac{\log \tau}{\log |\varepsilon^{(0)}|},$$

assuming that $|\varepsilon^{(0)}| < 1$. In other words, Newton's method requires about $\log_2 R$ iterations to satisfy the halting criterion. Similar reasoning shows that the secant method requires about $(\log_2 R)/(\log_2 \gamma)$ iterations.

Consider one "work unit" to be the computational effort needed to evaluate f, and let w signify the number of work units required to evaluate f'. Then Newton's method requires roughly $(1+w)\log_2 R$ work units to satisfy the halting criterion, while the secant method requires about $1\cdot(\log_2 R)/(\log_2 \gamma)$ work units. These two quantities are equal when $w = (1/\log_2 \gamma) - 1 \simeq 0.44042$. Therefore our rule of thumb is as follows: Use the secant method whenever the evaluation of f' requires more than 44 percent as much work as the evaluation of f.

Mathematical Details

The remaining task is to analyze finite-difference Newton methods (3.5-1) in general. The main theorem delineates when such methods converge, and the proof technique yields as corollaries some concrete results about convergence rates. To establish a precise convergence rate for the secant method, though, we must add hypotheses to prove a more specialized theorem.

We begin with an estimate relating the finite-difference analog D_m to the exact derivative of f.

LEMMA 3.12. *Let $f: (a,b) \to \mathbb{R}$ be differentiable, and suppose that f' satisfies a Lipschitz condition with Lipschitz constant $L > 0$. Call*

$$\mu := \inf_{x\in(a,b)} |f'(x)|.$$

There exists a number $\overline{h}_0 > 0$ such that, for every $h \in [-\overline{h}_0, \overline{h}_0]$,

$$\left|\frac{f(x+h)-f(x)}{h}\right| \geq \frac{\mu}{2},$$

so long as $x, x+h \in (a,b)$.

PROOF: We argue by contradiction. If no such number $\overline{h}_0$ exists, then we can pick a sequence $\{h_n\}$ in $\mathbb{R}$ with $h_n \to 0$ such that, for some $x \in (a,b)$,

$$\left|\frac{f(x+h_n)-f(x)}{h_n}\right| < \frac{\mu}{2},$$

for $n = 1, 2, 3, \ldots$. But then $|f'(x)| \leq \mu/2$, which is impossible. ∎

The main theorem is analogous to Theorem 3.9, in that it establishes *local* convergence. The salient requirement is that we pick the offsets h_m small enough so that the difference quotients D_m stay close in value to $f'(x^{(m)})$. The proof uses several ideas from the previous section.

THEOREM 3.13. *Let $f: (a,b) \to \mathbb{R}$ be differentiable and obey the following conditions:*

(i) *f has a zero $x^* \in (a,b)$.*

(ii) *$\mu := \inf_{x\in(a,b)} |f(x)| > 0$.*

(iii) *f' has Lipschitz constant $L > 0$ on (a,b).*

Let the sequence $\{h_m\}$ generate the finite-difference Newton method (3.5-1). Then there exist positive constants $\delta, \overline{h}$ such that, whenever the initial guess $x^{(0)} \in (x^* - \delta, x^* + \delta)$ and the generating sequence satisfies $|h_m| < \overline{h}$, each iterate $x^{(m)} \in (x^* - \delta, x^* + \delta)$. Moreover, $x^{(m)} \to x^*$ as $m \to \infty$.

PROOF: Call $\varepsilon^{(m)} := x^* - x^{(m)}$, and, using Lemma 3.12, choose $\overline{h}_0 > 0$ small enough to guarantee that

$$D_m := \frac{f(x^{(m)} + h) - f(x^{(m)})}{h} \geq \frac{\mu}{2},$$

whenever $|h| \leq \overline{h}_0$ and $x^{(m)}, x^{(m)} + h \in (a,b)$. Henceforth, we assume that $|h_m| \leq \overline{h}_0$.

We start by establishing the following claim: For any sequence $\{x^{(m)}\}$ of iterates generated by the scheme (3.5-1) and lying in the interval (a,b),

$$\left|\varepsilon^{(m+1)}\right| \leq \frac{L}{\mu}\left(\left|\varepsilon^{(m)}\right| + \left|h_m\right|\right)\left|\varepsilon^{(m)}\right|. \tag{3.5-4}$$

For proof observe that, since $f(x^*) = 0$,

$$\varepsilon^{(m+1)} = x^* - x^{(m+1)} = x^* - x^{(m)} + \frac{f(x^{(m)})}{D_m} - \frac{f(x^*)}{D_m}.$$

From this identity we deduce that

$$\begin{aligned} \left|\varepsilon^{(m+1)}\right| \leq \; & \underbrace{\left|\frac{1}{D_m}\right|}_{\text{(I)}}\Bigg[\underbrace{\left|f(x^*) - f(x^{(m)}) - f'(x^{(m)})\varepsilon^{(m)}\right|}_{\text{(II)}} \\ & + \underbrace{\left|f'(x^{(m)}) - D_m\right|}_{\text{(III)}}\left|\varepsilon^{(m)}\right|\Bigg]. \end{aligned} \tag{3.5-5}$$

By Lemma 3.12, (I) $\leq 2/\mu$. Also, Lemma 3.8 implies that (II) $\leq L|\varepsilon^{(m)}|^2/2$ and that (III) $\leq L|h_m|/2$. Substituting these estimates into the inequality (3.5-5) proves the claim.

Now we have the tools needed to establish the theorem. As in the proof of Theorem 3.9, let $\delta_0 := \min\{|x^* - a|, |x^* - b|\}$, and pick any $\theta \in (0,1)$. Define $\delta := \min\{\,\delta_0, \theta\mu/(2L)\,\}$ and $\overline{h} := \min\{\,\overline{h}_0, \delta\,\}$. Assume that $\{x^{(m)}\}$ is a sequence of iterates generated by the scheme (3.5-1), with $x^{(0)} \in (x^*-\delta, x^*+\delta)$ and $|h_m| \le \bar{h}$. It suffices to prove by induction on m that $|\varepsilon^{(m+1)}| \le \theta|\varepsilon^{(m)}|$.

When $m = 0$, the hypotheses ensure that $|\varepsilon^{(0)}| < \delta$ and $|h_0| \le \bar{h} \le \delta$, so the inequality (3.5-4) applies. We obtain

$$|\varepsilon^{(1)}| \le \frac{L}{\mu}\left(|\varepsilon^{(0)}| + |h_0|\right)|\varepsilon^{(0)}| \le \theta\,|\varepsilon^{(0)}|.$$

Now assume that $|\varepsilon^{(i+1)}| \le \theta|\varepsilon^{(i)}|$ for $i = 0, 1, \ldots, m-1$, so in particular $x^{(m)} \in (x^* - \delta, x^* + \delta)$ and $|\varepsilon^{(m)}| \le \delta$. The inequality (3.5-4) again yields

$$|\varepsilon^{(m+1)}| \le \frac{L}{\mu}\left(|\varepsilon^{(m)}| + |h_m|\right)|\varepsilon^{(m)}| \le \theta\,|\varepsilon^{(m)}|, \tag{3.5-6}$$

completing the induction and the proof. ∎

The estimate (3.5-6) shows that one can select a constant offset $h_m = h$ small enough to ensure that the scheme (3.5-2) converges linearly. For more sophisticated schemes, one can exploit properties of the sequences $\{h_m\}$ to refine the convergence estimates. For example:

COROLLARY 3.14. *If the hypotheses of Theorem 3.13 apply and $h_m \to 0$ as $m \to \infty$, then the finite-difference Newton method (3.5-1) converges superlinearly.*

PROOF: In the proof of Theorem 3.13, define

$$C_m := \frac{\theta L}{2\mu}\max\left\{|\varepsilon^{(m)}|, |h_m|\right\}.$$

The inequality (3.5-6) gives $|\varepsilon^{(m+1)}| \le C_m|\varepsilon^{(m)}|$, and $C_m \to 0$ as $m \to \infty$. ∎

In particular, the secant method converges superlinearly. Problem 8 asks for a proof of the following corollary and mentions an application:

COROLLARY 3.15. *If the hypotheses of Theorem 3.13 apply and there exists a constant $C > 0$ such that $|h_m| \le C|\varepsilon^{(m)}|$, then the finite-difference Newton method (3.5-1) converges quadratically.*

It remains to establish the precise convergence rate of the secant method. In preparation for this task, we review some elementary results from the theory of **divided differences**, summarized in Appendix A. Given a function $f: [a, b] \to \mathbb{R}$ and a set $\Delta := \{x_0, x_1, \ldots, x_n\}$ of distinct points in $[a, b]$, we define the divided differences of f on Δ inductively:

$$f[x_i] := f(x_i),$$

$$f[x_i, x_{i+1}, \ldots, x_{i+k}] := \frac{f[x_{i+1}, \ldots, x_{i+k}] - f[x_i, \ldots, x_{i+k-1}]}{x_{i+k} - x_i}.$$

For example,

$$f[x_{i-1}, x_i] := \frac{f(x_i) - f(x_{i-1})}{x_i - x_{i-1}},$$

$$f[x_{i-1}, x_i, x_{i+1}] := \frac{f[x_i, x_{i+1}] - f[x_{i-1}, x_i]}{x_{i+1} - x_{i-1}}.$$

Theorem A.1 asserts a useful fact about divided differences: *If $f \in C^n([a,b])$, then there exists a point $\zeta \in (a,b)$ such that*

$$f[x_0, x_1, \ldots, x_n] = \frac{f^{(n)}(\zeta)}{n!}. \tag{3.5-7}$$

This property of divided differences recalls the mean value theorem, which plays a central role in its proof. Equation (3.5-7) also suggests a connection between divided differences and differentiation. We use the equation in proving the following theorem.

THEOREM 3.16. *Let $f \in C^2([a,b])$, and suppose that*

(i) *$f(x^*) = 0$ for some point $x^* \in (a,b)$.*

(ii) *$\mu := \inf_{x\in[a,b]} |f'(x)| > 0$.*

(iii) *$\nu := \sup_{x\in[a,b]} |f''(x)| > 0$.*

If the secant method with initial guesses $x^{(0)}, x^{(1)} \in [a,b]$ converges to x^*, then the errors $\varepsilon^{(m)} := x^* - x^{(m)}$ obey an inequality of the form

$$\left|\varepsilon^{(m+1)}\right| \le C\left|\varepsilon^{(m)}\right|^{\gamma},$$

where $C > 0$ is a constant independent of m and $\gamma := (1+\sqrt{5})/2 \simeq 1.618$.

PROOF: By the definition of the secant method and Equation (3.5-7),

$$\begin{aligned}
-\varepsilon^{(m+1)} &= -\varepsilon^{(m)} - \frac{x^{(m)} - x^{(m-1)}}{f(x^{(m)}) - f(x^{(m-1)})} f(x^{(m)}) \\
&= \varepsilon^{(m)}\varepsilon^{(m-1)} \frac{f\left[x^{(m-1)}, x^{(m)}, x^*\right]}{f\left[x^{(m-1)}, x^{(m)}\right]} \\
&= \varepsilon^{(m)}\varepsilon^{(m-1)} \frac{f''(\zeta_2)}{2f'(\zeta_1)},
\end{aligned}$$

for some points $\zeta_1, \zeta_2 \in (a,b)$. Call $M := \nu/(2\mu) \neq 0$. We have

$$\left|\varepsilon^{(m+1)}\right| \le M\left|\varepsilon^{(m)}\right| \left|\varepsilon^{(m-1)}\right|. \tag{3.5-8}$$

To determine the order of convergence of the secant method, we seek constants $p, C > 0$ such that

$$\left|\varepsilon^{(m+1)}\right| \le C\left|\varepsilon^{(m)}\right|^p \le C^{p+1}\left|\varepsilon^{(m-1)}\right|^{p^2}.$$

According to the inequality (3.5-8), these constants exist provided that they also satisfy the inequality

$$\left|\varepsilon^{(m+1)}\right| \le MC\left|\varepsilon^{(m-1)}\right|^{p+1}.$$

Therefore, it suffices to find positive solutions C and p to the equations

$$C^{p+1} = MC, \qquad p^2 = p + 1.$$

The solutions are $C = M^{1/p}$ and $p = \gamma$. ∎

3.6 Successive Substitution: Several Variables

Motivation and Construction

Many of the methods that we have discussed for solving $f(x) = 0$ extend to methods for *systems* of nonlinear equations. These systems have the general form

$$\mathbf{f}(\mathbf{x}) := \begin{bmatrix} f_1(x_1, x_2, \ldots, x_n) \\ f_2(x_1, x_2, \ldots, x_n) \\ \vdots \\ f_n(x_1, x_2, \ldots, x_n) \end{bmatrix} = \mathbf{0}. \tag{3.6-1}$$

Here is a simple example with $n = 2$:

$$\begin{aligned} x_1^2 \sin x_2 &= 0, \\ \cos x_1 + x_2 - 1 &= 0. \end{aligned} \tag{3.6-2}$$

This section examines the method of successive substitution, introduced in Section 3.3, in the more general setting of Equation (3.6-1).

The multidimensional character of Equation (3.6-1) causes several difficulties. One of these is that simple, sure-fire methods like bisection have no straightforward extension to systems with $n > 1$. Therefore it is all the more crucial to understand the nature and approximate locations of zeros before launching a numerical scheme. On the other hand, intuition about systems with $n > 1$ can be hard won, especially since geometric reasoning grows more and more difficult as n increases. In many applications one can learn as much from analytic or physical considerations as from attempts to visualize the geometry of the problem.

The system (3.6-2), which is simple enough to allow both geometric and analytic reasoning, illustrates some of the features of nonlinear systems. Here the graph of $\mathbf{f}$ is the set of points $(x_1, x_2, f_1(x_1, x_2), f_2(x_1, x_2))$ in $\mathbb{R}^4$. We have no hope of plotting this graph to find the zeros of $\mathbf{f}$. However, we can examine the level sets $f_1(x_1, x_2) = 0$ and $f_2(x_1, x_2) = 0$ separately. The function $f_1(x_1, x_2) = x_1^2 \sin x_2$ vanishes along the line $x_1 = 0$ and along every line of the form $x_2 = k\pi$, for $k = 0, \pm 1, \pm 2, \ldots$. Figure 1 shows these lines as dashed curves. The function $f_2(x_1, x_2) = \cos x_1 + x_2 - 1$ vanishes on the curve $x_2 = 1 - \cos x_1$, plotted as a solid curve in Figure 1. The two level sets intersect at the points $(x_1, x_2) = (0, 2k\pi)$ and $(x_1, x_2) = (2, (2k+1)\pi)$, for $k = 0, \pm 1, \pm 2, \ldots$. These points are the zeros of the system (3.6-1).

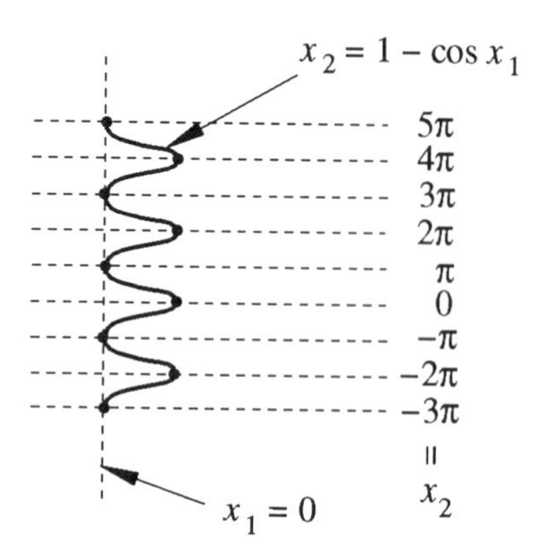

FIGURE 1. *Level sets $f_1 = 0$ (dashed curve) and $f_2 = 0$ (solid curve) for the system (3.6-2), showing the locations of the zeros of the system.*

In this case, the system of equations has infinitely many solutions. The best that we can hope for numerically is an iterative scheme that converges to one of the zeros $\mathbf{x}^* = (x_1^*, x_2^*)$, given an initial guess lying close to $\mathbf{x}^*$ but far from the other zeros. Also, the system changes dramatically with small changes in one of the parameters. Suppose that we replace $f_2(x_1, x_2) = \cos x_1 + x_2 - 1$ by the perturbed function $\hat{f}_2(x_1, x_2) := \cos x_1 + x_2 - (1 + \epsilon)$, where $|\epsilon| \ll 1$. Then the new system has two zeros near each point $(0, 2k\pi)$ when $\epsilon < 0$ but only one zero in the entire plane when $\epsilon > 0$. These features — nonuniqueness of solutions and sensitivity to small changes in paramters — commonly occur in more complicated systems and in systems having higher dimension, where analytic and geometric properties of the zeros may be quite obscure.

To solve systems of the form (3.6-1) by successive substitution, we construct an iteration function $\mathbf{\Phi}$ that has a fixed point $\mathbf{x}^* = \mathbf{\Phi}(\mathbf{x}^*)$ at the desired zero $\mathbf{x}^*$ of $\mathbf{f}$. Then, given an initial guess $\mathbf{x}^{(0)}$, we iterate using the algorithm

$$\mathbf{x}^{(m+1)} \leftarrow \mathbf{\Phi}(\mathbf{x}^{(m)}). \tag{3.6-3}$$

We wish to know under what circumstances, and how fast, $\mathbf{x}^{(m)} \to \mathbf{x}^*$ as $m \to \infty$. We discuss this issue, then examine an application.

Convergence Criteria

Much of the theory concerning the scheme (3.6-3) follows the conceptual paths established in Section 3.3 for the one-dimensional case. Given a few key definitions and appropriate changes in notation, one can mimic the proofs of theorems from that section to obtain convergence criteria for successive substitution in several variables. In the discussion below, we forgo formal proofs wherever such straightforward extensions are possible. Readers should verify that the extensions are indeed straightforward!

The main change necessary in multidimensional settings is to convert absolute values of various scalar quantities to norms of their vector counterparts. In what follows, $\|\cdot\|$ stands for any norm on $\mathbb{R}^n$. By Theorem 0.6, all norms on $\mathbb{R}^n$ are equivalent, and hence any condition that guarantees convergence of a sequence $\{\mathbf{x}^{(m)}\}$ in one norm suffices to guarantee convergence in any norm on $\mathbb{R}^n$. (See Problem 5, Chapter 0.)

We measure convergence rates in as in the one-dimensional theory:

DEFINITION. *Let $p \geq 1$. An iterative scheme that produces sequences $\{\mathbf{x}^{(m)}\}$ of approximations to $\mathbf{x}^* \in \mathbb{R}^n$* **converges with order** *p if there exists a constant $C > 0$ and an integer $M \geq 0$ such that*

$$\left\|\mathbf{x}^* - \mathbf{x}^{(m+1)}\right\| \leq C \left\|\mathbf{x}^* - \mathbf{x}^{(m)}\right\|^p$$

whenever $m \geq M$. If $p = 1$, we must have $0 \leq C < 1$, and we say that the scheme converges **linearly**. *If $p = 2$, the scheme converges* **quadratically**. *If*

$$\lim_{m\to\infty} \frac{\|\mathbf{x}^* - \mathbf{x}^{(m+1)}\|}{\|\mathbf{x}^* - \mathbf{x}^{(m)}\|^p} = C,$$

then we call C the **asymptotic error constant**. *The scheme* **converges superlinearly** *if there exists a sequence $\{C_m\}$ of positive real numbers such that $C_m \to 0$ and*

$$\left\|\mathbf{x}^* - \mathbf{x}^{(m+1)}\right\| \leq C_m \left\|\mathbf{x}^* - \mathbf{x}^{(m)}\right\|.$$

As in one dimension, we prefer schemes that converge superlinearly whenever they are available.

Central to the convergence of the scheme (3.6-3) is the following concept:

DEFINITION. *Let $S \subset \mathbb{R}^n$. A function $\mathbf{\Phi}: S \to \mathbb{R}^n$ satisfies a* **Lipschitz condition** *on S (with respect to the norm $\|\cdot\|$) if there exists a constant $L > 0$ such that, for any two points $\mathbf{x}, \mathbf{y} \in S$,*

$$\|\mathbf{\Phi}(\mathbf{x}) - \mathbf{\Phi}(\mathbf{y})\| \leq L\|\mathbf{x} - \mathbf{y}\|.$$

The greatest lower bound for such constants is the **Lipschitz constant** *for* $\mathbf{\Phi}$ *on* S. *If* $\mathbf{\Phi}$ *has Lipschitz constant* $L < 1$ *on* S, *then* $\mathbf{\Phi}$ *is a* **contraction** *on* S.

The next result serves as a multidimensional analog of Lemma 3.3, which estimates the distance between any two iterates in the sequence $\{\mathbf{x}^{(m)}\}$ generated by the scheme (3.6-3). The multidimensional case requires a slight change in the hypotheses: Instead of demanding that $\mathbf{\Phi}$ be a contraction on some interval containing the iterates, we require $\mathbf{\Phi}$ to be a contraction on an open set containing the iterates.

LEMMA 3.17. *Let* m, n *be positive integers, and let* $\{\mathbf{x}^{(m)}\}$ *be a sequence of iterates generated by the successive substitution scheme (3.6-3). Suppose that the iteration function* $\mathbf{\Phi}$ *has Lipschitz constant* $L < 1$ *on some open set containing every iterate* $\mathbf{x}^{(m)}$. *For* $j = 0, 1, \ldots, m$,

$$\left\|\mathbf{x}^{(m+n)} - \mathbf{x}^{(m)}\right\| \leq L^{m-j}\frac{1-L^n}{1-L}\left\|\mathbf{x}^{(j+1)} - \mathbf{x}^{(j)}\right\|. \qquad (3.6\text{-}4)$$

PROOF: The proof follows the argument used for Lemma 3.3. ∎

The basic convergence result is analogous to Theorem 3.6. It asserts, in effect, that contractions yield convergent iterative schemes. As in one dimension, $\mathbf{\Phi}$ need not be a contraction globally, so long as it is one in some region surrounding the fixed point $\mathbf{x}^*$. Once again, we must replace intervals by sets in $\mathbb{R}^n$. In this case, we use the multidimensional analogs of open intervals $(x^* - \delta, x^* + \delta)$, which are open balls:

$$\mathcal{B}_\delta(\mathbf{x}^*) := \left\{\mathbf{x} \in \mathbb{R}^n : \|\mathbf{x}^* - \mathbf{x}\| < \delta\right\}.$$

THEOREM 3.18. *Suppose that the iteration function* $\mathbf{\Phi}$ *has a fixed point* $\mathbf{x}^*$ *and that* $\mathbf{\Phi}$ *is a contraction on some ball* $\mathcal{B}_\delta(\mathbf{x}^*)$, *where* $\delta > 0$. *Let* $\mathbf{x}^{(0)}$ *be any initial guess lying in* $\mathcal{B}_\delta(\mathbf{x}^*)$, *and denote by* $\{\mathbf{x}^{(m)}\}$ *the sequence of iterates generated by the successive substitution scheme (3.6-3). Then each* $\mathbf{x}^{(m)} \in \mathcal{B}_\delta(\mathbf{x}^*)$, *and* $\mathbf{x}^{(m)} \to \mathbf{x}^*$ *as* $m \to \infty$.

PROOF: The proof follows the argument used for Theorem 3.6. ∎

It is also possible to prove a direct analog of the more powerful convergence result, Theorem 3.7. This result rests on hypotheses concerning the behavior of $\mathbf{\Phi}$ on a *closed* ball

$$\overline{\mathcal{B}_\delta(\mathbf{x}^{(0)})} := \left\{\mathbf{x} \in \mathbb{R}^n : \|\mathbf{x}^{(0)} - \mathbf{x}\| \leq \delta\right\}$$

centered at the initial guess $\mathbf{x}^{(0)}$.

THEOREM 3.19. *Let $\mathbf{\Phi}$ be a contraction with Lipschitz constant L on some closed ball $\overline{\mathcal{B}_\delta(\mathbf{x}^{(0)})}$ about an initial guess $x^{(0)}$. Suppose that the successive substitution scheme (3.6-3) satisfies the condition*

$$\left\|\mathbf{x}^{(1)} - \mathbf{x}^{(0)}\right\| = \left\|\mathbf{\Phi}(\mathbf{x}^{(0)}) - \mathbf{x}^{(0)}\right\| \leq (1-L)\delta.$$

Then

(i) *Each iterate $\mathbf{x}^{(m)} \in \overline{\mathcal{B}_\delta(\mathbf{x}^{(0)})}$.*

(ii) *The sequence $\{\mathbf{x}^{(m)}\}$ converges to a point $\mathbf{x}^* \in \overline{\mathcal{B}_\delta(\mathbf{x}^{(0)})}$.*

(iii) *The limit $\mathbf{x}^*$ is the unique fixed point of $\mathbf{\Phi}$ in $\overline{\mathcal{B}_\delta(\mathbf{x}^{(0)})}$.*

PROOF: The proof follows the same reasoning as that of Theorem 3.7. ∎

When $\mathbf{\Phi}$ is a contraction near $\mathbf{x}^*$, the corresponding successive substitution scheme converges at least linearly, assuming that we pick an appropriate initial guess. To see this, observe that since $\mathbf{x}^*$ is a fixed point of $\mathbf{\Phi}$,

$$\left\|\mathbf{x}^* - \mathbf{x}^{(m+1)}\right\| = \left\|\mathbf{\Phi}(\mathbf{x}^*) - \mathbf{\Phi}(\mathbf{x}^{(m)})\right\| \leq L\left\|\mathbf{x}^* - \mathbf{x}^{(m)}\right\|.$$

Heuristically, we expect each iteration to reduce the magnitude of the error $\varepsilon^{(m)} := \mathbf{x}^* - \mathbf{x}^{(m)}$ at least by the factor L.

Lemma 3.17 allows us to estimate this error at any iteration of a convergent scheme: Letting $n \to \infty$ in the inequality (3.6-4), we obtain

$$\lim_{n\to\infty}\left\|\mathbf{x}^{(m+n)} - \mathbf{x}^{(m)}\right\| = \left\|\mathbf{x}^* - \mathbf{x}^{(m)}\right\| \leq \frac{L^{m-j}}{1-L}\left\|\mathbf{x}^{(j+1)} - \mathbf{x}^{(j)}\right\|.$$

The two special cases $j = 0$ and $j = m-1$ yield the following a priori and a posteriori error estimates.

COROLLARY 3.20. *Under the hypotheses of Theorem 3.18, the iterates $\mathbf{x}^{(m)}$ generated using the successive substitution scheme (3.6-3) obey the error estimates*

(i) $\left\|\mathbf{x}^* - \mathbf{x}^{(m)}\right\| \leq \dfrac{L^m}{1-L}\left\|\mathbf{x}^{(1)} - \mathbf{x}^{(0)}\right\|,$

(ii) $\left\|\mathbf{x}^* - \mathbf{x}^{(m)}\right\| \leq \dfrac{L}{1-L}\left\|\mathbf{x}^{(m)} - \mathbf{x}^{(m-1)}\right\|.$

One final analogy with the one-dimensional case is the relationship between the Lipschitz constant L of the iteration function $\mathbf{\Phi}$ and its derivatives. This relationship has practical value, since it allows us to calculate L by inspecting the derivative of $\mathbf{\Phi}$. Before exploring this connection, we review the concept of differentiability of vector-valued functions. In the remainder of this section, $\Omega \subset \mathbb{R}^n$ is an open set, and $\mathbf{\Phi}$ has component functions $\Phi_1, \Phi_2, \ldots, \Phi_n$.

DEFINITION. *Let $\mathbf{\Phi}: \Omega \to \mathbb{R}^n$. The* **Jacobian matrix** *of $\mathbf{\Phi}$ at a point $\mathbf{x} \in \Omega$ is the matrix $\mathsf{J}_{\mathbf{\Phi}}(\mathbf{x}) \in \mathbb{R}^{n \times n}$ whose (i,j)th entry is*

$$j_{i,j}(\mathbf{x}) := \frac{\partial \Phi_i}{\partial x_j}(\mathbf{x}),$$

provided that this quantity exists.

Since the matrix entries $\partial \Phi_i / \partial x_j$ are functions of $\mathbf{x} = (x_1, x_2, \ldots, x_n)$, we regard $\mathsf{J}: \Omega \to \mathbb{R}^{n \times n}$.

We say that $\mathbf{\Phi}$ is **continuously differentiable** on Ω if each component function $\Phi_i \in C^1(\Omega)$. The following proposition generalizes the assertion in Section 3.3 that, for a continuously differentiable scalar function Φ, the Lipschitz constant L is an upper bound for $|\Phi'|$. The hypothesis of convexity allows us to apply Theorem 0.15, which is a multidimensional version of the mean value theorem.

PROPOSITION 3.21. *Let $\Omega \subset \mathbb{R}^n$ be open and convex, and let $\mathbf{\Phi}: \Omega \to \mathbb{R}^n$ be continuously differentiable on Ω. $\mathbf{\Phi}$ satisfies a Lipschitz condition (with respect to the norm $\|\cdot\|_\infty$) on Ω if there exists a constant $L > 0$ such that the entries of the Jacobian matrix for $\mathbf{\Phi}$ satisfy the inequality*

$$\left| \frac{\partial \Phi_i}{\partial x_j}(\mathbf{x}) \right| \leq \frac{L}{n},$$

for all $\mathbf{x} \in \Omega$.

PROOF: Let $\mathbf{x}, \mathbf{y} \in \Omega$. By Theorem 0.15, there exist points ζ_i, $i = 1, 2, \ldots, n$ on the line segment joining $\mathbf{x}$ and $\mathbf{y}$ such that

$$\Phi_i(\mathbf{x}) - \Phi_i(\mathbf{y}) = \nabla \Phi_i(\zeta_i) \cdot (\mathbf{x} - \mathbf{y}) = \sum_{j=1}^{n} \frac{\partial \Phi_i}{\partial x_j}(\zeta_i)(x_j - y_j).$$

Therefore,

$$\begin{aligned}
\|\mathbf{\Phi}(\mathbf{x}) - \mathbf{\Phi}(\mathbf{y})\|_\infty &= \max_{1\le i\le n} |\Phi_i(\mathbf{x}) - \Phi_i(\mathbf{y})| \\
&\le \sum_{j=1}^{n} \frac{L}{n}|x_j - y_j| \\
&\le \sum_{j=1}^{n} \frac{L}{n}\|\mathbf{x}-\mathbf{y}\|_\infty = L\|\mathbf{x}-\mathbf{y}\|_\infty. \qquad \blacksquare
\end{aligned}$$

An Application to Differential Equations

Problem 9 examines a routine application of the theory just presented. We close this section with a discussion of a more specialized application that arises in later chapters. Consider a system of ordinary differential equations having the form

$$\frac{d}{dt}\begin{bmatrix} u_1 \\ u_2 \\ \vdots \\ u_n \end{bmatrix} = \begin{bmatrix} \phi_1(u_1, u_2, \ldots, u_n) \\ \phi_2(u_1, u_2, \ldots, u_n) \\ \vdots \\ \phi_n(u_1, u_2, \ldots, u_n) \end{bmatrix}, \tag{3.6-5}$$

which we abbreviate as $d\mathbf{u}/dt = \boldsymbol{\phi}(\mathbf{u})$. It is common to think of t as time, although other independent variables occur in practice.

Chapter 7 develops techniques for approximating differential systems of this type via sets of algebraic equations. One of the simplest such techniques is to replace the unknown function $\mathbf{u}(t)$ by an approximate **grid function** $\mathbf{U}$, defined only on a discrete set $t = t_0, t_1 := t_0 + k, t_2 := t_0 + 2k, \ldots$ of "time levels." Viewing $\mathbf{U}_i := \mathbf{U}(t_i)$ as an approximation to $\mathbf{u}(t_i)$, we approximate the differential equation (3.6-5) by replacing derivatives with difference quotients:

$$\frac{\mathbf{U}_{i+1} - \mathbf{U}_i}{k} = \boldsymbol{\phi}(\mathbf{U}_{i+1}),$$

or

$$\mathbf{U}_{i+1} = \mathbf{U}_i + k\boldsymbol{\phi}(\mathbf{U}_{i+1}). \tag{3.6-6}$$

Computationally, we treat this equation as an updating scheme: Given an initial value $\mathbf{U}_0 := \mathbf{u}(t_0)$, we solve Equation (3.6-6) for $\mathbf{U}_1$, which we then employ as a known value in solving for $\mathbf{U}_2$, and so forth. At each step in the process, we must solve a possibly nonlinear system of equations having the form

$$\mathbf{x} = \mathbf{c} + k\boldsymbol{\phi}(\mathbf{x}),$$

where $\mathbf{x}$ is unknown and $\mathbf{c}$ is a constant vector. Thus Equation (3.6-6) gives rise to a successive substitution scheme

$$\mathbf{U}_{i+1}^{(m+1)} = \mathbf{U}_i + k\boldsymbol{\phi}(\mathbf{U}_{i+1}^{(m)}), \tag{3.6-7}$$

in which the iteration function is $\mathbf{\Phi}(\mathbf{x}) := \mathbf{c} + k\boldsymbol{\phi}(\mathbf{x})$.

According to Theorem 3.19, we expect this scheme to converge at each time level if $\mathbf{\Phi}$ is a contraction in a region surrounding $\mathbf{x}$. Proposition 3.22 permits us to verify this condition by checking whether the derivatives of $\mathbf{\Phi}$ obey bounds of the form

$$\left|\frac{\partial \Phi_i}{\partial x_j}\right| \leq \frac{\theta}{n},$$

for some $\theta \in (0,1)$. Suppose that each of the derivatives $\partial\phi_i/\partial x_j$ satisfies $|\partial\phi_i/\partial x_j| \leq M$ for some constant $M > 0$. Since $\partial\Phi_i/\partial x_j = k\partial\phi_i/\partial x_j$, $\mathbf{\Phi}$ is a contraction if $k \leq \theta/(2M)$, that is, if we pick a small enough time step.

As a concrete example, consider the two-dimensional system

$$\frac{d}{dt}\begin{bmatrix} u_1 \\ u_2 \end{bmatrix} = \begin{bmatrix} -(1+u_1^2)^{-1}\sin u_2 \\ u_1 + \exp(-u_2^2) \end{bmatrix}.$$

In this case, the discrete approximation takes the form

$$\mathbf{U}_{i+1} = \mathbf{U}_i + k\boldsymbol{\phi}(\mathbf{U}_{i+1}),$$

where

$$\begin{bmatrix} \phi_1(x_1,x_2) \\ \phi_2(x_1,x_2) \end{bmatrix} := \begin{bmatrix} -(1+x_1^2)^{-1}\sin x_2 \\ x_1 + \exp(-x_2^2) \end{bmatrix}.$$

The derivatives of $\boldsymbol{\phi}$ obey the following bounds:

$$\begin{aligned}
\left|\frac{\partial \phi_1}{\partial x_1}\right| &= \left|2x_1(1+x_1)^{-2}\sin x_2\right| \leq 1, \\
\left|\frac{\partial \phi_1}{\partial x_2}\right| &= \left|-(1+x_1^2)^{-1}\cos x_2\right| \leq 1, \\
\left|\frac{\partial \phi_2}{\partial x_1}\right| &= 1, \\
\left|\frac{\partial \phi_2}{\partial x_2}\right| &= \left|-2x_2\exp(-x_2^2)\right| \leq \sqrt{2}e^{-1/2} \simeq 0.8578.
\end{aligned}$$

Therefore we take $M = 1$. Since $n = 2$ in this case, the successive substitution scheme (3.6-7) for updating $\mathbf{U}_i$ converges for any time step $k \in (0, \frac{1}{2})$.

3.7 Newton's Method: Several Variables

Motivation and Construction

As with the numerical solution of single equations, we hope to solve systems like $\mathbf{f}(\mathbf{x}) = \mathbf{0}$ using methods that converge superlinearly. Prototypical is the multidimensional extension of Newton's method.

One way to construct this method is to draw a formal analogy with the one-dimensional scheme (3.4-2), which we rewrite as follows:

$$(i)\ \text{Solve } f'(x^{(m)})\delta^{(m+1)} = -f(x^{(m)}) \text{ for } \delta^{(m+1)}.$$

$$(ii)\ x^{(m+1)} \leftarrow x^{(m)} + \delta^{(m+1)}.$$

In the multidimensional case, the iterates $\mathbf{x}^{(m)}$ and increments $\boldsymbol{\delta}^{(m)}$ belong to $\mathbb{R}^n$, and the natural analog of f' is the Jacobian matrix J_f of $\mathbf{f}$. These observations suggest the following steps, starting with an initial guess $\mathbf{x}^{(0)}$:

$$(i)\ \text{Solve } \mathsf{J}_f(\mathbf{x}^{(m)})\,\boldsymbol{\delta}^{(m+1)} = -\mathbf{f}(\mathbf{x}^{(m)}) \text{ for } \boldsymbol{\delta}^{(m+1)}.$$

$$(ii)\ \mathbf{x}^{(m+1)} \leftarrow \mathbf{x}^{(m)} + \boldsymbol{\delta}^{(m+1)}. \tag{3.7-1}$$

For this analogy to make sense, $\mathbf{f}$ must be differentiable, with J_f invertible, on some neighborhood $\Omega \subset \mathbb{R}^n$ of the sought zero $\mathbf{x}^*$.

As a concrete example, consider the system (3.6-2):

$$\begin{aligned} f_1(x_1, x_2) := x_1^2 \sin x_2 &= 0, \\ f_2(x_1, x_2) := \cos x_1 + x_2 - 1 &= 0. \end{aligned}$$

This set of nonlinear equations has a solution at $\mathbf{x}^* = (x_1^*, x_2^*)^\top = (0,0)^\top$. The Jacobian matrix for the function $\mathbf{f} = (f_1, f_2)^\top$ at a point $\mathbf{x} = (x_1, x_2)^\top$ is

$$\mathsf{J}_f(x_1, x_2) = \begin{bmatrix} 2x_1 \sin x_2 & x_1^2 \cos x_2 \\ -\sin x_1 & 1 \end{bmatrix}.$$

If we adopt the initial guess $\mathbf{x}^{(0)} = (x_1^{(0)}, x_2^{(0)})^\top = (\pi/2, \pi/2)^\top$, the first iteration of Newton's method requires that we solve the linear system

$$\begin{bmatrix} \pi \sin(\pi/2) & (\pi^2/4)\cos(\pi/2) \\ -\sin(\pi/2) & 1 \end{bmatrix} \begin{bmatrix} \delta_1^{(1)} \\ \delta_2^{(1)} \end{bmatrix} = -\begin{bmatrix} (\pi/2)^2 \sin(\pi/2) \\ \cos(\pi/2) + \pi/2 - 1 \end{bmatrix},$$

that is,

$$\begin{bmatrix} \pi & 0 \\ -1 & 1 \end{bmatrix} \begin{bmatrix} \delta_1^{(1)} \\ \delta_2^{(1)} \end{bmatrix} = -\begin{bmatrix} \pi^2/4 \\ \pi/2 - 1 \end{bmatrix}.$$

We find that

$$\begin{bmatrix} \delta_1^{(1)} \\ \delta_2^{(1)} \end{bmatrix} = \begin{bmatrix} -\pi/4 \\ 1 - \pi/4 \end{bmatrix},$$

and therefore the next iterate is

$$\mathbf{x}^{(1)} = \begin{bmatrix} x_1^{(1)} \\ x_2^{(1)} \end{bmatrix} = \begin{bmatrix} \pi/2 \\ \pi/2 \end{bmatrix} + \begin{bmatrix} -\pi/4 \\ 1 - \pi/4 \end{bmatrix} = \begin{bmatrix} \pi/4 \\ 1 + \pi/4 \end{bmatrix}.$$

Figure 1 shows the iterates $\mathbf{x}^{(0)}$ and $\mathbf{x}^{(1)}$ along with the exact solution $\mathbf{x}^* = \mathbf{0}$.

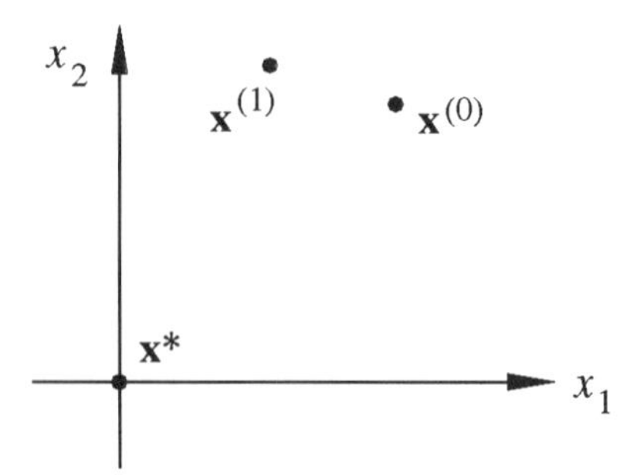

FIGURE 1. *Two iterates generated by Newton's method for the system (3.6-2).*

Practical Considerations

Step (*i*) of the method (3.7-1) calls for the solution of an $n \times n$ linear system $\mathsf{J}_f(\mathbf{x}^{(m)})\boldsymbol{\delta}^{(m+1)} = -\mathbf{f}(\mathbf{x}^{(m)})$ at each iteration. When n is large, this task can require a great deal of computational effort. In designing codes, it pays to minimize the work devoted to solving linear systems involving the Jacobian matrix.

One simple way to do this is to compute $\mathsf{J}_f(\mathbf{x}^{(0)})$ and then use it instead of $\mathsf{J}_f(\mathbf{x}^{(m)})$ in subsequent iterations. This tactic allows one to exploit the work of numerically "inverting" J_f for several iterations, instead of inverting a new matrix at every iteration. Heuristically, we expect the tactic to work in problems where $\mathsf{J}_f(\mathbf{x})$ varies slowly with $\mathbf{x}$.

Another approach is to exploit any special structure that J_f may have. For example, certain optimization problems lead to nonlinear systems for which the Jacobian matrix is symmetric and positive definite. Consider an open set $\Omega \subset \mathbb{R}^n$ and a function $\varphi \in C^2(\Omega)$. A point $\mathbf{x}^* \in \Omega$ is a **local minimum** for φ if there is some ball $\mathcal{B}_\epsilon(\mathbf{x}^*) \subset \Omega$ in which $\varphi(\mathbf{x}^*) \leq \varphi(\mathbf{x})$ for every $\mathbf{x} \in \mathcal{B}_\epsilon(\mathbf{x}^*)$. Schemes for minimizing such functions often use the following fact: The local minima for φ are points $\mathbf{x}^* \in \Omega$ for which $\nabla\varphi(\mathbf{x}^*) = \mathbf{0}$. Thus, by identifying $\mathbf{f}(\mathbf{x}) := \nabla\varphi(\mathbf{x})$, we can find local minima of φ by solving a system of the form $\mathbf{f}(\mathbf{x}) = \mathbf{0}$. This observation also allows us to find local maxima of φ, since these points are local minima for the function $-\varphi$.

The Jacobian matrix J_f of $\mathbf{f}$ in such problems is the Hessian matrix $\mathsf{H}_\varphi(\mathbf{x})$ of φ. As reviewed in Section 0.4, the (i,j)th entry of this matrix is

$$h_{i,j}(\mathbf{x}) := \frac{\partial^2\varphi}{\partial x_i \partial x_j}(\mathbf{x}),$$

and $\mathsf{H}_\varphi(\mathbf{x})$ is therefore symmetric for any $\mathbf{x} \in \Omega$ when $\varphi \in C^2(\Omega)$. Moreover, local minima of φ correspond to points where $\nabla\varphi = \mathbf{0}$ *and* H_φ is positive definite. When solving this type of problem, we can therefore employ special methods for symmetric, positive definite matrices in the execution of step (i) of the scheme (3.7-1).

We move now to a broader concern. The multidimensional version of Newton's method suffers from a limitation that is familiar from the one-dimensional case: The iterates $\mathbf{x}^{(m)}$ typically converge to a zero $\mathbf{x}^*$ of $\mathbf{f}$ only for good initial guesses $\mathbf{x}^{(0)}$. This caveat is more distressing in several dimensions, since we have no direct analog of the bisection method of Section 3.2 to provide slow but sure progress from poor initial guesses toward good ones.

There is a simple idea, however, that can often extend the set of initial guesses leading to convergent iterative sequences $\{\mathbf{x}^{(m)}\}$. The idea has its origin in the observation that a zero $\mathbf{x}^*$ of $\mathbf{f}$ is a minimum for the real-valued function $\|\mathbf{f}\|_2^2$. To find this minimum starting with an initial guess $\mathbf{x}^{(0)}$, we can regard the increments $\boldsymbol{\delta}^{(m+1)}$ generated by Newton's method as indicating *directions* along which to search for a local minimum of $\|\mathbf{f}\|_2^2$. If $\|\mathbf{f}(\mathbf{x}^{(m)} + \boldsymbol{\delta}^{(m+1)})\|_2^2 \geq \|\mathbf{f}(\mathbf{x}^{(m)})\|_2^2$, then we can try repeatedly halving the increment until we achieve a reduction in the value of $\|\mathbf{f}\|_2^2$. This "damping" strategy is clearly fallible, since $\mathbf{x}^{(0)}$ might already be a local minimum for $\|\mathbf{f}\|_2^2$ that lies far from an actual zero. Problem 13 examines this issue.

The following algorithm executes the damping scheme:

ALGORITHM 3.4 (DAMPED NEWTON METHOD). *Given an initial guess* $\mathbf{x}^{(0)}$, *a tolerance* $\tau > 0$, *and a maximum number* $k_{\max}$ *of allowable increment halvings, the following steps compute a sequence* $\{\mathbf{x}^{(m)}\}$ *of iterates using the damped Newton method.*

1. $m \leftarrow 0$.
2. If $\|\mathbf{f}(\mathbf{x}^{(m)})\|_2^2 \geq \tau$ then:
3. Solve $\mathsf{J}_f(\mathbf{x}^{(m)})\, \boldsymbol{\delta}^{(m+1)} = -\mathbf{f}(\mathbf{x}^{(m)})$ for $\boldsymbol{\delta}^{(m+1)}$.
4. $k \leftarrow 0$.
5. If $\|\mathbf{f}(\mathbf{x}^{(m)} + \boldsymbol{\delta}^{(m+1)})\|_2^2 \geq \|\mathbf{f}(\mathbf{x}^{(m)})\|_2^2$, then:
6. $k \leftarrow k + 1$.
7. If $k > k_{\max}$ then stop; initial guess $\mathbf{x}^{(0)}$ fails.
8. $\boldsymbol{\delta}^{(m+1)} \leftarrow \frac{1}{2}\boldsymbol{\delta}^{(m+1)}$.
9. Go to 5.
10. End if.
11. $\mathbf{x}^{(m+1)} \leftarrow \mathbf{x}^{(m)} + \boldsymbol{\delta}^{(m+1)}$.
12. $m \leftarrow m + 1$.

13. Go to 2.

14. End if.

15. End.

Ideally, the increment halvings in this algorithm keep the iterates $\mathbf{x}^{(m)}$ within a reasonable distance of the sought zero $\mathbf{x}^*$, until finally the algorithm produces an iterate that allows Newton's method to converge without further damping.

Another class of techniques, called **continuation methods**, also deserve mention. The idea is to connect the problem $\mathbf{f}(\mathbf{x}) = \mathbf{0}$, whose solution is unknown, by a continuous path to a problem $\mathbf{f}_0(\mathbf{x}) = \mathbf{0}$ whose solution $\mathbf{x}_0^*$ we know. We then solve a succession of problems along this path, eventually arriving at a problem so close to the original one that its solution furnishes a good intial guess to $\mathbf{x}^*$. Figure 2 illustrates this idea.

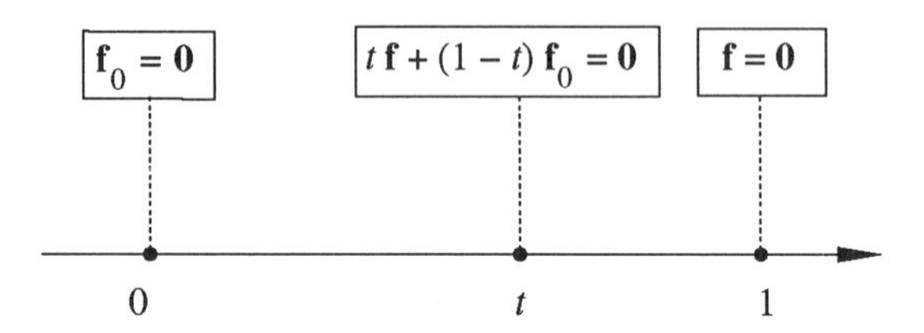

FIGURE 2. *Conceptual picture of a continuation method.*

To be more concrete, observe that we can use any initial guess $\mathbf{x}^{(0)}$ to define $\mathbf{f}_0(\mathbf{x}) := \mathbf{f}(\mathbf{x}) - \mathbf{f}(\mathbf{x}^{(0)})$, which obviously has $\mathbf{x}^{(0)}$ as a zero. Define a one-parameter family $\mathbf{F}_t$ of vector-valued functions by

$$\mathbf{F}_t(\mathbf{x}) := t\,\mathbf{f}(\mathbf{x}) + (1-t)\,\mathbf{f}_0(\mathbf{x}), \qquad t \in [0,1].$$

Clearly $\mathbf{F}_0(\mathbf{x}) = \mathbf{f}_0(\mathbf{x})$, and $\mathbf{F}_1(\mathbf{x}) = \mathbf{f}(\mathbf{x})$. If $\mathbf{f}$ is "tame" enough, then we can regard the nonlinear equation $\mathbf{F}_t(\mathbf{x}) = \mathbf{0}$ as having a solution $\mathbf{x}^*(t)$ that depends continuously on the parameter t.

Now construct a grid $\{0 = t_0, t_1, \ldots, t_{n-1}, t_n = 1\}$ on $[0,1]$. Since $\mathbf{x}^*(0) := \mathbf{x}^{(0)}$ is the exact solution to $\mathbf{F}_0(\mathbf{x}) = \mathbf{0}$, it is presumably close to the solution $\mathbf{x}^*(t_1)$ of $\mathbf{F}_{t_1}(\mathbf{x}) = \mathbf{0}$. Therefore, employing $\mathbf{x}^{(0)}$ as an initial guess, use Newton's method to solve for an approximation to $\mathbf{x}^*(t_1)$. Then, employing this vector as an initial guess, use Newton's method to solve the equation $\mathbf{F}_{t_2}(\mathbf{x}) = \mathbf{0}$. Proceeding in this way, we arrive at the penultimate problem $\mathbf{F}_{t_{n-1}}(\mathbf{x}) = \mathbf{0}$, whose approximate solution serves as initial guess for the original problem $\mathbf{f}(\mathbf{x}) = \mathbf{F}_1(\mathbf{x}) = \mathbf{0}$.

This heuristic raises many questions. We refer readers to Ortega and Rheinboldt [3] for more detail.

In many applications, there is no easy way to evaluate the entries $\partial f_i/\partial x_j$ of the Jacobian matrix J_f exactly. Here, paralleling the one-dimensional case, one can use difference quotients to approximate the partial derivatives. This approach replaces the Jacobian matrix $\mathsf{J}_f(\mathbf{x}^{(m)})$ by an approximation $\hat{\mathsf{J}}_m$, whose (i,j)th entry has the form

$$\frac{f_i(\mathbf{x}^{(m)}+h_m\mathbf{e}_j)-f_i(\mathbf{x}^{(m)})}{h_m}.$$

Here, $\{h_m\}$ is a sequence of small offsets, and $\mathbf{e}_j$ denotes the jth unit basis vector, all of whose entries are 0 except for the jth, which is 1. The following finite-difference Newton method results:

$$\begin{aligned} &(i)\ \text{Solve } \hat{\mathsf{J}}_m\boldsymbol{\delta}^{(m+1)} = -\mathbf{f}(\mathbf{x}^{(m)}) \text{ for } \boldsymbol{\delta}^{(m+1)}. \\ &(ii)\ \mathbf{x}^{(m+1)} \leftarrow \mathbf{x}^{(m)} + \boldsymbol{\delta}^{(m+1)}. \end{aligned} \tag{3.7-2}$$

As in the one-dimensional case examined in Section 3.5, the performance of this scheme depends strongly upon how one chooses the offsets h_m. We show later in this section that it is possible to pick these offsets so that the iterative scheme (3.7-2) converges at least linearly.

Mathematical Details

We explore two theoretical issues: We analyze the convergence of Newton's method, and we establish conditions under which finite-difference Newton methods converge. The former topic is standard; the latter may be of interest in certain applications.

For the remainder of this section, $\Omega \subset \mathbb{R}^n$, and $\|\cdot\|$ signifies an arbitrary norm on $\mathbb{R}^n$ or its subordinate matrix norm on $\mathbb{R}^{n\times n}$. Also, we often adopt the hypothesis that the Jacobian matrix J_f of a function $\mathbf{f}:\Omega \to \mathbb{R}^n$ has a Lipschitz constant $L > 0$ on Ω. This assertion means that, for any $\mathbf{x},\mathbf{y} \in \Omega$,

$$\|\mathsf{J}_f(\mathbf{x}) - \mathsf{J}_f(\mathbf{y})\| \le L\|\mathbf{x}-\mathbf{y}\|.$$

Here, the matrix norm appearing on the left side of the inequality is the matrix norm subordinate to the vector norm appearing on the right.

Newton's method for solving the nonlinear system $\mathbf{f}(\mathbf{x}) = \mathbf{0}$ has the explicit form

$$\mathbf{x}^{(m+1)} = \mathbf{x}^{(m)} - \mathsf{J}_f^{-1}(\mathbf{x}^{(m)})\mathbf{f}(\mathbf{x}^{(m)}). \tag{3.7-3}$$

We begin the analysis by proving a multidimensional version of Lemma 3.8.

LEMMA 3.22. *Let $\Omega \subset \mathbb{R}^n$ be convex and open. Assume that $\mathbf{f}:\Omega \to \mathbb{R}^n$ is differentiable and that its Jacobian matrix J_f has Lipschitz constant $L > 0$. Then*

$$\|\mathbf{f}(\mathbf{x}) - \mathbf{f}(\mathbf{y}) - \mathsf{J}_f(\mathbf{y})(\mathbf{x}-\mathbf{y})\| \le \frac{L}{2}\|\mathbf{x}-\mathbf{y}\|^2. \tag{3.7-4}$$

This lemma establishes how well the affine model $\mathbf{f}(\mathbf{y}) + \mathsf{J}_f(\mathbf{y})(\mathbf{x}-\mathbf{y})$ approximates $\mathbf{f}(\mathbf{x})$ near the point $\mathbf{y}$.

PROOF: Pick any points $\mathbf{x}, \mathbf{y} \in \Omega$. Since Ω is convex, the line segment

$$\left\{\mathbf{y} + t(\mathbf{x}-\mathbf{y}) \in \mathbb{R}^n : 0 \le t \le 1\right\}$$

lies entirely in Ω. Define $\boldsymbol{\psi}: [0,1] \to \mathbb{R}^n$ by setting $\boldsymbol{\psi}(t) := \mathbf{f}(\mathbf{y} + t(\mathbf{x}-\mathbf{y}))$. The function $\boldsymbol{\psi}$ is differentiable since $\mathbf{f}$ is, and for any $t \in [0,1]$ we have

$$\begin{aligned}
\|\boldsymbol{\psi}'(t) - \boldsymbol{\psi}'(0)\| &= \|\mathsf{J}_f\left(\mathbf{y} + t(\mathbf{x}-\mathbf{y})\right)(\mathbf{x}-\mathbf{y}) - \mathsf{J}_f(\mathbf{y})(\mathbf{x}-\mathbf{y})\| \\
&\le \|\mathsf{J}_f\left(\mathbf{y} + t(\mathbf{x}-\mathbf{y})\right) - \mathsf{J}_f(\mathbf{y})\|\,\|\mathbf{x}-\mathbf{y}\| \\
&\le L\|t\mathbf{x} - t\mathbf{y}\|\,\|\mathbf{x}-\mathbf{y}\| = Lt\|\mathbf{x}-\mathbf{y}\|^2.
\end{aligned}$$

Using this inequality and the fundamental theorem of calculus, we estimate the left side of (3.7-4):

$$\begin{aligned}
\|\mathbf{f}(\mathbf{x}) - \mathbf{f}(\mathbf{y}) - \mathsf{J}_f(\mathbf{y})(\mathbf{x}-\mathbf{y})\| &= \|\boldsymbol{\psi}(1) - \boldsymbol{\psi}(0) - \boldsymbol{\psi}'(0)\| \\
&= \left\| \int_0^1 \left[\boldsymbol{\psi}'(t) - \boldsymbol{\psi}'(0)\right] dt \right\| \\
&\le \int_0^1 \left\|\boldsymbol{\psi}'(t) - \boldsymbol{\psi}'(0)\right\| dt \\
&\le L\|\mathbf{x}-\mathbf{y}\|^2 \int_0^1 t\, dt \\
&= \frac{L}{2}\|\mathbf{x}-\mathbf{y}\|^2.
\end{aligned}$$

This is the required bound. ∎

The next lemma bounds the distance between successive iterates in Newton's method. The bound plays a crucial role in the proof of the main convergence theorem, but unfortunately it requires a list of hypotheses that exert rather strong control over the Jacobian matrix of the function $\mathbf{f}$. The hypotheses appear again in the main theorem.

LEMMA 3.23. *Let $\Omega \subset \mathbb{R}^n$ be open and convex, and let $\mathbf{f}: \overline{\Omega} \to \mathbb{R}^n$ be differentiable with Jacobian matrix J_f. Assume further that J_f obeys the following hypotheses:*

(i) *$\mathsf{J}_f(\mathbf{x})$ is invertible for all $\mathbf{x} \in \Omega$.*

(ii) J_f *has Lipschitz constant* $L > 0$ *on* Ω.

(iii) *There is a constant* $M > 0$ *such that* $\|\mathsf{J}_f^{-1}(\mathbf{x})\| \leq M$ *for all* $\mathbf{x} \in \Omega$.

(iv) *It is possible to choose an initial guess* $\mathbf{x}^{(0)} \in \Omega$ *such that the quantity* $B := \|\mathsf{J}_f^{-1}(\mathbf{x}^{(0)})\mathbf{f}(\mathbf{x}^{(0)})\|$ *obeys the inequality* $\theta := \frac{1}{2}BLM < 1$.

If Newton's method with initial guess $\mathbf{x}^{(0)}$ *generates an iterative sequence* $\{\mathbf{x}^{(m)}\}$ *such that each* $\mathbf{x}^{(m)} \in \Omega$, *then for* $j = 0, 1, 2, \ldots$,

$$\left\|\mathbf{x}^{(j+1)} - \mathbf{x}^{(j)}\right\| \leq B\theta^{2^j - 1} \to 0 \qquad \textit{as} \quad j \to \infty.$$

PROOF: We use induction on j. The identity (3.7-3) yields

$$\left\|\mathbf{x}^{(1)} - \mathbf{x}^{(0)}\right\| = \left\|\mathbf{x}^{(0)} - \mathsf{J}_f^{-1}(\mathbf{x}^{(0)})\,\mathbf{f}(\mathbf{x}^{(0)}) - \mathbf{x}^{(0)}\right\| = B,$$

by definition of B. Hence the conclusion holds for $j = 0$. If it holds for $j = i - 1$, then we can proceed similarly and apply the hypothesis (iii) to obtain

$$\left\|\mathbf{x}^{(i+1)} - \mathbf{x}^{(i)}\right\| = \left\|-\mathsf{J}_f^{-1}(\mathbf{x}^{(i)})\,\mathbf{f}(\mathbf{x}^{(i)})\right\| \leq M\left\|\mathbf{f}(\mathbf{x}^{(i)})\right\|.$$

We now subtract $\mathbf{0}$, in the form $\mathbf{f}(\mathbf{x}^{(i-1)}) + \mathsf{J}_f(\mathbf{x}^{(i-1)})\,(\mathbf{x}^{(i)} - \mathbf{x}^{(i-1)})$, from the quantity inside the norm on the right. This trick, together with the hypothesis (ii) and Lemma 3.22, yield

$$\begin{aligned}\left\|\mathbf{x}^{(i+1)} - \mathbf{x}^{(i)}\right\| &\leq M\left\|\mathbf{f}(\mathbf{x}^{(i)}) - \mathbf{f}(\mathbf{x}^{(i-1)}) - \mathsf{J}_f(\mathbf{x}^{(i-1)})\,(\mathbf{x}^{(i)} - \mathbf{x}^{(i-1)})\right\| \\ &\leq \frac{LM}{2}\left\|\mathbf{x}^{(i)} - \mathbf{x}^{(i-1)}\right\|^2.\end{aligned}$$

Applying the inductive hypothesis to the norm on the right, we get

$$\left\|\mathbf{x}^{(i+1)} - \mathbf{x}^{(i)}\right\| \leq \frac{LM}{2}\left(B\theta^{2^{i-1}-1}\right)^2 = B\theta^{2^i - 1}.$$

This completes the induction. ∎

Now we come to the main convergence theorem for Newton's method.

THEOREM 3.24. *Let* $\Omega \subset \mathbb{R}^n$ *be open and convex. Assume that* $\mathbf{f}\colon \overline{\Omega} \to \mathbb{R}^n$ *is continuous on* $\overline{\Omega}$ *and differentiable on* Ω *with Jacobian matrix* J_f. *Assume also that the hypotheses* (i) *through* (iv) *of Lemma 3.23 apply. Call* $\delta := B/(1-\theta)$. *If the ball* $\mathcal{B}_\delta(\mathbf{x}^{(0)}) \subset \Omega$, *then the sequence* $\{\mathbf{x}^{(m)}\}$ *generated by Newton's method with initial guess* $\mathbf{x}^{(0)}$ *satisfies the following conditions:*

(A) *Each* $\mathbf{x}^{(m)} \in \mathcal{B}_\delta(\mathbf{x}^{(0)})$.

(B) *There is a point* $\mathbf{x}^* \in \overline{\mathcal{B}_\delta(\mathbf{x}^{(0)})}$ *such that* $\mathbf{x}^{(m)} \to \mathbf{x}^*$ *as* $m \to \infty$.

(C) $\mathbf{f}(\mathbf{x}^*) = \mathbf{0}$.

(D) *There is a constant* $C > 0$ *such that* $\|\mathbf{x}^* - \mathbf{x}^{(m+1)}\| \le C\|\mathbf{x}^* - \mathbf{x}^{(m)}\|^2$ *for* $m = 0, 1, 2, \ldots$.

Conclusion (A) guarantees that the sequence $\{\mathbf{x}^{(m)}\}$ generated by Equation (3.7-3) is well defined; (B) and (C) assert that Newton's method converges to a zero $\mathbf{x}^*$ of $\mathbf{f}$, and (D) states that the convergence is quadratic.

PROOF: We prove (A) by induction on m. The fact that $\mathbf{x}^{(0)} \in \mathcal{B}_\delta(\mathbf{x}^{(0)})$ is obvious. Assume that $\mathbf{x}^{(i)} \in \mathcal{B}_\delta(\mathbf{x}^{(0)})$ for $i = 0, 1, \ldots, m$. We have

$$\mathbf{x}^{(m+1)} - \mathbf{x}^{(0)} = \sum_{i=0}^{m} \left(\mathbf{x}^{(i+1)} - \mathbf{x}^{(i)}\right),$$

so the triangle inequality and Lemma 3.23 imply that

$$\begin{aligned} \left\|\mathbf{x}^{(m+1)} - \mathbf{x}^{(0)}\right\| \le \sum_{i=0}^{m} \left\|\mathbf{x}^{(i+1)} - \mathbf{x}^{(i)}\right\| &\le B\sum_{i=1}^{m} \theta^{2^i-1} \\ &< B\sum_{i=0}^{\infty} \theta^i = \frac{B}{1-\theta} = \delta. \end{aligned}$$

Therefore $\mathbf{x}^{(m+1)} \in \mathcal{B}_\delta(\mathbf{x}^{(0)})$, and we have finished the induction.

To prove (B), it suffices to show that $\{\mathbf{x}^{(m)}\}$ is a Cauchy sequence. Once again, we use Lemma 3.23: If $n > m$, then

$$\begin{aligned} \left\|\mathbf{x}^{(n)} - \mathbf{x}^{(m)}\right\| &\le \sum_{i=m}^{n-1} \left\|\mathbf{x}^{(i+1)} - \mathbf{x}^{(i)}\right\| \\ &\le B\,\theta^{2^m-1} \sum_{i=0}^{\infty} \left(\theta^{2^m}\right)^i = \frac{B\theta^{2^m-1}}{1-\theta^{2^m}}. \end{aligned}$$

The last quantity on the right tends to 0 as $m \to \infty$. Consequently, given any $\epsilon > 0$, we can make $\|\mathbf{x}^{(n)} - \mathbf{x}^{(m)}\| < \epsilon$ by taking m sufficiently large. It follows that the sequence $\{\mathbf{x}^{(m)}\}$ converges to some point $\mathbf{x}^* \in \overline{\mathcal{B}_\delta(\mathbf{x}^{(0)})}$.

For (C), observe that continuity of $\mathbf{f}$ implies that $\mathbf{f}(\mathbf{x}^{(m)}) \to \mathbf{f}(\mathbf{x}^*)$ as $m \to \infty$, so it is enough to demonstrate that $\mathbf{f}(\mathbf{x}^{(m)}) \to \mathbf{0}$ as $m \to \infty$. By Equation (3.7-3),

$$\begin{aligned} \left\|\mathbf{f}(\mathbf{x}^{(m)})\right\| &= \left\|-\mathsf{J}_f(\mathbf{x}^{(m)})\,(\mathbf{x}^{(m+1)} - \mathbf{x}^{(m)})\right\| \\ &\le \left\|\mathsf{J}_f(\mathbf{x}^{(m)})\right\| \left\|\mathbf{x}^{(m+1)} - \mathbf{x}^{(m)}\right\|. \end{aligned}$$

Since $\|\mathbf{x}^{(m+1)} - \mathbf{x}^{(m)}\| \to 0$ as $m \to \infty$, the proof of (C) hinges on whether we can bound the growth of the factor $\|\mathsf{J}_f(\mathbf{x}^{(m)})\|$. Using the triangle inequality in the form (0.3-1), the Lipschitz condition (*ii*) and conclusion (A), we obtain the estimate

$$\begin{aligned}\left\|\mathsf{J}_f(\mathbf{x}^{(m)})\right\| - \left\|\mathsf{J}_f(\mathbf{x}^{(0)})\right\| &\leq \left\|\mathsf{J}_f(\mathbf{x}^{(m)}) - \mathsf{J}_f(\mathbf{x}^{(0)})\right\| \\ &\leq L\left\|\mathbf{x}^{(m)} - \mathbf{x}^{(0)}\right\| \leq L\delta.\end{aligned}$$

Therefore, $\|\mathsf{J}_f(\mathbf{x}^{(m)})\| \leq L\delta + \|\mathsf{J}_f(\mathbf{x}^{(0)})\|$, and we have established (C).

Finally we prove (D). Using Equation (3.7-3), the invertibility hypothesis (*i*), and the fact that $\mathbf{f}(\mathbf{x}^*) = \mathbf{0}$, we find that

$$\begin{aligned}\left\|\mathbf{x}^{(m+1)} - \mathbf{x}^*\right\| &= \left\|\mathbf{x}^{(m)} - \mathbf{x}^* - \mathsf{J}_f^{-1}(\mathbf{x}^{(m)})\mathbf{f}(\mathbf{x}^{(m)})\right\| \\ &= \left\|\mathsf{J}_f^{-1}(\mathbf{x}^{(m)})\left[-\mathsf{J}_f(\mathbf{x}^{(m)})(\mathbf{x}^* - \mathbf{x}^{(m)}) - \mathbf{f}(\mathbf{x}^{(m)})\right]\right\| \\ &\leq \underbrace{\left\|\mathsf{J}_f^{-1}(\mathbf{x}^{(m)})\right\|}_{(\text{I})}\underbrace{\left\|\mathbf{f}(\mathbf{x}^*) - \mathbf{f}(\mathbf{x}^{(m)}) - \mathsf{J}_f(\mathbf{x}^{(m)})(\mathbf{x}^* - \mathbf{x}^{(m)})\right\|}_{(\text{II})}.\end{aligned}$$

But (I) $\leq M$ by hypothesis (*iii*), and (II) $\leq \frac{1}{2}L\|\mathbf{x}^* - \mathbf{x}^{(m)}\|^2$ by Lemma 3.22. Conclusion (D) follows, with $C = LM/2$. ∎

The convergence analysis for finite-difference Newton methods in several variables is analogous to the analysis given in Section 3.5 for the one-dimensional case. In preparation for the main theorem, we state three lemmas. The first one establishes how well we can expect finite-difference analogs to approximate the Jacobian matrix. In the following, Ω_h denotes the collection of all points $\mathbf{x} \in \Omega$ such that $\mathbf{x} + h\mathbf{e}_j \in \Omega$ for every standard basis vector $\mathbf{e}_j$.

LEMMA 3.25. *Let $\Omega \subset \mathbb{R}^n$ be open and convex, and let $\mathbf{f}: \Omega \to \mathbb{R}^n$ be continuously differentiable. Suppose that the Jacobian matrix J_f of $\mathbf{f}$ has Lipschitz constant $L > 0$ on Ω. For $\mathbf{x} \in \Omega_h$, define $\hat{\mathsf{J}}(\mathbf{x}) \in \mathbb{R}^{n \times n}$ to be the matrix whose (i,j)th entry is*

$$\frac{f_i(\mathbf{x} + h\mathbf{e}_j) - f_i(\mathbf{x})}{h}.$$

Then for any $\mathbf{x} \in \Omega$

$$\left\|\hat{\mathsf{J}}(\mathbf{x}) - \mathsf{J}_f(\mathbf{x})\right\|_1 \leq \frac{L|h|}{2}. \tag{3.7-5}$$

This lemma deviates from habit by specifying the norm, $\|\cdot\|_1$, in which the inequality holds. This choice is actually a matter of convenience: The proofs of later results turn out to be simpler when we use a norm in which $\|\mathbf{e}_j\| = 1$.

PROOF: The jth column of the matrix $\hat{\mathsf{J}}(\mathbf{x}) - \mathsf{J}_f(\mathbf{x})$ is

$$\frac{\mathbf{f}(\mathbf{x}+h\mathbf{e}_j) - \mathbf{f}(\mathbf{x})}{h} - \mathsf{J}_f(\mathbf{x})\mathbf{e}_j = \frac{1}{h}\left[\mathbf{f}(\mathbf{x}+h\mathbf{e}_j) - \mathbf{f}(\mathbf{x}) - \mathsf{J}_f(\mathbf{x})h\mathbf{e}_j\right].$$

Taking norms and applying Lemma 3.22, we obtain

$$\left\|\frac{\mathbf{f}(\mathbf{x}+h\mathbf{e}_j) - \mathbf{f}(\mathbf{x})}{h} - \mathsf{J}_f(\mathbf{x})\mathbf{e}_j\right\|_1 \leq \frac{1}{|h|}\frac{L}{2}|h|^2\|\mathbf{e}_j\|_1^2 = \frac{L|h|}{2}.$$

But the matrix norm $\|\hat{\mathsf{J}}(\mathbf{x}) - \mathsf{J}_f(\mathbf{x})\|_1$ is the maximum of the quantity on the left, taken over the column indices $j = 1, 2, \ldots, n$. The inequality (3.7-5) follows. ∎

Finite-difference Newton methods essentially use a "perturbed" version of the true Jacobian matrix of $\mathbf{f}$. One question that arises in the analysis of the methods is whether we can deduce the nonsingularity of the "perturbed Jacobian" from the nonsingularity of the exact version. The next lemma is useful in settling issues of this type.

LEMMA 3.26. *Let $\|\cdot\|$ be any norm on $\mathbb{R}^{n\times n}$ that is subordinate to a norm on $\mathbb{R}^n$. If $\mathsf{E} \in \mathbb{R}^{n\times n}$ has norm $\|\mathsf{E}\| < 1$, then the matrix $\mathsf{I} - \mathsf{E}$ is nonsingular, and*

$$\|(\mathsf{I} - \mathsf{E})^{-1}\| \leq \frac{1}{1 - \|\mathsf{E}\|}. \tag{3.7-6}$$

PROOF: To show that $\mathsf{I} - \mathsf{E}$ is nonsingular, it suffices to prove that $\mathbf{x} = \mathbf{0}$ is the only solution to the equation $(\mathsf{I} - \mathsf{E})\mathbf{x} = \mathbf{0}$. For any $\mathbf{x} \in \mathbb{R}^n$, the version (0.3-1) of the triangle inequality implies that

$$\begin{aligned}\|(\mathsf{I} - \mathsf{E})\mathbf{x}\| = \|\mathbf{x} - \mathsf{E}\mathbf{x}\| &\geq \|\mathbf{x}\| - \|\mathsf{E}\|\,\|\mathbf{x}\| \\ &= (1 - \|\mathsf{E}\|)\,\|\mathbf{x}\|.\end{aligned}$$

But $1 - \|\mathsf{E}\| > 0$, so $\|(\mathsf{I} - \mathsf{E})\mathbf{x}\| = 0$ only if $\mathbf{x} = \mathbf{0}$, as desired. To establish the inequality (3.7-6), note that

$$\begin{aligned}1 = \|(\mathsf{I} - \mathsf{E})(\mathsf{I} - \mathsf{E})^{-1}\| &= \|(\mathsf{I} - \mathsf{E})^{-1} - \mathsf{E}(\mathsf{I} - \mathsf{E})^{-1}\| \\ &\geq \|(\mathsf{I} - \mathsf{E})^{-1}\| - \|\mathsf{E}\|\,\|(\mathsf{I} - \mathsf{E})^{-1}\| \\ &= (1 - \|\mathsf{E}\|)\,\|(\mathsf{I} - \mathsf{E})^{-1}\|.\end{aligned}$$

Since $1 - \|\mathsf{E}\| > 0$, we can divide through by this quantity to obtain the inequality (3.7-6). ∎

COROLLARY 3.27 (PERTURBATION LEMMA). *If* $\mathsf{A} \in \mathbb{R}^{n\times n}$ *is nonsingular and* $\mathsf{B} \in \mathbb{R}^{n\times n}$ *satisfies the inequality* $\|\mathsf{A}^{-1}(\mathsf{B}-\mathsf{A})\| < 1$ *for some subordinate matrix norm* $\|\cdot\|$, *then* B *is nonsingular. Moreover,*

$$\|\mathsf{B}^{-1}\| \le \frac{\|\mathsf{A}^{-1}\|}{1-\|\mathsf{A}^{-1}(\mathsf{B}-\mathsf{A})\|}.$$

PROOF: This is Problem 10. ∎

We now have the tools needed to prove the central convergence result for finite-difference Newton methods.

THEOREM 3.28. *Let* $\Omega \subset \mathbb{R}^n$ *be open and convex. Suppose that* $\mathbf{f}: \Omega \to \mathbb{R}^n$ *is continuously differentiable and that* $\mathbf{f}$ *and its Jacobian matrix* J_f *obey the following conditions:*

(i) $\mathbf{f}$ *has a zero* $\mathbf{x}^* \in \Omega$.

(ii) $\mathsf{J}_f(\mathbf{x})$ *is nonsingular for every* $\mathbf{x} \in \Omega$, *and there is a constant* $M > 0$ *such that* $\|\mathsf{J}_f^{-1}(\mathbf{x})\| \le M$.

(iii) J_f *has Lipschitz constant* $L > 0$ *on* Ω.

Let the real sequence $\{h_m\}$ *generate the finite-difference Newton method (3.7-2). Then there exist positive constants* δ *and* $\bar{h}$ *such that, whenever the initial guess* $\mathbf{x}^{(0)} \in \mathcal{B}_\delta(\mathbf{x}^*)$ *and the generating sequence satisfies* $|h_m| < \bar{h}$, *each iterate* $\mathbf{x}^{(m)} \in \mathcal{B}_\delta(\mathbf{x}^*)$. *Moreover,* $\mathbf{x}^{(m)} \to \mathbf{x}^*$ *as* $m \to \infty$.

PROOF: By the equivalence of norms on $\mathbb{R}^n$, it suffices to argue using the norm $\|\cdot\|_1$. Pick δ and $\bar{h}$ small enough to guarantee that $\mathcal{B}_\delta(\mathbf{x}^*) \subset \Omega$ and $\delta + \bar{h} \le 1/(2LM)$. Our first task is to show that the finite-difference analog $\hat{\mathsf{J}}_m$ to the Jacobian matrix is nonsingular whenever $\mathbf{x}^{(m)} \in \mathcal{B}_\delta(\mathbf{x}^*)$. Observe that

$$\left\|\mathsf{J}_f^{-1}(\mathbf{x}^*)\left[\hat{\mathsf{J}}_m - \mathsf{J}_f(\mathbf{x}^*)\right]\right\|_1 \le M\left\|\hat{\mathsf{J}}_m - \mathsf{J}_f(\mathbf{x}^{(m)}) + \mathsf{J}_f(\mathbf{x}^{(m)}) - \mathsf{J}_f(\mathbf{x}^*)\right\|_1$$

$$\le M\Big[\underbrace{\left\|\hat{\mathsf{J}}_m - \mathsf{J}_f(\mathbf{x}^{(m)})\right\|_1}_{\text{(I)}} + \underbrace{\left\|\mathsf{J}_f(\mathbf{x}^{(m)}) - \mathsf{J}_f(\mathbf{x}^*)\right\|_1}_{\text{(II)}}\Big].$$

Lemma 3.25 implies that (I) $\le L|h_m|/2 < L\bar{h}/2$, and the Lipschitz condition (iii) implies that (II) $< L\delta$. Therefore,

$$\begin{aligned}\left\|\mathsf{J}_f^{-1}(\mathbf{x}^*)\left[\hat{\mathsf{J}}_m - \mathsf{J}_f(\mathbf{x}^*)\right]\right\|_1 &\le M\left(\tfrac{1}{2}L\bar{h} + L\delta\right)\\ &< LM(\bar{h}+\delta) \le \tfrac{1}{2}.\end{aligned}$$

The perturbation lemma now guarantees that $\hat{\mathsf{J}}_m$ is nonsingular and that $\|\hat{\mathsf{J}}_m^{-1}\|_1 \leq 2M$. In particular, each new iterate $\mathbf{x}^{(m+1)}$ is well defined whenever $\mathbf{x}^{(m)} \in \mathcal{B}_\delta(\mathbf{x}^*)$.

Next we prove by induction on m that each iterate $\mathbf{x}^{(m)} \in \mathcal{B}_\delta(\mathbf{x}^*)$. By hypothesis, $\mathbf{x}^{(0)} \in \mathcal{B}_\delta(\mathbf{x}^*)$. Assume that $\mathbf{x}^{(m)} \in \mathcal{B}_\delta(\mathbf{x}^*)$, and call $\boldsymbol{\varepsilon}^{(m)} := \mathbf{x}^* - \mathbf{x}^{(m)}$. Paralleling the proof of Theorem 3.13, we note that

$$\begin{aligned}
\boldsymbol{\varepsilon}^{(m+1)} &= \mathbf{x}^* - \mathbf{x}^{(m)} + \hat{\mathsf{J}}_m^{-1}\mathbf{f}(\mathbf{x}^{(m)}) - \hat{\mathsf{J}}_m^{-1}\mathbf{f}(\mathbf{x}^*) \\
&= -\hat{\mathsf{J}}_m^{-1}\Big\{\mathbf{f}(\mathbf{x}^*) - \mathbf{f}(\mathbf{x}^{(m)}) - \mathsf{J}_f(\mathbf{x}^{(m)})\boldsymbol{\varepsilon}^{(m)} \\
&\qquad + \left[\mathsf{J}_f(\mathbf{x}^{(m)}) - \hat{\mathsf{J}}_m\right]\boldsymbol{\varepsilon}^{(m)}\Big\}.
\end{aligned}$$

Taking norms gives

$$\begin{aligned}
\left\|\boldsymbol{\varepsilon}^{(m+1)}\right\|_1 &\leq \underbrace{\left\|\hat{\mathsf{J}}_m^{-1}\right\|_1}_{(\text{III})}\Big[\underbrace{\left\|\mathbf{f}(\mathbf{x}^*) - \mathbf{f}(\mathbf{x}^{(m)}) - \mathsf{J}_f(\mathbf{x}^{(m)})\boldsymbol{\varepsilon}^{(m)}\right\|_1}_{(\text{IV})} \\
&\qquad + \underbrace{\left\|\mathsf{J}_f(\mathbf{x}^{(m)}) - \hat{\mathsf{J}}_m\right\|_1\left\|\boldsymbol{\varepsilon}^{(m)}\right\|_1}_{(\text{V})}\Big].
\end{aligned}$$

But we have already shown that (III) $\leq 2M$, and Lemma 3.22 and the inductive hypothesis ensure that

$$(\text{IV}) \leq \frac{L}{2}\left\|\boldsymbol{\varepsilon}^{(m)}\right\|_1^2 < \frac{L\delta}{2}\left\|\boldsymbol{\varepsilon}^{(m)}\right\|_1.$$

Also, by Lemma 3.25, (V) $\leq \frac{1}{2}L\bar{h}\|\boldsymbol{\varepsilon}^{(m)}\|_1$. It follows that

$$\left\|\boldsymbol{\varepsilon}^{(m+1)}\right\|_1 \leq LM(\delta + \bar{h})\left\|\boldsymbol{\varepsilon}^{(m)}\right\|_1 \leq \tfrac{1}{2}\left\|\boldsymbol{\varepsilon}^{(m)}\right\|_1. \tag{3.7-7}$$

Therefore $\mathbf{x}^{(m+1)} \in \mathcal{B}_\delta(\mathbf{x}^*)$, and the induction is complete. The last inequality also shows that $\mathbf{x}^{(m)} \to \mathbf{x}^*$ as $m \to \infty$. ∎

The inequality (3.7-7) asserts that properly constructed finite-difference Newton methods converge at least linearly in some neighborhood of a zero $\mathbf{x}^*$. In fact, more is possible. Enthusiastic readers should formulate and prove conditions on the sequence $\{h_m\}$ of offsets for which the corresponding finite-difference Newton methods converge superlinearly and quadratically. Corollaries 3.14 and 3.15 furnish reasonable guides.

Further Remarks

We close the chapter with a cursory look at **Broyden's method**. Two observations motivate this scheme. First, by analogy with the one-dimensional case, we expect the finite-difference Newton method (3.7-2) to converge superlinearly when the offsets $h_m \to 0$ as $m \to \infty$. Second, when $\mathbf{f}$ is expensive to evaluate, much of the work associated with the scheme (3.7-2) occurs in the computation of the Jacobian analogs $\hat{\mathsf{J}}_m$.

Broyden's method addresses the first observation by choosing $\hat{\mathsf{J}}_m$ in a fashion reminiscent of the secant method. To start the method, one typically chooses $\hat{\mathsf{J}}_0$ to be either the exact Jacobian $\mathsf{J}_f(\mathbf{x}^{(0)})$ or some finite-difference analog of it. Given $\hat{\mathsf{J}}_{m-1}$, the method defines a matrix $\hat{\mathsf{J}}_m$ such that

$$\hat{\mathsf{J}}_m(\mathbf{x}^{(m)} - \mathbf{x}^{(m-1)}) = \mathbf{f}(\mathbf{x}^{(m)}) - \mathbf{f}(\mathbf{x}^{(m-1)}).$$

This "secant-like" condition dictates the action of the matrix $\hat{\mathsf{J}}_m$ only on the subspace of $\mathbb{R}^n$ containing vectors proportional to $\mathbf{x}^{(m)} - \mathbf{x}^{(m-1)}$, so it does not define $\hat{\mathsf{J}}_m$ uniquely. To complete the definition, we further stipulate that $\hat{\mathsf{J}}_m\mathbf{z} = \hat{\mathsf{J}}_{m-1}\mathbf{z}$ for any vector $\mathbf{z} \in \mathbb{R}^n$ that is orthogonal to $\mathbf{x}^{(m)} - \mathbf{x}^{(m-1)}$. In other words, the procedure for updating $\hat{\mathsf{J}}_{m-1}$ to $\hat{\mathsf{J}}_m$ has no effect on those vectors. This stipulation appeals more to a desire for simplicitly than to a respect for the actual behavior of $\mathbf{f}$. In fact, if we define $\mathbf{d}_m := \mathbf{x}^{(m)} - \mathbf{x}^{(m-1)}$ and $\mathbf{y}_m := \mathbf{f}(\mathbf{x}^{(m)}) - \mathbf{f}(\mathbf{x}^{(m-1)})$, then the updated matrix

$$\hat{\mathsf{J}}_m := \hat{\mathsf{J}}_{m-1} + \frac{\mathbf{y}_m - \hat{\mathsf{J}}_{m-1}\mathbf{d}_m}{\|\mathbf{d}_m\|_2^2}\mathbf{d}_m^\top \tag{3.7-8}$$

satisfies the conditions mentioned above. Problem 14 asks for verification.

We pause to comment on the notation in Equation (3.7-8). For two nonzero vectors $\mathbf{a} = (a_1, a_2, \ldots, a_n)^\top$ and $\mathbf{b} = (b_1, b_2, \ldots, b_n)^\top$, the product $\mathbf{ab}^\top$ denotes the following matrix in $\mathbb{R}^{n\times n}$:

$$\mathbf{ab}^\top := \begin{bmatrix} a_1b_1 & \cdots & a_1b_n \\ \vdots & & \vdots \\ a_nb_1 & \cdots & a_nb_n \end{bmatrix}.$$

Every column of this matrix is a multiple of the vector $\mathbf{a}$; therefore, the columns of the matrix $\mathbf{ab}^\top$ span a subspace of $\mathbb{R}^n$ having dimension 1. In other words, the **rank** of the matrix $\mathbf{ab}^\top$ is 1. Equation (3.7-8) defines $\hat{\mathsf{J}}_m$ as a **rank-one update** of the approximate Jacobian matrix $\hat{\mathsf{J}}_{m-1}$ used in the previous iteration.

The updating formula (3.7-8) circumvents most of the computational effort involved in computing a new analog to the Jacobian matrix at each iteration. However, in practice the rank-one updates may not adequately capture the changes in the nonlinear function $\mathbf{f}$ for more than a few iterations, and it may be necessary to revert to an ordinary finite-difference Jacobian periodically to keep the method convergent. For an analysis of Broyden's method and more details concerning its implementation, we refer readers to Dennis and Schnabel ([1], Chapter 8).

3.8 Problems

PROBLEM 1. Chapter 1 discusses piecewise polynomial approximations to functions, in which the coefficients used to compute an approximation to $f(\overline{x})$ depend on which interval $[x_i, x_{i+1}]$ of a grid $\{x_0, x_1, \ldots, x_N\}$ contains the point $\overline{x}$. The naive way to determine $[x_i, x_{i+1}]$ is to test the intervals in order, for $i = 1, 2, \ldots$, until we reach the first value of i for which $\overline{x} < x_i$. This algorithm takes $\mathcal{O}(N)$ iterations to find the interval that contains $\overline{x}$. A faster technique, called **logarithmic searching**, uses an integer version of bisection:

1. $i_{\text{left}} \leftarrow 0$, $i_{\text{right}} \leftarrow N$.
2. If $i_{\text{left}} - i_{\text{right}} > 1$ then:
3. $\quad i_{\text{mid}} \leftarrow \text{int}\,[(i_{\text{right}} + i_{\text{left}})/2]$.
4. $\quad$ If $x_{i_{\text{mid}}} > \overline{x}$ then:
5. $\quad\quad i_{\text{right}} \leftarrow i_{\text{mid}}$.
6. $\quad$ Else:
7. $\quad\quad i_{\text{left}} \leftarrow i_{\text{mid}}$.
8. $\quad$ End if.
9. $\quad$ Go to 2.
10. End if.
11. $i \leftarrow i_{\text{left}}$.
12. End.

Here, $\text{int}\,(x)$ denotes the largest integer that is less than or equal to x. How many iterations do you expect this algorithm to take?

PROBLEM 2. The polynomial $f(x) = x^3 - 3x + 1$ has three real roots x_1^*, x_2^*, x_3^*. For each root, devise a successive substitution scheme $x_i^{(m+1)} \leftarrow \Phi(x_i^{(m)})$ that converges to x_i^*. Using appropriate initial guesses $x_i^{(0)}$, demonstrate both theoretically and computationally that the scheme converges with order $p \geq 1$.

PROBLEM 3. Prove that, if $\Phi: [a,b] \to \mathbb{R}$ satisfies a Lipschitz condition on $[a,b]$, then Φ is continuous on $[a,b]$. Find a function that satisfies a Lipschitz condtion on $[a,b]$ but that does not belong to $C^1([a,b])$. Therefore, the Lipschitz condition is at least as strong as continuity, but it is weaker than continuous differentiability. Does continuity of Φ on a set $S \subset \mathbb{R}$ guarantee that Φ satisfies a Lipschitz condition on S?

PROBLEM 4. Suppose that you have three iterative schemes that generate the following sequences of iterates, respectively:

$$\text{(A)} \quad \{2^{-m}\} \qquad \text{(B)} \quad \left\{2^{-2^m}\right\} \qquad \text{(C)} \quad \left\{2^{-m^2}\right\}.$$

Using this evidence, characterize the convergence rates of the schemes.

PROBLEM 5.

(A) Using the line of reasoning leading to the estimate (3.4-3), state and prove a theorem regarding quadratic convergence in Newton's method near simple roots.

(B) It is possible to generalize this idea to produce successive substitution schemes that converge with any order $p = 2, 3, \ldots$. The idea is to construct an iteration function Φ such that $\Phi'(x^*) = \Phi''(x^*) = \cdots = \Phi^{(p-1)}(x^*) = 0$, making sure in the process that $|\Phi^{(p)}(x^*)|$ remains appropriately bounded. State and prove such a theorem.

(C) In Newton's method, the iteration function is $\Phi(x) = x - f(x)/f'(x)$. By analyzing Φ' near x^*, show that Newton's method applied to a function $f \in C^2([a,b])$ converges linearly, but no faster, near a root having multiplicity 2 or greater.

PROBLEM 6. A sequence $\{x_m\}$ of real numbers is **bounded** if there is a number $M > 0$ such that $|x_m| \leq M$ for every index m. The sequence is **monotonic** if it is either **nondecreasing**, that is, $x_m \leq x_{m+1}$ for all m, or **nonincreasing**, that is, $x_{m+1} \leq x_m$ for all m. Prove that every bounded, nondecreasing sequence of real numbers has a limit. (Hint: the limit is $\sup x_m$.) The proof that every bounded, nondecreasing sequence of real numbers has a limit is similar.

PROBLEM 7. Derive an error estimate analogous to the inequality (3.4-5) for the case when x^* is a root of multiplicity $q > 1$.

PROBLEM 8.

(A) Prove Corollary 3.15.

(B) Prove that the conclusion of Corollary 3.15 remains valid for functions $f \in C^1([a,b])$ if we replace the condition $|h_m| \le C|\varepsilon^{(m)}|$ by $|h_m| \le C|f(x^{(m)})|$. (The choice $h_m = f(x^{(m)})$ yields **Steffensen's method**.)

PROBLEM 9. Use successive substitution to solve the following system of equations:

$$x^2 + y^2 = x,$$

$$x^2 - y^2 = y.$$

Discuss convergence, order of convergence, and error estimates.

PROBLEM 10. Prove Corollary 3.27.

PROBLEM 11. Define an iteration function Φ by $\Phi(x) = x - qf(x)/f'(x)$. Prove the following: If $f \in C^3(\mathbb{R})$ has a zero x^* of multiplicity q, then there exists $\delta > 0$ such that the iterative scheme $x^{(m+1)} = \Phi(x^{(m)})$ converges with order $p = 2$ for any initial guess $x^{(0)} \in (x^* - \delta, x^* + \delta)$.

PROBLEM 12. Write a computer program to solve the system (3.6-2) using an arbitrary initial guess $\mathbf{x}^{(0)}$. Through computational experiments, characterize regions of the plane $\mathbb{R}^2$ according to the zeros, if any, to which initial guesses in the regions converge.

PROBLEM 13. Let $\Omega \subset \mathbb{R}^n$ be open and convex, and assume that $\mathbf{f}: \Omega \to \mathbb{R}^n$ is continuously differentiable. Assume further that the Jacobian matrix $\mathsf{J}_f(\mathbf{x})$ is nonsingular and has Lipschitz constant $L > 0$ on Ω. Suppose that $\mathbf{x} \in \Omega$ with $\mathbf{f}(\mathbf{x}) \neq \mathbf{0}$, and define

$$\mathbf{y} := \mathbf{x} - \omega\, \mathsf{J}_f^{-1}(\mathbf{x})\, \mathbf{f}(\mathbf{x}).$$

Prove that there exists $\overline{\omega} > 0$ such that $\|\mathbf{f}(\mathbf{y})\|_2^2 < \|\mathbf{f}(\mathbf{x})\|_2^2$ whenever $\omega < \overline{\omega}$.

PROBLEM 14.

(A) Let $\hat{\mathsf{J}}_m \in \mathbb{R}^{n\times n}$ be the Broyden update defined in Equation (3.7-8). Prove that $\hat{\mathsf{J}}_m(\mathbf{x}^{(m)} - \mathbf{x}^{(m-1)}) = \mathbf{f}(\mathbf{x}^{(m)}) - \mathbf{f}(\mathbf{x}^{(m-1)})$ and that $\hat{\mathsf{J}}_m \mathbf{z} = \hat{\mathsf{J}}_{m-1}\mathbf{z}$ whenever $\mathbf{z}$ is orthogonal to $\mathbf{x}^{(m)} - \mathbf{x}^{(m-1)}$.

(B) Prove the **Sherman-Morrison theorem**: If $\mathsf{A} \in \mathbb{R}^{n\times n}$ is nonsingular and $\mathbf{a}, \mathbf{b} \in \mathbb{R}^n$ satisfy the condition $\mathbf{b}^\top \mathsf{A}^{-1}\mathbf{a} \neq -1$, then the matrix $\mathsf{A} + \mathbf{a}\mathbf{b}^\top \in \mathbb{R}^{n\times n}$ is nonsingular, and

$$(\mathsf{A} + \mathbf{a}\mathbf{b}^\top)^{-1} = \mathsf{A}^{-1} - \frac{\mathsf{A}^{-1}\mathbf{a}\mathbf{b}^\top\mathsf{A}^{-1}}{1 + \mathbf{a}\mathbf{b}^\top}.$$

(c) Comment on the applicability of the Sherman-Morrison formula to Broyden's method.

PROBLEM 15. Fixed points are quite common: Prove that every continuous function $\Phi: [a, b] \to [a, b]$ has a fixed point. (Hint: Show that $g(x) := x - \Phi(x)$ has a zero in $[a, b]$ by considering the values of $g(a)$ and $g(b)$.)

3.9 References

1. J.E. Dennis, Jr. and R.B. Schnabel, *Numerical Methods for Unconstrained Optimization and Nonlinear Equations,* Prentice-Hall, Englewood Cliffs, NJ, 1983.

2. P. Henrici, *Elements of Numerical Analysis*, Wiley, New York, 1964.

3. J.M. Ortega and W.C. Rheinboldt, *Iterative Solution of Nonlinear Equations in Several Variables*, Academic Press, New York, 1970.

4. J.H. Press, B.P. Flannery, S.A. Teukolsky, and W.T. Vetterling, *Numerical Recipes: The Art of Scientific Computing*, Cambridge University Press, Cambridge, U.K., 1986.

5. J.H. Wilkinson, "The evaluation of the zeros of ill-conditioned polynomials. Part I," *Numer. Math. 1* (1959), 150-180.

Chapter 4

Iterative Methods for Linear Systems

4.1 Introduction

Although iterating may seem most natural in nonlinear settings, where direct solution techniques typically are not available, it has tremendous applicability to linear systems as well. The discussion of band matrices in Section 2.4 hints at this utility. Many applications lead to large band matrices in which the bands themselves are sparse. Direct methods often use computer memory and arithmetic extravagantly in such applications, while iterative schemes offer opportunities for greater efficiency.

For example, finite-difference approximations of certain partial differential equations yield block-tridiagonal matrices:

$$
\mathsf{A} = \begin{array}{c} \\ m \\ m \\ \vdots \\ \\ \\ \end{array}
\begin{array}{c} \begin{array}{ccccc} m & m & \cdots & & \end{array} \\
\begin{pmatrix}
\mathsf{T}_1 & \mathsf{D}_1 & & & \\
\mathsf{D}_2 & \mathsf{T}_2 & \mathsf{D}_2 & & \\
& & \ddots & & \\
& & \mathsf{D}_{n-1} & \mathsf{T}_{n-1} & \mathsf{D}_{n-1} \\
& & & \mathsf{D}_n & \mathsf{T}_n
\end{pmatrix} \end{array}. \qquad (4.1\text{-}1)
$$

Typically, each block $\mathsf{T}_i \in \mathbb{R}^{m \times m}$ is tridiagonal, while each of the off-diagonal blocks $\mathsf{D}_i \in \mathbb{R}^{m \times m}$ is diagonal. The sparseness of such systems is an attractive consequence of the approximations that generate them.

For concreteness, consider the unit square $\Omega = (0,1) \times (0,1) \subset \mathbb{R}^2$ and the following boundary-value problem for the Laplace equation:

$$\begin{aligned} -\frac{\partial^2 u}{\partial x^2}(x,y) - \frac{\partial^2 u}{\partial y^2}(x,y) = 0, \qquad & (x,y) \in \Omega; \\ u(x,y) = u_\partial(x,y), \qquad & (x,y) \in \partial\Omega. \end{aligned} \tag{4.1-2}$$

Here, $\partial\Omega$ denotes the boundary of Ω, and the known function u_∂ defines the boundary values of the unknown function u. Given the uniform grid on Ω drawn in Figure 1, standard finite-difference approximations convert the derivatives in the Laplace equation to the algebraic analogs

$$\begin{aligned} \frac{\partial^2 u}{\partial x^2}(x_k, y_l) &\simeq \frac{u_{k-1,l} - 2u_{k,l} + u_{k+1,l}}{h^2}, \\ \frac{\partial^2 u}{\partial y^2}(x_k, y_l) &\simeq \frac{u_{k,l-1} - 2u_{k,l} + u_{k,l+1}}{h^2}, \end{aligned}$$

where h stands for the distance between adjacent grid lines and $u_{k,l}$ denotes the approximate value of $u(x_k, y_l)$. (Chapter 8 discusses this approximation in detail.) Substituting these analogs into the Laplace equation yields

$$4u_{k,l} - (u_{k-1,l} + u_{k,l-1} + u_{k,l+1} + u_{k+1,l}) = 0. \tag{4.1-3}$$

By assigning known values of $u_\partial(x_k, y_l)$ to corresponding values $u_{k,l}$, we arrive at the following block-tridiagonal system:

$$\begin{bmatrix} 4 & -1 & 0 & -1 & & & & & \\ -1 & 4 & -1 & 0 & -1 & & & & \\ 0 & -1 & 4 & 0 & 0 & -1 & & & \\ -1 & 0 & 0 & 4 & -1 & 0 & -1 & & \\ & -1 & 0 & -1 & 4 & -1 & 0 & -1 & \\ & & -1 & 0 & -1 & 4 & 0 & 0 & -1 \\ & & & -1 & 0 & 0 & 4 & -1 & 0 \\ & & & & -1 & 0 & -1 & 4 & -1 \\ & & & & & -1 & 0 & -1 & 4 \end{bmatrix} \begin{bmatrix} u_{1,1} \\ u_{2,1} \\ u_{3,1} \\ u_{1,2} \\ u_{2,2} \\ u_{3,2} \\ u_{1,3} \\ u_{2,3} \\ u_{3,3} \end{bmatrix} = \begin{bmatrix} u_{1,0} + u_{0,1} \\ u_{2,0} \\ u_{3,0} + u_{4,1} \\ u_{0,2} \\ 0 \\ u_{4,2} \\ u_{0,3} + u_{1,4} \\ u_{2,4} \\ u_{3,4} + u_{4,3} \end{bmatrix} \tag{4.1-4}$$

where the vector on the right contains known boundary values. We denote this system more briefly as $\mathbf{Au} = \mathbf{b}$. For finer grids on the square Ω, the maximum number of nonzero entries on each row of the matrix for this problem remains constant at five, and many more zeros appear inside the nonzero band.

(Throughout this chapter we use the symbol $\mathbf{u}$, instead of $\mathbf{x}$, for the vector of unknowns. Thus the system to be solved is $\mathbf{Au} = \mathbf{b}$. This change in notation reflects the close connection between the iterative methods discussed and approximations to differential equations, in which u commonly denotes the unknown function.)

Direct methods based on banded LU factorization tend to preserve the block-tridiagonal structure but fill in the originally sparse blocks. This fill-in

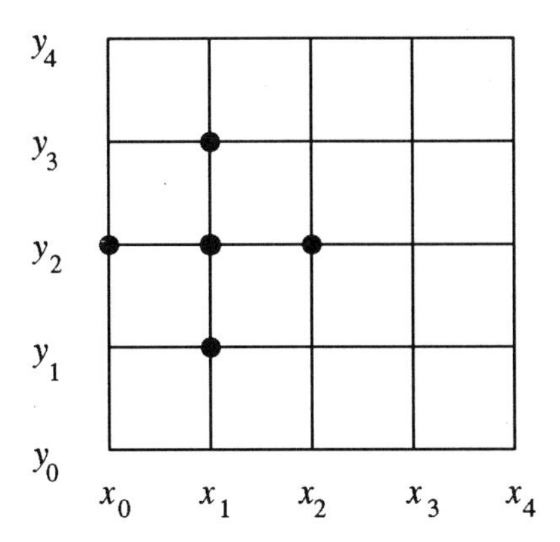

FIGURE 1. *A simple grid on the unit square, showing how the standard finite-difference approximation to the Laplace equation couples unknown nodal values* $u_{i,j}$.

demands more storage space than the original zero structure, and the new nonzero entries increase the operation counts in subsequent uses of forward and backward substitution. Direct methods thus forfeit some of the properties that make finite-difference approximations so attractive.

The idea behind iterative methods is to generate a sequence $\{\mathbf{u}^{(m)}\}$ of approximate solution vectors, each computed using arithmetic involving only the nonzero entries of A. Since the methods avoid fill-in, the computational cost per iteration stays small. The main goal is to guarantee that $\mathbf{u}^{(m)} \to \mathbf{u}$ rapidly.

In this chapter, we explore the rudiments of two distinct ideas. The first is **relaxation**, that is, the successive adjustment of small subsets of unknowns using tentative values for the other unknowns. The second idea is **searching**. Here, one constructs successive approximations $\mathbf{u}^{(m)}$ in multidimensional Euclidean space by marching in directions determined by optimization criteria.

Before proceeding, we review the algebraic setting and some notation. Earlier chapters discuss linear algebraic methods in the context of the real vector spaces $\mathbb{R}^n$. In the present chapter, we sometimes work in the more general vector space $\mathbb{C}^n$, which contains all n-tuples of elements from the set $\mathbb{C}$ of complex numbers. There corresponds to $\mathbb{C}^n$ an associated space of linear transformations represented by elements of $\mathbb{C}^{n\times n}$, the set of all $n \times n$ matrices whose entries belong to $\mathbb{C}$.

The elementary algebra of $\mathbb{C}^n$ extends that of $\mathbb{R}^n$. The main differences arise in the computation of norms. Recall that the magnitude of a complex number $z = x + iy$, with $x, y \in \mathbb{R}$, is $|z| := \sqrt{\overline{z}z}$, where $\overline{z} := x - iy$ denotes the **complex conjugate** of z. The definitions of various norms on $\mathbb{C}^n$ follow accordingly: If $\mathbf{z} = (z_1, z_2, \ldots, z_n)^\top$, then

$$\begin{aligned}
\|\mathbf{z}\|_1 &:= \sum_{j=1}^{n} \sqrt{z_j \bar{z}_j}; \\
\|\mathbf{z}\|_2 &:= \left(\sum_{j=1}^{n} z_j \bar{z}_j \right)^{1/2}; \\
\|\mathbf{z}\|_\infty &:= \max_{1 \le j \le n} \left\{ \sqrt{\bar{z}_j z_j} \right\}.
\end{aligned}$$

As in $\mathbb{R}^n$, for any norm $\|\cdot\|$ on $\mathbb{C}^n$ there is a subordinate matrix norm $\|\cdot\|: \mathbb{C}^{n \times n} \to \mathbb{R}$ by the equation

$$\|\mathsf{A}\| := \sup_{\|\mathbf{z}\| \neq 0} \frac{\|\mathsf{A}\mathbf{z}\|}{\|\mathbf{z}\|}. \tag{4.1-5}$$

Familiar facts about norms on $\mathbb{R}^{n \times n}$ extend to the complex setting: Any matrix norm on $\mathbb{C}^{n \times n}$ is continuous, and all norms on $\mathbb{C}^{n \times n}$ are equivalent.

Some economy of notation is available if we generalize the transpose of a real matrix. For a matrix $\mathsf{A} \in \mathbb{C}^{m \times n}$, the **Hermitian transpose** (or **conjugate transpose**) of A is

$$\mathsf{A}^* := \bar{\mathsf{A}}^\top \in \mathbb{C}^{n \times m}.$$

Obviously, $\mathsf{A}^* = \mathsf{A}^\top$ when the entries of A are all real. If we think of vectors $\mathbf{z} \in \mathbb{C}^n$ as column vectors, or elements of $\mathbb{C}^{n \times 1}$, then the Hermitian conjugate of $\mathbf{z}$ is $\mathbf{z}^*$, the row vector whose entries are the complex conjugates of the entries of $\mathbf{z}$. The following identities are easy to check:

(*i*) $\|\mathbf{z}\|_2 = \sqrt{\mathbf{z}^* \mathbf{z}}$.

(*ii*) $(\mathsf{AB})^* = \mathsf{B}^* \mathsf{A}^*$.

(*iii*) $(\mathsf{A}^*)^* = \mathsf{A}$.

Finally, the inner product of two vectors $\mathbf{w}, \mathbf{z} \in \mathbb{C}^n$ is $\mathbf{w}^* \mathbf{z}$. As in any inner-product space, two vectors in $\mathbb{C}^n$ are orthogonal if their inner product vanishes.

4.2 Conceptual Foundations

Rigorous analysis of iterative methods for linear systems relies on several theoretical results. We devote this section to a review of these results. However, we use most of the concepts developed here only in discussing mathematical details. For readers intent on skipping the theory, only the basic facts about eigenvalues and eigenvectors, culminating below in Proposition 4.1, are needed in later chapters.

Suppose that an iterative scheme for a system $\mathbf{A}\mathbf{u} = \mathbf{b}$ produces a sequence $\{\mathbf{u}^{(m)}\}$ of iterates. It is crucial to determine conditions under which the errors $\boldsymbol{\varepsilon}^{(m)} := \mathbf{u} - \mathbf{u}^{(m)} \to 0$ as $m \to \infty$. As Section 4.3 shows, with many iterative schemes we can associate matrices G that relate errors at successive iterations in the following way:

$$\boldsymbol{\varepsilon}^{(m+1)} = \mathsf{G}\boldsymbol{\varepsilon}^{(m)} = \cdots = \mathsf{G}^{m+1}\boldsymbol{\varepsilon}^{(0)}.$$

We therefore analyze such repeated applications of the matrix G, both in terms of norms and in terms of the action of G on specific vectors.

We begin with the action of G.

DEFINITION. *Let $\mathsf{G} \in \mathbb{C}^{n\times n}$. A number $\lambda \in \mathbb{C}$ is an* **eigenvalue** *of G if there exists a nonzero vector $\boldsymbol{v} \in \mathbb{C}^n$ such that $\mathsf{G}\boldsymbol{v} = \lambda\boldsymbol{v}$. In this case, $\boldsymbol{v}$ is an* **eigenvector** *associated with λ. The set $\sigma(\mathsf{G})$ of all eigenvalues of G is the* **spectrum** *of G.*

Thus G "stretches" an eigenvector by a (possibly complex) factor, namely the associated eigenvalue. For a given eigenvalue, the associated eigenvector is not unique: If $\boldsymbol{v}$ is an eigenvector associated with eigenvalue λ, then so is $\alpha\boldsymbol{v}$ for any nonzero constant $\alpha \in \mathbb{C}$.

A number λ is an eigenvalue of $\mathsf{G} \in \mathbb{C}^{n\times n}$ if and only if there is a nonzero solution $\boldsymbol{v} \in \mathbb{C}^n$ to the linear system $(\mathsf{G} - \lambda\mathsf{I})\boldsymbol{v} = \mathbf{0}$, where I is the identity matrix. Such a solution exists if and only if $\det(\mathsf{G} - \lambda\mathsf{I}) = 0$. Since the **characteristic polynomial** $\det(\mathsf{G} - \lambda\mathsf{I})$ has degree n in λ, G has n eigenvalues, counted according to their multiplicities as zeros. The multiplicity of an eigenvalue λ as a zero of the characteristic polynomial of G is the **algebraic multiplicity** of λ. Thus, finding the eigenvalues of a matrix is equivalent to the nonlinear problem of finding the zeros of a polynomial. When n is large, this equivalence is more useful theoretically than computationally, as we explore in Chapter 5.

When do two matrices have the same spectrum?

DEFINITION. *The matrices $\mathsf{G}, \mathsf{H} \in \mathbb{C}^{n\times n}$ are* **similar** *if there exists a nonsingular matrix $\mathsf{S} \in \mathbb{C}^{n\times n}$ such that $\mathsf{H} = \mathsf{S}\mathsf{G}\mathsf{S}^{-1}$. The mapping $\mathsf{A} \mapsto \mathsf{S}\mathsf{A}\mathsf{S}^{-1}$ is a* **similarity transformation**.

It is nearly obvious that similarity is an equivalence relation. More to the point, similarity transformations preserve eigenvalues:

PROPOSITION 4.1. *If* $\mathsf{G}, \mathsf{H} \in \mathbb{C}^{n\times n}$ *are similar, then* G *and* H *have the same spectrum.*

PROOF: The two matrices have the same characteristic polynomial:

$$\begin{aligned}\det(\lambda\mathsf{I}-\mathsf{H}) = \det(\lambda\mathsf{I}-\mathsf{SGS}^{-1}) &= \det[\mathsf{S}(\lambda\mathsf{I}-\mathsf{G})\mathsf{S}^{-1}] \\ &= \det\mathsf{S}^{-1}\det(\lambda\mathsf{I}-\mathsf{G})\det\mathsf{S} \\ &= \det(\lambda\mathsf{I}-\mathsf{G}). \qquad \blacksquare\end{aligned}$$

If $\lambda \in \sigma(\mathsf{G})$ has associated eigenvector $\boldsymbol{v}$, then the eigenvector of SGS^{-1} associated with λ is $\mathsf{S}\boldsymbol{v}$.

As in $\mathbb{R}^{n\times n}$, if G is upper or lower triangular, then its eigenvalues are just the diagonal entries of G. Thus one can easily determine the spectra of upper and lower triangular matrices by inspection. Remarkably, every matrix in $\mathbb{C}^{n\times n}$ is similar to an upper triangular matrix. We now explore this fact.

DEFINITION. *A matrix* $\mathsf{U} \in \mathbb{C}^{n\times n}$ *is* **unitary** *if* $\mathsf{U}^*\mathsf{U} = \mathsf{I}$. *Two matrices* $\mathsf{G}, \mathsf{H} \in \mathbb{C}^{n\times n}$ *are* **unitarily similar** *if there is a unitary matrix* $\mathsf{U} \in \mathbb{C}^{n\times n}$ *such that* $\mathsf{H} = \mathsf{U}^*\mathsf{GU}$.

All unitary matrices are nonsingular, since the definition implies that $\mathsf{U}^{-1} = \mathsf{U}^*$. Therefore, matrices that are unitarily similar have the same eigenvalues. A unitary matrix is a square array whose columns, considered as individual vectors $\mathbf{v}_1, \mathbf{v}_2, \ldots, \mathbf{v}_n \in \mathbb{C}^n$, form an orthonormal basis for $\mathbb{C}^n$. This fact follows from the observation that U has n columns and from the equivalence between the requirement $\mathsf{U}^*\mathsf{U} = \mathsf{I}$ and the orthonormality condition

$$\mathbf{v}_j^*\mathbf{v}_k = \begin{cases} 1, & \text{if} \quad k = j; \\ 0, & \text{if} \quad k \neq j. \end{cases} \tag{4.2-1}$$

THEOREM 4.2 (SCHUR NORMAL FORM). *Every matrix* $\mathsf{G} \in \mathbb{C}^{n\times n}$ *is unitarily similar to an upper triangular matrix in* $\mathbb{C}^{n\times n}$.

PROOF: We use induction on n, the case $n = 1$ being trivial. Assume that any matrix in $\mathbb{C}^{(n-1)\times(n-1)}$ is unitarily similar to an upper triangular matrix, and let $\mathsf{G} \in \mathbb{C}^{n\times n}$. Suppose that λ_1 is an eigenvalue of G associated with an eigenvector $\mathbf{v}_1$, which we assume to be normalized so that $\|\mathbf{v}_1\|_2 = 1$. Pick an orthonormal basis $\{\mathbf{v}_1, \mathbf{v}_2, \ldots, \mathbf{v}_n\}$ for $\mathbb{C}^n$ (One can do this, for example, using the Gram-Schmidt procedure; see Strang [6].) Now construct a matrix $\mathsf{V} \in \mathbb{C}^{n\times n}$ by taking its columns to be the vectors $\mathbf{v}_1, \mathbf{v}_2, \ldots, \mathbf{v}_n$. We denote V in block form as $[\mathbf{v}_1\ \mathbf{v}_2\ \cdots\ \mathbf{v}_n]$.

The matrix V has two salient properties. First, it is unitary, since the vectors $\mathbf{v}_j$ form an orthonormal set. Second,

$$\mathsf{V}^*\mathsf{G}\mathsf{V} = \mathsf{V}^*[\lambda_1\mathbf{v}_1 \quad \mathsf{G}\mathbf{v}_2 \quad \cdots \quad \mathsf{G}\mathbf{v}_n] = \begin{matrix} \\ 1 \\ n-1 \end{matrix} \begin{matrix} \quad 1 \qquad n-1 \\ \begin{pmatrix} \lambda_1 & \mathbf{g}^* \\ \mathbf{0} & \mathsf{G}_1 \end{pmatrix} \end{matrix},$$

where $\mathbf{g} \in \mathbb{C}^{n-1}$ and $\mathsf{G}_1 \in \mathbb{C}^{(n-1)\times(n-1)}$.

The next objective is to convert the block G_1 to upper triangular form. By the inductive hypothesis, there is a unitary matrix $\mathsf{U}_1 \in \mathbb{C}^{(n-1)\times(n-1)}$ such that

$$\mathsf{U}_1^*\mathsf{G}_1\mathsf{U}_1 = \begin{bmatrix} \lambda_2 & \star & \cdots & \star \\ & \lambda_3 & & \vdots \\ & & \ddots & \star \\ & & & \lambda_n \end{bmatrix},$$

where $\lambda_2, \lambda_3, \ldots, \lambda_n$ are the eigenvalues of G_1 and the symbols "$\star$" stand for entries that may be nonzero. Define $\mathsf{U} \in \mathbb{C}^{n\times n}$ by setting

$$\mathsf{U} = \mathsf{V}\begin{pmatrix} 1 & \mathbf{0}^* \\ \mathbf{0} & \mathsf{U}_1 \end{pmatrix}.$$

It is easy to check that U is unitary. Moreover,

$$\begin{aligned} \mathsf{U}^*\mathsf{G}\mathsf{U} &= \begin{pmatrix} 1 & \mathbf{0}^* \\ \mathbf{0} & \mathsf{U}_1^* \end{pmatrix}\mathsf{V}^*\mathsf{G}\mathsf{V}\begin{pmatrix} 1 & \mathbf{0}^* \\ \mathbf{0} & \mathsf{U}_1 \end{pmatrix} \\ &= \begin{pmatrix} 1 & \mathbf{0}^* \\ \mathbf{0} & \mathsf{U}_1^* \end{pmatrix}\begin{pmatrix} \lambda_1 & \mathbf{g}^* \\ \mathbf{0} & \mathsf{G}_1 \end{pmatrix}\begin{pmatrix} 1 & \mathbf{0}^* \\ \mathbf{0} & \mathsf{U}_1 \end{pmatrix} \\ &= \begin{pmatrix} 1 & \mathbf{0}^* \\ \mathbf{0} & \mathsf{U}_1^* \end{pmatrix}\begin{pmatrix} \lambda_1 & \mathbf{g}^*\mathsf{U}_1 \\ \mathbf{0} & \mathsf{G}_1\mathsf{U}_1 \end{pmatrix} = \begin{pmatrix} \lambda_1 & \mathbf{g}^*\mathsf{U}_1 \\ \mathbf{0} & \mathsf{U}_1^*\mathsf{G}_1\mathsf{U}_1 \end{pmatrix}. \end{aligned}$$

The last matrix is upper triangular. ∎

The Schur normal form facilitates an estimate of the spectral radius $\varrho(\mathsf{G})$ in terms of $\|\mathsf{G}\|$. From the definition (4.1-5) it follows that $|\lambda|\,\|\boldsymbol{v}\| \le \|\mathsf{G}\|\,\|\boldsymbol{v}\|$ for any eigenvector $\boldsymbol{v}$, and hence $\varrho(\mathsf{G}) \le \|\mathsf{G}\|$ for any subordinate matrix norm $\|\cdot\|$. More is true:

THEOREM 4.3. *Let $\mathsf{G} \in \mathbb{C}^{n\times n}$. For any $\epsilon > 0$ there exists a subordinate matrix norm $\|\cdot\|: \mathbb{C}^{n\times n} \to \mathbb{R}$ such that $\varrho(\mathsf{G}) \le \|\mathsf{G}\| \le \varrho(\mathsf{G}) + \epsilon$.*

(The norm $\|\cdot\|$ that works depends upon G and ϵ.) This theorem is pivotal in the analysis of G^m, so its long proof is worth the effort. We use the following simple lemma, whose proof is Problem 1:

LEMMA 4.4. *The equation* $\|\mathbf{z}\|_{\mathrm{II}} := \|\mathsf{M}\mathbf{z}\|_{\mathrm{I}}$ *defines a norm* $\|\cdot\|_{\mathrm{II}}$ *on* $\mathbb{C}^n$ *whenever* $\|\cdot\|_{\mathrm{I}}$ *is a norm on* $\mathbb{C}^n$ *and* $\mathsf{M} \in \mathbb{C}^{n\times n}$ *is nonsingular.*

PROOF OF THEOREM 4.3: We need only find a subordinate matrix norm that satisfies the inequality $\|\mathsf{G}\| \leq \varrho(\mathsf{G}) + \epsilon$. We begin by defining a suitable norm on $\mathbb{C}^n$. By Theorem 4.2, there is a unitary matrix $\mathsf{U} \in \mathbb{C}^{n\times n}$ such that the matrix UGU^{-1} is upper triangular and has the eigenvalues $\lambda_1, \lambda_2, \ldots, \lambda_n$ of G as its diagonal entries. Decompose this matrix as follows:

$$\mathsf{UGU}^{-1} = \mathsf{L} + \mathsf{T},$$

where L is diagonal with diagonal entries $\lambda_1, \lambda_2, \ldots, \lambda_n$ and T is upper triangular. Pick any $\delta > 0$, and construct the diagonal matrix

$$\mathsf{D} := \begin{bmatrix} 1 & & & \\ & \delta^{-1} & & \\ & & \ddots & \\ & & & \delta^{1-n} \end{bmatrix}.$$

The matrix $\mathsf{C} := \mathsf{D}(\mathsf{L}+\mathsf{T})\mathsf{D}^{-1}$ has three properties of interest. First, since $\mathsf{DLD}^{-1} = \mathsf{L}$, $\mathsf{C} = \mathsf{L} + \mathsf{E}$, where $\mathsf{E} := \mathsf{DTD}^{-1}$ is upper triangular. Second, the entries of E and T stand in the relationship $e_{i,j} = t_{i,j}\delta^{j-i}$ for $i < j$. In other words, $e_{i,j} = \mathcal{O}(\delta)$, which we can make arbitrarily small in magnitude by choosing δ small. Third, from the equations

$$\mathsf{U}^{-1}\mathsf{D}^{-1}\mathsf{CDU} = \mathsf{U}^{-1}\mathsf{D}^{-1}\mathsf{D}(\mathsf{L}+\mathsf{T})\mathsf{D}^{-1}\mathsf{DU} = (\mathsf{U}^{-1}\mathsf{L}+\mathsf{T})\mathsf{U} = \mathsf{G}$$

there follows the identity

$$\mathsf{CDU} = \mathsf{DUG}. \tag{4.2-2}$$

We now define the desired vector norm. For any $\mathbf{z} \in \mathbb{C}^n$, let

$$\|\mathbf{z}\| := \|\mathsf{DU}\mathbf{z}\|_2 = \sqrt{\mathbf{z}^*\mathsf{U}^*\mathsf{D}^*\mathsf{DU}\mathbf{z}}.$$

Lemma 4.4 ensures that $\|\cdot\|$ is indeed a norm, since the matrix DU has inverse $\mathsf{U}^*\mathsf{D}^{-1}$. To demonstrate that the matrix norm subordinate to $\|\cdot\|$ fulfills the conclusions of the theorem, it suffices to establish the following fact: For any $\mathbf{z} \in \mathbb{C}^n$ with $\|\mathbf{z}\| = 1$, $\|\mathsf{G}\mathbf{z}\| \leq \varrho(\mathsf{G}) + \mathcal{O}(\delta)$. Given this inequality, we can make $\|\mathsf{G}\| \leq \varrho(\mathsf{G}) + \epsilon$ by choosing δ small enough.

From equation (4.2-2),

$$\|\mathsf{G}\mathbf{z}\| = \|\mathsf{DUG}\mathbf{z}\|_2 = \|\mathsf{CDU}\mathbf{z}\|_2.$$

Now let $\mathbf{w} := \mathsf{DU}\mathbf{z}$. We have

$$\begin{aligned} \|\mathsf{G}\mathbf{z}\| = \|\mathsf{C}\mathbf{w}\|_2 &= \sqrt{\mathbf{w}^*\mathsf{C}^*\mathsf{C}\mathbf{w}} \\ &= \sqrt{\mathbf{w}^*(\mathsf{L}^*+\mathsf{E}^*)(\mathsf{L}+\mathsf{E})\mathbf{w}} = \sqrt{\mathbf{w}^*(\mathsf{L}^*\mathsf{L}+\mathsf{B})\mathbf{w}}, \end{aligned}$$

where the matrix $\mathsf{B} := \mathsf{L}^*\mathsf{E}+\mathsf{E}^*\mathsf{L}+\mathsf{E}^*\mathsf{E}$ has entries that are $\mathcal{O}(\delta)$ in magnitude. Therefore,

$$\begin{aligned} \|\mathsf{Gz}\| &\leq \sqrt{[\varrho(\mathsf{G})]^2\mathbf{w}^*\mathbf{w} + n^2\mathcal{O}(\delta)\mathbf{w}^*\mathbf{w}} \\ &= \sqrt{\mathbf{w}^*\mathbf{w}}\sqrt{[\varrho(\mathsf{G})]^2 + \mathcal{O}(\delta)} \\ &= \|\mathbf{w}\|_2\,[\varrho(\mathsf{G}) + \mathcal{O}(\delta)]\,. \end{aligned}$$

(In the last step, we use the binomial series $(1+\alpha)^{1/2} = 1 + \frac{1}{2}\alpha - \frac{1}{8}\alpha^2 + \frac{1}{16}\alpha^3 + \mathcal{O}(\alpha^4)$.) But $\|\mathbf{w}\|_2 = \|\mathsf{DUz}\|_2 = \|\mathbf{z}\| = 1$, so $\|\mathsf{Gz}\| \leq \varrho(\mathsf{G}) + \mathcal{O}(\delta)$, as desired. ∎

Now we examine the repeated application of a matrix G.

DEFINITION. *A matrix* $\mathsf{G} \in \mathbb{C}^{n\times n}$ *is* **convergent** *if* $\lim_{m\to\infty} \mathsf{G}^m = 0$.

As shown in the next section, convergent matrices are the keys to convergent iterative schemes. The following theorem characterizes convergent matrices.

THEOREM 4.5. *Let* $\mathsf{G} \in \mathbb{C}^{n\times n}$. *The following are equivalent:*

(*i*) G *is convergent.*

(*ii*) $\lim_{m\to\infty} \|\mathsf{G}^m\| = 0$ *for some matrix norm* $\|\cdot\|$.

(*iii*) $\varrho(\mathsf{G}) < 1$.

PROOF: We first prove that (*i*) and (*ii*) are equivalent, then prove that (*ii*) and (*iii*) are equivalent. To show that (*i*) implies (*ii*), we recall that matrix norms $\|\cdot\|$ are continuous. Therefore, for any matrix norm, $\|\mathsf{G}^m\| \to 0$ whenever $\mathsf{G}^m \to 0$.

To show that (*ii*) implies (*i*), assume that $\|\mathsf{G}^m\| \to 0$ as $m \to \infty$. Since all norms on $\mathbb{C}^{n\times n}$ are equivalent, there exists a constant $M > 0$ such that $0 \leq \|\mathsf{G}^m\|_\infty \leq M\|\mathsf{G}^m\|$ for every m. Therefore, if $\|\mathsf{G}^m\| \to 0$ as $m \to \infty$, then the maximum row sum

$$\max_i \left\{ \sum_{j=1}^n |g_{i,j}| \right\}$$

of A^m tends to 0 as $m \to \infty$ as well. This fact implies that each entry of G^m tends to 0. We have established that (*i*) and (*ii*) are equivalent.

To demonstrate that (*ii*) implies (*iii*), observe that λ^m is an eigenvalue of G^m if and only if λ is an eigenvalue of G (see Problem 2). If the eigenvalues of G are $\lambda_1, \lambda_2, \ldots, \lambda_n$, then

$$\varrho(\mathsf{G}^m) = \max_j\{|\lambda_j^m|\} = \left(\max_j\{|\lambda_j|\}\right)^m = [\varrho(\mathsf{G})]^m.$$

Now assume that $||\mathsf{G}^m|| \to 0$ as $m \to \infty$. We have

$$0 \leq [\varrho(\mathsf{G})]^m = \varrho(\mathsf{G}^m) \leq ||\mathsf{G}^m|| \to 0 \quad \text{as} \quad m \to \infty,$$

so $\lim_{m\to\infty}[\varrho(\mathsf{G})]^m = 0$. This statement holds only if $\varrho(\mathsf{G}) < 1$.

Finally, we prove that (*iii*) implies (*ii*). Assume that $\varrho(\mathsf{G}) < 1$, so that $\varrho(\mathsf{G}) < 1-\epsilon$ for some $\epsilon > 0$. By Theorem 4.3, there exists a subordinate matrix norm $||\cdot||$ such that $||\mathsf{G}|| \leq \varrho(\mathsf{G})+\epsilon < 1$. From the inequality $||\mathsf{AB}|| \leq ||\mathsf{A}||\,||\mathsf{B}||$ we now conclude that $0 \leq ||\mathsf{G}^m|| \leq ||\mathsf{G}||^m \to 0$ as $m \to \infty$. ∎

4.3 Matrix-Splitting Techniques

Motivation and Construction

One important strategy for solving a linear system $\mathsf{A}\mathbf{u} = \mathbf{b}$ involves splitting the matrix A as $\mathsf{A} = \mathsf{B} + (\mathsf{A} - \mathsf{B})$. Given such a splitting and an initial guess $\mathbf{u}^{(0)}$, we can iterate on the system

$$\mathsf{B}\mathbf{u}^{(m+1)} = (\mathsf{B} - \mathsf{A})\mathbf{u}^{(m)} + \mathbf{b}. \tag{4.3-1}$$

This strategy hinges on two properties: First, it must be significantly easier to solve systems involving B than to solve those involving A. Second, the iterates $\mathbf{u}^{(m)}$ should converge quickly to the exact answer $\mathbf{u}$. Formally, this second requirement means the following: If $\boldsymbol{\varepsilon}^{(m)} := \mathbf{u} - \mathbf{u}^{(m)}$ denotes the error at iteration level m, then we want $||\boldsymbol{\varepsilon}^{(m)}|| \to 0$ as $m \to \infty$, as rapidly as possible. In practice, of course, we want to reach an iterate $\mathbf{u}^{(m)}$ that lies within some prescribed distance of $\mathbf{u}$ after a fairly small number of iterations.

Several well known iterative schemes use matrix splittings based on the decomposition

$$\mathsf{A} = \mathsf{L} + \mathsf{D} + \mathsf{U},$$

where D is the diagonal part of A, L is the lower triangular part, and U is the upper triangular part. More specifically, D, L, and U have the following entries:

$$d_{i,j} = \begin{cases} a_{i,j}, & \text{if} \quad i = j, \\ 0, & \text{if} \quad i \neq j; \end{cases}$$

$$l_{i,j} = \begin{cases} a_{i,j}, & \text{if} \quad i > j, \\ 0, & \text{if} \quad i \leq j; \end{cases}$$

$$u_{i,j} = \begin{cases} a_{i,j}, & \text{if} \quad i < j, \\ 0, & \text{if} \quad i \geq j. \end{cases}$$

(L and U as defined here differ from the upper and lower triangular *factors* discussed in Chapter 2.)

If the diagonal part D has no zeros among its diagonal entries, then it is certainly easy to invert: D^{-1} is the diagonal matrix whose (i, i)th entry is

$d_{i,i}^{-1}$. This observation suggests the matrix splitting $\mathsf{B} = \mathsf{D}$, $\mathsf{A} - \mathsf{B} = \mathsf{L} + \mathsf{U}$. In matrix form, the following iterative scheme for $\mathsf{A}\mathbf{u} = \mathbf{b}$ results:

$$\begin{aligned} \mathbf{u}^{(m+1)} &= -\mathsf{D}^{-1}(\mathsf{L} + \mathsf{U})\mathbf{u}^{(m)} + \mathsf{D}^{-1}\mathbf{b} \\ &= (\mathsf{I} - \mathsf{D}^{-1}\mathsf{A})\mathbf{u}^{(m)} + \mathsf{D}^{-1}\mathbf{b}. \end{aligned} \tag{4.3-2}$$

This scheme is the **Jacobi method**.

The matrix form (4.3-2) is not the most computationally useful form for the iterative scheme. To get a form more amenable to computer programming, consider the original linear system $\mathsf{A}\mathbf{u} = \mathbf{b}$ written in component form:

$$\sum_{j=1}^{n} a_{i,j} u_j = b_i, \quad i = 1, 2, \ldots, n.$$

We obtain the Jacobi method by solving the ith equation for the unknown u_i, then "lagging" the other unknowns u_j, $j \neq i$, by one iteration:

$$u_i^{(m+1)} = -\frac{1}{a_{i,i}} \sum_{j \neq i} a_{i,j} u_j^{(m)} + \frac{b_i}{a_{i,i}}, \quad i = 1, 2, \ldots, n. \tag{4.3-3}$$

For example, for the finite-difference approximation (4.1-3) to the Laplace equation, the Jacobi method yields the iterative equations

$$u_{k,l}^{(m+1)} = \frac{1}{4}\left(u_{k-1,l}^{(m)} + u_{k,l-1}^{(m)} + u_{k,l+1}^{(m)} + u_{k+1,l}^{(m)}\right).$$

(Even though we retain the double subscripts for the unknowns $u_{i,j}$ in this context, we regard these unknowns as entries of a single vector, say $\mathbf{u} = (u_{1,1}, u_{1,2}, \ldots, u_{n,n})^{\top}$.) When the expression on the right calls for a value of $u_{i,j}$ associated with a node (x_i, y_j) located on the boundary of the grid, it is necessary to substitute a known boundary value for $u_{i,j}$. In programming the Jacobi method, one simply inserts equations of this form inside a loop running over all indices of the unknowns. One execution of the loop, passing through all index values, thus constitutes one iteration. The method requires the storage of two vectors of iterates: one for the "old" values associated with iterative level m and one for the "new" iterates associated with level $m + 1$.

The Jacobi method admits a modification in which we multiply the correction vector $\mathbf{u}^{(m+1)} - \mathbf{u}^{(m)}$ by a "damping factor" ω before adding it to the previous iterate $\mathbf{u}^{(m)}$. This **damped Jacobi method** has the following matrix form:

$$\overline{\mathbf{u}} := (\mathsf{I} - \mathsf{D}^{-1}\mathsf{A})\mathbf{u}^{(m)} + \mathsf{D}^{-1}\mathbf{b},$$

$$\mathbf{u}^{(m+1)} = \mathbf{u}^{(m)} + \omega(\overline{\mathbf{u}} - \mathbf{u}^{(m)}).$$

Eliminating $\overline{\mathbf{u}}$ gives

$$\mathbf{u}^{(m+1)} = (\mathsf{I} - \omega\mathsf{D}^{-1}\mathsf{A})\mathbf{u}^{(m)} + \omega\mathsf{D}^{-1}\mathbf{b}. \tag{4.3-4}$$

Clearly, the choice $\omega = 1$ yields the original Jacobi method. For the finite-difference equations (4.1-3), the method reduces to the iterative equations

$$u_{k,l}^{(m+1)} = u_{k,l}^{(m)} + \omega \left\{ \frac{1}{4} \left[u_{k-1,l}^{(m)} + u_{k,l-1}^{(m)} + u_{k,l+1}^{(m)} + u_{k+1,l}^{(m)} \right] - u_{k,l}^{(m)} \right\}.$$

For this application, the damped Jacobi method converges for $0 < \omega \leq 1$ (see Problem 3). We revisit this method briefly at the end of this section.

One can refine the Jacobi method in another way. Observe that new iterates $u_i^{(m+1)}$ become available for use during the course of each iteration. Since $u_i^{(m+1)}$ is presumably closer to the exact value u_i than $u_i^{(m)}$, we might use the newer value as soon as we have computed it inside the iterative loop, instead of waiting for the next execution of the loop. The **Gauss-Seidel method** results:

$$u_i^{(m+1)} = -\frac{1}{a_{i,i}} \left(\sum_{j<i} a_{i,j} \underbrace{u_j^{(m+1)}}_{\text{known}} + \sum_{j>i} a_{i,j} u_j^{(m)} \right) + \frac{b_i}{a_{i,i}}. \tag{4.3-5}$$

The values $u_j^{(m+1)}$ for $j < i$ are known from earlier passes through the loop during the same iteration. In coding this scheme, it is not necessary to store separate vectors for iterative levels m and $m+1$. We actually need just one vector for the unknowns, since we can overwrite its entries with new values as soon as they become available.

The Gauss-Seidel method has the following matrix representation:

$$\mathbf{u}^{(m+1)} = -(\mathsf{L} + \mathsf{D})^{-1}\mathsf{U}\mathbf{u}^{(m)} + (\mathsf{L} + \mathsf{D})^{-1}\mathbf{b}.$$

In the matrix splitting, $\mathsf{B} = \mathsf{L} + \mathsf{D}$, $\mathsf{A} - \mathsf{B} = \mathsf{U}$.

Unlike the Jacobi method, the Gauss-Seidel method generates iterates that depend on the order that we assign to the unknowns. Consider the model problem (4.1-1). In the **lexicographic** ordering of the variables $u_{i,j}$, the unknown vector is

$$(u_{1,1}, u_{1,2}, u_{1,3}, u_{2,1}, u_{2,2}, u_{2,3}, u_{3,1}, u_{3,2}, u_{3,3})^{\mathsf{T}}.$$

With this arrangement, the Gauss-Seidel iteration has the form

$$u_{k,l}^{(m+1)} = \frac{1}{4} \left(u_{k-1,l}^{(m+1)} + u_{k,l-1}^{(m+1)} + u_{k,l+1}^{(m)} + u_{k+1,l}^{(m)} \right),$$

since the unknowns $u_{k-1,l}$ and $u_{k,l-1}$ appear earlier in the lexicographic ordering than $u_{k,l}$.

In **red-black** ordering, the unknowns $u_{k,l}$ for which $k + l$ is even come before those for which $k + l$ is odd, and the ordering within each of these two subsets is lexicographic. (Imagine the grid drawn in Figure 1 of Section 4.1 as a checkerboard, with nodes alternately colored black and red according to whether $k+l$ is even or odd.) Thus the unknown vector in the model equations (4.1-1) is as follows:

$$(\underbrace{u_{1,1}, u_{1,3}, u_{2,2}, u_{3,1}, u_{3,3}}_{\text{black}}, \underbrace{u_{1,2}, u_{2,1}, u_{2,3}, u_{3,2}}_{\text{red}})^{\mathsf{T}}.$$

In the red-black ordering for the problem (4.1-1), the Gauss-Seidel equations have the following form when $u_{k,l}$ is a "black" unknown:

$$u_{k,l}^{(m+1)} = \frac{1}{4}\left(u_{k-1,l}^{(m)} + u_{k,l-1}^{(m)} + u_{k,l+1}^{(m)} + u_{k+1,l}^{(m)}\right), \quad k+l \ \text{ even.} \qquad (4.3\text{-}6a)$$

All of the unknowns appearing on the right are "red" and therefore appear later in the vector of unknowns. For the red unknowns, the Gauss-Seidel equation is

$$u_{k,l}^{(m+1)} = \frac{1}{4}\left(u_{k-1,l}^{(m+1)} + u_{k,l-1}^{(m+1)} + u_{k,l+1}^{(m+1)} + u_{k+1,l}^{(m+1)}\right), \quad k+l \ \text{ odd.} \quad (4.3\text{-}6b)$$

All of the unknowns on the right are black and therefore already have been assigned values at the new iteration level $m+1$.

Like the Jacobi method, the Gauss-Seidel method also admits a modification based on adjustments to the correction vector. This method is known as **successive overrelaxation (SOR)**. To describe it, let us rewrite the Gauss-Seidel method as follows:

$$\begin{aligned} u_i^{(m+1)} &= u_i^{(m)} + \delta_i^{(m+1)}, \\ \delta_i^{(m+1)} &:= -\sum_{j<i} \frac{a_{i,j}}{a_{i,i}} u_j^{(m+1)} - \sum_{j>i} \frac{a_{i,j}}{a_{i,i}} u_j^{(m)} + \frac{b_i}{a_{i,i}} - u_i^{(m)}, \end{aligned}$$

the quantity $\delta_i^{(m+1)}$ being the correction. In SOR, one replaces the scheme $u_i^{(m+1)} = u_i^{(m)} + \delta_i^{(m+1)}$ by the iterative equation

$$u_i^{(m+1)} = u_i^{(m)} + \omega\delta_i^{(m+1)},$$

choosing the **overrelaxation parameter** ω to speed convergence. Clearly, the choice $\omega = 1$ yields the Gauss-Seidel method.

From an algorithmic viewpoint, the SOR equations have the form

$$u_i^{(m+1)} = (1-\omega)u_i^{(m)} + \omega\left(-\sum_{j<i} \frac{a_{i,j}}{a_{i,i}} u_j^{(m+1)} - \sum_{j>i} \frac{a_{i,j}}{a_{i,i}} u_j^{(m)} + \frac{b_i}{a_{i,i}}\right). \quad (4.3\text{-}7)$$

For our model problem (4.1-1), when we order the unknowns lexicographically, SOR reduces to the the following scheme:

$$u_{i,j}^{(m+1)} = (1-\omega)u_{i,j}^{(m)} + \frac{\omega}{4}\left(u_{i-1,j}^{(m+1)} + u_{i,j-1}^{(m+1)} + u_{i,j+1}^{(m)} + u_{i+1,j}^{(m)}\right).$$

The matrix form for SOR is as follows:

$$\mathbf{u}^{(m+1)} = \underbrace{\left(\omega^{-1}\mathsf{D}+\mathsf{L}\right)^{-1}}_{\mathsf{B}^{-1}} \underbrace{\left[\left(\omega^{-1}-1\right)\mathsf{D}-\mathsf{U}\right]}_{\mathsf{B}-\mathsf{A}} \mathbf{u}^{(m)} + \left(\omega^{-1}\mathsf{D}+\mathsf{L}\right)^{-1}\mathbf{b}, \quad (4.3\text{-}8)$$

with the matrix splitting $\mathsf{A} = \mathsf{B} + (\mathsf{A} - \mathsf{B})$ indicated.

Some theory is available to guide the choice of overrelaxation parameter ω in certain common applications. We devote Section 4.4 to this topic.

The Jacobi and Gauss-Seidel methods are examples of **relaxation** methods. This terminology reflects an analogy with certain mechanical systems, such as elastic membranes, that give rise to linear systems involving symmetric, positive definite matrices A. As we explore in Section 4.5, the solution $\mathbf{u}$ of the linear system $\mathsf{A}\mathbf{u} = \mathbf{b}$ for such matrices corresponds to the minimum of the function $F(\mathbf{v}) = \frac{1}{2}\mathbf{v}^{\top}\mathsf{A}\mathbf{v} + \mathbf{b}^{\top}\mathbf{v}$, which for the mechanical systems in question represents the energy. In this analogy, iterates $\mathbf{u}^{(m)} \neq \mathbf{u}$ represent configurations of the system that do not quite minimize energy, and the iterative adjustment of the vectors $\mathbf{u}^{(m)}$ toward the vector $\mathbf{u}$ corresponds to "relaxation" of the mechanical system to its equilibrium state ([5], Section 38).

Practical Considerations

The most important practical questions regarding the Jacobi and Gauss-Seidel methods concern their convergence. The matrix-splitting viewpoint yields both general and specific convergence criteria for the methods. Both of the methods fall under the rubric of **stationary iterative methods**, the general form of which is

$$\mathbf{u}^{(m+1)} = \mathsf{G}\mathbf{u}^{(m)} + \mathbf{k}, \qquad (4.3\text{-}9)$$

where G is a constant **iteration matrix** and $\mathbf{k}$ is a constant vector. For matrix-splitting methods for the system $\mathsf{A}\mathbf{u} = \mathbf{b}$, the iteration matrix is $\mathsf{G} = \mathsf{I} - \mathsf{B}^{-1}\mathsf{A}$, and $\mathbf{k} = \mathsf{B}^{-1}\mathbf{b}$. In particular, for the Jacobi method,

$$\mathsf{G}_J := -\mathsf{D}^{-1}(\mathsf{L}+\mathsf{U}) = \mathsf{I} - \mathsf{D}^{-1}\mathsf{A}, \qquad \mathbf{k}_J := \mathsf{D}^{-1}\mathbf{b},$$

while for the Gauss-Seidel method

$$\mathsf{G}_1 := -(\mathsf{L}+\mathsf{D})^{-1}\mathsf{U}, \qquad \mathbf{k}_1 := (\mathsf{L}+\mathsf{D})^{-1}\mathbf{b}.$$

A stationary iterative scheme for $\mathsf{A}\mathbf{u} = \mathbf{b}$ is **consistent** if $\mathbf{u} = \mathsf{G}\mathbf{u} + \mathbf{k}$, that is, if the solution to the linear system is a fixed point of the iteration. It **converges** if $\mathbf{u} - \mathbf{u}^{(m)} := \boldsymbol{\varepsilon}^{(m)} \to \mathbf{0}$ for any initial guess $\mathbf{u}^{(0)}$. Later in this section we prove the following general convergence criterion:

> *A consistent stationary iterative scheme (4.3-9) for* $\mathsf{A}\mathbf{u} = \mathbf{b}$ *converges if and only if* $\varrho(\mathsf{G}) < 1$.

In other words, the iterative scheme converges from any initial guess $\mathbf{u}^{(0)}$ precisely when all eigenvalues of G have magnitude less than 1.

While this criterion furnishes a general characterization of convergence, it refers to the spectrum of the matrix G, which is typically difficult to ascertain. By using the fact that $\varrho(\mathsf{G}) \leq ||\mathsf{G}||$ for any subordinate matrix norm $||\cdot||$ (see Problem 9, Chapter 2), we deduce the following corollary of the general criterion:

> *If* $||\mathsf{G}|| < 1$ *for some subordinate matrix norm, then stationary iterative schemes using* G *as the iteration matrix converge.*

This criterion gives only a sufficient condition for convergence, but it can be more convenient to check, since such subordinate matrix norms as $||\cdot||_1$ (the "maximum column sum") and $||\cdot||_\infty$ (the "maximum row sum") are easy to compute.

If one can find a subordinate matrix norm in which $||\mathsf{G}|| < 1$, then one can use it to estimate the error $\boldsymbol{\varepsilon}^{(m)} = \mathbf{u} - \mathbf{u}^{(m)}$ at each iteration in terms of the difference $\mathbf{u}^{(m)} - \mathbf{u}^{(m-1)}$ between successive iterates. By consistency, $\boldsymbol{\varepsilon}^{(m)} = \mathsf{G}\mathbf{u}+\mathbf{k}-\mathsf{G}\mathbf{u}^{(m-1)}-\mathbf{k} = \mathsf{G}\boldsymbol{\varepsilon}^{(m-1)}$. Also, $\boldsymbol{\varepsilon}^{(m)} = \mathbf{u}^{(m+1)} - \mathbf{u}^{(m)} - \mathbf{u}^{(m+1)} + \mathbf{u} = \mathsf{G}(\mathbf{u}^{(m)} - \mathbf{u}^{(m-1)}) - \mathsf{G}(\mathbf{u}^{(m)} - \mathbf{u})$. Combining these two observations and taking norms, we get

$$||\boldsymbol{\varepsilon}^{(m)}|| \leq ||\mathsf{G}||\,||\mathbf{u}^{(m)} - \mathbf{u}^{(m-1)}|| + ||\mathsf{G}||\,||\boldsymbol{\varepsilon}^{(m)}||,$$

or, assuming that $||\mathsf{G}|| < 1$,

$$||\boldsymbol{\varepsilon}^{(m)}|| \leq \frac{||\mathsf{G}||}{1 - ||\mathsf{G}||}||\mathbf{u}^{(m)} - \mathbf{u}^{(m-1)}||. \tag{4.3-10}$$

The inequality (4.3-10) bounds the unknown error in terms that are computable at each iteration. Thus one can iterate until the quantity on the right falls below a prescribed error tolerance, being assured that the error at that iteration level is no larger in norm.

Still, the criterion $||\mathsf{G}|| < 1$ can be frustrating in its failure to identify convergent schemes. For example, for the finite-difference system (4.1-4) the Jacobi iteration matrix has $||\mathsf{G}_J||_1 = ||\mathsf{G}_J||_\infty = 1$, so neither of these norms allows one to conclude that the Jacobi method converges for this problem. Yet the Jacobi method does converge for this system of equations. In fact, it converges for essentially all systems arising from elementary difference approximations to analogs of the boundary-value problem (4.1-2). Observations of this sort have led numerical analysts to develop a fairly sophisticated body of theory concerning matrix splitting methods. We investigate some of this theory later in this section.

In practice, it is important for an iterative scheme to converge rapidly. The rate at which $||\boldsymbol{\varepsilon}^{(m)}|| \to 0$ depends upon the spectral radius of the iteration matrix G, according to an argument that we now sketch heuristically. As we have seen, $\boldsymbol{\varepsilon}^{(m)} = \mathsf{G}^m\boldsymbol{\varepsilon}^{(0)}$. Consider the case when $\mathsf{G} \in \mathbb{R}^{n\times n}$ has n

linearly independent eigenvectors $\boldsymbol{v}_1, \boldsymbol{v}_2, \ldots, \boldsymbol{v}_n$ with associated eigenvalues $\lambda_1, \lambda_2, \ldots, \lambda_n \neq 0$, respectively, ordered so that $|\lambda_1| \leq |\lambda_2| \leq \cdots \leq |\lambda_n|$. In this case we can expand the initial error as a sum of n linearly independent eigenvectors, $\boldsymbol{\varepsilon}^{(0)} = \boldsymbol{v}_1 + \boldsymbol{v}_2 + \cdots + \boldsymbol{v}_n$. (We absorb the scalar constants into the eigenvectors.) Hence,

$$\boldsymbol{\varepsilon}^{(1)} = \mathsf{G}\boldsymbol{\varepsilon}^{(0)} = \sum_{i=1}^{n} \mathsf{G}\boldsymbol{v}_i = \sum_{i=1}^{n} \lambda_i \boldsymbol{v}_i.$$

Similarly,

$$\boldsymbol{\varepsilon}^{(2)} = \sum_{i=1}^{n} \lambda_i^2 \boldsymbol{v}_i,$$

and so forth. In general,

$$\boldsymbol{\varepsilon}^{(m)} = \sum_{i=1}^{n} \lambda_i^m \boldsymbol{v}_i,$$

or

$$\lambda_n^{-m} \boldsymbol{\varepsilon}^{(m)} = \sum_{i=1}^{n} \left(\frac{\lambda_i}{\lambda_n} \right)^m \boldsymbol{v}_i.$$

In the sum on the right side of this last equation, all terms tend to $\mathbf{0}$ as $m \to \infty$, except those associated with eigenvalues λ_i for which $|\lambda_i| = |\lambda_n| = \varrho(\mathsf{G})$.

Thus, as $m \to \infty$, the error $\boldsymbol{\varepsilon}^{(m)}$ tends to a superposition of eigenvectors whose associated eigenvalues have magnitude $\varrho(\mathsf{G})$. Therefore, as $m \to \infty$, the ratio $\|\boldsymbol{\varepsilon}^{(m+1)}\|/\|\boldsymbol{\varepsilon}^{(m)}\|$ tends to $\varrho(\mathsf{G})$. In other words, $\varrho(\mathsf{G})$ measures the asymptotic factor by which the scheme reduces errors at each iteration. Over the course of k iterations, the error reduction is, asymptotically, $[\varrho(\mathsf{G})]^k = \|\boldsymbol{\varepsilon}^{(m+k)}\|/\|\boldsymbol{\varepsilon}^{(m)}\|$. Hence the number k of iterations needed to obtain a value of $1/10$ for this ratio — thereby gaining one decimal digit of accuracy in the iterative approximation to $\mathbf{u}$ — is $k = -1/\log_{10} \varrho(\mathsf{G})$. This reasoning motivates the following:

DEFINITION. *For a stationary iterative scheme* $\mathbf{u}^{(m+1)} = \mathsf{G}\mathbf{u}^{(m)} + \mathbf{k}$, *the number* $R(\mathsf{G}) := -\log_{10} \varrho(\mathsf{G})$ *is the* **convergence rate**.

Large values of $R(\mathsf{G})$ correspond to rapid convergence.

The argument just given requires some modification in cases when $\mathsf{G} \in \mathbb{R}^{n \times n}$ does not have n linearly independent eigenvectors. The heuristic is a useful one, however, and we encounter it again in the next section and in the development of numerical methods for solving eigenvalue problems. It is in the latter context, in Chapter 5, that we examine the changes needed to generalize the argument.

One unfortunate aspect of the Jacobi and Gauss-Seidel methods is that they tend to converge slowly for large problems. A sample problem helps to illustrate the qualitative behavior of the convergence rate as the order of the matrix equation increases.

Consider the boundary-value problem (4.1-2). Instead of fixing the grid size at 4×4, as in Figure 1 of Section 4.1, consider the grid size to be variable at $N \times N$, so that the mesh size $h = 1/N$. The finite-difference approximations still lead to algebraic approximations of the form (4.1-3), but now the matrix equation corresponding to Equation (4.1-4) has size $(N-1)^2$. The system matrix $\mathsf{A} \in \mathbb{R}^{(N-1)^2 \times (N-1)^2}$ is now block-tridiagonal with block size $(N-1) \times (N-1)$, and the unknown vector $\mathbf{u} \in \mathbb{R}^{(N-1)^2}$ has entries $u_{k,l}$, where $k, l = 1, 2, \ldots, N-1$ are indices associated with interior nodes of the grid. The aim is to determine the convergence rate of the Jacobi method applied to this problem. In particular, we wish to assess the behavior of $R(\mathsf{G}_J)$ as $h \to 0$, this limit corresponding to more accurate finite-difference analogs and also to larger matrix problems.

The system matrix A has $(N-1)^2$ eigenvectors: one associated with each of the ordered pairs (p, q), $p, q = 1, 2, \ldots N-1$. Problem 3 asks for proof that a typical eigenvector has entries $\sin(pk\pi h)\sin(ql\pi h)$, $k, l = 1, 2, \ldots, N-1$, and is associated with the eigenvalue

$$\mu_{p,q} = 4\left[\sin^2\left(\frac{p\pi h}{2}\right) + \sin^2\left(\frac{q\pi h}{2}\right)\right].$$

The smallest eigenvalue of A is the one for which $p = q = 1$. By the Taylor theorem,

$$\mu_{\min} = 8\sin^2\left(\frac{\pi h}{2}\right) = 8\left[\left(\frac{\pi h}{2}\right)^2 - \frac{1}{3}\left(\frac{\pi h}{2}\right)^4 + \cdots\right] = 2\pi^2 h^2 + \mathcal{O}(h^4).$$

The largest eigenvalue corresponds to the indices $p = q = N-1$:

$$\mu_{\max} = 8\sin^2\left(\frac{(N-1)\pi h}{2}\right) = 8 - 8\sin^2\left(\frac{\pi h}{2}\right).$$

As an aside, we mention that A is symmetric and positive definite, so its condition number is

$$\operatorname{cond}_2(\mathsf{A}) = \frac{\mu_{\max}}{\mu_{\min}} = \mathcal{O}(h^{-2}) \qquad \text{as} \quad h \to 0. \tag{4.3-11}$$

Therefore the algebraic approximation to the partial differential equation becomes more poorly conditioned as we refine the grid. This phenomenon typifies discretizations of differential equations.

Given this knowledge of the spectrum of A, we can determine the spectrum of the Jacobi iteration matrix G_J. Since in this case

$$\mathsf{G}_J = -\mathsf{D}^{-1}(\mathsf{L} + \mathsf{U}) = \mathsf{I} - \mathsf{D}^{-1}\mathsf{A} = \mathsf{I} - \tfrac{1}{4}\mathsf{A},$$

the eigenvalues of G_J have the form $\lambda_{p,q} = 1 - \mu_{p,q}/4$, where $\mu_{p,q}$ ranges over the eigenvalues of A. That is,

$$\begin{aligned}\lambda_{p,q} &= 1 - \sin^2\left(\frac{p\pi h}{2}\right) - \sin^2\left(\frac{q\pi h}{2}\right)\\ &= \tfrac{1}{2}[\cos(p\pi h) + \cos(q\pi h)], \quad p, q = 1, 2, \ldots N-1.\end{aligned}$$

The largest of these values corresponds to the indices $p = q = 1$, so

$$\varrho(\mathsf{G}_J) = |\cos(\pi h)| = 1 - \frac{(\pi h)^2}{2} + \mathcal{O}(h^4), \quad \text{as} \quad h \to 0.$$

The convergence rate of the Jacobi method for this system of equations is therefore

$$R(\mathsf{G}_J) = -\log_{10}\left(1 - \frac{(\pi h)^2}{2} + \mathcal{O}(h^4)\right) = \frac{(\pi h^2)}{2} + \mathcal{O}(h^4).$$

Consequently, as $h \to 0$, the convergence rate of the Jacobi method tends to 0 at the same rate as h^2. For very fine grids (small h), the convergence can be excruciatingly slow.

The Gauss-Seidel method behaves in a qualitatively similar way. However, for this sample problem and analogous ones, $R(\mathsf{G}_1)$ is larger than $R(\mathsf{G}_J)$, so the Gauss-Seidel method converges somewhat more rapidly. In the next section we show that, in fairly typical circumstances, $R(\mathsf{G}_1) = 2R(\mathsf{G}_J)$.

One more practical aspect is worth mentioning. While the Gauss-Seidel method typically converges more rapidly, the Jacobi method is more amenable to parallel computations. Since the equation (4.3-3) for the updated value of u_i does not require us to know updated values for any of the other unknowns u_j, we can compute updated values for all of the unknowns $u_1, u_2, \ldots, u_n$ simultaneously, sending the equation for each update to a separate processor on a parallel-processing machine. In contrast, the Gauss-Seidel method requires knowledge of updated values of u_j *within* each iteration, so the scheme is not generally amenable to parallel processing. In special cases, though, some parallelism may be available. For example, with the red-black ordering of Equations (4.3-6), the finite-difference approximations to the Laplace equation allow one to process all of the updates for black unknowns concurrently, then to process all of the updates for red unknowns concurrently.

Mathematical Details

We now derive rigorous convergence criteria for the Jacobi and Gauss-Seidel schemes. While several results are available almost immediately, they are disappointing in that they do not apply to analogs of the finite-difference system (4.1-4). Since such systems constitute an important setting for matrix-splitting schemes, we devote some attention to the refinements needed to accommodate them.

Fundamental is the connection between iterative convergence and the spectrum of the iteration matrix G.

THEOREM 4.6. *A consistent, stationary iterative scheme (4.3-9) converges if and only if* $\varrho(\mathsf{G}) < 1$.

PROOF: First assume that the iterative scheme converges. Thus, for any initial guess $\mathbf{u}^{(0)}$, $\boldsymbol{\varepsilon}^{(m)} := \mathbf{u} - \mathbf{u}^{(m)} \to 0$ as $m \to \infty$. As observed earlier, consistency implies that $\boldsymbol{\varepsilon}^{(m)} = \mathsf{G}^m \boldsymbol{\varepsilon}^{(0)}$. Let λ be an eigenvalue of G, and choose the initial guess $\mathbf{u}^{(0)}$ so that $\boldsymbol{\varepsilon}^{(0)} = \mathbf{u} - \mathbf{u}^{(0)}$ is an eigenvector associated with λ. By hypothesis, $\boldsymbol{\varepsilon}^{(m)} = \mathsf{G}^m \boldsymbol{\varepsilon}^{(0)} = \lambda^m \boldsymbol{\varepsilon}^{(0)} \to 0$ as $m \to \infty$. This fact implies that $|\lambda| < 1$, and since λ is an arbitrary eigenvalue of G we conclude that $\varrho(\mathsf{G}) < 1$.

Now assume that $\varrho(\mathsf{G}) < 1$. By Theorem 4.5, G is convergent, so

$$0 \leq \|\boldsymbol{\varepsilon}^{(m)}\| = \|\mathsf{G}^m \boldsymbol{\varepsilon}^{(0)}\| \leq \|\mathsf{G}^m\| \, \|\boldsymbol{\varepsilon}^{(0)}\| \to 0,$$

as $m \to \infty$. Therefore $\boldsymbol{\varepsilon}^{(m)} \to \mathbf{0}$, for arbitrary initial guess $\mathbf{u}^{(0)}$. ∎

We have already mentioned the following consequence:

COROLLARY 4.7. *If there is a subordinate matrix norm* $\|\cdot\|$ *in which* $\|\mathsf{G}\| < 1$, *then the iterative scheme (4.3-6) converges.*

These observations yield concrete convergence criteria for the Jacobi and Gauss-Seidel methods. From an intuitive viewpoint, the Jacobi method treats a linear system $\mathsf{A}\mathbf{u} = \mathbf{b}$ as if the "lagged" terms, associated with off-diagonal entries of A, are in some way less influential than those associated with diagonal entries. The following convergence criterion formalizes this intuition. Recall that a matrix $\mathsf{A} \in \mathbb{R}^{n \times n}$ is strictly (row) diagonally dominant if

$$|a_{i,i}| > \sum_{j \neq i} |a_{i,j}|, \qquad i = 1, 2, \ldots, n.$$

THEOREM 4.8. *The Jacobi method for* $\mathsf{A}\mathbf{u} = \mathbf{b}$ *converges whenever* A *is strictly diagonally dominant.*

PROOF: Given the decomposition $\mathsf{A} = \mathsf{L} + \mathsf{D} + \mathsf{U}$, the Jacobi iteration matrix is $\mathsf{G}_J := -\mathsf{D}^{-1}(\mathsf{L} + \mathsf{U})$. Hence by strict diagonal dominance we have

$$\|\mathsf{G}_J\|_\infty = \|-\mathsf{D}^{-1}(\mathsf{L} + \mathsf{U})\|_\infty = \max_{1 \leq i \leq n} \frac{1}{|a_{i,i}|} \sum_{j \neq i} |a_{i,j}| < 1. \qquad ∎$$

The same criterion also applies to the Gauss-Seidel method:

THEOREM 4.9. *The Gauss-Seidel method for* $\mathsf{A}\mathbf{u} = \mathbf{b}$ *converges whenever* A *is strictly diagonally dominant.*

The proof, however, is more involved.

PROOF: Rewrite the hypothesis of strict diagonal dominance as follows:

$$r := \max_{1 \le i \le n} \sum_{j \ne i} \left| \frac{a_{i,j}}{a_{i,i}} \right| < 1.$$

We show that the sequence $\{\varepsilon^{(m)}\}$ of errors obeys the inequality $\|\varepsilon^{(m)}\|_\infty \le r\|\varepsilon^{(m-1)}\|_\infty$, from which it follows that $\|\varepsilon^{(m)}\|_\infty \to 0$ as $m \to \infty$. First notice that

$$\begin{aligned} u_i^{(m)} &= -\sum_{j<i} \frac{a_{i,j}}{a_{i,i}} u_j^{(m)} - \sum_{j>i} \frac{a_{i,j}}{a_{i,i}} u_j^{(m-1)} + \frac{b_i}{a_{i,i}}, \\ u_i &= -\sum_{j<i} \frac{a_{i,j}}{a_{i,i}} u_j - \sum_{j>i} \frac{a_{i,j}}{a_{i,i}} u_j + \frac{b_i}{a_{i,i}}. \end{aligned}$$

Subtracting the first of these identities from the second yields

$$\varepsilon_i^{(m)} = -\sum_{j<i} \frac{a_{i,j}}{a_{i,i}} \varepsilon_j^{(m)} - \sum_{j>i} \frac{a_{i,j}}{a_{i,i}} \varepsilon_j^{(m-1)}.$$

We now show that $|\varepsilon_i^{(m)}| \le r\|\varepsilon^{(m-1}\|_\infty$, proceeding by induction on the index i. When $i = 1$, the triangle inequality yields

$$\begin{aligned} |\varepsilon_1^{(m)}| &\le \sum_{j>1} \left| \frac{a_{1,j}}{a_{1,1}} \right| \left| \varepsilon_j^{(m-1)} \right| \\ &\le \|\varepsilon^{(m-1)}\|_\infty \sum_{j>1} \left| \frac{a_{1,j}}{a_{1,1}} \right| \le r\|\varepsilon^{(m-1)}\|_\infty. \end{aligned}$$

Now assume that $|\varepsilon_j^{(m)}| \le r|\varepsilon_j^{(m-1)}|$ whenever $j < i$. Applying the triangle inequality, the inductive hypothesis, and the fact that $0 \le r < 1$, we find that

$$\begin{aligned} |\varepsilon_i^{(m)}| &\le \sum_{j<i} \left| \frac{a_{i,j}}{a_{i,i}} \right| \left| \varepsilon_j^{(m)} \right| + \sum_{j>i} \left| \frac{a_{i,j}}{a_{i,i}} \right| \left| \varepsilon_j^{(m-1)} \right| \\ &\le r\|\varepsilon^{(m-1)}\|_\infty \sum_{j<i} \left| \frac{a_{i,j}}{a_{i,i}} \right| + \|\varepsilon^{(m-1)}\|_\infty \sum_{j>i} \left| \frac{a_{i,j}}{a_{i,i}} \right| \\ &\le \|\varepsilon^{(m-1)}\|_\infty \sum_{j \ne i} \left| \frac{a_{i,j}}{a_{i,i}} \right| \le r\|\varepsilon^{(m-1)}\|_\infty. \end{aligned}$$

Therefore, $|\varepsilon_i^{(m)}| \leq r\|\boldsymbol{\varepsilon}^{(m-1)}\|_\infty$ for each i, and $\|\boldsymbol{\varepsilon}^{(m)}\|_\infty \leq r\|\boldsymbol{\varepsilon}^{(m-1)}\|_\infty$, as desired. ∎

While these theorems confirm the intuitive view that the Jacobi and Gauss-Seidel schemes rely on diagonal dominance, they do not guarantee convergence of finite-difference approximations to the Laplace equation and its analogs. Indeed, the block-tridiagonal matrix in Equation (4.1-4) is not strictly diagonally dominant. We can relax the condition that A be strictly diagonally dominant by adopting an additional hypothesis.

DEFINITION. *A matrix* $\mathsf{A} \in \mathbb{R}^{n\times n}$ *is* **reducible** *if there exists a permutation matrix* $\mathsf{P} \in \mathbb{R}^{n\times n}$ *such that*

$$\mathsf{PAP}^\top = \begin{matrix} & p & n-p \\ p & \multicolumn{2}{c}{\multirow{2}{*}{}} \end{matrix} \quad \begin{array}{c} \\ p \\ n-p \end{array} \begin{array}{c} \begin{array}{cc} p & n-p \end{array} \\ \begin{pmatrix} \mathsf{B}_{1,1} & \mathsf{B}_{1,2} \\ 0 & \mathsf{B}_{2,2} \end{pmatrix} \end{array}. \tag{4.3-12}$$

If A *is not reducible, it is* **irreducible***.*

As we demonstrate shortly, irreducibility serves as the additional hypothesis needed to weaken the condition of strict diagonal dominance. Recall from Section 2.2 that permutation matrices have as their columns the standard unit basis vectors $\mathbf{e}_1, \mathbf{e}_2, \ldots, \mathbf{e}_n$, arranged in some order. To see what kind of transformation the matrix $\mathsf{PAP}^\top$ effects, observe that the mapping $\mathsf{A} \mapsto \mathsf{PA}$ permutes the rows of A, and then the mapping $\mathsf{PA} \mapsto (\mathsf{PA})\mathsf{P}^\top$ applies the same permutation to the columns of PA.

If A is reducible, then one can reorder the equations and unknowns (that is, the rows and columns) in the linear system $\mathsf{A}\mathbf{u} = \mathbf{b}$ to get the block structure

$$\begin{array}{c} \\ p \\ n-p \end{array} \begin{array}{c} \begin{array}{cc} p & n-p \end{array} \\ \begin{pmatrix} \mathsf{B}_{1,1} & \mathsf{B}_{1,2} \\ 0 & \mathsf{B}_{2,2} \end{pmatrix} \end{array} \begin{array}{c} \\ \begin{pmatrix} \mathbf{u}_1 \\ \mathbf{u}_2 \end{pmatrix} \end{array} \begin{array}{c} \\ = \end{array} \begin{array}{c} \\ \begin{pmatrix} \mathbf{b}_1 \\ \mathbf{b}_2 \end{pmatrix} \end{array}.$$

We can rewrite this system as

$$\begin{aligned} \mathsf{B}_{1,1}\mathbf{u}_1 + \mathsf{B}_{1,2}\mathbf{u}_2 &= \mathbf{b}_1, \\ \mathsf{B}_{2,2}\mathbf{u}_2 &= \mathbf{b}_2. \end{aligned}$$

Thus the unknowns stored in the block vector $\mathbf{u}_2$ are independent of the entries of $\mathbf{b}$ stored in the block $\mathbf{b}_1$. In the context of Equation (4.1-4), such a block partitioning implies that some of the unknowns associated with interior nodes of the grid are independent of some of the boundary values. This conclusion violates a physically important property of the Laplace equation and its analogs: All interior values depend upon all of the boundary values. On the strength of this idea, we expect matrices arising from reasonable difference schemes such as Equation (4.1-3) to be irreducible.

It is also worthwhile to develop computable characterizations of irreducibility. The following characterization concerns the zero structure of A. We denote by $\mathcal{N}$ the index set $\{1, 2, \ldots, n\}$.

PROPOSITION 4.10. *A matrix* $\mathsf{A} \in \mathbb{R}^{n \times n}$ *is reducible if and only if there exists a nonempty, proper subset* I *of* $\mathcal{N}$ *such that* $a_{i,j} = 0$ *whenever* $i \in I$ *and* $j \notin I$.

PROOF: Assume that A is reducible, and let P be the permutation matrix that reduces A to the form prescribed in Equation (4.3-12), with zeros in the first p entries of rows $p+1$ through n. The set I in this case is the set of indices of rows of A mapped to rows $p+1, p+2, \ldots, n$ under the transformation $\mathsf{A} \mapsto \mathsf{PAP}^{\top}$.

Now assume that $I = \{i_1, i_2, \ldots, i_{n-p}\}$ is a set of indices, with $\emptyset \neq I \neq \mathcal{N}$, such that $a_{i,j} = 0$ whenever $i \in I$ and $j \notin I$. Let P be the permutation associated with any permutation on $(1, 2, \ldots, n)$ that maps $\{i_1, i_2, \ldots, i_{n-p}\}$ onto the set $\{p+1, p+2, \ldots, n\}$. One can easily check that $\mathsf{PAP}^{\top}$ has the block structure shown in Equation (4.3-12). ∎

The following example helps make the preceding argument concrete. Suppose that $\mathsf{A} \in \mathbb{R}^{4 \times 4}$ has the following zero structure:

$$\mathsf{A} = \begin{bmatrix} \star & \star & \star & \star \\ 0 & \star & 0 & \star \\ \star & \star & \star & \star \\ 0 & \star & 0 & \star \end{bmatrix}.$$

Here, $I = \{2, 4\}$. A simple calculation shows that either of the permutations $2 \mapsto 3 \mapsto 2$ or $2 \mapsto 4 \mapsto 3 \mapsto 2$ will accomplish the reduction of A to the form (4.3-12). The corresponding permutation matrices are, respectively,

$$\mathsf{P} = \begin{bmatrix} 1 & 0 & 0 & 0 \\ 0 & 0 & 1 & 0 \\ 0 & 1 & 0 & 0 \\ 0 & 0 & 0 & 1 \end{bmatrix} \quad \text{or} \quad \begin{bmatrix} 1 & 0 & 0 & 0 \\ 0 & 0 & 1 & 0 \\ 0 & 0 & 0 & 1 \\ 0 & 1 & 0 & 0 \end{bmatrix}.$$

The characterization of reducibility in terms of zero structure leads to an interesting and useful pictorial method for checking irreducibility. To develop this technique, we introduce some terminology and another theorem.

DEFINITION. *Let* $\mathsf{A} \in \mathbb{R}^{n \times n}$ *and* $i, j \in \mathcal{N}$. *An* A-**chain** *for* (i, j) *is a sequence* $\{i, i_1, i_2, \ldots, i_k, j\} \subset \mathcal{N}$ *such that* $0 \notin \{a_{i,i_1}, a_{i_1,i_2}, \ldots a_{i_k,j}\}$.

THEOREM 4.11. *A matrix* $\mathsf{A} \in \mathbb{R}^{n \times n}$ *is irreducible if and only if, for any indices* $i \neq j$, *there exists an* A-*chain for* (i, j).

PROOF: First assume that A is irreducible, and pick $i \in \mathcal{N}$. Define

$$J := \Big\{k \in \mathcal{N} \;:\; \text{there exists an A-chain for} \quad (i,k)\Big\}.$$

We establish by contradiction that $J = \mathcal{N}$. Observe that $J \neq \emptyset$, since otherwise $a_{i,1} = a_{i,2} = \cdots = a_{i,n} = 0$, in which case A is reducible by the permutation $i \mapsto n \mapsto i$. Now assume that $\mathcal{N}\backslash J \neq \emptyset$, so that for some index j there is no A-chain for (i,j). We claim that $a_{l,m} = 0$ whenever $l \in J$ and $m \in \mathcal{N}\backslash J$, contradicting the irreducibility of A by Proposition 4.10. To justify the claim, note that there exists an A-chain $(i, i_1, \ldots, i_k, l)$ for (i,l), by construction. If the claim is false, so that $a_{l,m} \neq 0$, then $(i, i_1, \ldots, i_k, l, m)$ is an A-chain for (i,m), which is impossible since $m \in \mathcal{N}\backslash J$. Hence the claim is valid.

Now suppose that there exists an A-chain for every pair (i,j) of indices with $i \neq j$. Again we argue by contradiction: If A is reducible, then there exists a nonempty, proper subset $I \subset \mathcal{N}$ such that $a_{i,j} = 0$ whenever $i \in I$ and $j \in \mathcal{N}\backslash I$. Choose $i \in I$ and $j \in \mathcal{N}\backslash I$, and let $(i, i_1, i_2, \ldots, i_k, j)$ be an A-chain for (i,j) guaranteed by the hypothesis. The fact that $a_{i,i_1} \neq 0$ implies that $i_i \in I$; the fact that $a_{i_1,i_2} \neq 0$ implies that $i_2 \in I$, and so forth, so that eventually the fact that $a_{i_k,j} \neq 0$ implies that $j \in I$, a contradiction. Therefore A must be irreducible. ∎

To interpret this theorem graphically, suppose that $A \in \mathbb{R}^{n \times n}$, and consider points $P_1, P_2, \ldots, P_n$ in the plane. If the matrix entry $a_{i,j} \neq 0$, then draw an arrow whose tail is P_i and whose head is P_j. This construction produces a **directed graph** for the set $\{P_1, P_2, \ldots, P_n\}$. According to Theorem 4.11, A is irreducible if and only if, for any pair of indices $i, j \in \mathcal{N}$, there exists a path of the form $P_i \to P_{i_1} \to P_{i_2} \to \cdots \to P_{i_k} \to P_j$ in the directed graph. For example, Figure 1 shows that the matrix in the finite-difference approximation (4.1-4) to the Laplace equation is irreducible.

It is now possible to establish convergence of the Jacobi method when the matrix satisfies the following relaxed version of diagonal dominance:

DEFINITION. *A matrix* $A \in \mathbb{R}^{n \times n}$ *is* **irreducibly (row) diagonally dominant** *if all of the following three conditions hold:*

(i) *A is irreducible.*

(ii) $|a_{i,i}| \geq \sum_{j \neq i} |a_{i,j}|$ *for* $i = 1, 2, \ldots, n$.

(iii) *There is at least one row index k for which* $|a_{k,k}| > \sum_{j \neq k} |a_{k,j}|$.

The matrix in Equation (4.1-4) is irreducibly diagonally dominant, even though it is not strictly diagonally dominant.

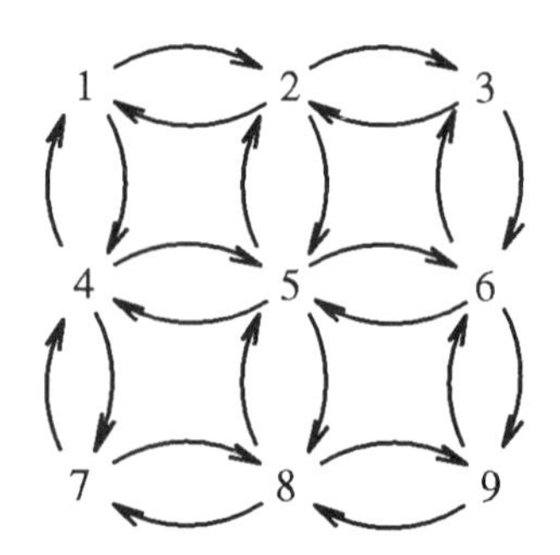

FIGURE 1. *Directed graph for the finite-difference approximation (4.1-4) to the Laplace equation, showing that the matrix for this problem is irreducible.*

In proving the convergence theorem for irreducibly diagonally dominant matrices, it is useful to compare vectors and matrices entrywise.

DEFINITION. *Let* $\mathbf{v} = (v_1, v_2, \ldots, v_n)^\top$ *and* $\mathbf{w} = (w_1, w_2, \ldots, w_n)^\top$. *Then*

(*i*) $\mathbf{v} \prec \mathbf{w}$ *if* $v_i < w_i$ *for every index* i.

(*ii*) $\mathbf{v} \preceq \mathbf{w}$ *if* $v_i \leq w_i$ *for every index* i.

(*iii*) $\mathbf{v} \precneqq \mathbf{w}$ *if* $\mathbf{v} \preceq \mathbf{w}$ *and* $\mathbf{v} \neq \mathbf{w}$.

(*iv*) $|\mathbf{v}| := (|v_1|, |v_2|, \ldots, |v_n|)^\top$.

THEOREM 4.12. *If* $\mathsf{A} \in \mathbb{R}^{n\times n}$ *is irreducibly diagonally dominant, then the Jacobi method for the linear system* $\mathsf{A}\mathbf{u} = \mathbf{b}$ *converges.*

PROOF: First observe that the Jacobi iteration matrix G_J for A is irreducible whenever A is. To show that $\varrho(\mathsf{G}_J) < 1$, it suffices to demonstrate that $|\mathsf{G}_J|^n\mathbf{1} \prec \mathbf{1}$, where $\mathbf{1} := (1, 1, \ldots 1)^\top \in \mathbb{R}^n$, that is,

$$\mathbf{0} \prec \mathbf{1} - |\mathsf{G}_J|^n\mathbf{1}. \tag{4.3-13}$$

For, in this case, $\|\,|\mathbf{G}_J|\,\|_\infty < 1$, and we have

$$[\varrho(\mathsf{G}_J)]^n = \varrho(\mathsf{G}_J^n) \leq \|\mathsf{G}_J^n\|_\infty \leq \|\,|\mathsf{G}_J|^n\,\|_\infty < 1,$$

from which the inequality $\varrho(\mathsf{G}_J) < 1$ follows.

To establish the inequality (4.3-13), notice that the conditions (*ii*) and (*iii*) in the definition of irreducible diagonal dominance imply that $|\mathsf{G}_J|\mathbf{1} \precneqq \mathbf{1}$. Therefore,

$$|\mathsf{G}_J|^n\mathbf{1} \preceq |\mathsf{G}_J|^{n-1}\mathbf{1} \preceq \cdots \preceq |\mathsf{G}_J|\mathbf{1} \precneqq \mathbf{1},$$

and hence

$$\mathbf{0} \npreceq \underbrace{\mathbf{1} - |\mathsf{G}_J|\mathbf{1}}_{\mathbf{d}_1} \preceq \underbrace{\mathbf{1} - |\mathsf{G}_J|^2\mathbf{1}}_{\mathbf{d}_2} \preceq \cdots \preceq \underbrace{\mathbf{1} - |\mathsf{G}_J|^n\mathbf{1}}_{\mathbf{d}_n}. \tag{4.3-14}$$

We complete the proof by showing that the vector $\mathbf{d}_n$ defined in this last chain of inequalities has n nonzero entries. We argue by contradiction: Assume that $\mathbf{d}_n$ has fewer than n nonzero entries. Since the vector $\mathbf{d}_1$ has at least one nonzero entry, it follows from the inequalities (4.3-14) that there is some index $k \in \mathcal{N}$ for which $\mathbf{d}_{k-1}$ and $\mathbf{d}_k$ have the same number p of nonzero entries. These entries must occur at the same indices. Now let $\mathsf{P} \in \mathbb{R}^{n\times n}$ be a permutation matrix such that

$$\begin{aligned} \mathsf{P}\mathbf{d}_{k-1} &= (\alpha_1, \alpha_2, \ldots, \alpha_p, 0, \ldots, 0)^\top &= \begin{bmatrix} \boldsymbol{\alpha} \\ \mathbf{0} \end{bmatrix}, \\ \mathsf{P}\mathbf{d}_k &= (\beta_1, \beta_2, \ldots, \beta_p, 0, \ldots, 0)^\top &= \begin{bmatrix} \boldsymbol{\beta} \\ \mathbf{0} \end{bmatrix}, \end{aligned}$$

where $\mathbf{0} \npreceq \boldsymbol{\alpha}, \boldsymbol{\beta} \in \mathbb{R}^p$. We have

$$|\mathsf{G}_J|\mathbf{d}_{k-1} = |\mathsf{G}_J|\mathbf{1} - |\mathsf{G}_J|^k\mathbf{1} \preceq \mathbf{1} - |\mathsf{G}_J|^k\mathbf{1} = \mathbf{d}_k,$$

which implies that

$$\mathsf{P}|\mathsf{G}_J|\mathsf{P}^\top\mathsf{P}\mathbf{d}_{k-1} \preceq \mathsf{P}\mathbf{d}_k = \begin{bmatrix} \boldsymbol{\beta} \\ \mathbf{0} \end{bmatrix}.$$

Rewrite this relationship in block form:

$$\underbrace{\begin{matrix} p \\ n-p \end{matrix}\overset{\begin{matrix} p & \quad n-p \end{matrix}}{\begin{pmatrix} |\mathsf{H}_{1,1}| & |\mathsf{H}_{1,2}| \\ |\mathsf{H}_{2,1}| & |\mathsf{H}_{2,2}| \end{pmatrix}}}_{\mathsf{P}|\mathsf{G}_J|\mathsf{P}^\top} \underbrace{\begin{pmatrix} \boldsymbol{\alpha} \\ \mathbf{0} \end{pmatrix}}_{\mathsf{P}\mathbf{d}_{k-1}} = \begin{pmatrix} |\mathsf{H}_{1,1}|\boldsymbol{\alpha} \\ |\mathsf{H}_{2,1}|\boldsymbol{\alpha} \end{pmatrix} \preceq \underbrace{\begin{pmatrix} \boldsymbol{\beta} \\ \mathbf{0} \end{pmatrix}}_{\mathsf{P}\mathbf{d}_k}.$$

Therefore, $|\mathsf{H}_{2,1}| = \mathsf{0}$. But if this is true, then the matrix $\mathsf{P}\mathsf{G}_J\mathsf{P}^\top$ has the block structure specified in Equation (4.3-12), and thus G_J must be reducible. This conclusion contradicts our hypotheses, completing the proof. ∎

Using a similar argument, one can prove that the Gauss-Seidel method converges whenever A is irreducibly diagonally dominant. Instead, we show in the next section that $\varrho(\mathsf{G}_1) < \varrho(\mathsf{G}_J)$ under reasonable hypotheses, and in these cases the Gauss-Seidel method converges whenever the Jacobi method does.

Further Remarks: Multigrid Methods

We conclude this section with a brief discussion of multigrid methods. We merely sketch some of the main ideas here. For a good introduction to multigrid methods, we recommend Briggs [2], whose treatment guides the discussion.

Consider, for simplicity, a one-dimensional boundary-value problem of the form

$$-u''(x) = f(x), \qquad u(0) = u(1) = 0. \tag{4.3-15}$$

The simplest finite-difference approximation to this problem uses the algebraic analog

$$-\frac{u_{i-1} - 2u_i + u_{i+1}}{h^2} = f(x_i),$$

where the values $x_i = ih = i/N$ are nodes in a uniform grid on the interval $[0,1]$, and the values u_i represent unknown approximate values of $u(x_i)$. This discrete approximation yields the following linear system:

$$\underbrace{\begin{bmatrix} 2 & -1 & & & \\ -1 & 2 & -1 & & \\ & & \ddots & & \\ & & -1 & 2 & -1 \\ & & & -1 & 2 \end{bmatrix}}_{\mathsf{A}} \underbrace{\begin{bmatrix} u_1 \\ u_2 \\ \vdots \\ u_{N-2} \\ u_{N-1} \end{bmatrix}}_{\mathbf{u}} = \underbrace{\begin{bmatrix} h^2 f(x_1) \\ h^2 f(x_2) \\ \vdots \\ h^2 f(x_{N-2}) \\ h^2 f(x_{N-1}) \end{bmatrix}}_{\mathbf{f}}. \tag{4.3-16}$$

Even though it is possible to solve this tridiagonal system efficiently using direct methods, the one-dimensional character of the problem makes it easier to illustrate the concepts behind the multigrid method, which finds its most important applications in multidimensional settings.

The matrix A has the following eigenvalues and corresponding eigenvectors:

$$\mu_k = 2 - 2\cos\left(\frac{k\pi}{N}\right) = 4\sin^2\left(\frac{k\pi}{2N}\right),$$

$$\boldsymbol{v}_k = \begin{bmatrix} \sin(k\pi/N) \\ \sin(2k\pi/N) \\ \vdots \\ \sin((N-1)k\pi/N) \end{bmatrix},$$

where $k = 1, 2, \ldots, N-1$. From a geometric viewpoint, the eigenvectors $\boldsymbol{v}_k$ corresponding to small values of the **wavenumber** k have entries $\sin(jk\pi/N)$ that vary smoothly with the index j. Entries of eigenvectors corresponding to large wavenumbers are more oscillatory as functions of j. Figure 2 shows plots of the entries of $\boldsymbol{v}_1$ and $\boldsymbol{v}_{16}$ versus the index j for $N = 32$, illustrating the difference between "smooth" eigenvectors and "oscillatory" ones.

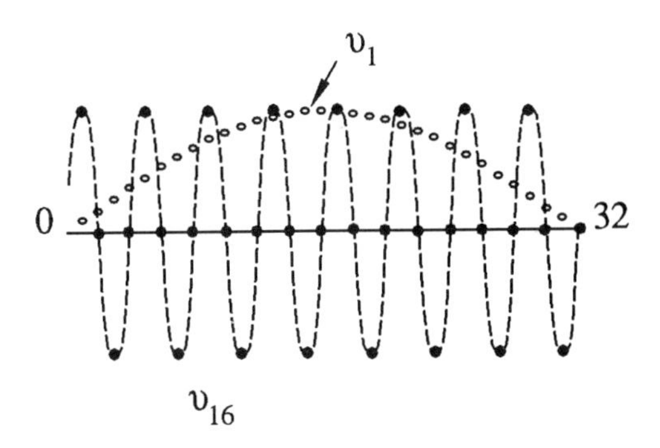

FIGURE 2. *Plots of the jth entries of the eigenvectors corresponding to wavenumbers $k = 1$ and $k = 16$ for the matrix in Equation (4.3-16). Here, $N = 32$.*

The eigenvectors of A are linearly independent and hence form a basis for $\mathbb{R}^{N-1}$. Thus, given an iterate $\mathbf{u}^{(m)}$ for the solution $\mathbf{u}$ of Equation (4.3-16), we can expand the error as

$$\boldsymbol{\varepsilon}^{(m)} := \mathbf{u} - \mathbf{u}^{(m)} = c_1\boldsymbol{v}_1 + c_2\boldsymbol{v}_2 + \cdots + c_{N-1}\boldsymbol{v}_{N-1}.$$

If the iteration matrix for a given iterative scheme is G, then the error at the next iteration is

$$\boldsymbol{\varepsilon}^{(m+1)} = \mathsf{G}\boldsymbol{\varepsilon}^{(m)} = \mathsf{G}c_1\boldsymbol{v}_1 + \mathsf{G}c_2\boldsymbol{v}_2 + \cdots + \mathsf{G}c_{N-1}\boldsymbol{v}_{N-1}.$$

It is instructive to ask what effect a given iterative scheme has on each of the **error modes** $c_k\boldsymbol{v}_k$ at a typical iteration level m.

Consider the damped Jacobi method (4.3-4). For the system (4.3-16), this scheme gives

$$u_j^{(m+1)} = u_j^{(m)} + \omega\left\{\frac{1}{2}\left[u_{j+1}^{(m)} + u_{j-1}^{(m)} + h^2 f(x_j)\right] - u_j^{(m)}\right\}.$$

The iteration matrix in this case has the form $\mathsf{G}_D := \mathsf{I} - \omega\mathsf{D}^{-1}\mathsf{A}$, and therefore it has eigenvalues

$$\lambda_k = 1 - \frac{\omega}{2}\mu_k = 1 - 2\omega\sin^2\left(\frac{k\pi}{2N}\right), \quad k = 1, 2, \ldots, N-1.$$

One easily checks that the corresponding eigenvectors of G_D are precisely the eigenvectors $\boldsymbol{v}_k$ of A. Therefore, using the expansion of the iterative error in terms of the eigenvectors of A, we have

$$\boldsymbol{\varepsilon}^{(m+1)} = \sum_{k=1}^{N-1} \mathsf{G}_D c_k\boldsymbol{v}_k = \sum_{k=1}^{N-1} \lambda_k c_k\boldsymbol{v}_k.$$

Thus the damped Jacobi method damps each error mode $c_k \boldsymbol{v}_k$ by the factor λ_k at each iteration.

The observation that relaxation schemes tend to damp certain error modes faster than others furnishes the heuristic for multigrid methods. To wit, instead of using one iterative scheme to damp all error modes, we regard the scheme as a *selective* damper, modifying it to act on modes associated with different wavenumber ranges. In the case of the damped Jacobi method, the error modes that undergo preferential damping depend upon the choice of the parameter ω. One interesting choice is $\omega = 2/3$. Figure 3 plots the value of the eigenvalue λ_k versus the wavenumber k for this case. For the "oscillatory" modes — those corresponding to wavenumbers in the interval $N/2 \leq k \leq N-1$ — the eigenvalues satisfy $|\lambda_k| < 1/3$, and consequently the scheme damps these modes rapidly. In contrast, for the "smooth" modes — those for which $k < N/2$ — the damping occurs more slowly. If an iterative method preferentially damps the oscillatory modes of error vectors, we call it a **smoother**.

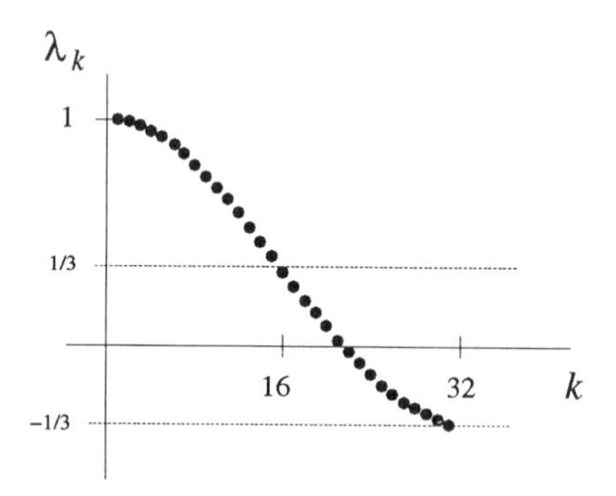

FIGURE 3. *Eigenvalues λ_k versus wavenumber k for the damped Jacobi method with $\omega = 2/3$, showing that the most effective damping occurs for the large wavenumbers associated with oscillatory error modes.*

A simple example helps to justify this terminology. Consider the discrete approximation to the boundary-value problem (4.3-15) with $f(x) = 0$ for all x. For the corresponding linear system (4.3-16), the solution vector is $\mathbf{u} = \mathbf{0}$, so the error at any iteration level m is just $\boldsymbol{\varepsilon}^{(m)} = -\mathbf{u}^{(m)}$. Let us examine the behavior of the damped Jacobi method, with $\omega = 2/3$, for the case when the initial guess $\mathbf{u}^{(0)} = \boldsymbol{v}_1 + 0.4\boldsymbol{v}_{16}$, a superposition of a smooth error mode and an oscillatory one. Figure 4(a) shows the graph of this initial guess. Figure 4(b) shows the graph of the iterate $\mathbf{u}^{(8)}$. Notice that after eight iterations there remains a large contribution from the smooth error mode, but the remaining contribution from the oscillatory mode is much smaller. The main effect of iterating here is to *smooth* the error.

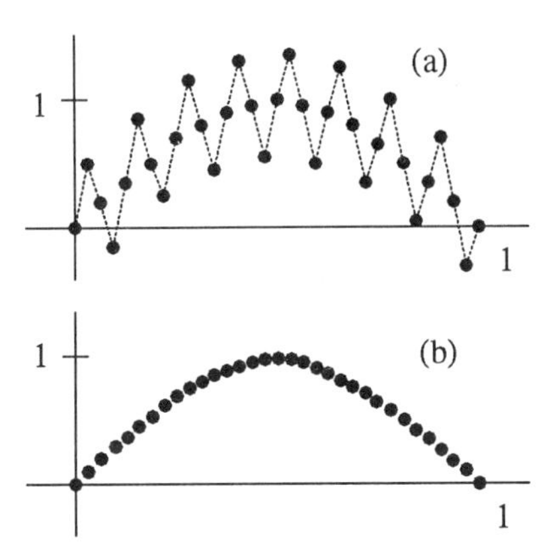

FIGURE 4. *(a) Graph of the initial guess* $\mathbf{u}^{(0)} = \boldsymbol{v}_1 + 0.4\boldsymbol{v}_{16}$ *for the discrete approximation to the boundary-value problem* $u'' = 0$, $u(0) = u(1) = 0$, *with* $N = 32$. *(b) Graph of* $\mathbf{u}^{(8)}$ *for the same problem.*

Another crucial idea in the multigrid approach is that errors that are smooth on fine grids typically appear to be more oscillatory on coarser grids. Consider, for example, the eigenvector $\boldsymbol{v}_4$ on the grid with $N = 32$, along with the projection of $\boldsymbol{v}_4$ on a coarser grid with $N = 8$. The vector is smooth ($k < N/2$) on the fine grid, but it is oscillatory with respect to the coarse grid. We therefore expect a smoother to be more effective at damping $\boldsymbol{v}_4$ if we transfer the iterations to coarse grids than if we keep iterating on the fine one. As an added benefit, the iterations on the coarser grids are cheaper than those on the fine grid, since the former involve fewer unknowns.

Let us summarize these observations: By transferring a problem between fine and coarse grids, we can rapidly damp all of the error modes using the smoother *and* reap computational benefits associated with frequent visits to coarse grids.

This remark suggests the following iterative strategy: Iterate a few times on a fine grid using the smoother. The error $\boldsymbol{\varepsilon}^{(m)}$ obeys the **error equation**, $\mathsf{A}\boldsymbol{\varepsilon}^{(m)} = \mathbf{r}^{(m)}$, where $\mathbf{r}^{(m)} := \mathbf{f} - \mathsf{A}\mathbf{u}^{(m)}$ denotes the **residual**. Then map this error equation to a coarser grid, where the error appears more oscillatory. Apply several iterations of the smoother again, then map to a still coarser grid. Eventually we will have mapped the problem to a grid so coarse that the associated matrix equation is small enough to solve directly. After solving on the coarsest grid, correct the solution on successively finer grids until we have reached the finest grid. This orchestration of the smoother with intergrid transfers is called a **V-cycle**; Figure 5 gives a schematic picture.

To be more concrete, assume that we have a nested sequence of grids,

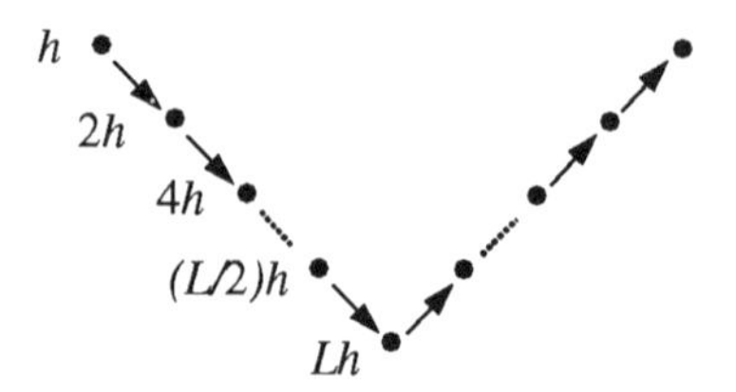

FIGURE 5. *Schematic diagram of a V-cycle on a sequence of nested grids.*

each of which has mesh size twice that of the next finer grid. Thus we have grids of mesh size $h, 2h, 4h, \ldots, (L/2)h, Lh$, where L is an integer power of 2. To map the error equation from fine grids to coarse grids, we use **restriction operators** I_{lh}^{2lh}. For example, to map a vector $\mathbf{v}^h$ defined on the finest grid to a vector $\mathbf{v}^{2h}$ defined on the grid having mesh size $2h$, we might set $\mathbf{v}_{2h} = \mathsf{I}_h^{2h}\mathbf{v}_h$, where I_h^{2h} denotes the projection operator that directly transfers every other entry of $\mathbf{v}^h$ to an entry of $\mathbf{v}^{2h}$:

$$v_j^{2h} = v_{2j}^h.$$

To transfer from coarse grids to fine grids, we use **prolongation operators** $\mathsf{I}_{lh}^{(l/2)h}$, the most common of which employ linear interpolation. For example, we might define $\mathbf{v}^h = \mathsf{I}_{2h}^h\mathbf{v}^{2h}$ as follows:

$$v_j^h = \begin{cases} v_{j/2}^{2h}, & \text{for} \quad j \text{ even}, \\ \frac{1}{2}(v_{(j+1)/2}^{2h} + v_{(j-1)/2}^{2h}), & \text{for} \quad j \text{ odd}. \end{cases}$$

The following algorithm outlines a V-cycle. The notation

$$\mathbf{u}^h \leftarrow \text{SMOOTH}^r\,(\mathsf{A}^h\mathbf{u}^h = \mathbf{f}^h\,;\,\mathbf{u}^{(0)})$$

means, "apply r iterations of the smoother to the linear system associated with the grid having mesh size h, using $\mathbf{u}^{(0)}$ as initial guess." The notation

$$\boldsymbol{\varepsilon}^{Lh} \leftarrow \text{SOLVE}\,(\mathsf{A}^{Lh}\boldsymbol{\varepsilon}^{Lh} = \mathbf{r}^{Lh})$$

means, "use a direct method to solve the error equation on the coarsest grid."

ALGORITHM 4.1 (V-CYCLE). *Given a smoother, a nested sequence of grids having mesh sizes $h, 2h, 4h, \ldots, Lh$, and a set of restriction and prolongation operators I_{Jh}^{Kh}, the following algorithm implements one multigrid V-cycle for the linear system $\mathsf{A}^h\mathbf{u}^h = \mathbf{f}^h$ associated with the finest grid.*

1. $\mathbf{u}^h \leftarrow \text{SMOOTH}^r\,(\mathsf{A}^h\mathbf{u}^h = \mathbf{f}^h\,;\,\mathbf{u}^{(0)})$.

2. $\mathbf{r}^h \leftarrow \mathbf{f}^h - \mathbf{A}^h\mathbf{u}^h$.

3. $\mathbf{r}^{2h} \leftarrow \mathsf{I}_h^{2h}\mathbf{r}^h$.

4. For $l = 2, 4, 8, \ldots, L/2$:

5. $\quad \boldsymbol{\varepsilon}^{lh} \leftarrow \text{SMOOTH}^r\ (\mathsf{A}^{lh}\boldsymbol{\varepsilon}^{lh} = \mathbf{r}^{lh}\ ;\ \mathbf{0})$.

6. $\quad \mathbf{r}^{lh} \leftarrow \mathsf{A}^{lh}\boldsymbol{\varepsilon}^{lh}$.

7. $\quad \mathbf{r}^{2lh} \leftarrow \mathsf{I}_{lh}^{2lh}\mathbf{r}^{lh}$.

8. $\quad l \leftarrow 2l$.

9. $\boldsymbol{\varepsilon}^{Lh} \leftarrow \text{SOLVE}\ (\mathsf{A}^{Lh}\boldsymbol{\varepsilon}^{Lh} = \mathbf{r}^{Lh})$.

10. For $l = L, L/2, \ldots, 2$:

11. $\quad \boldsymbol{\varepsilon}^{(l/2)h} \leftarrow \boldsymbol{\varepsilon}^{(l/2)h} + \mathsf{I}_{lh}^{(l/2)h}\boldsymbol{\varepsilon}^{lh}$.

12. $\quad \boldsymbol{\varepsilon}^{(l/2)h} \leftarrow \text{SMOOTH}^r(\mathsf{A}^{(l/2)h}\boldsymbol{\varepsilon}^{(l/2)h} = \mathbf{r}^{(l/2)h}\ ;\ \boldsymbol{\varepsilon}^{(l/2)h})$.

13. $\mathbf{u}^h \leftarrow \mathbf{u}^h + \mathsf{I}_{2h}^h\boldsymbol{\varepsilon}^{2h}$.

14. $\mathbf{u}^h \leftarrow \text{SMOOTH}^r(\mathsf{A}^h\mathbf{u}^h = \mathbf{f}^h\ ;\ \mathbf{u}^h)$.

Many other formulations and orchestrations are possible; see Briggs [2].

4.4 Successive Overrelaxation

Motivation

We devote this section to the method of successive overrelaxation (SOR) introduced in Section 4.3. Recall the algorithmic form of the method for a linear system $\mathsf{A}\mathbf{u} = \mathbf{b}$:

$$u_i^{(m+1)} = (1-\omega)u_i^{(m)} + \omega\left(-\sum_{j<i}\frac{a_{i,j}}{a_{i,i}}u_j^{(m+1)} - \sum_{j>i}\frac{a_{i,j}}{a_{i,i}}u_j^{(m)} + \frac{b_i}{a_{i,i}}\right).$$

These equations correspond to the matrix form

$$\mathbf{u}^{(m+1)} = -\underbrace{\left(\omega^{-1}\mathsf{D}+\mathsf{L}\right)^{-1}}_{\mathsf{B}^{-1}}\underbrace{\left[\left(1-\omega^{-1}\right)\mathsf{D}+\mathsf{U}\right]}_{\mathsf{A}-\mathsf{B}}\mathbf{u}^{(m)} + \left(\omega^{-1}\mathsf{D}+\mathsf{L}\right)^{-1}\mathbf{b},$$

with the splitting $\mathsf{A} = \mathsf{B} + (\mathsf{A}-\mathsf{B})$ as indicated. A key question is how to choose the overrelaxation parameter ω.

The theory developed in this section applies to analogs of the model problem (4.1-1), that is, to discretizations of the Laplace operator or its generalizations. Matrices in such problems are typically symmetric and positive definite. Moreover, they often are amenable to analysis more delicate than that available for general symmetric, positive definite matrices. This analysis helps motivate schemes for speeding the convergence of SOR by proper choice of ω. It also further elucidates the relationship between the Jacobi and Gauss-Seidel methods discussed in the previous section.

Practical Considerations

We begin by reviewing the key mathematical results, which we prove later. Throughout, we use the following notation for various iteration matrices associated with the linear system $\mathsf{A}\mathbf{u} = \mathbf{b}$:

$$\begin{aligned} \mathsf{G}_J &:= -\mathsf{D}^{-1}(\mathsf{L}+\mathsf{U}) && \text{Jacobi;} \\ \mathsf{G}_1 &:= -(\mathsf{D}+\mathsf{L})^{-1}\mathsf{U} && \text{Gauss-Seidel;} \\ \mathsf{G}_\omega &:= -(\omega^{-1}\mathsf{D}+\mathsf{L})^{-1}[(1-\omega^{-1})\mathsf{D}+\mathsf{U}] && \text{SOR.} \end{aligned}$$

The Ostrowski-Reich theorem asserts that, when $\mathsf{A} \in \mathbb{R}^{n\times n}$ is symmetric and positive definite, SOR converges whenever $0 < \omega < 2$. We prove this theorem shortly. The theorem is disappointing, since it fails to specify how to choose particular values of ω that yield the most rapid convergence. Figure 1, showing how $\varrho(\mathsf{G}_\omega)$ varies with ω for a typical difference approximation to the Laplace operator, illustrates the problem. The graph indicates that the spectral radius of G_ω attains a minimum at a value, denoted ω_{opt}, located between 1 and 2. Outside a fairly small neighborhood of ω_{opt}, $\varrho(\mathsf{G}_\omega)$ assumes values much closer to 1. Therefore, unless we select a value of ω close to ω_{opt}, we forfeit much of SOR's potential benefit.

Determining ω_{opt} is difficult in general. However, if the system matrix A enjoys certain additional properties, then one can identify ω_{opt}, at least analytically. Let us review the results of this theory and construct an algorithm for using it computationally, saving rigorous proofs for later. We start with some definitions.

The iterates generated by the Gauss-Seidel method and SOR depend upon the ordering of the equations and unknowns. The following definition identifies a useful class of orderings.

DEFINITION. *A matrix* $\mathsf{A} \in \mathbb{R}^{n\times n}$ *is* **consistently ordered** *if, for all nonzero values of* α*, the eigenvalues of the matrix*

$$\mathsf{G}(\alpha) := \alpha\mathsf{D}^{-1}\mathsf{L} + \alpha^{-1}\mathsf{D}^{-1}\mathsf{U}$$

are independent of α.

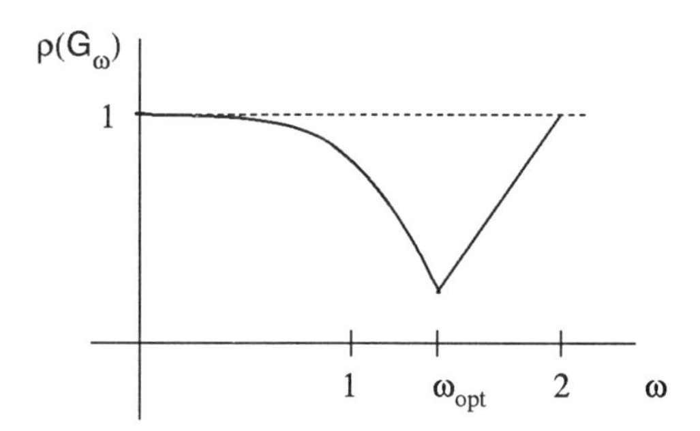

FIGURE 1. *Typical plot of the spectral radius of the iteration matrix for* SOR *versus the overrelaxation parameter* ω, *showing a minimum value at* ω_{opt}.

Certain block tridiagonal matrices are consistently ordered. In particular, consider the block tridiagonal structure

$$
\mathsf{A} \;=\; \begin{matrix} m \\ m \\ \vdots \\ \\ \\ \end{matrix} \overset{\begin{matrix} m & & m & & \cdots \end{matrix}}{\begin{pmatrix} \mathsf{D}_1 & \mathsf{U}_1 & & & \\ \mathsf{L}_2 & \mathsf{D}_2 & \mathsf{U}_2 & & \\ & & \ddots & & \\ & & \mathsf{L}_{n-1} & \mathsf{D}_{n-1} & \mathsf{U}_{n-1} \\ & & & \mathsf{L}_n & \mathsf{D}_n \end{pmatrix}}. \tag{4.4-1}
$$

Such matrices are consistently ordered whenever all of the diagonal blocks D_i have nonzero diagonal parts, that is, whenever $\operatorname{diag}(\mathsf{D}_i) \neq 0$. To see this, observe that $\mathsf{G}(\alpha)$ in this case has the same eigenvalues as the similar matrix $\mathsf{Q}^{-1}\mathsf{G}(\alpha)\mathsf{Q}$, where

$$
\mathsf{Q} := \begin{pmatrix} \mathsf{I} & & & & \\ & \alpha\mathsf{I} & & & \\ & & \ddots & & \\ & & & \alpha^{n-2}\mathsf{I} & \\ & & & & \alpha^{n-1}\mathsf{I} \end{pmatrix}.
$$

Problem 5 asks for verification that $\det(\lambda\mathsf{I} - \mathsf{Q}^{-1}\mathsf{G}(\alpha)\mathsf{Q}) = \det(\lambda\mathsf{I} + \mathsf{L} + \mathsf{U})$, so that $\mathsf{G}(\alpha)$ has the same eigenvalues as $\mathsf{G}(1)$. In particular, the finite-difference Laplace matrix in Equation (4.1-4) is consistently ordered.

The next definition concerns the zero structure of the system matrix A.

DEFINITION. *A matrix* $\mathsf{A} \in \mathbb{R}^{n\times n}$ *has* **property A** *if there is a permutation matrix* $\mathsf{P} \in \mathbb{R}^{n\times n}$ *such that* $\mathsf{PAP}^\top$ *has the following block structure:*

$$
\mathsf{PAP}^\top = \begin{pmatrix} \mathsf{D}_1 & \mathsf{M}_1 \\ \mathsf{M}_2 & \mathsf{D}_2 \end{pmatrix}. \tag{4.4-2}
$$

Here, D_1, D_2 are square, diagonal matrices, not necessarily having the same size.

If A has property A, then the permutation matrix P in Equation (4.4-2) effects the following transformation on the Jacobi iteration matrix G_J:

$$PG_JP^T = -P(D^{-1}L)P^T - P(D^{-1}U)P^T = \begin{pmatrix} 0 & -D_1^{-1}M_1 \\ -D_2^{-1}M_2 & 0 \end{pmatrix}.$$

The finite-difference approximation to the Laplace operator, introduced in Equation (4.1-3), possesses property A. For example, if we reorder the equations and unknowns in the matrix equation (4.1-4) under the red-black ordering of Figure 2, then we obtain the following matrix equation:

$$\begin{bmatrix} 4 & & & & & -1 & -1 & 0 & 0 \\ & 4 & & & & -1 & 0 & -1 & 0 \\ & & 4 & & & -1 & -1 & -1 & -1 \\ & & & 4 & & 0 & -1 & 0 & -1 \\ & & & & 4 & 0 & 0 & -1 & -1 \\ -1 & -1 & -1 & 0 & 0 & 4 & & & \\ -1 & 0 & -1 & -1 & 0 & & 4 & & \\ 0 & -1 & -1 & 0 & -1 & & & 4 & \\ 0 & 0 & -1 & -1 & -1 & & & & 4 \end{bmatrix} \begin{bmatrix} u_{1,3} \\ u_{3,3} \\ u_{2,2} \\ u_{1,1} \\ u_{3,1} \\ u_{2,3} \\ u_{1,2} \\ u_{3,2} \\ u_{2,1} \end{bmatrix} = \begin{bmatrix} u_{0,3} + u_{1,4} \\ u_{3,4} + u_{4,3} \\ 0 \\ u_{1,0} + u_{0,1} \\ u_{3,0} + u_{4,1} \\ u_{2,4} \\ u_{0,2} \\ u_{4,2} \\ u_{2,0} \end{bmatrix}.$$

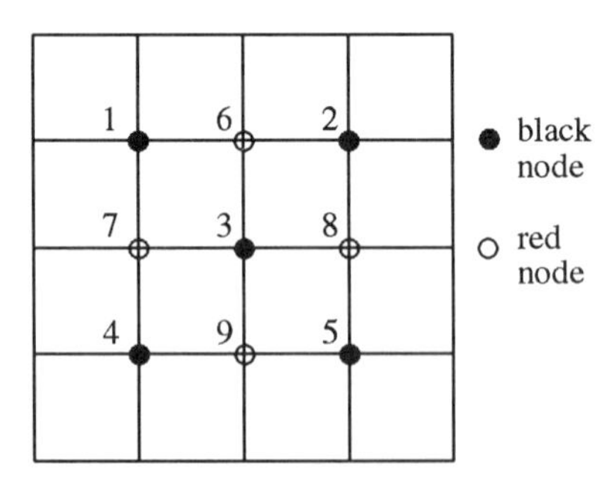

FIGURE 2. *Red-black numbering scheme for the nodes in the grid shown in Figure 1 of Section 4.1*

The following is a summary of important facts about SOR, proved below:

(i) If $A \in \mathbb{R}^{n \times n}$ is consistently ordered and has property A, then

$$\varrho(G_1) = [\varrho(G_J)]^2.$$

In terms of the convergence rates $R(G) := -\log_{10} \varrho(G)$ defined in Section 4.3, this identity implies that $R(G_1) = 2R(G_J)$. In other words, when the Jacobi method converges for such matrices, the Gauss-Seidel method converges twice as fast.

(ii) If, in addition, A is symmetric and positive definite, then the graph of $\varrho(\mathsf{G}_\omega)$ versus ω has the form shown in Figure 1. The optimal choice of overrelaxation parameter is

$$\omega_{\text{opt}} = \frac{2}{1+\sqrt{1-[\varrho(\mathsf{G}_J)]^2}} = \frac{2}{1+\sqrt{1-\varrho(\mathsf{G}_1)}}.$$

(iii) With this choice, $\varrho(\mathsf{G}_{\omega_{\text{opt}}}) = \omega_{\text{opt}} - 1$.

The main difficulty in applying these results lies in the fact that exact knowledge of $\varrho(\mathsf{G}_J)$ (or $\varrho(\mathsf{G}_1)$) is typically unavailable. Even without this knowledge, though, one can devise schemes that generate better approximations to ω_{opt} during the iterations. Thus, by adjusting the value of ω as the iterations proceed, one can steer the scheme toward its fastest convergence rate. Let us sketch the basic ideas behind one such approach, using heuristics developed in Section 4.3.

Consider the case when the Gauss-Seidel iteration matrix $\mathsf{G}_1 \in \mathbb{R}^{n\times n}$ has distinct eigenvalues $\lambda_1, \lambda_2, \ldots, \lambda_n$ that we can number so that $|\lambda_1| \leq |\lambda_2| \leq \cdots \leq |\lambda_{n-1}| < |\lambda_n|$. The distinctness of the eigenvalues implies that the corresponding eigenvectors $\boldsymbol{v}_1, \boldsymbol{v}_2, \ldots, \boldsymbol{v}_n$ of G_1 form a basis for $\mathbb{R}^n$. Let $\{\mathbf{u}^{(m)}\}$ signify a sequence of iterates generated by the scheme $\mathbf{u}^{(m+1)} = \mathsf{G}_1\mathbf{u}^{(m)} + \mathbf{k}$, starting with some initial guess $\mathbf{u}^{(0)}$. Denote by $\mathbf{d}_m := \mathbf{u}^{(m)} - \mathbf{u}^{(m-1)}$ the difference between successive iterates. Since the iteration equation is linear, $\mathbf{d}_{m+1} = \mathsf{G}_1\mathbf{d}_m$.

Now expand the first difference vector $\mathbf{d}_1$ as a linear combination of the eigenvectors $\boldsymbol{v}_i$. Since eigenvectors are unique only up to constant multiples, we write this expansion as

$$\mathbf{d}_1 = \sum_{i=1}^{n} \boldsymbol{v}_i.$$

It follows that

$$\begin{aligned}
\mathbf{d}_2 &= \mathsf{G}_1\mathbf{d}_1 &&= \sum_{i=1}^{n} \lambda_i \boldsymbol{v}_i,\\
&\vdots \\
\mathbf{d}_{m+1} &= \mathsf{G}_1\mathbf{d}_{m-1} &&= \sum_{i=1}^{n} \lambda_i^m \boldsymbol{v}_i.
\end{aligned}$$

Hence,

$$\lambda_n^{-m}\mathbf{d}_{m+1} = \sum_{i=1}^{n} \left(\frac{\lambda_i}{\lambda_n}\right)^m \boldsymbol{v}_i,$$

the last sum tending to $\boldsymbol{v}_n$ as $m \to \infty$, since $|\lambda_i/\lambda_n| < 1$ when $i \neq n$. In other words, as the iterations progress, the difference vector $\mathbf{d}_m$ approaches an eigenvector associated with the dominant eigenvalue λ_n.

To make use of this observation, define the **Rayleigh quotient**

$$r_m := \frac{(\mathsf{G}_1 \mathbf{d}_m)^\top \mathbf{d}_m}{\mathbf{d}_m^\top \mathbf{d}_m} = \frac{\mathbf{d}_{m+1}^\top \mathbf{d}_m}{\mathbf{d}_m^\top \mathbf{d}_m}.$$

Since $\mathsf{G}_1 \mathbf{d}_m \to \lambda_n \mathbf{d}_m$ as $m \to \infty$, $|r_m| \to |\lambda_n| = \varrho(\mathsf{G}_1)$ as $m \to \infty$. We therefore expect the Rayleigh quotient of successive differences to furnish a good estimate of $\varrho(\mathsf{G}_1)$ after several iterations.

This heuristic leads to the following strategy for estimating ω_{opt}. Begin the SOR iterations with $\omega = 1$, that is, start with the Gauss-Seidel scheme. While $\omega = 1$, use the differences $\mathbf{d}_m$ between successive iterates to compute the Rayleigh quotients r_m. After several iterations, r_m approaches a constant, which we adopt as an estimate of $\varrho(\mathsf{G}_1)$. As soon as $|r_m - r_{m-1}|$ is small, set

$$\omega = \frac{2}{1 + \sqrt{1 - |r_m|}} \simeq \omega_{\text{opt}}.$$

The next algorithm implements this strategy. The notation "$\mathbf{u} \leftarrow \mathsf{G}_\omega \mathbf{u} + \mathbf{k}$" means, "execute one iteration of SOR with the current value of ω."

ALGORITHM 4.2 (SOR WITH NEARLY OPTIMAL ω). *Let $\mathsf{A} \in \mathbb{R}^{n \times n}$ be consistently ordered and have property A. The following algorithm solves the linear system $\mathsf{A}\mathbf{u} = \mathbf{b}$ via SOR, using $\mathbf{u}^{(0)}$ as initial guess. The algorithm uses the first few iterations to estimate ω_{opt}. The parameter τ is a positive tolerance on the difference between successive estimates of the spectral radius $\varrho(\mathsf{G}_1)$ of the Gauss-Seidel method. The parameter $\epsilon > 0$ is a tolerance on the norm of the error $\mathbf{u} - \mathbf{u}^{(m)}$.*

1. $\mathbf{u} \leftarrow \mathbf{u}^0$.
2. $\mathbf{d} \leftarrow \mathbf{0}$.
3. $r \leftarrow 1$.
4. $r_{\text{old}} \leftarrow 0$.
5. $\omega \leftarrow 1$.
6. $\mathbf{u}_{\text{old}} \leftarrow \mathbf{u}$.
7. If $\|\mathbf{u} - \mathbf{u}_{\text{old}}\| \geq [(1 - \|\mathsf{G}_\omega\|)/\|\mathsf{G}\|]\epsilon$ then:
8. $\quad\quad \mathbf{u} \leftarrow \mathsf{G}_\omega \mathbf{u} + \mathbf{k}$.
9. $\quad\quad r_{\text{old}} \leftarrow r$.
10. $\quad\quad \mathbf{d}_{\text{old}} \leftarrow \mathbf{d}$.
11. $\quad\quad \mathbf{d} \leftarrow \mathbf{u} - \mathbf{u}_{\text{old}}$.

12. $\quad r \leftarrow \mathbf{d}^{\mathsf{T}}\mathbf{d}_{\text{old}}/\|\mathbf{d}_{\text{old}}\|_2^2$.
13. $\quad$ If $|r - r_{\text{old}}| < \tau$ then:
14. $\quad\quad \omega \leftarrow 2/(1+\sqrt{1-r})$.
15. $\quad$ End if.
16. End if.
17. End.

Mathematical Details

The Ostrowski-Reich theorem gives the basic convergence result for SOR. Its proof relies on the following lemma:

LEMMA 4.13. *If* $\mathsf{A} \in \mathbb{R}^{n\times n}$ *is symmetric and positive definite, then its diagonal entries are positive.*

PROOF: This is Problem 7.

THEOREM 4.14 (OSTROWSKI-REICH). *If* $\mathsf{A} \in \mathbb{R}^{n\times n}$ *is symmetric and positive definite, then* $\varrho(\mathsf{G}_\omega) < 1$ *whenever* $0 < \omega < 2$.

PROOF: If $\mathsf{A} = \mathsf{L} + \mathsf{D} + \mathsf{U}$ is the decomposition of A into lower triangular, diagonal, and upper triangular parts, then the fact that A is symmetric and positive definite implies that $\mathsf{U} = \mathsf{L}^*$. By Lemma 4.13, D has diagonal entries $d_{i,i} > 0$. Therefore, in the matrix splitting $\mathsf{A} = \mathsf{B} + (\mathsf{A} - \mathsf{B})$ that gives rise to SOR, the matrix $\mathsf{B} = \omega^{-1}\mathsf{D} + \mathsf{L}$ has eigenvalues $d_{i,i}/\omega > 0$. In particular, B is nonsingular.

Three observations are relevant. First, let $\mathsf{Q} := \mathsf{A}^{-1}(2\mathsf{B} - \mathsf{A})$. Then the matrix $\mathsf{Q} + \mathsf{I} = 2\mathsf{A}^{-1}\mathsf{B}$ is nonsingular, and it is easy to check that

$$(\mathsf{Q} - \mathsf{I})(\mathsf{Q} + \mathsf{I})^{-1} = \mathsf{I} - \mathsf{B}^{-1}\mathsf{A} = \mathsf{G}_\omega.$$

Second,

$$\mathsf{B} + \mathsf{B}^* - \mathsf{A} = \omega^{-1}\mathsf{D} + \mathsf{L} + \omega^{-1}\mathsf{D} + \mathsf{L}^* - \mathsf{A} = (2\omega^{-1} - 1)\mathsf{D},$$

which is clearly symmetric and positive definite since $(2\omega^{-1} - 1)d_{i,i} > 0$. Third, all eigenvalues of Q have positive real parts. To see this, suppose that μ is an eigenvalue of Q, with $\mathsf{Q}\mathbf{y} = \mathsf{A}^{-1}(2\mathsf{B} - \mathsf{A})\mathbf{y} = \mu\mathbf{y}$. Then $\mathbf{y}^*\mathsf{Q}^* = \overline{\mu}\mathbf{y}^*$, and since $\mathsf{A}^* = \mathsf{A}$ we have

$$\begin{aligned} \mathbf{y}^*(2\mathsf{B} - \mathsf{A})\mathbf{y} &= \mu\mathbf{y}^*\mathsf{A}\mathbf{y}, \\ \mathbf{y}^*(2\mathsf{B}^* - \mathsf{A})\mathbf{y} &= \overline{\mu}\mathbf{y}^*\mathsf{A}\mathbf{y}. \end{aligned}$$

Adding these two equations gives

$$2\mathbf{y}^*(\mathsf{B}+\mathsf{B}^*-\mathsf{A})\mathbf{y} = 2\Re(\mu)\mathbf{y}^*\mathsf{A}\mathbf{y},$$

where $\Re(\mu)$ denotes the real part of μ. Therefore, following the second observation, we have

$$\Re(\mu) = \frac{\mathbf{y}^*(\mathsf{B}+\mathsf{B}^*-\mathsf{A})\mathbf{y}}{\mathbf{y}^*\mathsf{A}\mathbf{y}} > 0.$$

Now let λ be an eigenvalue of G_ω with associated eigenvector $\boldsymbol{v} \neq \mathbf{0}$. We wish to show that $|\lambda| < 1$. We have

$$\mathsf{G}_\omega \boldsymbol{v} = (\mathsf{Q}-\mathsf{I})(\mathsf{Q}+\mathsf{I})^{-1}\boldsymbol{v} = \lambda \boldsymbol{v},$$

which we can rewrite as

$$(\mathsf{Q}-\mathsf{I})\mathbf{v} = \lambda(\mathsf{Q}+\mathsf{I})\mathbf{v}$$

by making the substitution $\mathbf{v} := (\mathsf{Q}+\mathsf{I})^{-1}\boldsymbol{v}$. Thus, $(1-\lambda)\mathsf{Q}\mathbf{v} = (1+\lambda)\mathbf{v}$. We claim that $\lambda \neq 1$. Otherwise, we would have $0 \cdot \mathsf{Q}\mathbf{v} = 2\mathbf{v}$, which would imply that $\mathbf{v} = (\mathsf{Q}+\mathsf{I})^{-1}\boldsymbol{v} = \mathbf{0}$, that is, that $\boldsymbol{v} = \mathbf{0}$. This is impossible, since we chose $\boldsymbol{v} \neq \mathbf{0}$.

The claim established, we observe that

$$\mathsf{Q}\mathbf{v} = \frac{1+\lambda}{1-\lambda}\mathbf{v},$$

which implies that $(1+\lambda)/(1-\lambda)$ is an eigenvalue of Q. From this fact it follows that

$$\Re\left(\frac{1+\lambda}{1-\lambda}\right) = \Re\left(\frac{(1+\lambda)(1-\overline{\lambda})}{|1-\lambda|^2}\right) > 0.$$

As a consequence,

$$\Re\left(1 + 2i\Im(\lambda) - |\lambda|^2\right) = 1 - |\lambda|^2 > 0,$$

where $\Im(\lambda)$ denotes the imaginary part of λ. Therefore, $1-|\lambda|^2 > 0$, so $|\lambda| < 1$. ∎

To gain further insight into how to choose ω, we examine certain spectral implications of property A and consistent ordering. Consider first the spectrum of the Jacobi iteration matrix G_J.

THEOREM 4.15. *Suppose that $\mathsf{A} \in \mathbb{R}^{n\times n}$ has property A and that the diagonal entries of A are all nonzero. Then $-\lambda$ is an eigenvalue of G_J whenever λ is.*

PROOF: It suffices to show that $\det(\lambda\mathsf{I}-\mathsf{G}_J) = 0$ if and only if $\det(-\lambda\mathsf{I}-\mathsf{G}_J) = 0$. Let P be the permutation matrix that effects the similarity transformation

(4.4-2). Since $\mathsf{P}^\top = \mathsf{P}^{-1}$ for any permutation matrix,

$$\begin{aligned}
\det(\lambda \mathsf{I} - \mathsf{G}_J) &= (\det \mathsf{P}) \det(\lambda \mathsf{I} - \mathsf{G}_J)(\det \mathsf{P}^\top) \\
&= \det(\lambda \mathsf{I} - \mathsf{P}\mathsf{G}_J\mathsf{P}^\top) \\
&= \det \begin{pmatrix} \lambda \mathsf{I} & -\mathsf{D}_1^{-1}\mathsf{M}_1 \\ -\mathsf{D}_2^{-1}\mathsf{M}_2 & \lambda \mathsf{I} \end{pmatrix} \\
&= \det(\lambda \mathsf{I} + \mathsf{P}\mathsf{G}_J\mathsf{P}^\top) \\
&= \det(\lambda \mathsf{I} + \mathsf{G}_J) = \pm \det(-\lambda \mathsf{I} - \mathsf{G}_J),
\end{aligned}$$

depending on whether n is even or odd. In either case, $\det(\lambda \mathsf{I} - \mathsf{G}_J) = 0$ if and only if $\det(-\lambda \mathsf{I} - \mathsf{G}_J) = 0$. ∎

Theorem 4.15 furnishes a key ingredient in the proof of the following relationship between the spectra of G_J and G_ω.

THEOREM 4.16. *Let $\mathsf{A} \in \mathbb{R}^{n\times n}$ be consistently ordered and have property A, and let ω be any nonzero real number. Then the following assertions are true:*

(A) *If λ_J is an eigenvalue of G_J and $\lambda_\omega \in \mathbb{C}$ satisfies the equation*

$$(\lambda_\omega + \omega - 1)^2 = \lambda_\omega \omega^2 \lambda_J^2, \tag{4.4-3}$$

then λ_ω is an eigenvalue of G_ω.

(B) *If $\lambda_\omega \neq 0$ is an eigenvalue of G_ω and $\lambda_J \in \mathbb{C}$ satisfies Equation (4.4-3), then λ_J is an eigenvalue of G_J.*

PROOF: Let $\mathsf{A} = \mathsf{L} + \mathsf{D} + \mathsf{U}$ be the standard decomposition into lower triangular, diagonal, and upper triangular parts. We begin the proof of (A) with the observation that $\mathsf{I} - \omega\mathsf{D}^{-1}\mathsf{L}$ is lower triangular with unit diagonal entries, so $\det(\mathsf{I} - \omega\mathsf{D}^{-1}\mathsf{L}) = 1$. Therefore, for any $\omega \in \mathbb{R}$,

$$\begin{aligned}
\det(\lambda \mathsf{I} - \mathsf{G}_\omega) &= \det(\mathsf{I} - \omega\mathsf{D}^{-1}\mathsf{L}) \det(\lambda \mathsf{I} - \mathsf{G}_\omega) \\
&= \det\left((\mathsf{I} - \omega\mathsf{D}^{-1}\mathsf{L}) \left\{\lambda \mathsf{I} - (\mathsf{I} - \omega\mathsf{D}^{-1}\mathsf{L})^{-1} \left[(1-\omega)\mathsf{I} + \omega\mathsf{D}^{-1}\mathsf{U}\right]\right\}\right) \\
&= \det\left[\lambda \mathsf{I} - \lambda\omega\mathsf{D}^{-1}\mathsf{L} - (1-\omega)\mathsf{I} - \omega\mathsf{D}^{-1}\mathsf{U}\right] \\
&= \det\left[(\lambda + \omega - 1)\mathsf{I} - \lambda\omega\mathsf{D}^{-1}\mathsf{L} - \omega\mathsf{D}^{-1}\mathsf{U}\right].
\end{aligned} \tag{4.4-4}$$

Assume that λ_J is an eigenvalue of G_J and that λ_ω is a solution to Equation (4.4-3). We show that λ_ω is an eigenvalue of G_ω by confirming that $\det(\lambda_\omega \mathsf{I} - \mathsf{G}_\omega) = 0$.

There are two cases. First, when $\lambda_\omega = 0$, Equation (4.4-3) reduces to the simpler equation $(\omega - 1)^2 = 0$, which implies that $\omega = 1$. Equation (4.4-4) now yields

$$\begin{aligned} \det(\lambda_\omega \mathsf{I} - \mathsf{G}_\omega) &= \det\left(\pm\lambda_\omega^{1/2}\lambda_J - \lambda_\omega \mathsf{D}^{-1}\mathsf{L} - \mathsf{D}^{-1}\mathsf{U}\right) \\ &= \det(-\mathsf{D}^{-1}\mathsf{U}) = 0, \end{aligned}$$

since the matrix $\mathsf{D}^{-1}\mathsf{U}$ is upper triangular with zero diagonal entries. In the second case, when $\lambda_\omega \neq 0$, Equation (4.4-3) implies that $\lambda_\omega + \omega - 1 = \pm\lambda_\omega^{1/2}\omega\lambda_J$. Since Theorem 4.15 guarantees that $-\lambda_J$ is an eigenvalue of G_J whenever λ_J is, we can assume that $\lambda_\omega + \omega - 1 = \lambda_\omega^{1/2}\omega\lambda_J$ without loss of generality. In this case Equation (4.4-4) gives

$$\begin{aligned} \det(\lambda_\omega \mathsf{I} - \mathsf{G}_\omega) &= \det\left[\lambda_\omega^{1/2}\omega\lambda_J \mathsf{I} - \lambda_\omega^{1/2}\omega\left(\lambda_\omega^{1/2}\mathsf{D}^{-1}\mathsf{L} + \lambda_\omega^{-1/2}\mathsf{D}^{-1}\mathsf{U}\right)\right] \\ &= (\lambda_\omega^{1/2}\omega)^n \det\left[\lambda_J \mathsf{I} - (\lambda_\omega^{1/2}\mathsf{D}^{-1}\mathsf{L} + \lambda_\omega^{-1/2}\mathsf{D}^{-1}\mathsf{U})\right]. \end{aligned} \qquad (4.4\text{-}5)$$

The last determinant vanishes when $\lambda_J \in \sigma(\lambda_\omega^{1/2}\mathsf{D}^{-1}\mathsf{L} + \lambda_\omega^{-1/2}\mathsf{D}^{-1}\mathsf{U})$. But, in the notation adopted for the definition of consistent ordering, this matrix is simply $\mathsf{G}(\lambda_\omega^{1/2})$. Consequently the hypothesis that A is consistently ordered guarantees that the last determinant in Equation (4.4-5) vanishes. This concludes the proof of (A).

The proof of (B) is shorter. Suppose that λ_ω is a nonzero eigenvalue of G_ω. Since $-\lambda_J$ is an eigenvalue of G_J whenever λ_J is, we need only show that λ_J is an eigenvalue of G_J whenever $\lambda_\omega + \omega - 1 = \lambda_\omega^{1/2}\omega\lambda_J$. But Equation (4.4-5) and the fact that A is consistently ordered imply that

$$\begin{aligned} \det(\lambda_J \mathsf{I} - \mathsf{G}_J) &= \det\left[\lambda_J \mathsf{I} - (\lambda_\omega^{1/2}\mathsf{D}^{-1}\mathsf{L} + \lambda_\omega^{-1/2}\mathsf{D}^{-1}\mathsf{U})\right] \\ &= \det(\lambda_\omega \mathsf{I} - \mathsf{G}_\omega) = 0. \end{aligned}$$

Therefore, λ_J is an eigenvalue of G_J. ∎

This theorem has a corollary that relates the convergence rates of the Jacobi and Gauss-Seidel schemes:

COROLLARY 4.17. *If* $\mathsf{A} \in \mathbb{R}^{n\times n}$ *is consistently ordered and has property A, then*

$$\varrho(\mathsf{G}_1) = [\varrho(\mathsf{G}_J)]^2. \qquad (4.4\text{-}6)$$

PROOF: In the case $\omega = 1$, Equation (4.4-3) collapses to $\lambda_\omega^2 = \lambda_\omega\lambda_J^2$. Hence either 0 is the only eigenvalue of both G_J and G_1, in which case the corollary is trivially true, or else G_J has nonzero eigenvalues $\pm\lambda_J$. According to Theorem 4.16, to each of these eigenvalues of G_J there corresponds an eigenvalue of G_1 having the form λ^2. ∎

In terms of the convergence rate defined in Section 4.3, this corollary asserts that $R(G_1) := -\log_{10}\varrho(G_1) = -2\log_{10}\varrho(G_J) = R(G_J)$. In other words, if the original system matrix A is consistently ordered and has property A, then the Gauss-Seidel method converges twice as fast as the Jacobi method whenever the latter converges.

Finally, we analyze the SOR parameter ω. The goal is to establish the value $\omega_{\text{opt}} \in (0,2)$ that minimizes $\varrho(G_\omega)$ and hence gives the fastest convergence of the iterative scheme. The analysis yields a value for ω_{opt} in terms of the spectral radius of the Jacobi iteration matrix G_J, which is related to the spectral radius of the Gauss-Seidel matrix G_1 by Corollary 4.17. The analysis also produces a value for $\varrho(G_{\omega_{\text{opt}}})$ and promotes graphic insight into how the convergence rate of SOR varies with the choice of ω.

THEOREM 4.18. *Let $A \in \mathbb{R}^{n\times n}$ be consistently ordered and have property A. If the Jacobi iteration matrix G_J has real eigenvalues and spectral radius $\varrho(G_J) < 1$, then the spectral radius $\varrho(G_\omega)$ of* SOR *assumes its minimum value when*

$$\omega = \omega_{\text{opt}} := \frac{2}{1+\sqrt{1-[\varrho(G_J)]^2}}. \tag{4.4-7}$$

In this case, $\varrho(G_{\omega_{\text{opt}}}) = \omega_{\text{opt}} - 1$.

Before proving the theorem, we remark that the eigenvalues of G_J are all real whenever A is symmetric and positive definite. Also, Section 4.3 establishes widely applicable conditions under which $\varrho(G_J) < 1$. We conclude that Theorem 4.18 applies to finite-difference approximations of the form (4.4-3) to the Laplace operator.

PROOF: Denote the eigenvalues of G_J by $\pm\mu_1, \pm\mu_2, \ldots, \pm\mu_M$, with the indexing chosen so that $0 \le \mu_1 < \mu_2 < \cdots < \mu_M = \varrho(G_J)$. Corresponding to each nonzero eigenvalue μ_j and each choice of ω we have a pair $\lambda_j^\pm$ of eigenvalues of G_ω. For the moment, choose a particular eigenvalue μ_j of G_J and a fixed value for ω and consider the curves $\ell(\lambda)$ and $q_j(\lambda)$ defined by the relations

$$\ell(\lambda) := \frac{1}{\omega}(\lambda+\omega-1), \qquad q_j(\lambda) := \pm\sqrt{\lambda}\mu_j. \tag{4.4-8}$$

According to Theorem 4.16, the eigenvalues $\lambda_j^\pm$ of G_ω that correspond to $\pm\mu_j$ are the ordinates of the two points where the graphs of $\ell(\lambda)$ and $q_j(\lambda)$ intersect, as drawn in Figure 3. Now let ω increase between the values 0 and 1. In the case $\mu_j = 0$, Equation (4.4-3) implies that $|\lambda_j^\pm| = |\omega - 1|$, which decreases monotonically. Otherwise, if $\mu_j \neq 0$, then the graph of the line ℓ rotates clockwise about the point $(1,1)$, starting as a vertical line when

$\omega = 0$ and ending with slope 1 when $\omega = 1$. During this rotation, λ_j^+ and λ_j^- both decrease monotonically. Since this reasoning holds for any choice of the eigenvalues μ_j, we conclude that $\varrho(\mathrm{G}_\omega)$ decreases monotonically as ω slides from 0 to 1. Therefore, of all values $\omega \in (0, 1]$, the choice $\omega = 1$ yields the fastest convergence available via SOR.

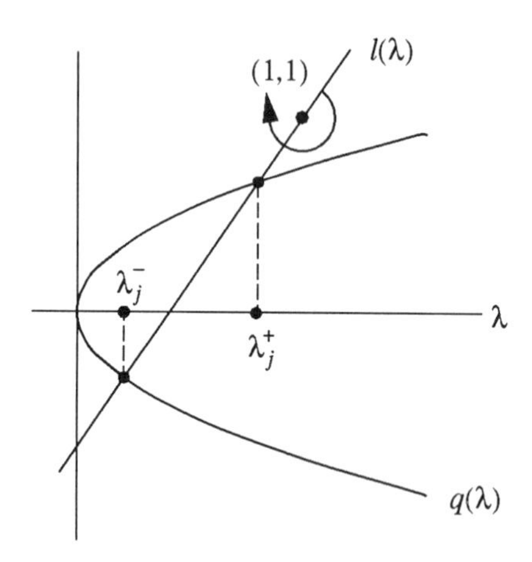

FIGURE 3. *Graphs of the relations $\ell(\lambda)$ and $q_j(\lambda)$ defined in Equation (4.4-8), shown for the case $0 < \omega \leq 1$. The arrow indicates the direction of rotation of the line $\ell(\lambda)$ that results when the overrelaxation parameter ω increases, in the case when $\mu_j \neq 0$.*

Now let ω increase between 1 and 2. If $\mu_j = 0$, then $|\lambda_j^\pm| = \omega - 1$. If $\mu_j \neq 0$, then λ_j^- increases while λ_j^+ decreases, as illustrated in Figure 4, until ω reaches a value where $\lambda_j^+ = \lambda_j^-$. At this value of ω the graph of $\ell(\lambda)$ is tangent to that of $q(\lambda)$, that is, $\ell(\lambda) = q(\lambda)$ and $\ell'(\lambda) = q'(\lambda)$. The first of these conditions implies that $\lambda + \omega - 1 = \lambda^{1/2}\mu_j\omega$. The second implies that $\lambda^{1/2} = \frac{1}{2}\mu_j\omega$. Together, these two conditions imply that $\frac{1}{4}\mu_j^2\omega^2 - \omega + 1 = 0$. By solving this quadratic equation for ω and imposing the constraint $\omega < 2$, we see that tangency occurs when

$$\omega = \omega_j := \frac{2}{1 + \sqrt{1 - \mu_j^2}}.$$

As ω increases beyond ω_j, the equation $\ell(\lambda) = q(\lambda)$ has no real roots. In this case, λ_i^- and λ_j^+ are complex conjugates, and one easily checks that $|\lambda_j^\pm| = \omega - 1$.

Let us summarize. When $\omega = \omega_j$, both of the eigenvalues $\lambda_j^\pm$ have magnitude $\omega_j - 1$, so $\varrho(\mathrm{G}_{\omega_j}) \geq \omega_j - 1$. When $\omega < \omega_j$, the geometry of Figure 4 indicates that $|\lambda_j^-| < \omega_j - 1 < |\lambda_j^+|$, so in these cases $\varrho(\mathrm{G}_\omega) > \omega_j - 1$. For

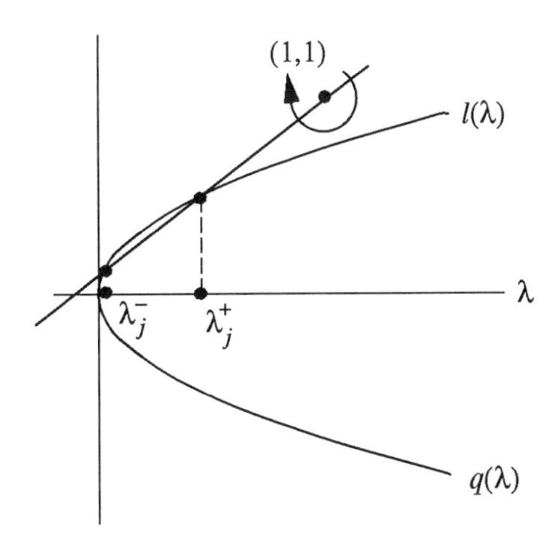

FIGURE 4. *Graphs of the relations $\ell(\lambda)$ and $q(\lambda)$, shown for the case $\mu_j \neq 0$ and $1 \leq \omega < 2$. The arrow shows the direction of rotation of the line $\ell(\lambda)$ when ω increases.*

$\omega > \omega_j$, the eigenvalues $\lambda_j^{\pm}$ of G_ω have magnitude $\omega - 1 > \omega_j - 1$, so in this case also $\varrho(G_\omega) > \omega_j - 1$.

Now consider what happens as μ_j ranges over the eigenvalues of G_J. It is enough to consider the nonnegative values of μ_j. If $\mu_j = 0$, then $|\lambda_j^{\pm}| = |\omega - 1|$. Suppose that the nonzero eigenvalues of G_J are $\pm\mu_m, \pm\mu_{m+1}, \ldots, \pm\mu_M$. As ω increases from 1 to 2, the eigenvalues $\lambda_m^{\pm}$ corresponding to μ_m become complex first, with magnitude $\omega_m - 1$. Then the eigenvalues $\lambda_{m+1}^{\pm}$ become complex with magnitude $\omega_{m+1} - 1$, and at this point $|\lambda_m^{\pm}| = \omega_m - 1$, too. This process continues until the eigenvalues $\lambda_M^{\pm}$ become complex with magnitude $\omega_M - 1$, and for $\omega_M \leq \omega < 2$ all eigenvalues of G_ω have magnitude $\omega - 1$. We conclude that

$$\omega_{\text{opt}} = \omega_M = \frac{2}{1 + \sqrt{1 - [\varrho(G_J)]^2}},$$

and that $\varrho(G_{\omega_{\text{opt}}}) = \omega_{\text{opt}} - 1$. ∎

COROLLARY 4.19. *Under the hypotheses of Theorem 4.18,*

$$\omega_{\text{opt}} = \frac{2}{1 + \sqrt{1 - \varrho(G_1)}}.$$

The graphic reasoning in the proof of Theorem 4.18 furnishes a qualitative picture of how $\varrho(G_\omega)$ varies with ω. As shown in Figure 1, $\varrho(G_\omega)$ decreases monotonically as ω ranges from 0 to ω_{opt}. At ω_{opt}, the graph reaches a cusp. For $\omega > \omega_{\text{opt}}$, all eigenvalues of G_ω have magnitude $\omega - 1$, so the graph of $\varrho(G_\omega)$ increases with unit slope in this region. Since the graph is steepest above the region of the ω-axis immediately to the left of ω_{opt}, SOR typically converges more rapidly for choices of ω slightly larger that ω_{opt} than for values slightly smaller than ω_{opt}.

Further Remarks

The strategy for estimating ω_{opt} in Algorithm 4.2 is a crude one. It relies on a technique called the **power method** for estimating the spectral radius of a matrix. As we discuss in Chapter 5, the power method can converge quite slowly, and in these cases Algorithm 4.1 spends many iterations using the Gauss-Seidel method before switching to the much faster SOR method. A variety of more sophisticated techniques exist for estimating ω_{opt}, some of them based on the use of Chebyshev polynomials, an application of which we discuss in the next section. We refer interested readers to Chapter 9 of Hageman and Young [4].

4.5 The Conjugate-Gradient Method

Motivation and Construction

We turn now to an iterative method based on *searching*. The idea behind the conjugate-gradient method is to find the solution of a linear system $\mathsf{A}\mathbf{u} = \mathbf{b}$ by searching for the solution of an equivalent minimization problem. The following proposition serves as the conceptual basis for the method:

PROPOSITION 4.20. *Let* $\mathsf{A} \in \mathbb{R}^{n\times n}$ *be symmetric and positive definite. Then* $\mathsf{A}\mathbf{u} = \mathbf{b}$ *if and only if the vector* $\mathbf{u}$ *minimizes the real-valued function* $F\colon \mathbb{R}^n \to \mathbb{R}$ *defined by* $F(\mathbf{v}) := \frac{1}{2}\mathbf{v}^\top\mathsf{A}\mathbf{v} - \mathbf{b}^\top\mathbf{v}$. *The minimum value of* F *is* $F(\mathsf{A}^{-1}\mathbf{b}) = -\frac{1}{2}\mathbf{b}^\top\mathsf{A}^{-1}\mathbf{b}$.

PROOF: This is Problem 11(b). ∎

As mentioned briefly in Section 4.3, in some applications the function F represents the energy of a mechanical system, such as an elastic membrane, and thus the solution of the matrix equation $\mathsf{A}\mathbf{u} = \mathbf{b}$ corresponds to the lowest energy state, or equilibrium, of the system.

The relationship between the linear system $\mathsf{A}\mathbf{u} = \mathbf{b}$ and the minimization of F furnishes a geometric interpretation as well as a physical one. Given any vector $\mathbf{v} \in \mathbb{R}^n$, the corresponding residual $\mathbf{r} := \mathbf{b} - \mathbf{A}\mathbf{v}$ is the negative of the gradient of F at $\mathbf{v}$; in symbols,

$$\mathbf{r} = \mathbf{b} - \mathbf{A}\mathbf{v} = -\nabla F(\mathbf{v}) := -\left(\frac{\partial F}{\partial x_1}(\mathbf{v}), \frac{\partial F}{\partial x_2}(\mathbf{v}), \ldots, \frac{\partial F}{\partial x_n}(\mathbf{v})\right).$$

Thus the residual vector $\mathbf{r}$ points in the direction of the steepest descent of F, as illustrated for the case $n = 2$ in Figure 1.

To find the minimum of F, imagine traveling downhill on the hypersurface formed by the graph of F, proceeding in steps. In the first step, we start at an initial guess $\mathbf{u}^{(0)}$ (that is, at the point $(\mathbf{u}_0, F(\mathbf{u}_0))$ on the hypersurface) and travel in the direction of $\mathbf{r}_0 := \mathbf{b} - \mathsf{A}\mathbf{u}^{(0)}$. We stop at the lowest point along the line defined by $\mathbf{u}^{(0)} + \alpha\mathbf{r}_0$, with α variable. Call this new position $\mathbf{u}^{(1)}$, and find a new search direction. By continuing to travel "downhill" along a sequence of search directions, we expect to arrive at successively better approximations $\mathbf{u}^{(1)}, \mathbf{u}^{(2)}, \ldots$ to the minimum $\mathbf{u}$.

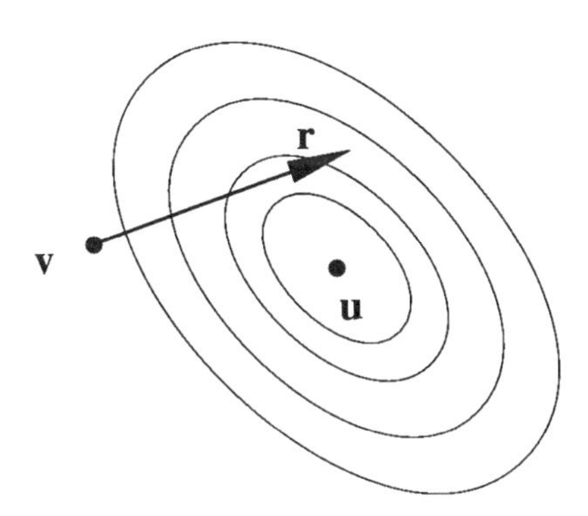

FIGURE 1. *Level sets of a function $F: \mathbb{R}^n \to \mathbb{R}$ whose minimization corresponds to the solution of a linear system. The point* $\mathbf{u}$ *represents the minimum, while the residual vector* $\mathbf{r}$ *associated with any other vector* $\mathbf{v}$ *points in the direction of steepest descent of F.*

This picture in mind, we envision an iterative scheme of the following structure for finding the minimum of F: Given an initial guess $\mathbf{u}^{(0)} \in \mathbb{R}^n$,

(A) Let $\mathbf{r}_0 = \mathbf{b} - \mathsf{A}\mathbf{u}^{(0)}$ (the direction of steepest descent).

(B) Let $\mathbf{p}_0 = \mathbf{r}_0$ (the initial search direction).

Then for $m = 1, 2, 3, \ldots$, perform the following steps:

(C) Find $\alpha_{m-1} \in \mathbb{R}$ such that $\mathbf{u}^{(m-1)} + \alpha_{m-1}\mathbf{p}_{m-1}$ minimizes F over the line $\mathbf{u}^{(m-1)} + \alpha\mathbf{p}_{m-1}$.

(D) Let $\mathbf{u}^{(m)} = \mathbf{u}^{(m-1)} + \alpha_{m-1}\mathbf{p}_{m-1}$ (the new iterate).

(E) Let $\mathbf{r}_m = \mathbf{b} - \mathsf{A}\mathbf{u}^{(m)}$ $(= -\nabla F(\mathbf{u}^{(m)})$, the new residual).

(F) Find a new search direction $\mathbf{p}_m$.

(G) $m \leftarrow m + 1$; go to (C).

To specify the scheme uniquely, we must define steps (C) and (F). In doing so, we use some new terminology:

DEFINITION. *Given a symmetric, positive definite matrix* $\mathsf{A} \in \mathbb{R}^{n\times n}$ *and two vectors* $\mathbf{v}, \mathbf{w} \in \mathbb{R}^n$, *the* **energy inner product** *associated with* A *is*

$$\langle \mathbf{v}, \mathbf{w} \rangle_A := \mathbf{v}^\top \mathsf{A}\mathbf{w}.$$

The function $\langle \cdot, \cdot \rangle_A$ satisfies the axioms for real inner products: For any $\mathbf{v}, \mathbf{w}, \mathbf{z} \in \mathbb{R}^n$ and any $c_1, c_2 \in \mathbb{R}$,

(i) $\langle \mathbf{v}, \mathbf{w} \rangle_A = \langle \mathbf{w}, \mathbf{v} \rangle_A$ (symmetry);

(ii) $\langle c_1\mathbf{v} + c_2\mathbf{w}, \mathbf{z} \rangle_A = c_1\langle \mathbf{v}, \mathbf{z} \rangle_A + c_2\langle \mathbf{w}, \mathbf{z} \rangle_A$ (linearity);

(iii) $\langle \mathbf{v}, \mathbf{v} \rangle_A \geq 0$, and $\langle \mathbf{v}, \mathbf{v} \rangle_A = 0$ only if $\mathbf{v} = \mathbf{0}$ (positive definiteness).

We leave verification for Problem 10. An important consequence of these axioms is that the quantity

$$||\mathbf{v}||_A := \sqrt{\langle \mathbf{v}, \mathbf{v} \rangle_A}$$

defines a norm on $\mathbb{R}^n$, called the **energy norm**.

Defining step (C) in the skeletal algorithm above is straightforward, if we know how to compute the **search directions** $\mathbf{p}_m$. Given vectors $\mathbf{u}^{(m)}$ and $\mathbf{p}_m$, we determine α_m by minimizing F over all vectors of the form $\mathbf{u}^{(m)}+\alpha\mathbf{p}_m$, where α ranges over $\mathbb{R}$. This set of vectors forms a line in $\mathbb{R}^n$, as Figure 2 suggests. To minimize F along this line, we regard F as a function of α and find where $dF/d\alpha = 0$:

$$\begin{aligned}\frac{d}{d\alpha}F(\mathbf{u}^{(m)} + \alpha\mathbf{p}_m) &= \nabla F(\mathbf{u}^{(m)} + \alpha\mathbf{p}_m)\frac{d}{d\alpha}(\mathbf{u}^{(m)} + \alpha\mathbf{p}_m) \\ &= [\mathsf{A}(\mathbf{u}^{(m)} + \alpha\mathbf{p}_m) - \mathbf{b}]^\top \mathbf{p}_m \\ &= [(\mathsf{A}\mathbf{u}^{(m)} - \mathbf{b})^\top + \alpha\mathbf{p}_m^\top \mathsf{A}^\top]\mathbf{p}_m.\end{aligned}$$

Since $\mathsf{A}\mathbf{u}^{(m)} - \mathbf{b} = -\mathbf{r}_m$ and $\mathsf{A}^\top = \mathsf{A}$,

$$\frac{d}{d\alpha}F(\mathbf{u}^{(m)} + \alpha\mathbf{p}_m) = -\mathbf{r}_m^\top\mathbf{p}_m + \alpha\langle \mathbf{p}_m, \mathbf{p}_m \rangle_A.$$

This quantity vanishes when

$$\alpha = \alpha_m := \frac{\mathbf{r}_m^\top \mathbf{p}_m}{||\mathbf{p}_m||_A^{\,2}}.$$

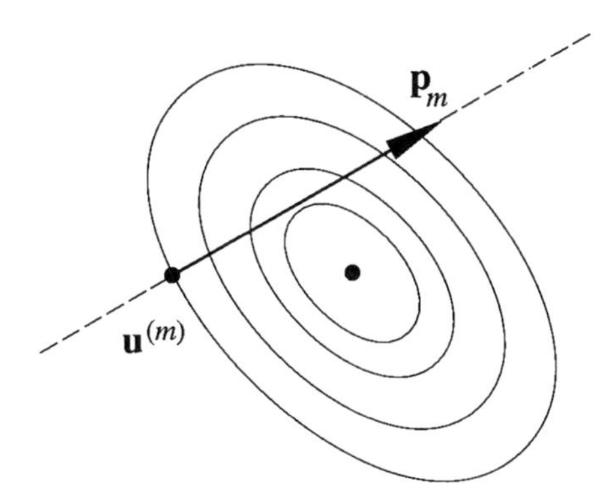

FIGURE 2. *The one-dimensional ray consisting of all vectors in $\mathbb{R}^n$ having the form $\mathbf{u}^{(m)} + \alpha\mathbf{p}_m$, where $\mathbf{u}^{(m)}$ and $\mathbf{p}_m$ are fixed and α ranges over $\mathbb{R}$.*

Defining step (F) requires some judgment. Naively, one might be tempted to choose the residuals $\mathbf{r}_m = \mathbf{b} - \mathbf{A}\mathbf{u}^{(m)}$ as the search directions, since they point in the directions of steepest descent of the function F. The resulting iterative scheme, called the **method of steepest descent**, turns out to be a poor one for many functions F. Figure 3 illustrates the difficulty for a function $F: \mathbb{R}^2 \to \mathbb{R}$ whose graph is a long, narrow valley. In this case, the direction $\mathbf{r}_m$ of steepest descent from a given point $\mathbf{u}^{(m)}$ may differ markedly from the direction pointing toward the bottom of the valley. Moreover, by traveling along the direction of steepest descent until one reaches the minimum of F along the line $\mathbf{u}^{(m)} + \alpha\mathbf{r}_m$, the method of steepest descent locates the next iterate $\mathbf{u}^{(m+1)}$ at a point where that line is tangent to the level sets of F. Consequently, the next search direction $\mathbf{r}^{(m+1)}$, being orthogonal to the level sets at $\mathbf{u}^{(m+1)}$, must be orthogonal to the old search direction $\mathbf{r}_m$. This geometry often causes the iterative scheme to take many short, inefficient switchbacks down to the valley floor, when a better choice of search directions could lead to a more direct descent.

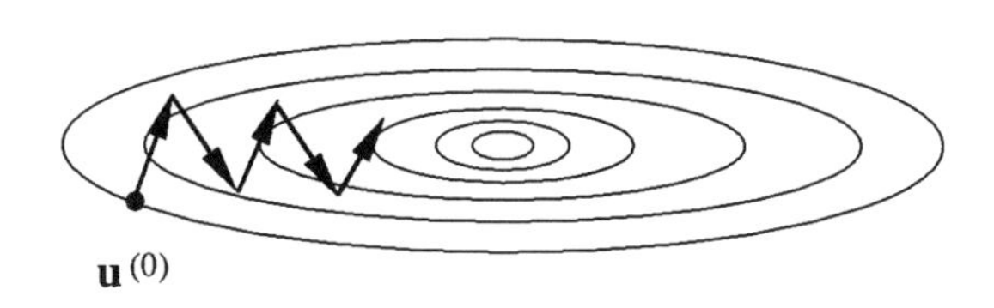

FIGURE 3. *Iterative behavior of the method of steepest descent when the graph of $F: \mathbb{R}^n \to \mathbb{R}$ is a long, narrow valley.*

It is possible to choose better search directions by stipulating that new search directions respect the progress made by previous ones. As explained above, choosing the initial search direction $\mathbf{p}_0 := \mathbf{r}_0$ guarantees descent along the direction opposite $\nabla F(\mathbf{u}^{(0)})$. The next iterate $\mathbf{u}^{(1)}$ lies at a point where the old search direction $\mathbf{p}_0$ is orthogonal to $\nabla F(\mathbf{u}^{(1)}) = \mathsf{A}\mathbf{u}^{(1)} - \mathbf{b}$. We write this latter condition as $\mathbf{p}_0^\top(\mathsf{A}\mathbf{u}^{(1)} - \mathbf{b}) = 0$. The next search direction $\mathbf{p}_1$ should be one along which ∇F *remains* orthogonal to $\mathbf{p}_0$, so that minimization along the line $\mathbf{u}^{(1)} + \alpha\mathbf{p}^{(1)}$ does not "undo" the progress made in searching along the direction $\mathbf{p}_0$. Thus we require that

$$\mathbf{p}_0^\top \nabla F(\mathbf{u}^{(1)} + \alpha\mathbf{p}_1) = \mathbf{p}_0^\top[\mathsf{A}(\mathbf{u}^{(1)} + \alpha\mathbf{p}_1) - \mathbf{b}] = \alpha\mathbf{p}_0^\top\mathsf{A}\mathbf{p}_1 = 0.$$

Each subsequent iteration follows the same logic. Instead of choosing the residuals $\mathbf{r}_m$ as search directions $\mathbf{p}_m$, we select $\mathbf{p}_m$ so that

$$\mathbf{p}_m^\top\mathsf{A}\mathbf{p}_{m-1} = \langle\mathbf{p}_m, \mathbf{p}_{m-1}\rangle_A = 0,$$

a condition that we describe by saying that $\mathbf{p}_m$ is A-**conjugate** to $\mathbf{p}_{m-1}$.

To specify the search directions concretely, let us find a vector of the form

$$\mathbf{p}_m = \mathbf{r}_m + \beta_{m-1}\mathbf{p}_{m-1}, \tag{4.5-1a}$$

choosing the parameter β_{m-1} to force $\mathbf{p}_m$ to be A-conjugate to $\mathbf{p}_{m-1}$. The conjugacy condition is $\langle\mathbf{r}_m + \beta_{m-1}\mathbf{p}_{m-1}, \mathbf{p}_{m-1}\rangle_A = 0$, which we can solve for β_{m-1} to get

$$\beta_{m-1} = -\frac{\langle\mathbf{r}_m, \mathbf{p}_{m-1}\rangle_A}{\|\mathbf{p}_{m-1}\|_A^2}. \tag{4.5-1b}$$

As we prove later, this choice of β_{m-1} forces the new search direction $\mathbf{p}_m$ to be A-conjugate, not just to $\mathbf{p}_{m-1}$, but also to *all* previous search directions $\mathbf{p}_0, \mathbf{p}_1, \ldots, \mathbf{p}_{m-1}$.

The choices of α_{m-1} and β_{m-1} specify the conjugate-gradient algorithm completely. Before listing the algorithm explicitly, though, it is useful to make one more observation for efficiency's sake. By left-multiplying the relationship $\mathbf{u}^{(m+1)} = \mathbf{u}^{(m)} + \alpha_m\mathbf{p}_m$ by A and subtracting both sides from $\mathbf{b}$, we arrive at the identity

$$\mathbf{r}_{m+1} = \mathbf{r}_m - \alpha_m\mathsf{A}\mathbf{p}_m. \tag{4.5-2}$$

Thus we can update the residual cheaply at each iteration, since the product $\mathsf{A}\mathbf{p}_m$ is already available from the calculation of α_m. The steps in the algorithm make this observation apparent:

ALGORITHM 4.3 (BASIC CONJUGATE-GRADIENT ALGORITHM). *Let* $\mathsf{A} \in \mathbb{R}^{n\times n}$ *be symmetric and positive definite. The following algorithm solves the equation* $\mathsf{A}\mathbf{u} = \mathbf{b}$ *using the method of conjugate gradients, starting with initial guess* $\mathsf{u}^{(0)}$.

1. $\mathbf{r}^{(0)} \leftarrow \mathbf{b} - \mathsf{A}\mathbf{u}^{(0)}$.
2. $\mathbf{p}^{(0)} \leftarrow \mathbf{r}^{(0)}$.
3. For $m = 1, 2, \ldots, n$:
4. $\quad \alpha_{m-1} \leftarrow \mathbf{r}_{m-1}^{\top}\mathbf{p}_{m-1}/\|\mathbf{p}_{m-1}\|_A^2$.
5. $\quad \mathbf{u}^{(m)} \leftarrow \mathbf{u}^{(m-1)} + \alpha_{m-1}\mathbf{p}_{m-1}$.
6. $\quad \mathbf{r}_m \leftarrow \mathbf{r}_{m-1} - \alpha_{m-1}\mathsf{A}\mathbf{p}_{m-1}$.
7. $\quad \mathbf{p}_m \leftarrow \mathbf{r}_m - \dfrac{\langle \mathbf{r}_m, \mathbf{p}_{m-1}\rangle_A}{\|\mathbf{p}_{m-1}\|_A^2}\mathbf{p}_{m-1}$.
8. Next m.
9. End.

Several features of this algorithm deserve comment. First, it requires relatively few arithmetic operations per iteration. Step 4 requires the calculation of two inner products in $\mathbb{R}^n$ and the matrix-vector product $\mathsf{A}\mathbf{p}_{m-1}$. The latter calculation is efficient if one encodes a matrix multiplication routine that exploits any sparse structure in A. Steps 5 through 7 require only multiplication by scalars and vector additions in $\mathbb{R}^n$, since the products $\mathsf{A}\mathbf{p}_{m-1}$ and $\|\mathbf{p}_{m-1}\|_A^2$ are already known from step 4.

Second, the algorithm as written requires at most n iterations to converge. In theory, the fact that each search direction $\mathbf{p}_m$ is A-conjugate to all previous search directions implies that, after n iterations, the algorithm will have minimized F along all possible directions in $\mathbb{R}^n$. Consequently the iterate $\mathbf{u}^{(n)}$ must be the exact solution. This theoretical observation typically has little practical import, however. One reason is that the accumulation of arithmetic errors destroys the mutual conjugacy of the search directions. A more salient reason is that, when n is large, we would prefer that the algorithm yield accurate iterates $\mathbf{u}^{(m)}$ in substantially *fewer* than n iterations. We now explore this possibility.

Practical Considerations

The utility of the method of conjugate gradients hinges on the speed with which it converges. We show later in this section that the error $\boldsymbol{\varepsilon}_m := \mathbf{u} - \mathbf{u}^{(m)}$ at the mth iteration obeys a bound of the form

$$\|\boldsymbol{\varepsilon}_m\|_A \leq 2\|\boldsymbol{\varepsilon}_0\|_A \left[\frac{\sqrt{\mathrm{cond}_2(\mathsf{A})} - 1}{\sqrt{\mathrm{cond}_2(\mathsf{A})} + 1}\right]^m. \tag{4.5-3}$$

This estimate suggests that the method converges slowly when the matrix A is ill conditioned. In particular, the basic conjugate-gradient method can perform poorly when A arises from approximations to partial differential equations on grids having small mesh size h. Indeed, Equation (4.3-11) indicates that $\mathrm{cond}_2(\mathsf{A}) = \mathcal{O}(h^{-2})$ in a typical application involving the Laplace operator. On the other hand, when $\mathrm{cond}_2(\mathsf{A})$ is only slightly larger than 1, the method converges rapidly.

These observations motivate the use of **preconditioners**.. The idea is to replace the original system $\mathsf{A}\mathbf{u} = \mathbf{b}$ by an equivalent problem $\tilde{\mathsf{A}}\mathbf{y} = \tilde{\mathbf{b}}$ in which the new matrix $\tilde{\mathsf{A}}$ has a smaller condition number than A.

In the original problem, we seek $\mathbf{u} \in \mathbb{R}^n$ such that $\mathsf{A}\mathbf{u} = \mathbf{b}$, which is equivalent to demanding that $\mathbf{u}$ minimize the function $F(\mathbf{v}) = \frac{1}{2}\mathbf{v}^\top\mathsf{A}\mathbf{v} - \mathbf{b}^\top\mathbf{v}$. Suppose that $\mathsf{B} \in \mathbb{R}^{n\times n}$ is nonsingular, and consider the change of variables $\mathbf{z} := \mathsf{B}\mathbf{v}$. We define a new function $G\colon \mathbb{R}^n \to \mathbb{R}$ as follows:

$$\begin{aligned} G(\mathbf{z}) := F(\mathbf{v}) &= \tfrac{1}{2}(\mathsf{B}^{-1}\mathbf{z})^\top\mathsf{A}(\mathsf{B}^{-1}\mathbf{z}) - \mathbf{b}^\top\mathsf{B}^{-1}\mathbf{z} \\ &= \tfrac{1}{2}\mathbf{z}^\top\underbrace{\mathsf{B}^{-\top}\mathsf{A}\mathsf{B}^{-1}}_{\tilde{\mathsf{A}}}\mathbf{z} - (\underbrace{\mathsf{B}^{-\top}\mathbf{b}}_{\tilde{\mathbf{b}}})^\top\mathbf{z}, \end{aligned}$$

where $\mathsf{B}^{-\top} := (\mathsf{B}^{-1})^\top$. The matrix $\tilde{\mathsf{A}}$ defined above is symmetric and positive definite, just as A is. With these definitions of $\tilde{\mathsf{A}}$ and $\tilde{\mathbf{b}}$, the new problem, equivalent to the original one, is to find $\mathbf{y} \in \mathbb{R}^n$ such that $\tilde{\mathsf{A}}\mathbf{y} = \tilde{\mathbf{b}}$. Solving this linear system is equivalent to demanding that $\mathbf{y}$ minimize the function $G(\mathbf{z}) = \frac{1}{2}\mathbf{z}^\top\tilde{\mathsf{A}}\mathbf{z} - \tilde{\mathbf{b}}^\top\mathbf{z}$.

Within this framework, consider the following strategy: Find a matrix B such that $\tilde{\mathsf{A}} := \mathsf{B}^{-\top}\mathsf{A}\mathsf{B}^{-1}$ is better conditioned than A, then apply the method of conjugate gradients to the transformed system $\tilde{\mathsf{A}}\mathbf{y} = \tilde{\mathbf{b}}$, then set $\mathbf{u} = \mathsf{B}^{-1}\mathbf{y}$. We defer for a moment the issue of how to choose B.

A straightforward translation of the skeletal algorithm developed earlier to the problem $\tilde{\mathsf{A}}\mathbf{y} = \tilde{\mathbf{b}}$ yields the following: Given an initial guess $\mathbf{y}^{(0)}$,

(A) Let $\mathbf{s}_0 = \tilde{\mathbf{b}} - \tilde{\mathsf{A}}\mathbf{y}^{(0)}$ (the direction of steepest descent).

(B) Let $\mathbf{q}_0 = \mathbf{s}_0$ (the initial search direction).

Then for $m = 1, 2, 3, \ldots$ perform the following steps:

(C) $\tilde{\alpha}_{m-1} \leftarrow \mathbf{s}_{m-1}^\top\mathbf{q}_{m-1}/\|\mathbf{q}_{m-1}\|_{\tilde{A}}^2$.

(D) Let $\mathbf{y}^{(m)} = \mathbf{y}^{(m-1)} + \tilde{\alpha}_{m-1}\mathbf{q}_{m-1}$.

(E) $\mathbf{s}_m \leftarrow \mathbf{s}_{m-1} - \tilde{\alpha}_{m-1}\tilde{\mathsf{A}}\mathbf{q}_{m-1}$ (the new residual).

(F) $\mathbf{q}_m \leftarrow \mathbf{s}_m - \dfrac{\langle \mathbf{s}_m, \mathbf{q}_{m-1}\rangle_{\tilde{A}}}{\|\mathbf{q}_{m-1}\|_{\tilde{A}}}\mathbf{q}_{m-1}$.

(G) $m \leftarrow m + 1$; go to (C).

It is useful to recast this scheme as a simpler modification of the original scheme. Using the relationships $\mathbf{u}^{(m)} = \mathsf{B}^{-1}\mathbf{y}^{(m)}$, $\mathbf{r}_m = \mathsf{B}^{\top}\mathbf{s}_m$, $\mathbf{p}_m = \mathsf{B}^{-1}\mathbf{q}_m$, and $\langle \mathbf{v}, \mathbf{w} \rangle_{\tilde{A}} = \langle \mathsf{B}^{-1}\mathbf{v}, \mathsf{B}^{-1}\mathbf{w} \rangle_A$, we obtain

ALGORITHM 4.4 (PRECONDITIONED CONJUGATE-GRADIENT ALGORITHM). *Let* $\mathsf{A} \in \mathbb{R}^{n\times n}$ *be symmetric and positive definite. The following algorithm solves the equation* $\mathsf{A}\mathbf{u} = \mathbf{b}$ *using the method of conjugate gradients, preconditioned by the invertible matrix* $\mathsf{B} \in \mathbb{R}^{n\times n}$. *The initial guess is* $\mathsf{u}^{(0)}$, *and* $\tau > 0$ *is a convergence tolerance on the norm of the residual* $\mathbf{r}_m = \mathbf{b} - \mathsf{A}\mathbf{u}^{(m)}$.

1. $m \leftarrow 0$.
2. $\mathbf{r}^{(0)} \leftarrow \mathbf{b} - \mathsf{A}\mathbf{u}^{(0)}$.
3. $\mathbf{p}^{(0)} \leftarrow \mathbf{r}^{(0)}$.
4. $\alpha_{m-1} \leftarrow \mathbf{r}_{m-1}^{\top}\mathbf{p}_{m-1}/\|\mathbf{p}_{m-1}\|_A^2$.
5. $\mathbf{u}^{(m)} \leftarrow \mathbf{u}^{(m-1)} + \alpha_{m-1}\mathbf{p}_{m-1}$.
6. $\mathbf{r}_m \leftarrow \mathbf{r}_{m-1} - \alpha_{m-1}\mathsf{A}\mathbf{p}_{m-1}$.
7. If $\|\mathbf{r}_m\|_2 \geq \tau$ then:
8. Solve $\mathsf{B}^{\top}\mathsf{B}\mathbf{z}_m = \mathbf{r}_m$ for $\mathbf{z}_m$.
9. $\mathbf{p}_m \leftarrow \mathbf{z}_m - \dfrac{\langle \mathbf{z}_m, \mathbf{p}_{m-1} \rangle_A}{\|\mathbf{p}_{m-1}\|_A^2}\mathbf{p}_{m-1}$.
10. $m \leftarrow m + 1$.
11. Go to 4.
12. End if.
13. End.

This preconditioned algorithm is similar to the basic version presented in Algorithm 4.3. The main differences occur in step 8, where one must solve a linear system involving the **preconditioner** $\mathsf{B}^{\top}\mathsf{B}$, and in step 9, which uses the result of step 8 in computing the new search direction.

Crucial to the effectiveness of preconditioned conjugate gradients is the choice of preconditioner $\mathsf{B}^{\top}\mathsf{B}$. This matter involves as much art as science. The desired properties for preconditioners are in a sense mutually conflicting. The requirement that $\text{cond}_2(\mathsf{B}^{-\top}\mathsf{A}\mathsf{B}^{-1})$ be significantly smaller than $\text{cond}_2(\mathsf{A})$ suggests that the choice $\mathsf{B}^{\top}\mathsf{B} = \mathsf{A}$ would be ideal, since in this case $\text{cond}_2(\mathsf{B}^{-\top}\mathsf{A}\mathsf{B}^{-1}) = \text{cond}_2(\mathsf{I}) = 1$. On the other hand, the computational requirements of step 8 make it desirable that the linear system $\mathsf{B}^{\top}\mathsf{B}\mathbf{z}_m = \mathbf{r}_m$

be easy to solve. This observation suggests that the choice $\mathsf{B}^\mathsf{T}\mathsf{B} = \mathsf{I}$ would be ideal.

Reasonable choices for preconditioners involve compromises between these two extremes. The idea is to choose $\mathsf{B}^\mathsf{T}\mathsf{B}$ to be a "simple" matrix that possesses much of the spectral structure of the original matrix A. One simple example of a preconditioner that is actually effective in some cases is $\mathsf{B}^\mathsf{T}\mathsf{B} = \text{diag}(\mathsf{A})$, that is, the diagonal part of A. Heuristically, we expect this choice to be most appropriate when A has large diagonal entries and small off-diagonal entries.

More sophisticated preconditioners often involve "partial" or "incomplete" factorizations of A. One popular example is the incomplete Cholesky decomposition. Consider the block-tridiagonal matrices that arise from difference approximations to analogs of the Laplace operator. Figure 4 shows the zero structure of such a matrix A and the zero structure of its Cholesky triangle C. As discussed in Section 2.4, fill-in typically destroys the sparseness within the nonzero band of A. As a consequence, computing C requires the calculation of nonzero entries in positions where A has zeros, and solving linear systems involving C requires arithmetic operations on these entries. Figure 4 also illustrates the idea behind the incomplete Cholesky decomposition. Here, one computes only the entries of C that correspond to nonzero entries of A. The resulting matrix $\hat{\mathsf{C}}$ obviously cannot be the correct Cholesky triangle for A, but $\hat{\mathsf{C}}^\mathsf{T}\hat{\mathsf{C}}$ preserves some of the essential structure of A. It is also inexpensive to compute, and we can solve systems of the form $\hat{\mathsf{C}}^\mathsf{T}\hat{\mathsf{C}}\mathbf{z}_m = \mathbf{r}_m$ by relatively cheap forward and backward substitutions.

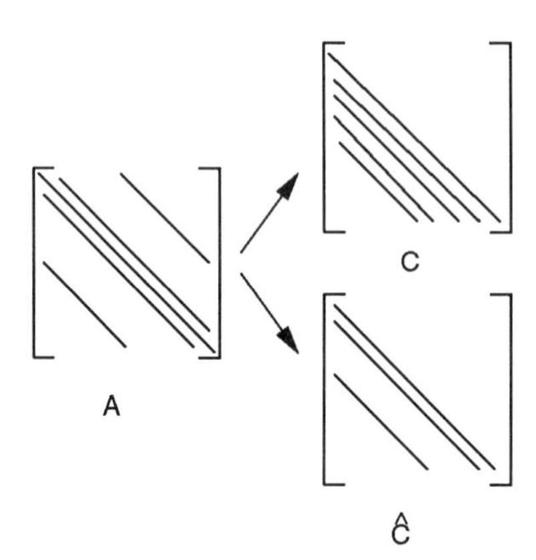

FIGURE 4. *Zero structures of block-tridiagonal matrices arising from a difference approximation to a Laplace-like operator, its Cholesky triangle, and the sparse preconditioner computed using incomplete Cholesky decomposition.*

Mathematical Details

The main theoretical tasks are to examine the geometry of the method in $\mathbb{R}^n$ and to justify the error estimate (4.5-3).

We begin with geometry. Part of the motivation for our choice of search directions $\mathbf{p}_m$ is to speed convergence by steering iterates $\mathbf{u}^{(m)}$ downhill on the graph of F. In this regard, the notion of A-conjugacy is crucial. However, we also want search direction to be independent of all previous search directions, in some sense. We now show a connection between A-conjugacy of the search directions and the orthogonality of the residuals $\mathbf{r}_m = \mathbf{b} - \mathsf{A}\mathbf{u}^{(m)}$.

We need the following lemma:

LEMMA 4.21. *Let $\mathsf{A} \in \mathbb{R}^{n\times n}$ be symmetric and positive definite. Let $\{\mathbf{p}_m\}$ be a sequence of search directions in $\mathbb{R}^n$ given by Equations (4.5-1), let $\mathbf{u}^{(m)}$ be the iterates generated by Algorithm 4.3 with initial guess $\mathbf{u}^{(0)} \in \mathbb{R}^n$, and denote by $\mathbf{r}_m$ the residual $\mathbf{b} - \mathsf{A}\mathbf{u}^{(m)}$. For $m = 0, 1, 2, \ldots$, we have*

$$\operatorname{span}\left\{\mathbf{p}_0, \mathbf{p}_1, \ldots, \mathbf{p}_m\right\} = \operatorname{span}\left\{\mathbf{r}_0, \mathbf{r}_1, \ldots, \mathbf{r}_m\right\} = \operatorname{span}\left\{\mathbf{r}_0, \mathsf{A}\mathbf{r}_0, \ldots, \mathsf{A}^m\mathbf{r}_0\right\}.$$

PROOF: We use induction on m. When $m = 0$, the proposition is trivial, since $\mathbf{p}_0 = \mathbf{r}_0$. Assume that the proposition holds for $m = k$. From Equation (4.5-2) we have

$$\mathbf{r}_{k+1} = \mathbf{r}_k - \alpha_k \mathsf{A}\mathbf{p}_k,$$

and the inductive hypothesis ensures that $\mathbf{p}_k \in \operatorname{span}\{\mathbf{r}_0, \mathsf{A}\mathbf{r}_0, \ldots, \mathsf{A}^k\mathbf{r}_0\}$, so $\mathsf{A}\mathbf{p}_k \in \operatorname{span}\{\mathbf{r}_0, \mathsf{A}\mathbf{r}_0, \ldots, \mathsf{A}^{k+1}\mathbf{r}_0\}$. In particular, the linear combination $\mathbf{r}_{k+1} - \mathbf{r}_k \in \operatorname{span}\{\mathbf{r}_0, \mathsf{A}\mathbf{r}_0, \ldots, \mathsf{A}^{k+1}\mathbf{r}_0\}$. This observation and the inductive hypothesis together imply that

$$\operatorname{span}\left\{\mathbf{r}_0, \mathbf{r}_1, \ldots, \mathbf{r}_{k+1}\right\} \subset \operatorname{span}\left\{\mathbf{r}_0, \mathsf{A}\mathbf{r}_0, \ldots, \mathsf{A}^{k+1}\mathbf{r}_0\right\}. \tag{4.5-4}$$

The inductive hypothesis also implies that $\mathsf{A}^k\mathbf{r}_0 \in \operatorname{span}\{\mathbf{p}_0, \mathbf{p}_1, \ldots, \mathbf{p}_k\}$, so $\mathsf{A}^{k+1}\mathbf{r}_0 \in \operatorname{span}\{\mathsf{A}\mathbf{p}_0, \mathsf{A}\mathbf{p}_1, \ldots, \mathsf{A}\mathbf{p}_k\}$. Therefore, by Equation (4.5-2) and the fact that $\mathbf{p}_0 = \mathbf{r}_0$, we have $\mathsf{A}^{k+1}\mathbf{r}_0 \in \operatorname{span}\{\mathbf{r}_0, \mathbf{r}_1, \ldots, \mathbf{r}_{k+1}\}$. It follows that

$$\operatorname{span}\left\{\mathbf{r}_0, \mathsf{A}\mathbf{r}_0, \ldots, \mathsf{A}^{k+1}\mathbf{r}_0\right\} \subset \operatorname{span}\left\{\mathbf{r}_0, \mathbf{r}_1, \ldots, \mathbf{r}_{k+1}\right\}. \tag{4.5-5}$$

The relationships (4.5-4) and (4.5-5) together imply that

$$\operatorname{span}\left\{\mathbf{r}_0, \mathsf{A}\mathbf{r}_0, \ldots, \mathsf{A}^{k+1}\mathbf{r}_0\right\} = \operatorname{span}\left\{\mathbf{r}_0, \mathbf{r}_1, \ldots, \mathbf{r}_{k+1}\right\}.$$

The fact that $\operatorname{span}\{\mathbf{p}_0, \mathbf{p}_1, \ldots, \mathbf{p}_{k+1}\} = \operatorname{span}\{\mathbf{r}_0, \mathbf{r}_1, \ldots, \mathbf{r}_{k+1}\}$ follows easily from the fact that $\mathbf{p}_0 = \mathbf{r}_0$ and the identity $\mathbf{p}_m = \mathbf{r}_m + \beta_{m-1}\mathbf{p}_{m-1}$. This completes the induction. ∎

The spaces in this lemma have a name:

DEFINITION. *The subspace*

$$\mathcal{K}_m := \operatorname{span}\left\{\mathbf{r}_0, \mathbf{r}_1, \ldots, \mathbf{r}_{m-1}\right\}$$

of $\mathbb{R}^n$ *is the mth* **Krylov subspace** *associated with the matrix* A *and the initial guess* $\mathbf{u}^{(0)}$.

Now we establish the geometric relationships between the search directions.

THEOREM 4.22. *If* $\mathsf{A} \in \mathbb{R}^{n\times n}$ *is symmetric and positive definite, then*

(A) *The search directions defined in Equations (4.5-1) are mutually* A-*conjugate, that is,* $\langle \mathbf{p}_k, \mathbf{p}_m\rangle_A = 0$ *whenever* $l \neq m$.

(B) *The residuals* $\mathbf{r}_m = \mathbf{b} - \mathbf{A}\mathbf{u}^{(m)}$ *in the conjugate gradient method are mutually orthogonal, that is,* $\mathbf{r}_l^\top \mathbf{r}_m = 0$ *whenever* $l \neq m$.

These relationships hold when the arithmetic used is exact. As mentioned earlier, roundoff errors can destroy conjugacy and orthogonality in computational practice.

PROOF: We use induction on the size of the indices l, m. When $l, m \leq 1$, the propositions (A) and (B) are true by construction of the algorithm. Assume that the propositions are true whenever $l, m \leq k$, for some integer $k \geq 1$. We begin the argument by establishing the following claim: For $m = 0, 1, \ldots, k$, $\mathbf{r}_{k+1}^\top \mathbf{p}_m = 0$. For proof, observe that, by Lemma 4.21,

$$\operatorname{span}\left\{\mathbf{p}_0, \mathbf{p}_1, \ldots, \mathbf{p}_m\right\} = \operatorname{span}\left\{\mathbf{r}_0, \mathbf{r}_1, \ldots, \mathbf{r}_m\right\}.$$

It follows from this identity and the inductive hypothesis that $\mathbf{r}_l^\top \mathbf{p}_m = 0$ whenever $m = 0, 1, \ldots, k-1$. Therefore, Equation (4.5-2) yields

$$\mathbf{r}_{k+1}^\top \mathbf{p}_m = (\mathbf{r}_k - \alpha_k \mathsf{A}\mathbf{p}_k)^\top \mathbf{p}_m = \mathbf{r}_k^\top \mathbf{p}_m - \alpha_k \langle \mathbf{p}_k, \mathbf{p}_m\rangle_A.$$

The first term on the right vanishes by the observation just made, and the second vanishes by the inductive hypothesis, so $\mathbf{r}_{k+1}^\top \mathbf{p}_m = 0$ for $m = 0, 1, \ldots, k-1$. It remains to examine the case $m = k$. From Equation (4.5-2),

$$\mathbf{r}_{k+1}^\top \mathbf{p}_k = \mathbf{r}_k^\top \mathbf{p}_k - \alpha_k \langle \mathbf{p}_k, \mathbf{p}_k\rangle_A = -\frac{d}{d\alpha} F(\mathbf{u}^{(k)} + \alpha_k \mathbf{p}_k) = 0,$$

by choice of α_k. This establishes the claim.

Since Lemma 4.21 guarantees that $\mathbf{r}_m \in \text{span}\,\{\mathbf{p}_0, \mathbf{p}_1, \ldots, \mathbf{p}_k\}$ for $m = 0, 1, \ldots, k$, we have $\mathbf{r}_{k+1}^{\mathsf{T}}\mathbf{r}_m = 0$ for $m = 0, 1, \ldots k$. Thus proposition (B) holds for $l, m \leq k+1$. The argument for proposition (A) rests on the identity $\mathbf{p}_{k+1} = \mathbf{r}_{k+1} + \beta_k \mathbf{p}_k$. By Equation (4.5-2), we have $\mathsf{A}\mathbf{p}_m \in \text{span}\,\{\mathbf{r}_0, \mathbf{r}_1, \ldots, \mathbf{r}_{m+1}\}$, and this fact together with proposition (B) imply that $\langle \mathbf{r}_{k+1}, \mathbf{p}_m \rangle_A = 0$ for $m = 0, 1, \ldots, k-1$. Therefore,

$$\langle \beta_k \mathbf{p}_k - \mathbf{p}_{k+1}, \mathbf{p}_m \rangle_A = 0, \qquad m = 0, 1, \ldots, k-1.$$

The first term $\langle \beta_k \mathbf{p}_k, \mathbf{p}_m \rangle_A$ in this energy inner product vanishes by the inductive hypothesis, so $\langle \mathbf{p}_{k+1}, \mathbf{p}_m \rangle_A = 0$ for $m = 0, 1, \ldots, k-1$. Since the method of constructing new search directions ensures that $\langle \mathbf{p}_{k+1}, \mathbf{p}_k \rangle_A = 0$, proposition (A) holds for $l, m \leq k+1$, and the induction is complete. ∎

As a consequence of this theorem, the Krylov subspace

$$\mathcal{K}_n = \text{span}\left\{\mathbf{r}_0, \mathbf{r}_1, \ldots, \mathbf{r}_{n-1}\right\}$$

has dimension n, and hence $\mathcal{K}_n = \mathbb{R}^n$. This observation has the following consequence:

COROLLARY 4.23. *If* $\mathsf{A} \in \mathbb{R}^{n \times n}$ *is symmetric and positive definite, then for some index* $m \leq n$ *the residual* $\mathbf{r}_m$ *generated by the conjugate-gradient method vanishes.*

In other words, for some $m \leq n$, $\mathbf{u}^{(m)}$ is the exact solution to $\mathsf{A}\mathbf{u} = \mathbf{b}$. Theoretically, then, the conjugate-gradient algorithm terminates, yielding the exact solution, after at most n iterations.

The fact that $\mathbf{r}_m = \mathbf{0}$ for some $m \leq n$ is of little practical interest, for two reasons. First, as mentioned, errors associated with machine arithmetic can destroy the conjugacy and orthogonality relationships established in Theorem 4.22. Hence $\mathbf{r}_m$ may *never* vanish in actual calculations. Second, in many settings n is a large number — 10^4 or even much larger — and for efficiency's sake we would like to stop iterating after a much smaller number of iterations — say, 10 or fewer. This practical desire motivates the derivation of error estimates for the conjugate-gradient method.

Before embarking on this project, note that one can reformulate the conjugate-gradient method so that the initial guess $\mathbf{u}^{(0)} = \mathbf{0}$. Given a linear system $\mathsf{A}\mathbf{v} = \mathbf{c}$ and an arbitrary initial guess $\mathbf{v}^{(0)}$, we simply apply the method to the system $\mathsf{A}\mathbf{u} = \mathbf{b}$, where $\mathbf{u} := \mathbf{v} - \mathbf{v}^{(0)}$, $\mathbf{b} := \mathbf{c} - \mathsf{A}\mathbf{v}^{(0)}$, and $\mathbf{u}^{(0)} := \mathbf{0}$. The iteration then generates iterates $\mathbf{u}^{(1)}, \mathbf{u}^{(2)}, \ldots$, which correspond to the iterates $\mathbf{v}^{(m)} = \mathbf{v}^{(0)} + \mathbf{u}^{(m)}$ of the original system. In the transformed system, the initial residual is $\mathbf{r}_0 = \mathbf{b}$, and the mth Krylov subspace is

$$\mathcal{K}_m = \text{span}\left\{\mathbf{r}_0, \mathsf{A}\mathbf{r}_0, \ldots, \mathsf{A}^{m-1}\mathbf{r}_0\right\} = \text{span}\left\{\mathbf{b}, \mathsf{A}\mathbf{b}, \ldots, \mathsf{A}^{m-1}\mathbf{b}\right\}.$$

The error estimate of interest arises from a crucial variational principle:

THEOREM 4.24. *If* $\mathsf{A} \in \mathbb{R}^{n\times n}$ *is symmetric and positive definite, then the mth step of the conjugate-gradient method, using* $\mathbf{u}^{(0)} = \mathbf{0}$ *as initial guess, minimizes the error over* $\mathcal{K}_m$ *in the following sense:*

$$\|\varepsilon_m\|_A = \|\mathbf{u} - \mathbf{u}^{(m)}\|_A = \min_{\mathbf{v}\in\mathcal{K}_m} \|\mathbf{u} - \mathbf{v}\|_A .$$

PROOF: This is Problem 12.

This theorem leads to a more concrete error estimate in terms of the energy norm $\|\cdot\|_A$. We denote by Π_m the collection of all polynomials having degree at most m.

COROLLARY 4.25. *Let* $\mathsf{A} \in \mathbb{R}^{n\times n}$ *be symmetric and positive definite, with eigenvalues* $\lambda_1, \lambda_2, \ldots, \lambda_n$. *Then the errors* $\varepsilon_m = \mathbf{u} - \mathbf{u}^{(m)}$ *in the conjugate-gradient method for* $\mathsf{A}\mathbf{u} = \mathbf{b}$, *with initial guess* $\mathbf{u}^{(0)} = \mathbf{0}$, *obey the bound*

$$\|\varepsilon_m\|_A \leq \|\varepsilon_0\|_A \max_j |p(\lambda_j)|,$$

for any polynomial $p \in \Pi_m$ *such that* $p(0) = 1$.

PROOF: We have seen that $\mathcal{K}_m = \operatorname{span}\{\mathbf{b}, \mathsf{A}\mathbf{b}, \ldots, \mathsf{A}^{m-1}\mathbf{b}\}$. This space is just the set of all vectors of the form $p(\mathsf{A})\mathbf{b}$, where $p \in \Pi_{m-1}$. From Theorem 4.24 and the fact that $\varepsilon_0 = \mathbf{u}$, it follows that

$$\begin{aligned}
\|\varepsilon_m\|_A &= \min\Big\{\|\mathbf{u} - p(\mathsf{A})\mathbf{b}\|_A : p \in \Pi_{m-1}\Big\} \\
&= \min\Big\{\|\varepsilon_0 - p(\mathsf{A})\mathsf{A}\varepsilon_0\|_A : p \in \Pi_{m-1}\Big\} \\
&\leq \min\Big\{\|\varepsilon_0\|_A \|\mathsf{I} - p(\mathsf{A})\mathsf{A}\|_A : p \in \Pi_{m-1}\Big\} \\
&= \min\Big\{\|\varepsilon_0\|_A \|p(\mathsf{A})\|_A : p \in \Pi_m \quad \text{and} \quad p(0) = 1\Big\}.
\end{aligned}$$

Thus $\|\varepsilon_m\|_A \leq \|p(\mathsf{A})\|_A \|\varepsilon_0\|_A$ for every polynomial $p \in \Pi_m$ such that $p(0) = 1$. Problem 13 asks for verification that $\|p(\mathsf{A})\|_A = \max_j |p(\lambda_j)|$, where λ_j ranges over the eigenvalues of A. This observation completes the proof. ∎

Denote the smallest and largest eigenvalues of A as $\lambda_{\min}$ and $\lambda_{\max}$, respectively. (These numbers are real and positive.) Corollary 4.25 implies that

$$\|\varepsilon_m\|_A \leq \|\varepsilon_0\|_A \sup\Big\{|p(z)| : \lambda_{\min} \leq z \leq \lambda_{\max}\Big\}, \qquad (4.5\text{-}6)$$

where p can be any polynomial in Π_m for which $p(0) = 1$. Think of the inequality (4.5-6) as a family of estimates, one for each $p \in \Pi_m$. For a *sharp* estimate, we should take p to be a polynomial in Π_m satisfying the following two conditions:

(*i*) $p(0) = 1$.

(*ii*) Among all polynomials of degree exactly m, p has the smallest excursion from 0.

There indeed exists such a polynomial. To construct it, we employ the Chebyshev polynomials discussed in Appendix C. The Chebyshev polynomial of degree m is defined as follows:

$$T_m(z) := \frac{1}{2}\left[\left(z + \sqrt{z^2-1}\right)^m + \left(z - \sqrt{z^2-1}\right)^m\right], \quad m = 0, 1, 2, \ldots.$$

We prove several properties of the Chebyshev polynomials in Appendix C. For example, T_m has m zeros and $m+1$ extrema in the interval $[-1, 1]$. Moreover, T_m has the form

$$T_m(z) = 2^{m-1}z^m + \alpha_{m-1}z^{m-1} + \alpha_{m-2}z^{m-2} + \cdots + \alpha_0,$$

for some real coefficients $\alpha_0, \alpha_1, \ldots, \alpha_{m-1}$. Among all polynomials p of degree m having this form, T_m minimizes the value of $||p||_\infty = \sup_{z\in[-1,1]} |p(z)|$. In fact, $||T_m||_\infty = 1$. Figure 5 shows the graph of T_6 on the interval $[-1, 1]$.

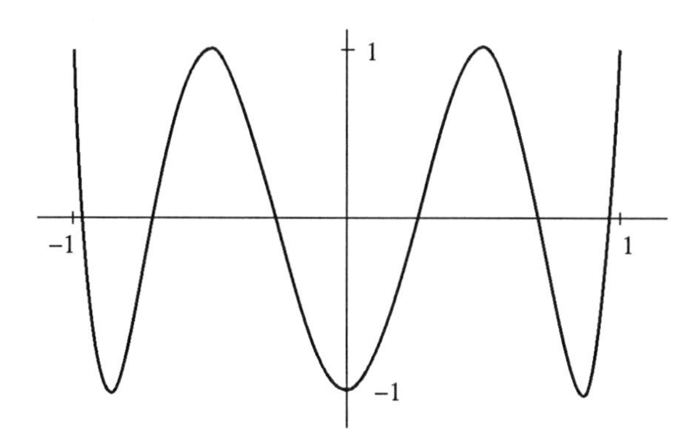

FIGURE 5. *Graph of the Chebyshev polynomial $T_6(z)$ on the interval* $[-1, 1]$.

For our application, we must shift T_m, so that it minimizes $||p||_\infty$ over the interval $[\lambda_{\min}, \lambda_{\max}]$ containing the eigenvalues of A, and then rescale, so that the resulting polynomial $\tilde{T}_m$ has the value 1 at $z = 0$. Define

$$\tilde{T}_m(z) := \frac{T_m\left(1 - 2\dfrac{z - \lambda_{\min}}{\lambda_{\max} - \lambda_{\min}}\right)}{T_m\left(1 + \dfrac{2\lambda_{\min}}{\lambda_{\max} - \lambda_{\min}}\right)}.$$

Thus $\tilde{T}_m$ has m zeros and $m + 1$ extrema in the interval $[\lambda_{\min}, \lambda_{\max}]$, and $\tilde{T}_m(0) = 1$. Figure 6 shows $\tilde{T}_6(z)$ when $\lambda_{\min} = 1$ and $\lambda_{\max} = 10$. Problem 15 asks for proof that, among all polynomials p of degree m with $p(0) = 1$, $\tilde{T}_m$ minimizes the value of

$$||p||_\infty = \sup_{z \in [\lambda_{\min}, \lambda_{\max}]} |p(z)|.$$

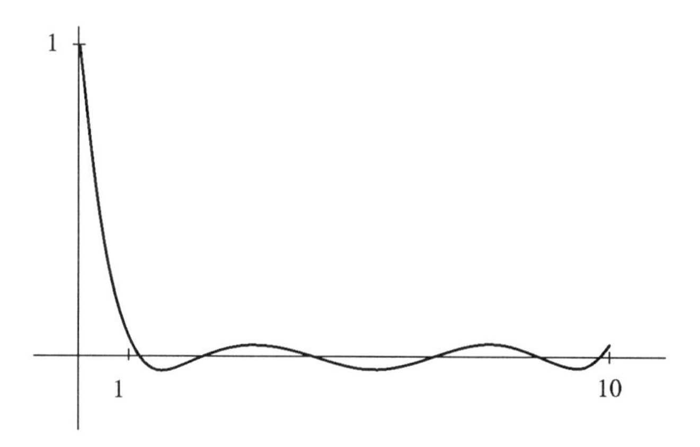

FIGURE 6. *Graph of the scaled, shifted Chebyshev polynomial* $\tilde{T}_6(z)$ *on the interval* $[0, 10]$.

Using $\tilde{T}_m$, we deduce a sharp estimate for the error $||\varepsilon_m||_A$. As a consequence of Corollary 4.25,

$$||\varepsilon_m||_A \le ||\tilde{T}_m||_\infty ||\varepsilon_0||_A.$$

To estimate the factor $||\tilde{T}_m||_\infty$, assume that $\lambda_{\max} > \lambda_{\min}$ and observe that

$$\begin{aligned} |\tilde{T}_m(z)| &= \frac{\left|T_m\left(1 - 2\dfrac{z - \lambda_{\min}}{\lambda_{\max} - \lambda_{\min}}\right)\right|}{\left|T_m\left(1 + \dfrac{2\lambda_{\min}}{\lambda_{\max} - \lambda_{\min}}\right)\right|} \\ &\le \frac{1}{\left|T_m\left(\dfrac{\lambda_{\max} + \lambda_{\min}}{\lambda_{\max} - \lambda_{\min}}\right)\right|}. \end{aligned}$$

The denominator in this last expression is

$$\left|T_m\left(\frac{\lambda_{\max}+\lambda_{\min}}{\lambda_{\max}-\lambda_{\min}}\right)\right| = \frac{1}{2}\left|\left(\frac{\sqrt{\lambda_{\max}}+\sqrt{\lambda_{\min}}}{\sqrt{\lambda_{\max}}-\sqrt{\lambda_{\min}}}\right)^m + \left(\frac{\sqrt{\lambda_{\max}}-\sqrt{\lambda_{\min}}}{\sqrt{\lambda_{\max}}+\sqrt{\lambda_{\min}}}\right)^m\right|.$$

The second of the two terms on the right being positive, we have

$$|\tilde{T}_m(z)| \le 2\left(\frac{\sqrt{\lambda_{\max}}-\sqrt{\lambda_{\min}}}{\sqrt{\lambda_{\max}}+\sqrt{\lambda_{\min}}}\right)^m.$$

Since A is symmetric and positive definite, $\lambda_{\max}/\lambda_{\min} = \mathrm{cond}_2(\mathsf{A})$, and therefore

$$\|\varepsilon_m\|_A \le 2\|\varepsilon_0\|_A \left[\frac{\sqrt{\mathrm{cond}_2(\mathsf{A})}-1}{\sqrt{\mathrm{cond}_2(\mathsf{A})}+1}\right]^m.$$

This estimate motivates our earlier discussion about preconditioning.

Further Remarks

Our discussion of the conjugate-gradient method barely scratches the surface, especially in light of the remarkable growth in research in this and related techniques in the past 15 years. One major research direction that we do not even mention above is the development of gradient-based search algorithms for linear systems involving nonsymmetric matrices. These methods typically lack some of the efficiency associated with the standard preconditioned conjugate-gradient schemes, but they have broader utility. Among the standard references in this literature are papers by Axelsson [1] and and Eisenstat et al. [3].

4.6 Problems

PROBLEM 1. Prove Lemma 4.4.

PROBLEM 2. Let $\mathsf{G} \in \mathbb{C}^{n\times n}$. Prove that λ is an eigenvalue of G if and only if λ^m is an eigenvalue of G^m. (Hint: $a^m - b^m = (a-b)(a^{m-1} + a^{m-2}b + \cdots + b^{m-1})$.)

PROBLEM 3. Consider the difference approximation (4.1-3) to the boundary-value problem (4.1-2) on a two-dimensional grid with nodes $(x_k, y_l) = (kh, lh)$, where $h = 1/n$ and $k, l = 0, 1, \ldots, n$. Show that the vector $\boldsymbol{v} \in \mathbb{R}^{(n-1)^2}$, with entries

$$v_{k,l} = \sin(pk\pi h)\sin(ql\pi h),$$

is an eigenvector for the system matrix for $p, q = 1, 2, \ldots, n-1$ and that the corresponding eigenvalue is $\mu = 4[\sin^2(p\pi h/2) + \sin^2(q\pi h/2)]$.

PROBLEM 4. Show that the iteration matrix for the damped Jacobi method applied to Equations (4.1-2) has the form $\mathsf{G}_D = \mathsf{I} - \frac{1}{4}\omega\mathsf{A}$. Using the results of Problem 3, determine the eigenvalues of G_D. Show that the damped Jacobi method converges for this problem when $0 < \omega \leq 1$.

PROBLEM 5. Prove that the block tridiagonal matrix in Equation (4.4-1) is consistently ordered whenever the diagonal blocks D_i satisfy the condition $\text{diag}(\mathsf{D}_i) \neq 0$.

PROBLEM 6. Write computer programs to solve the boundary-value problem (4.1-2) on the unit square $(0,1) \times (0,1)$, allowing for an $N \times N$ uniform grid having variable size N. Use the approximating finite-difference equations (4.1-3) and the following boundary values:

$$u(x,0) = u(x,1) = 1, \qquad 0 < x < 1;$$

$$u(0,y) = 2, \quad u(1,y) = 3, \qquad 0 < y < 1.$$

Demonstrate computationally that the convergence rate of the Gauss-Seidel method is twice that of the Jacobi method. Investigate the convergence rates as N varies.

PROBLEM 7. Prove Lemma 4.13.

PROBLEM 8. **Line successive overrelaxation** (LSOR) is similar to SOR, except that one solves for an entire line (for example, l =constant) of nodal unknowns $u_{k,l}$ in each step of every iteration. The defining equations for the Poisson problem

$$\frac{\partial^2 u}{\partial x^2} + \frac{\partial^2 u}{\partial y^2} = \sin x \sin y,$$

$$u(0,y) = u(\pi,y) = u(x,0) = u(x,\pi) = 0,$$

are as follows:

$$\overline{u}_{k,l} = \tfrac{1}{4}(\overline{u}_{k+1,l} + \overline{u}_{k-1,l} + u_{k,l+1}^{(m)} + u_{k,l-1}^{(m+1)} + h^2 \sin x_k \sin y_l),$$

$$u_{k,l}^{(m+1)} = \omega\overline{u}_{k,l} + (1-\omega)u_{k,l}^{(m)}.$$

In the first equation, the barred quantities are intermediate unknowns, analogous to the intermediate variables solved for in the Gauss-Seidel step of ordinary SOR. After determining them, one uses them in the second equation to correct the old iterative values using a relaxation parameter ω. Write a program to solve the Poisson problem using LSOR on a grid with 10 cells on a side and various values of $\omega \in (1,2)$.

PROBLEM 9. Assume that $\mathsf{A} \in \mathbb{R}^{n\times n}$ is singular, so that the system $\mathsf{A}\mathbf{u} = \mathbf{b}$ does not have a unique solution. Consider a matrix splitting scheme $\mathsf{A} =$

$\mathsf{B} + (\mathsf{A} - \mathsf{B})$, where B is nonsingular. Show that the corresponding iterative scheme does not converge.

PROBLEM 10. Verify that $\langle \cdot, \cdot \rangle_A$ is an inner product whenever the matrix $\mathsf{A} \in \mathbb{R}^{n \times n}$ is symmetric and positive definite.

PROBLEM 11. Prove that the kth step of the conjugate-gradient method for $\mathsf{A}\mathbf{u} = \mathbf{b}$ produces an iterate $\mathbf{u}^{(k)}$ that minimizes $||\mathbf{u} - \mathbf{v}||_A$ over all vectors $\mathbf{v}$ in the Krylov subspace $\mathcal{K}_m = \text{span}\{\mathbf{p}_0, \mathbf{p}_1, \ldots, \mathbf{p}_{k-1}\}$. (The vectors $\mathbf{p}_i$ are the search directions. Assume that $\mathbf{u}_0 = \mathbf{0}$.) (Hint: The identity $\mathsf{A}\mathbf{u} = \mathbf{b}$ implies that $\langle \varepsilon_m, \varepsilon_m \rangle_A = \mathbf{b}^\top \mathsf{A}^{-1}\mathbf{b} + 2F(\mathbf{u}^{(m)})$.)

PROBLEM 12.

(A) Show that $A \in \mathbb{R}^{n \times n}$ is positive definite if and only if A^{-1} is.

(B) Let $\mathsf{A} \in \mathbb{R}^{n \times n}$ be symmetric and positive definite, and let $\mathbf{b} \in \mathbb{R}^n$. Show that $\mathsf{A}\mathbf{u} = \mathbf{b}$ if and only if $\mathbf{u}$ minimizes the function $F(\mathbf{v}) = \frac{1}{2}\mathbf{v}^\top \mathsf{A}\mathbf{v} - \mathbf{b}^\top \mathbf{v}$.

PROBLEM 13. Let $\mathsf{A} \in \mathbb{R}^{n \times n}$ be symmetric and positive definite, and let p be a polynomial. Show that $||p(\mathsf{A})||_A = \max_j |p(\lambda_j)|$, where λ_j ranges over the eigenvalues of A.

PROBLEM 14. Consider the matrix equation $\mathsf{A}\mathbf{u} = \mathbf{b}$ and the preconditioner $\mathsf{B}^\top \mathsf{B}$, where

$$\mathsf{A} = \begin{bmatrix} 5 & 1 \\ 1 & 1 \end{bmatrix}, \quad \mathbf{b} = \begin{bmatrix} 1 \\ 1 \end{bmatrix}, \quad \mathsf{B}^\top \mathsf{B} = \begin{bmatrix} 5 & 0 \\ 0 & 1 \end{bmatrix}.$$

Take $\mathbf{u}_0 = (0, 0)^\top$ as an initial guess. Sketch the level curves of the function $F(\mathbf{v}) = \frac{1}{2}\mathbf{v}^\top \mathsf{A}\mathbf{v} - \mathbf{b}^\top \mathbf{v}$ and the iterates $\mathbf{u}^{(1)}, \mathbf{u}^{(2)}$ for the conjugate-gradient method. Do the same for the preconditioned-conjugate gradient method.

PROBLEM 15. Prove that, among all polynomials p of degree m with $p(0) = 1$, the scaled, shifted Chebyshev polynomial $\tilde{T}_m$ minimizes the value of

$$||p||_\infty = \sup_{z \in [\lambda_{\min}, \lambda_{\max}]} |p(z)|.$$

(Hint: $(p - T_m)(0) = 0$.)

4.7 References

1. O. Axelsson, "Conjugate gradient type methods for unsymmetric and inconsistent systems of linear equations," *Lin. Alg. & Its Applic. 29* (1980), 1–16.

2. W.L. Briggs, *A Multigrid Tutorial,* Society for Industrial and Applied Mathematics, Philadelphia, 1987.

3. S.C. Eisenstat, H.C. Elman, and M.H. Schultz, "Variational iterative methods for nonsymmetric systems of linear equations," *SIAM J. Numer. Anal. 20:2* (1983), 345–357.

4. L.A. Hageman and D.M.Young, *Applied Iterative Methods,* Academic Press, New York, 1981.

5. R.V. Southwell, *Relaxation Methods in Theoretical Physics,* Oxford University Press, Oxford, 1946.

6. G. Strang, *Linear Algebra and its Applications* (2nd ed.), Academic Press, New York, 1980.

Chapter 5

Eigenvalue Problems

Finding the eigenvalues of a square matrix is a difficult problem that arises in a wide variety of scientific applications. The problem amounts to a special case of the more general nonlinear problems considered in Chapter 3, and iterative methods can be distressingly ineffective. Fortunately, the problem's close connections with numerical linear algebra lend it enough structure to admit elegant and comparatively efficient solutions. This chapter presents two of the most important numerical techniques for solving eigenvalue problems: the power method and the QR method. We restrict attention to eigenvalues of real matrices, although much of the theory extends in natural ways to matrices with complex entries. Along the way, we examine numerical aspects of the QR decomposition of matrices, a useful topic in its own right. We begin with a review of basic facts about eigenvalue problems.

5.1 Basic Facts About Eigenvalues

Sections 2.5 and 4.2 introduce the basic definitions of eigenvalues and eigenvectors and the notion of similarity transformation. Recall that the spectrum of a square matrix A is the set $\sigma(\mathsf{A})$ of all eigenvalues of A and that the algebraic multiplicity of an eigenvalue λ is its multiplicity as a zero of the characteristic polynomial $\det(\mathsf{A} - \lambda\mathsf{I})$.

The following proposition lists several properties of eigenvalues not discussed earlier.

PROPOSITION 5.1. *Let* $\mathsf{A} \in \mathbb{R}^{n\times n}$.

(i) *If* A *is upper or lower triangular, then its eigenvalues are its diagonal entries.*

(ii) *If* $\lambda \in \sigma(\mathsf{A})$, *then* $\overline{\lambda} \in \sigma(\mathsf{A})$.

(iii) *If* $\lambda \in \sigma(\mathsf{A})$, *then* $\lambda \in \sigma(\mathsf{A}^\top)$.

(iv) *If λ is an eigenvalue of* A *associated with eigenvector* $\boldsymbol{v}$ *and* $p(x)$ *is any polynomial, then* $p(\lambda)$ *is an eigenvalue of* $p(\mathsf{A})$ *associated with eigenvector* $\boldsymbol{v}$.

PROOF: Assertion (*i*) follows from the observation that, if A is upper or lower triangular, then $\det(\mathsf{A}-\lambda\mathsf{I}) = (a_{1,1}-\lambda)(a_{2,2}-\lambda)\cdots(a_{n,n}-\lambda)$. Assertion (*ii*) is a direct consequence of the fact that λ is a zero of the characteristic polynomial of a real matrix. To prove (*iii*), observe that

$$\det(\mathsf{A}^\top - \lambda\mathsf{I}) = \det\left[(\mathsf{A}-\lambda\mathsf{I})^\top\right] = \det(\mathsf{A}-\lambda\mathsf{I}).$$

For (*iv*), it is easy to establish that $\mathsf{A}^k\boldsymbol{v} = \lambda^k\boldsymbol{v}$ whenever λ is an eigenvalue associated with eigenvector $\boldsymbol{v}$. Therefore, if $p(x) = c_0 + c_1x + \cdots + c_mx^m$, then

$$\begin{aligned} p(\mathsf{A})\boldsymbol{v} &= (c_0\mathsf{I} + c_1\mathsf{A} + \cdots + + c_m\mathsf{A}^m)\boldsymbol{v} \\ &= (c_0 + c_1\lambda + \cdots + c_m\lambda^m)\boldsymbol{v}, \end{aligned}$$

that is, $p(\mathsf{A})\boldsymbol{v} = p(\lambda)\boldsymbol{v}$. ∎

While part (*iii*) of this proposition describes a relationship between the eigenvectors of A and those of $\mathsf{A}^\top$, the relationship between the eigenvectors of the two matrices is not simple. It is possible (see Problem 1) to show that, if $\mathsf{A}\boldsymbol{v} = \lambda\boldsymbol{v}$ and $\mathsf{A}^\top\boldsymbol{\omega} = \mu\boldsymbol{\omega}$, with $\lambda \neq \mu$, then $\boldsymbol{\omega}^\top\boldsymbol{v} = 0$. However, this conclusion does not imply that $\boldsymbol{\omega}$ and $\boldsymbol{v}$ are orthogonal; they may be complex vectors, in which case $\boldsymbol{\omega}^\top\boldsymbol{v}$ does not constitute an inner product.

Recall that two square matrices A and B are similar, and thus have the same spectrum, if there is a nonsingular matrix S such that $\mathsf{B} = \mathsf{SAS}^{-1}$. A major theme of numerical eigenvalue calculations is the reduction of a general matrix $\mathsf{A} \in \mathbb{R}^{n\times n}$ to a similar matrix, whose eigenvalues are easy to compute, via similarity transformations. Section 5.4 exploits this strategy.

A salient question is whether the eigenvectors of a matrix $\mathsf{A} \in \mathbb{R}^{n\times n}$ span $\mathbb{C}^n$. Several simple matrices illustrate the range of possibilities. For example, the matrix

$$\mathsf{A}_1 = \begin{bmatrix} 1 & 1 & 1 \\ 0 & 1 & 1 \\ 0 & 0 & 1 \end{bmatrix}$$

has eigenvalue $\lambda = 1$, with algebraic multiplicity 3. All eigenvectors of this matrix have the form $\boldsymbol{v} = \alpha(1,0,0)^\top$, where α is a nonzero complex constant. In this case, the eigenvectors span a one-dimensional subspace of $\mathbb{C}^3$. In contrast, the matrix

$$\mathsf{A}_2 = \mathsf{I} = \begin{bmatrix} 1 & 0 & 0 \\ 0 & 1 & 0 \\ 0 & 0 & 1 \end{bmatrix}$$

also has eigenvalue $\lambda = 1$ with algebraic multiplicity 3, but for this matrix $(1,0,0)^\top, (0,1,0)^\top, (0,0,1)^\top$ are linearly independent eigenvectors. In this case the eigenvectors span $\mathbb{C}^3$. Finally, consider the matrix

$$\mathsf{A}_3 = \begin{bmatrix} 0 & 0 & 1 \\ 0 & 2 & 0 \\ 3 & 0 & 0 \end{bmatrix}.$$

This matrix has the following eigenvalue-eigenvector pairs:

$$\left(\sqrt{3}, \begin{bmatrix} 1 \\ 0 \\ \sqrt{3} \end{bmatrix}\right), \quad \left(2, \begin{bmatrix} 0 \\ 1 \\ 0 \end{bmatrix}\right), \quad \left(-\sqrt{3}, \begin{bmatrix} 1 \\ 0 \\ -\sqrt{3} \end{bmatrix}\right).$$

In this case, the eigenvalues are distinct, and the eigenvectors span $\mathbb{C}^3$.

For a given eigenvalue $\lambda \in \sigma(\mathsf{A})$, denote by $\mathcal{L}_A(\lambda)$ the span of the eigenvectors of A associated with λ. We call the dimension of $\mathcal{L}_A(\lambda)$ the **geometric multiplicity** of λ. In the examples just shown, the geometric multiplicity of the eigenvalue $\lambda = 1$ of A_1 is 1; the geometric multiplicity of $\lambda = 1$ as an eigenvalue of A_2 is 3. Each of the eigenvalues of A_3 has geometric multiplicity 1.

The space spanned by all eigenvectors of a matrix has close connections to the effectiveness of similarity transformations in simplifying its structure:

THEOREM 5.2. *A matrix $\mathsf{A} \in \mathbb{R}^{n\times n}$ has n linearly independent eigenvectors if and only if A is similar to a diagonal matrix.*

PROOF: Assume first that A has n linearly independent eigenvectors, which we denote as $\boldsymbol{v}_1, \boldsymbol{v}_2, \ldots, \boldsymbol{v}_n$. Let $\lambda_1, \lambda_2, \ldots, \lambda_n$ be the corresponding eigenvalues. Form the matrix $\mathsf{S} := [\boldsymbol{v}_1, \boldsymbol{v}_2, \ldots, \boldsymbol{v}_n]$. This matrix is nonsingular, since its columns are the linearly independent eigenvectors. Also,

$$\begin{aligned} \mathsf{AS} &= \mathsf{A}[\boldsymbol{v}_1, \boldsymbol{v}_2, \ldots, \boldsymbol{v}_n] \\ &= [\lambda_1\boldsymbol{v}_1, \lambda_2\boldsymbol{v}_2, \ldots, \lambda_n\boldsymbol{v}_n] = \mathsf{SD}, \end{aligned} \tag{5.1-1}$$

where $\mathsf{D} := \operatorname{diag}(\lambda_1, \lambda_2, \ldots, \lambda_n)$. It follows that $\mathsf{A} = \mathsf{SDS}^{-1}$.

Now assume that $\mathsf{A} = \mathsf{SDS}^{-1}$, where D is a diagonal matrix. Then $\mathsf{AS} = \mathsf{SD}$, and Equation (5.1-1) shows that the columns of S are linearly independent eigenvectors of S. ∎

Testing for linear independence of eigenvectors can be tedious, but in one important case it is unnecessary:

THEOREM 5.3. *Let $\mathsf{A} \in \mathbb{R}^{n\times n}$, and suppose that $\lambda_1, \lambda_2, \ldots, \lambda_k \in \sigma(\mathsf{A})$ are distinct. Then the corresponding eigenvectors $\boldsymbol{v}_1, \boldsymbol{v}_2, \ldots, \boldsymbol{v}_k$ are linearly independent.*

PROOF: We argue by contradiction. Let l be the largest integer such that $\boldsymbol{v}_1, \boldsymbol{v}_2, \ldots, \boldsymbol{v}_l$ are linearly independent, and assume that $l < k$. Then there exist constants $c_1, c_2, \ldots, c_{l+1}$, not all zero, such that

$$c_1\boldsymbol{v}_1 + c_2\boldsymbol{v}_2 + \cdots + c_{l+1}\boldsymbol{v}_{l+1} = \mathbf{0}. \tag{5.1-2}$$

Multiplying this equation by A and then by λ_{l+1} yields

$$\begin{array}{ccccccccc} c_1\lambda_1\boldsymbol{v}_1 & + & c_2\lambda_2\boldsymbol{v}_2 & + & \cdots & + & c_{l+1}\lambda_{l+1}\boldsymbol{v}_{l+1} & = & \mathbf{0}, \\ c_1\lambda_{l+1}\boldsymbol{v}_1 & + & c_2\lambda_{l+1}\boldsymbol{v}_2 & + & \cdots & + & c_{l+1}\lambda_{l+1}\boldsymbol{v}_{l+1} & = & \mathbf{0}. \end{array}$$

Subtracting the second of these equations from the first gives

$$c_1(\lambda_1 - \lambda_{l+1})\boldsymbol{v}_1 + c_2(\lambda_2 - \lambda_{l+1})\boldsymbol{v}_2 + \cdots + c_l(\lambda_l - \lambda_{l+1})\boldsymbol{v}_l = \mathbf{0}.$$

The linear independence of the vectors $\boldsymbol{v}_1, \boldsymbol{v}_2, \ldots, \boldsymbol{v}_l$ now implies that $c_j(\lambda_j - \lambda_{l+1}) = 0$, for $j = 1, 2, \ldots, l$. Since the eigenvalues are distinct, $c_1 = c_2 = \cdots = c_l = 0$. Therefore, by Equation (5.1-2), $c_{l+1} = 0$ too, contradicting the fact that not all of the constants c_j vanish. ∎

COROLLARY 5.4. *If* $\mathsf{A} \in \mathbb{R}^{n\times n}$ *has* n *distinct eigenvalues, then* A *is similar to a diagonal matrix.*

A special case of this corollary occurs when $|\lambda_1| > |\lambda_2| > \cdots > |\lambda_n|$. Here the eigenvalues λ_j must be real, since complex eigenvalues come in conjugate pairs having equal magnitude. In this case, then, A is similar to a real, diagonal matrix.

Often one can glean information about the location of eigenvalues in the complex plane without much computational effort. One of the most famous results along these lines is the following:

THEOREM 5.5 (GERSCHGORIN). *Let* $\mathsf{A} \in \mathbb{R}^{n\times n}$, *and define the closed disks*

$$D_i := \left\{ z \in \mathbb{C} \; : \; |z - a_{i,i}| \le \sum_{j\neq i} |a_{i,j}| \right\}. \tag{5.1-3}$$

Then

$$\sigma(\mathsf{A}) \subset \bigcup_{j=1}^{n} D_i.$$

We call the set D_i defined in Equation (5.1-3) the ith **Gerschgorin disk** of A. For the matrix A_3 discussed above, the Gerschgorin disks are

$$D_1 = \overline{\mathcal{B}_1(0)}, \qquad D_2 = \overline{\mathcal{B}_0(2)}, \qquad D_3 = \overline{\mathcal{B}_3(0)}.$$

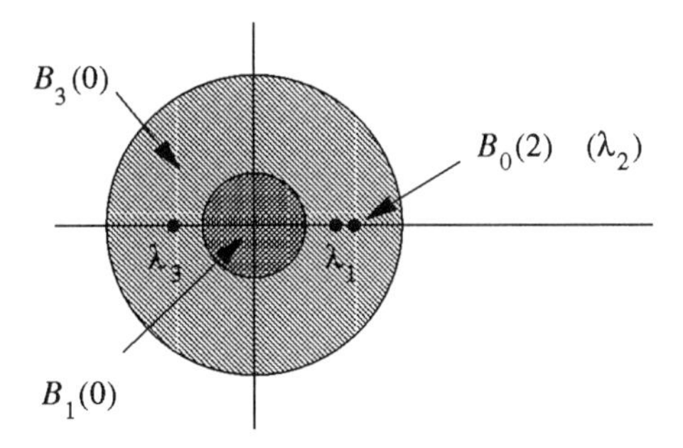

FIGURE 1. *Union of the Gerschgorin disks for the matrix* A_3*, showing the locations of its three eigenvalues.*

The union of these disks is $\overline{\mathcal{B}_3(0)}$. Figure 1 depicts this set along with the three eigenvalues of A_3.

PROOF: Let $\lambda \in \sigma(\mathsf{A})$, and pick a vector $\boldsymbol{v} \in \mathcal{L}_A(\lambda)$ with $\|\boldsymbol{v}\|_\infty = 1$. Thus, for some index k, $|v_k| = 1$. The fact that the kth component of $\mathsf{A}\boldsymbol{v}$ is λv_k implies that $\lambda v_k = a_{k,1}\boldsymbol{v}_1 + a_{k,2}\boldsymbol{v}_2 + \cdots + a_{k,n}\boldsymbol{v}_n$, and hence

$$(\lambda - a_{k,k})v_k = \sum_{j \neq k} a_{k,j}v_j.$$

From the triangle inequality we conclude that

$$|\lambda - a_{k,k}| \leq \sum_{j\neq k} |a_{k,j}||v_j| \leq \sum_{j \neq k} |a_{k,j}|.$$

It follows that $\lambda \in D_k$. ∎

The Gerschgorin theorem leads to another result that concerns perturbations of matrices whose spectra are known.

THEOREM 5.6. *Assume that* $\mathsf{A} \in \mathbb{R}^{n\times n}$ *is similar to a diagonal matrix* D *via the similarity transformation* $\mathsf{A} = \mathsf{SDS}^{-1}$*, and let* $\mathsf{E} \in \mathbb{R}^{n\times n}$*. Denote by* $\lambda_1, \lambda_2, \ldots, \lambda_n$ *the eigenvalues of* A*, and define the disks*

$$B_j := \left\{z \in \mathbb{C} \ : \ |z - \lambda_j| \leq \|\mathsf{E}\|_\infty \operatorname{cond}_\infty \mathsf{S}\right\}, \quad j = 1, 2, \ldots, n.$$

Then

$$\sigma(\mathsf{A} + \mathsf{E}) \subset \bigcup_{j=1}^{n} B_j.$$

PROOF: Observe that $\sigma(\mathsf{A}+\mathsf{E}) = \sigma(\mathsf{S}^{-1}(\mathsf{A}+\mathsf{E})\mathsf{S}) = \sigma(\mathsf{D}+\mathsf{S}^{-1}\mathsf{E}\mathsf{S})$. By Theorem 5.5, $\sigma(\mathsf{A}+\mathsf{E})$ lies in the union of Gerschgorin disks for $\mathsf{D}+\mathsf{S}^{-1}\mathsf{E}\mathsf{S}$. Call $\mathsf{F} := \mathsf{S}^{-1}\mathsf{E}\mathsf{S}$. By Gerschgorin's theorem,

$$\sigma(\mathsf{A}+\mathsf{E}) \subset \bigcup_{j=1}^{n} D_j,$$

where

$$D_j := \left\{ z \in \mathbb{C} \ : \ |z + (\lambda_j - f_{j,j})| \leq \sum_{k \neq j} |f_{j,k}| \right\}.$$

But, for $z \in D_j$,

$$\begin{aligned} |z - \lambda_j| &\leq |z - (\lambda_j + f_{j,j})| + |f_{j,j}| \leq |f_{j,j}| + \sum_{k \neq j} |f_{j,k}| \\ &\leq \|\mathsf{F}\|_\infty \leq \|\mathsf{S}^{-1}\|_\infty \|\mathsf{E}\|_\infty \|\mathsf{S}\|_\infty \\ &= \|\mathsf{E}\|_\infty \operatorname{cond}_\infty \mathsf{S}. \end{aligned}$$

It follows that $D_j \subset B_j$ and hence that

$$\sigma(\mathsf{A}+\mathsf{E}) \subset \bigcup_{j=1}^{n} B_j. \qquad \blacksquare$$

5.2 Power Methods

The power method is among the simplest of numerical techniques for finding eigenvalues. The overall idea has several limitations, and as a consequence the power method and its relatives are not the most common choices for robust numerical algorithms. Nevertheless, the scheme reveals some important aspects of the eigenvalue problem. It also serves as a backdrop for the more sophisticated approaches considered in Sections 5.3 and 5.4.

Motivation and Construction

Consider a matrix $\mathsf{A} \in \mathbb{R}^{n \times n}$ whose eigenvalues stand in the relationship $|\lambda_1| > |\lambda_2| \geq \cdots \geq |\lambda_n|$. In this case λ_1 must be real, or else $\overline{\lambda}_1$ would be a different eigenvalue having equal magnitude. We say that λ_1 is the **dominant eigenvalue** of A. Suppose also that the eigenvectors $\boldsymbol{v}_1, \boldsymbol{v}_2, \ldots, \boldsymbol{v}_n$ of A span $\mathbb{C}^n$. Thus a generic vector $\mathbf{z}^{(0)} \in \mathbb{C}^n$ has an expansion

$$\mathbf{z}^{(0)} = \alpha_1 \boldsymbol{v}_1 + \alpha_2 \boldsymbol{v}_2 + \cdots + \alpha_n \boldsymbol{v}_n.$$

We assume that $\alpha_1 \neq 0$. Indeed, a "random" choice of $\mathbf{z}^{(0)}$ will almost certainly have a nonzero component in the direction of $\boldsymbol{v}_1$, since the hyperplane span$\{\boldsymbol{v}_2, \boldsymbol{v}_3, \ldots, \boldsymbol{v}_n\}$ is an extremely small subset (having zero measure) of $\mathbb{C}^n$.

Multiply the vector $\mathbf{z}^{(0)}$ repeatedly by the matrix A, calling $\mathbf{z}^{(m)} := \mathsf{A}^m \mathbf{z}^{(0)}$:

$$\mathbf{z}^{(m)} = \sum_{j=1}^{n} \lambda_j^m \alpha_j \boldsymbol{v}_j = \lambda_1^m \left[\alpha_1 \boldsymbol{v}_1 + \sum_{j=2}^{n} \left(\frac{\lambda_j}{\lambda_1} \right)^m \alpha_j \boldsymbol{v}_j \right]. \tag{5.2-1}$$

Heuristically, as $m \to \infty$, $\mathbf{z}^{(m)}$ tends to a vector that is collinear with $\boldsymbol{v}_1$. In this way, taking successive powers of A yields information about its dominant eigenvalue λ_1.

It is possible to recover the dominant eigenvalue λ_1 from this iteration. Recall the Rayleigh quotient for $\mathbf{z}^{(m)}$, defined in Section 4.4:

$$\rho_m := \frac{(\mathbf{z}^{(m)})^\top \mathsf{A} \mathbf{z}^{(m)}}{(\mathbf{z}^{(m)})^\top \mathbf{z}^{(m)}}.$$

It is an easy exercise to show that $\rho_m \to \lambda_1$ as $m \to \infty$.

Two observations help to convert this reasoning to a workable algorithm. First, when $|\lambda_1| \neq 1$, Equation (5.2-1) suggests that the iterates $\mathbf{z}^{(m)}$ either grow without bound or shrink toward zero. Computationally, either occurrence can be disastrous, leading to machine overflow or loss of precision. To avoid such scaling problems, we normalize $\mathbf{z}^{(m)}$ at each step, setting

$$\mathbf{z}^{(m)} \leftarrow \frac{\mathsf{A}\mathbf{z}^{(m-1)}}{\|\mathsf{A}\mathbf{z}^{(m-1)}\|},$$

where $\|\cdot\|$ stands for any norm on $\mathbb{C}^n$.

Second, there may be cheap alternatives to the Rayleigh quotient. Let $F: \mathbb{C}^n \to \mathbb{C}$ be any linear function for which $F(\boldsymbol{v}_1) \neq 0$. An example might be $F(\mathbf{z}) = \mathbf{z}^\top \mathbf{e}_1$, where $\mathbf{e}_1 := (1, 0, \ldots, 0)^\top$. Then

$$\frac{F(\mathsf{A}\mathbf{z}^{(m)})}{F(\mathbf{z}^{(m)})} = \lambda_1 \frac{F(\boldsymbol{v}_1) + F(\boldsymbol{\varepsilon}^{(m+1)})}{F(\boldsymbol{v}_1) + F(\boldsymbol{\varepsilon}^{(m)})},$$

where

$$\boldsymbol{\varepsilon}^{(m)} := \sum_{j=2}^{n} \left(\frac{\lambda_j}{\lambda_1} \right)^m \alpha_j \boldsymbol{v}_j \to 0 \quad \text{as} \quad m \to \infty.$$

It follows that

$$\frac{F(\mathsf{A}\mathbf{z}^{(m)})}{F(\mathbf{z}^{(m)})} \to \lambda_1 \quad \text{as} \quad m \to \infty.$$

Incorporating these observations, we arrive at the following algorithm:

ALGORITHM 5.1 (POWER METHOD). *Given a matrix $\mathsf{A} \in \mathbb{R}^{n \times n}$, an initial vector $\mathbf{z} \in \mathbb{C}^n$ with $\|\mathbf{z}\| = 1$, a maximum number M of iterations, and a tolerance $\tau > 0$, the following algorithm implements the power method for finding the dominant eigenvalue of* A.

1. For $m = 1, 2, \ldots, M$:
2. $\quad \mathbf{z}_{\text{old}} \leftarrow \mathbf{z}$.
3. $\quad \mathbf{z} \leftarrow \mathsf{A}\mathbf{z}$.
4. $\quad \lambda \leftarrow F(\mathbf{z})/F(\mathbf{z}_{\text{old}})$.
5. $\quad \mathbf{z} \leftarrow \mathbf{z}/\|\mathbf{z}\|$.
6. $\quad$ If $\|\mathbf{z} - \mathbf{z}_{\text{old}}\| < \tau$ go to 8.
7. Next m.
8. End.

The most recently computed values of λ and $\mathbf{z}$ are the current approximations of the dominant eigenvalue λ_1 and an associated eigenvector, respectively. One can show (Problem 3) that the algorithm converges even when the dominant eigenvalue λ_1 is associated with several linearly independent eigenvectors.

Practical Considerations

What happens when there is no single dominant eigenvalue? What happens in the numerically similar case when $|\lambda_1| \simeq |\lambda_2|$? Finally, is there a reasonable strategy for finding *all* eigenvalues of a matrix A? We devote the remainder of this section to these issues.

Consider first the case when there is no single dominant eigenvalue of A, say $|\lambda_1| = |\lambda_2| = \cdots = |\lambda_s| > |\lambda_{s+1}| \geq \cdots \geq |\lambda_n|$, where not all of the numbers $\lambda_1, \lambda_2, \ldots, \lambda_s$ are equal. In this case the power method does not converge. Suppose, for example, that there are two dominant eigenvalues, occurring as a complex conjugate pair $\lambda_1 = re^{i\theta}$, $\overline{\lambda}_1 = re^{-i\theta}$ with associated eigenvectors $\boldsymbol{v}_1, \overline{\boldsymbol{v}}_1$. Then a generic initial vector $\mathbf{z}^{(0)}$ has the form

$$\mathbf{z}^{(0)} = \alpha_1 \boldsymbol{v}_1 + \alpha_2 \overline{\boldsymbol{v}}_1 + \sum_{j=3}^{n} \alpha_j \boldsymbol{v}_j.$$

Repeated multiplication by A yields

$$\mathsf{A}^m \mathbf{z}^{(0)} = r^m \left[\alpha_1 e^{im\theta} \boldsymbol{v}_1 + \alpha_2 e^{-im\theta} \overline{\boldsymbol{v}}_1 + \sum_{j=3}^{n} \alpha_j \left(\frac{\lambda_j}{r} \right)^m \boldsymbol{v}_j \right].$$

The terms inside the summation sign tend to zero as $m \to \infty$, but the first two terms survive. The presence of the complex exponential factors $e^{im\theta}$ and $e^{-im\theta}$ implies that, after the nondominant terms have effectively died off, the vector $\mathbf{z}^{(m)}$ exhibits oscillatory behavior. Wilkinson [3, Chapter 9] discusses this case more thoroughly.

When $|\lambda_1| \simeq |\lambda_2|$ but λ_1 is still dominant, the power method converges slowly. Equation (5.2-1) shows why: The term $(\lambda_2/\lambda_1)^m \alpha_2 \boldsymbol{v}_2$ decays slowly as $m \to \infty$. One strategy for avoiding this difficulty is to devise a variant of the power method that seeks the eigenvalue of A lying closest to a prescribed point in the complex plane. As Figure 1 illustrates, this approach exploits the fact that the distance between λ_1 and λ_2 can be large even when the two eigenvalues have nearly the same magnitude.

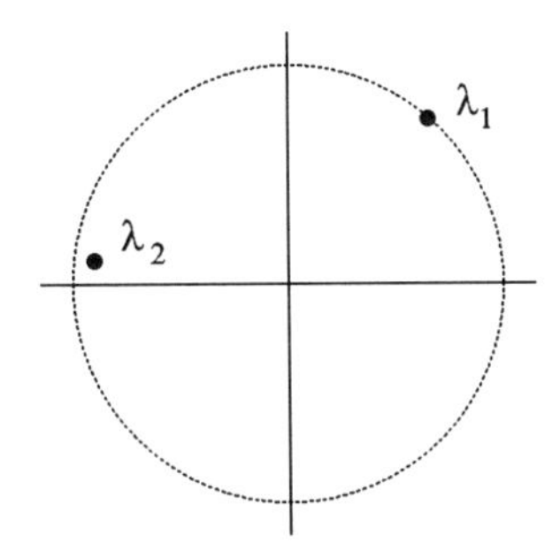

FIGURE 1. *Eigenvalues λ_1 and λ_2 that have nearly the same magnitude but lie far apart in the complex plane.*

The **inverse power method** facilitates this strategy. The following proposition motivates the method:

PROPOSITION 5.7. *If $\mathsf{A} \in \mathbb{R}^{n\times n}$ is nonsingular and $\lambda \in \sigma(\mathsf{A})$, then $\lambda^{-1} \in \sigma(\mathsf{A}^{-1})$.*

PROOF: This is an easy exercise. ∎

We use this idea to compute the eigenvalue of A that has smallest magnitude. Assume that $|\lambda_1| \geq |\lambda_2| \geq \cdots \geq |\lambda_{n-1}| > |\lambda_n| > 0$, so that $|\lambda_n^{-1}| > |\lambda_{n-1}^{-1}| \geq \cdots \geq |\lambda_1^{-1}|$. Applying the power method to A^{-1} yields a sequence $\mathbf{z}^{(m+1)} = \mathsf{A}^{-1}\mathbf{z}^{(m)}/\|\mathsf{A}^{-1}\mathbf{z}^{(m)}\|$ whose Rayleigh quotients converge to λ_n^{-1}.

In practice, we do not compute A^{-1} explicitly. Instead, at each iteration we solve the linear system $\mathsf{A}\mathbf{z} = \mathbf{z}^{(m)}$ for $\mathbf{z}$, setting $\mathbf{z}^{(m+1)} = \mathbf{z}/\|\mathbf{z}\|$. Since the matrix A remains fixed as we iterate, we can initiate the algorithm by computing the LU factorization for A once, so that each iteration requires only forward and backward substitution.

Think of the inverse power method as a scheme for finding the eigenvalue of A that lies closest to the origin. In this view, it is natural to seek the eigenvalue of A lying closest to a prescribed point $\mu \notin \sigma(\mathsf{A})$ by means of a **shift of origin**. Again, we assume that A is nonsingular. By Proposition

5.1, $\lambda - \mu$ is an eigenvalue of $\mathsf{A} - \mu\mathsf{I}$ whenever λ is an eigenvalue of A, so by Proposition 5.7 $(\lambda - \mu)^{-1}$ is an eigenvalue of $(\mathsf{A} - \mu\mathsf{I})^{-1}$. Hence, if λ is the eigenvalue of A lying closest to μ, then the shifted inverse power method,

$$\text{Solve} \quad (\mathsf{A} - \mu\mathsf{I})\mathbf{z} = \mathbf{z}^{(m-1)} \quad \text{for} \quad \mathbf{z}$$

$$\mathbf{z}^{(m)} \leftarrow \mathbf{z}/\|\mathbf{z}\|,$$

converges to an eigenvector of A associated with λ.

Finally, we turn to the problem of finding all eigenvalues of a matrix using the power method or one of its variants. The approach reviewed here is called **deflation** [1]. Having computed an eigenvalue-eigenvector pair $(\lambda_1, \boldsymbol{v}_1)$ of $\mathsf{A} \in \mathbb{R}^{n \times n}$, we construct a matrix $\mathsf{A}' \in \mathbb{R}^{(n-1)\times(n-1)}$ whose spectrum consists of the remaining eigenvalues $\lambda_2, \lambda_3, \ldots, \lambda_n$ of A.

One way to do this is use a similarity transformation. Suppose, for example, that $\mathsf{S} \in \mathbb{R}^{n \times n}$ is a nonsingular matrix such that $\boldsymbol{v}_1 = \mathsf{S}\mathbf{e}_1$, where $\mathbf{e}_1 = (1, 0, \ldots, 0)^\top$. By similarity, $\sigma(\mathsf{S}^{-1}\mathsf{AS}) = \sigma(\mathsf{A})$. Also,

$$(\mathsf{S}^{-1}\mathsf{AS})\mathbf{e}_1 = (\mathsf{S}^{-1}\mathsf{AS})\mathsf{S}^{-1}\boldsymbol{v}_1 = \lambda_1 \mathsf{S}^{-1}\boldsymbol{v}_1 = \lambda_1 \mathbf{e}_1.$$

Consequently $\mathsf{S}^{-1}\mathsf{AS}$ has the following structure:

$$\mathsf{S}^{-1}\mathsf{AS} = \begin{array}{c} \\ 1 \\ n-1 \end{array} \begin{array}{c} \begin{array}{cc} 1 & n-1 \end{array} \\ \begin{pmatrix} \lambda_1 & \mathbf{a}^\top \\ \mathbf{0} & \mathsf{A}' \end{pmatrix} \end{array}.$$

(Compare this construction with that used in the development of the Schur normal form in Theorem 4.2.) The $(n-1) \times (n-1)$ block A' now has eigenvalues $\lambda_2, \lambda_3, \ldots, \lambda_n$.

For concreteness, let $\boldsymbol{v}_1 = (v_1, v_2, \ldots, v_n)^\top$, where $v_1 \neq 1$. Define $\mathsf{S} := \mathsf{I} - 2\mathbf{w}\mathbf{w}^\top$, where

$$w_1 := \sqrt{\frac{1 - v_1}{2}}, \qquad w_j := -\frac{v_j}{2w_1}, \quad j = 2, 3, \ldots, n.$$

It is easy to check that $\mathsf{S}^{-1}\mathsf{AS}\mathbf{e}_1 = \lambda_1\mathbf{e}_1$ and hence that the first column of the matrix $\mathsf{S}^{-1}\mathsf{AS}$ is $(\lambda_1, 0, \ldots, 0)^\top$, as desired.

The power method with deflation is too clumsy to be a preferred technique in actual computations, even though it has some heuristic appeal. In the next two sections we discuss a far more useful approach.

5.3 The QR Method: Underlying Concepts

The most widely used method for computing eigenvalues employs a technique, called **QR decomposition**, that is useful more broadly. We devote this section to a description of the QR method and to a brief discussion of one of its uses outside the context of eigenvalue problems, namely the solution of least-squares problems. We focus on the matrices having real entries. However, most of the theory extends naturally to complex matrices, as we sketch at the end of the section.

QR Decomposition

The LU decomposition introduced in Chapter 2 factors a matrix $\mathsf{A} = \mathsf{LU}$, where L has a special structure and U is upper triangular. In particular, L is lower triangular with unit diagonal entries. The idea behind the QR decomposition is similar: We factor $\mathsf{A} = \mathsf{QR}$, where now R is upper triangular and again Q has a special structure. To wit,

DEFINITION. *A matrix* $\mathsf{Q} \in \mathbb{R}^{n \times n}$ *is* **orthogonal** *if* $\mathsf{Q}^\top \mathsf{Q} = \mathsf{I}$.

An equivalent characterization of an orthogonal matrix $\mathsf{Q} \in \mathbb{R}^{n \times n}$ is that its columns form an orthonormal set of vectors $\mathbf{q}_1, \mathbf{q}_2, \ldots, \mathbf{q}_n \in \mathbb{R}^n$. Orthogonal matrices are the real analogs of unitary matrices, introduced in Section 4.2: A matrix is orthogonal if and only if it is real and unitary.

Orthogonal matrices have several key properties. For any vector $\mathbf{x} \in \mathbb{R}^n$, $\|\mathsf{Q}\mathbf{x}\|_2^2 = (\mathsf{Q}\mathbf{x})^\top \mathsf{Q}\mathbf{x} = \mathbf{x}^\top \mathsf{Q}^\top \mathsf{Q}\mathbf{x} = \mathbf{x}^\top \mathsf{I}\mathbf{x} = \|\mathbf{x}\|_2^2$. Hence, as linear transformations, orthogonal matrices preserve Euclidean length. It follows that all eigenvalues of an orthogonal matrix Q have unit magnitude and hence that $|\det \mathsf{Q}| = 1$. It is easy to show that the product of two orthogonal matrices is also orthogonal; in fact, the set of orthogonal matrices in $\mathbb{R}^{n \times n}$ forms a group (Problem 12).

In principle, one can use the decomposition $\mathsf{A} = \mathsf{QR}$ of a nonsingular matrix A to solve linear systems. Since $\mathsf{Q}^{-1} = \mathsf{Q}^\top$, the system $\mathsf{A}\mathbf{x} = \mathbf{b}$ reduces to the upper triangular system $\mathsf{R}\mathbf{x} = \mathsf{Q}^\top \mathbf{b}$, which one can solve via backward substitution. However, as we demonstrate later, the QR decomposition typically requires more arithmetic than the LU decomposition, so this approach does not receive much use outside of specialized contexts.

Two classes of orthogonal matrices are especially useful. In $\mathbb{R}^{2 \times 2}$, these classes have simple geometric interpretations. The first class consists of **rotations**:

$$\mathsf{Q} = \begin{bmatrix} \cos\theta & \sin\theta \\ -\sin\theta & \cos\theta \end{bmatrix}.$$

Multiplying an arbitrary vector $\mathbf{x} \in \mathbb{R}^2$ by such a matrix produces a vector $\mathsf{Q}\mathbf{x}$ that lies at an angle θ to $\mathbf{x}$, as shown in Figure 1(a). The second class consists of **reflections**. These matrices have the form

$$\mathsf{Q} = \mathsf{I} - 2\mathbf{w}\mathbf{w}^\top,$$

where $\|\mathbf{w}\|_2 = 1$. The action of Q in this case is to reflect an arbitrary vector $\mathbf{x} \in \mathbb{R}^2$ across the line perpendicular to the unit vector $\mathbf{w}$, as illustrated in Figure 1(b).

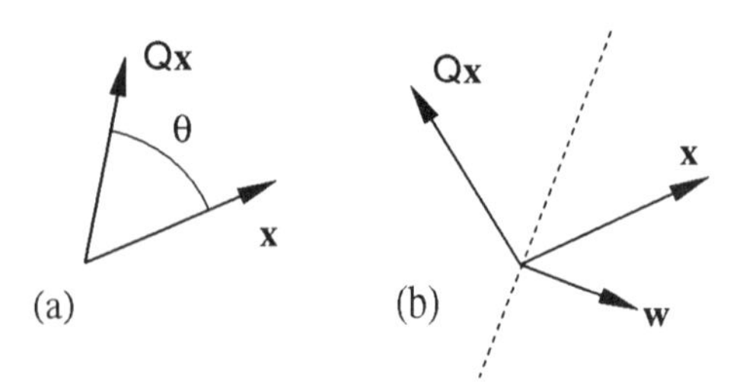

FIGURE 1. *(a) Effect of a rotation through angle θ on a vector $\mathbf{x} \in \mathbb{R}^2$. (b) Effect of a reflection of $\mathbf{x}$ across the line perpendicular to a vector $\mathbf{w}$.*

These classes generalize to $\mathbb{R}^{n\times n}$. The simplest extensions of rotations are **plane rotations** or **Givens transformations**. These have the form

$$\mathsf{Q} = \begin{bmatrix} 1 & & & & & & & & & \\ & \ddots & & & & & & & & \\ & & 1 & & & & & & & \\ & & & q_{i,i} & & & q_{i,j} & & & \\ & & & & 1 & & & & & \\ & & & & & \ddots & & & & \\ & & & & & & 1 & & & \\ & & & q_{j,i} & & & q_{j,j} & & & \\ & & & & & & & 1 & & \\ & & & & & & & & \ddots & \\ & & & & & & & & & 1 \end{bmatrix}, \tag{5.3-1}$$

where $q_{i,i} = q_{j,j} = \cos\theta$ and $q_{i,j} = -q_{j,i} = \sin\theta$. One can show (see Problem 4) that Givens transformations are orthogonal and that their action is to rotate a vector $\mathbf{x}$ through the angle θ in the (x_i, x_j)-plane, leaving the other coordinates x_k invariant.

Reflections extend in a straightforward way to **plane reflections** or **Householder transformations**. These have the same form in $\mathbb{R}^{n\times n}$ as in $\mathbb{R}^{2\times 2}$:

$$\mathsf{Q} = \mathsf{I} - 2\mathbf{w}\mathbf{w}^\top, \tag{5.3-2}$$

where $\|\mathbf{w}\|_2 = 1$. Problem 5 asks for proof that Householder transformations are symmetric, orthogonal matrices and that their action is to reflect vectors across the hyperplane in $\mathbb{R}^n$ that is perpendicular to $\mathbf{w}$. Symmetry and orthogonality together imply that, if Q is a Householder transformation, then $\mathsf{Q}^{-1} = \mathsf{Q}^\top = \mathsf{Q}$.

These two classes of matrices serve as tools in the construction of QR decompositions. Such a decomposition is possible for *every* matrix $\mathsf{A} \in \mathbb{R}^{n\times n}$:

THEOREM 5.8 (QR DECOMPOSITION). *If* $\mathsf{A} \in \mathbb{R}^{n\times n}$, *then there is an orthogonal matrix* $\mathsf{Q} \in \mathbb{R}^{n\times n}$ *and an upper triangular matrix* $\mathsf{R} \in \mathbb{R}^{n\times n}$ *such that* $\mathsf{A} = \mathsf{QR}$.

PROOF: The idea is to apply a succession of Householder transformations to A to triangularize it column by column. The result of these steps is R, and the Householder transformations determine Q.

The first step is representative: Let $\mathsf{A} = [\mathbf{a}_1, \mathbf{a}_2, \ldots, \mathbf{a}_n]$, where the vectors $\mathbf{a}_k$ denote columns of A. Define $\mathbf{w}_1 := \mathbf{v}_1/\|\mathbf{v}_1\|_2$, where

$$\alpha_1 := -\text{sgn}\,(a_{1,1})\,\|\mathbf{a}_1\|_2,$$

$$\mathbf{v}_1 := \mathbf{a}_1 - \alpha_1 \mathbf{e}_1,$$

$$\|\mathbf{v}_1\|_2 = \sqrt{2\alpha_1^2 - 2a_{1,1}\alpha_1}.$$

(The choice of sign in the definition of α_1 avoids numerically undesirable cancellation in the computation of $\|\mathbf{v}_1\|_2$.) By construction, $\mathbf{w}_1^\top \mathbf{w}_1 = 1$, so the matrix $\mathsf{H}_1 := \mathsf{I} - 2\mathbf{w}_1\mathbf{w}_1^\top$ is a Householder transformation. We claim that $\mathsf{H}_1\mathbf{a}_1 = (\alpha_1, 0, \ldots, 0)^\top$, so that the first column of $\mathsf{H}_1\mathsf{A}$ has only zeros below the diagonal entry. To see this fact, observe that

$$\begin{aligned} \mathsf{H}_1\mathbf{a}_1 &= \mathbf{a}_1 - 2\mathbf{w}_1\mathbf{w}_1^\top \mathbf{a}_1 \\ &= \mathbf{a}_1 - \frac{2\mathbf{v}_1}{\|\mathbf{v}_1\|_2^2}\mathbf{v}_1^\top \mathbf{a}_1. \end{aligned}$$

But $\mathbf{v}_1^\top \mathbf{a}_1 = \|\mathbf{a}_1\|_2^2 - \alpha_1 a_{1,1}$, so $\mathsf{H}_1\mathbf{a}_1$ collapses to

$$\mathbf{a}_1 - \frac{2}{\|\mathbf{v}_1\|_2^2}(\mathbf{v}_1^\top \mathbf{a}_1)\mathbf{v}_1 = (\alpha_1, 0, \ldots, 0)^\top,$$

establishing the claim.

(It is worth noting at this point that the quantity

$$\beta_1 := \frac{2}{\|\mathbf{v}_1\|_2^2} = \frac{1}{\alpha_1^2 - \alpha_1 a_{1,1}}$$

is more efficient to compute than $1/\|\mathbf{v}_1\|_2$, since the former does not require extraction of a square root. We exploit this observation in constructing an algorithm for QR decomposition.)

We now proceed by induction. After step $k-1$, we have reduced the original matrix A to a matrix that is upper triangular through the row and column indexed $k-1$. We now operate on the $(n-k+1)\times(n-k+1)$ submatrix that remains. At step k, let $\mathbf{w}_k := \mathbf{v}_k/\|\mathbf{v}_k\|_2$, where

$$\mathbf{v}_k := (0, \ldots, 0, a_{k,k} - \alpha_k, a_{k+1,k}, \ldots, a_{n,k}),$$

and

$$\alpha_k := -\mathrm{sgn}\,(a_{k,k})\,\|(0,\ldots,0,a_{k,k},a_{k+1,k},\ldots,a_{n,k})^\top\|_2,$$

$$\|\mathbf{v}_k\|_2 = \sqrt{2\alpha_k^2 - 2a_{k,k}\alpha_k} \quad (= \sqrt{2/\beta_k}).$$

Then $\mathsf{H}_k := \mathsf{I} - 2\mathbf{w}_k\mathbf{w}_k^\top$ is a Householder transformation as before, and the matrix $\mathsf{H}_k \cdots \mathsf{H}_2\mathsf{H}_1\mathsf{A}$ has zeros below the diagonal in columns $1, 2, \ldots, k$.

After step $n-1$, we have the upper triangular matrix

$$\mathsf{R} := \mathsf{H}_{n-1}\cdots\mathsf{H}_2\mathsf{H}_1\mathsf{A}.$$

Thus $\mathsf{A} = \mathsf{QR}$, where $\mathsf{Q} := \mathsf{H}_1^{-1}\mathsf{H}_2^{-1}\cdots\mathsf{H}_{n-1}^{-1} = \mathsf{H}_1\mathsf{H}_2\cdots\mathsf{H}_{n-1}$. The matrix Q is orthogonal since it is the product of orthogonal matrices. ∎

Problem 6 asks the reader to "walk through" the QR decomposition of a 3×3 matrix, identifying the plane reflections used at each stage.

The geometry used in the proof of the QR decomposition underlies a general-purpose algorithm:

ALGORITHM 5.2 (HOUSEHOLDER ORTHOGONALIZATION). *Given a matrix $\mathsf{A} \in \mathbb{R}^{n\times n}$, the following algorithm computes the* QR *decomposition of* A. *The algorithm overwrites* A *with the upper triangular matrix* R.

1. For $k = 1, 2, \ldots, n-1$:
2. $\quad\alpha_k \leftarrow -\mathrm{sgn}\,(a_{k,k})\,\|(0,\ldots,0,a_{k,k},a_{k+1,k},\ldots,a_{n,k})^\top\|_2$.
3. $\quad\mathbf{v}_k \leftarrow (0,\ldots,0,a_{k,k}-\alpha_k,a_{k+1,k},\ldots,a_{n,k})^\top$.
4. $\quad\beta_k \leftarrow 1/(\alpha_k^2 - \alpha_k a_{k,k})$.
5. $\quad a_{k,k} \leftarrow \alpha_k$.
6. $\quad$For $j = k+1, k+2, \ldots, n$:
7. $\quad\quad\mathbf{a}_j \leftarrow \mathbf{a}_j - \beta_k(\mathbf{v}_k^\top\mathbf{a}_j)\mathbf{v}_k$.
8. $\quad$Next j.
9. Next k.
10. End.

Let us count the arithmetic operations required in this algorithm. In the inner loop, steps 6 through 8, we must compute $\mathbf{v}_k^\top\mathbf{a}_j$ and $\mathbf{a}_j - \beta_k(\mathbf{v}_k^\top\mathbf{a}_j)\mathbf{v}_k$ for the $n-k$ columns indexed by j. The operation counts for these steps are, respectively,

$$(n-k+1)\ \text{multiplications} \quad + \quad (n-k)\ \text{additions}$$

and

$$(1+n-k+1) \quad \text{multiplications} \quad + \quad (n-k+1) \quad \text{additions}.$$

Thus a typical pass through the inner loop requires $(n-k)[4(n-k)+4]$ arithmetic operations. Since the arithmetic required in the inner loop is the dominant contribution to the overall computational effort, the total operation count is roughly

$$4\sum_{k=1}^{n-1}\left[(n-k)^2+(n-k)\right] = \tfrac{4}{3}n^3 + \mathcal{O}(n^2).$$

Thus Householder orthogonalization requires approximately twice as many operations as ordinary LU decomposition.

It is possible to build a similar procedure using Givens transformations instead of Householder transformations, but the resulting algorithm typically requires more arithmetic. We discuss an important class of exceptions in the next section.

Application to Least-Squares Problems

While the main purpose in developing QR decompositions is to compute eigenvalues, the technique has other significant applications. Among these is the solution of least-squares problems. Recall from Section 1.7 that solving such problems involves finding a solution to the normal equations,

$$\mathsf{A}^\top\mathsf{A}\mathbf{x} = \mathsf{A}^\top\mathbf{b}. \tag{5.3-3}$$

Here, $\mathsf{A} \in \mathbb{R}^{n\times m}$, with $n \geq m$. We assume that A has linearly independent columns, so that $\det(\mathsf{A}^\top\mathsf{A}) \neq 0$ and the system has a unique solution.

To apply the QR decomposition to this problem, we first extend the decomposition to nonsquare matrices:

PROPOSITION 5.9. *If $\mathsf{A} \in \mathbb{R}^{n\times m}$, with $n \geq m$, then there exists an orthogonal matrix $\mathsf{Q} \in \mathbb{R}^{n\times n}$ and an upper triangular matrix $\mathsf{R} \in \mathbb{R}^{n\times m}$ such that $\mathsf{A} = \mathsf{QR}$.*

In this case the QR decomposition has a zero structure exemplified in the following schematic:

$$\underbrace{\begin{bmatrix} \star & \star & \star \\ \star & \star & \star \\ \star & \star & \star \\ \star & \star & \star \end{bmatrix}}_{n\times m} = \underbrace{\begin{bmatrix} \star & \star & \star & \star \\ \star & \star & \star & \star \\ \star & \star & \star & \star \\ \star & \star & \star & \star \end{bmatrix}}_{n\times n}\underbrace{\begin{bmatrix} \star & \star & \star \\ & \star & \star \\ & & \star \\ & & \end{bmatrix}}_{n\times m}.$$

PROOF: Construct a square matrix $\mathsf{A}_\square := [\mathsf{A} \quad 0] \in \mathbb{R}^{n\times n}$ by appending $n-m$ columns of zeros to A. Then apply the Householder orthogonalization to $\mathsf{A}_\square$, truncating the outer loop in Algorithm 5.1 as follows:

1. For $k = 1, 2, \ldots, m-1$: ∎

This QR decomposition simplifies the normal equations (5.3-3) through the following set of equivalent linear systems:

$$\begin{aligned} \mathsf{A}^\mathsf{T}\mathsf{A}\mathbf{x} &= \mathsf{A}^\mathsf{T}\mathbf{b}, \\ \mathsf{R}^\mathsf{T}\mathsf{Q}^\mathsf{T}\mathsf{Q}\mathsf{R}\mathbf{x} &= \mathsf{R}^\mathsf{T}\mathsf{Q}^\mathsf{T}\mathbf{b}, \\ \mathsf{R}^\mathsf{T}\mathsf{R}\mathbf{x} &= \mathsf{R}^\mathsf{T}\mathsf{Q}^\mathsf{T}\mathbf{b}. \end{aligned} \tag{5.3-4}$$

Now consider the matrix $\mathsf{R}_\square \in \mathbb{R}^{m\times m}$ comprising the first m rows of R:

$$\mathsf{R} = \begin{bmatrix} \mathsf{R}_\square \\ 0 \end{bmatrix}.$$

The matrix $\mathsf{R}_\square$ is upper triangular, and $\mathsf{A}^\mathsf{T}\mathsf{A} = \mathsf{R}^\mathsf{T}\mathsf{R} = \mathsf{R}_\square^\mathsf{T}\mathsf{R}_\square$. Also, since $\mathsf{A}^\mathsf{T}\mathsf{A}$ is nonsingular, so is $\mathsf{R}_\square^\mathsf{T}\mathsf{R}_\square$. This fact implies that $\mathsf{R}_\square^\mathsf{T}$ is nonsingular, since if $\mathsf{R}_\square^\mathsf{T}$ has a zero eigenvalue then so does $\mathsf{R}_\square^\mathsf{T}\mathsf{R}_\square$, which is impossible.

Using these observations, we rewrite the normal equations in a form that is easy to solve. The right side of Equation (5.3-4) reduces to the vector

$$\mathsf{R}^\mathsf{T}\underbrace{\mathsf{Q}^\mathsf{T}\mathbf{b}}_{\mathbf{c}} = \begin{bmatrix} \mathsf{R}_\square^\mathsf{T} & 0 \end{bmatrix} \begin{bmatrix} \mathbf{c}_1 \\ \mathbf{c}_2 \end{bmatrix} = \mathsf{R}_\square^\mathsf{T}\mathbf{c}_1,$$

where $\mathbf{c}_1 \in \mathbb{R}^m$ contains the first m entries of the vector $\mathsf{Q}^\mathsf{T}\mathbf{b}$. Thus the normal equations (5.3-3) collapse to the system $\mathsf{R}_\square^\mathsf{T}\mathsf{R}_\square\mathbf{x} = \mathsf{R}_\square^\mathsf{T}\mathbf{c}$. But, because $\mathsf{R}_\square^\mathsf{T}$ is nonsingular, this system reduces to $\mathsf{R}_\square\mathbf{x} = \mathbf{c}_1$. In summary, having computed the QR decomposition of $\mathsf{A}^\mathsf{T}\mathsf{A}$, we easily convert the normal equations to a system that is solvable via backward substitution.

Of course, to construct the QR decomposition of $\mathsf{A}^\mathsf{T}\mathsf{A}$ to begin with costs more operations than are required to solve the normal equations (5.3-3) directly. The advantage of using the QR decomposition in least-squares systems consists not in the reduction in operation count but rather in the fact that the system $\mathsf{R}_\square\mathbf{x} = \mathbf{c}_1$ is typically better conditioned than the system (5.3-3). When A is a square matrix, roughly speaking, $\operatorname{cond}(\mathsf{A}^\mathsf{T}\mathsf{A})$ is the square of $\operatorname{cond}(\mathsf{A})$. Similar conditioning problems arise when A is not square. As an illustration of how poor conditioning can arise, consider the 4×3 matrix

$$\mathsf{A} = \begin{bmatrix} 1 & 1 & 1 \\ \epsilon & 0 & 0 \\ 0 & \epsilon & 0 \\ 0 & 0 & \epsilon \end{bmatrix}.$$

In this case,

$$\mathsf{A}^\top\mathsf{A} = \begin{bmatrix} 1+\epsilon^2 & 1 & 1 \\ 1 & 1+\epsilon^2 & 1 \\ 1 & 1 & 1+\epsilon^2 \end{bmatrix},$$

the eigenvalues of which are $\epsilon^2, \epsilon^2, 3+\epsilon^2$. Therefore

$$\mathrm{cond}_2(\mathsf{A}^\top\mathsf{A}) = \frac{3+\epsilon^2}{\epsilon^2} = \mathcal{O}(\epsilon^{-2}),$$

as $\epsilon \to 0$.

Further Remarks

The theory presented in this section extends to matrices $\mathsf{A} \in \mathbb{C}^{n\times n}$, provided we change terminology appropriately. For every matrix $\mathsf{A} \in \mathbb{C}^{n\times n}$, there exists a unitary matrix $\mathsf{Q} \in \mathbb{C}^{n\times n}$ and an upper triangular matrix $\mathsf{R} \in \mathbb{C}^{n\times n}$ such that $\mathsf{A} = \mathsf{QR}$. (Recall that Q is unitary if $\mathsf{Q}^*\mathsf{Q} = \mathsf{QQ}^* = \mathsf{I}$, where Q^* denotes the conjugate transpose, or Hermitian transpose, of Q.) The proof parallels that for real matrices: One uses Householder transformations of the form $\mathsf{H}_k = \mathsf{I} - 2\mathbf{w}\mathbf{w}^*$, where $\mathbf{w} \in \mathbb{C}^n$ with $\mathbf{w}^*\mathbf{w} = 1$, to triangularize A column by column.

QR decompositions are not unique. However, as the following proposition demonstrates, all QR decompositions of a nonsingular matrix are closely related, a fact that is useful for the theory developed in the next section.

PROPOSITION 5.10. *Let $\mathsf{A} \in \mathbb{R}^{n\times n}$ be nonsingular. Then the decomposition $\mathsf{A} = \mathsf{QR}$ is unique up to the choice of signs on the diagonal entries of R.*

The assertion means the following: Given two QR decompositions $\mathsf{A} = \mathsf{Q}_1\mathsf{R}_1$ and $\mathsf{A} = \mathsf{Q}_2\mathsf{R}_2$, there is a diagonal matrix U, whose diagonal entries are all ± 1, such that $\mathsf{R}_2 = \mathsf{U}\mathsf{R}_1$. Concomitantly, $\mathsf{Q}_2 = \mathsf{Q}_1\mathsf{U}^\top = \mathsf{Q}_1\mathsf{U}$.

PROOF: Assume that $\mathsf{A} = \mathsf{Q}_1\mathsf{R}_1 = \mathsf{Q}_2\mathsf{R}_2$. Since $|\det \mathsf{Q}_1| = |\det \mathsf{Q}_2| = 1$ and $\det \mathsf{A} \neq 0$ by hypothesis, $\det \mathsf{R}_1 \neq 0 \neq \det \mathsf{R}_2$. Consequently, neither R_1 nor R_2 has a zero among its eigenvalues, which are its diagonal entries. Consider the matrix $\mathsf{U} := \mathsf{Q}_2^\top\mathsf{Q}_1 = \mathsf{R}_2\mathsf{R}_1^{-1}$. Being the product of orthogonal matrices, U is orthogonal. Also, by the results of Problem 7, U is upper triangular. It follows that

$$\mathsf{U}^\top\mathsf{U} = \begin{bmatrix} u_{1,1} & & \\ \vdots & \ddots & \\ u_{1,n} & \cdots & u_{n,n} \end{bmatrix} \begin{bmatrix} u_{1,1} & \cdots & u_{1,n} \\ & \ddots & \vdots \\ & & u_{n,n} \end{bmatrix} = \mathsf{I}.$$

This equation can hold only if

$$\mathsf{U} = \begin{bmatrix} \pm 1 & & \\ & \ddots & \\ & & \pm 1 \end{bmatrix}.$$

Since $\mathsf{Q}_2\mathsf{R}_2 = \mathsf{Q}_2\mathsf{U}\mathsf{R}_1$, $\mathsf{R}_2 = \mathsf{U}\mathsf{R}_1$, as desired. ∎

The following corollary is immediate:

COROLLARY 5.11. *Any nonsingular matrix* $\mathsf{A} \in \mathbb{R}^{n\times n}$ *has a unique* QR *decomposition in which the diagonal entries of* R *are positive.*

Analogous reasoning shows that, for any nonsingular matrix $\mathsf{A} \in \mathbb{C}^{n\times n}$, the QR decomposition is unique up to multiplication by a diagonal matrix of the form

$$\mathsf{U} = \begin{bmatrix} e^{i\phi_1} & & & \\ & e^{i\phi_2} & & \\ & & \ddots & \\ & & & e^{i\phi_n} \end{bmatrix}, \tag{5.3-5}$$

where each $\phi_j \in \mathbb{R}$. Thus, given two QR decompositions $\mathsf{A} = \mathsf{Q}_1\mathsf{R}_1 = \mathsf{Q}_2\mathsf{R}_2$, we must have $\mathsf{Q}_2 = \mathsf{Q}_2\mathsf{U}^*$ and $\mathsf{R}_2 = \mathsf{U}\mathsf{R}_1$, for some matrix of the form (5.3-5).

5.4 The QR Method: Implementation

Motivation and Construction

QR decompositions form the core of the most widely used numerical technique for computing eigenvalues. This technique, the **QR method**, has several attractive features: It finds all eigenvalues of a matrix; its behavior in the presence of equal-magnitude eigenvalues is much tamer than that of the power method, and it simplifies when the matrix in question is symmetric [1].

The idea is both simple and seemingly mysterious. Given $\mathsf{A} \in \mathbb{R}^{n\times n}$, we decompose $\mathsf{A} = \mathsf{Q}\mathsf{R}$, where Q is orthogonal and R is upper triangular. We then form a new matrix $\mathsf{A}_1 := \mathsf{R}\mathsf{Q}$. The procedure continues by induction: Given $\mathsf{A}_m = \mathsf{Q}_m\mathsf{R}_m$, we form $\mathsf{A}_{m+1} := \mathsf{R}_m\mathsf{Q}_m$. While the procedure itself is straightforward, the result is not: For a "typical" matrix A with eigenvalues $\lambda_1, \lambda_2, \ldots, \lambda_n$,

$$\mathsf{A}_m \to \begin{bmatrix} \lambda_1 & \star & \cdots & \star \\ & \lambda_2 & & \vdots \\ & & \ddots & \star \\ & & & \lambda_n \end{bmatrix},$$

as $m \to \infty$. This result holds whenever $|\lambda_1| > |\lambda_2| > \cdots > |\lambda_n|$. In other cases, the sequence $\{\mathsf{A}_m\}$ still yields useful information about the spectrum of A; we discuss details later.

There are two sources of mystery in this procedure. First, what connection do the matrices A_m have with the spectrum of A? Second, why does the sequence A_m tend to such a convenient structure? The first question is easier to answer:

PROPOSITION 5.12. *The matrices* A_m *generated by the* QR *method are all similar to* A.

PROOF: The argument is by induction. Clearly $\mathsf{A}_0 := \mathsf{A}$ is similar to itself. To complete the induction, we need only show that A_{m+1} is similar to A_m. We have

$$\mathsf{A}_{m+1} = \mathsf{R}_m\mathsf{Q}_m = \mathsf{Q}_m^{-1}\mathsf{Q}_m\mathsf{R}_m\mathsf{Q}_m = \mathsf{Q}_m^{-1}\mathsf{A}_m\mathsf{Q}_m. \qquad \blacksquare$$

The second question is subtler; we discuss it later in this section.

The following example illustrates the QR method, showing iterates A_m generated for the matrix

$$\mathsf{A} = \begin{bmatrix} 2 & -1 & 0 \\ -1 & 2 & -1 \\ 0 & -1 & 2 \end{bmatrix}.$$

This matrix has eigenvalues 3.4142, 2.000, and 0.5858, accurate to four decimal places.

$$\begin{aligned}
\mathsf{A}_1 &= \begin{bmatrix} 2 & -1 & 0 \\ -1 & 2 & -1 \\ 0 & -1 & 2 \end{bmatrix} \\
&= \underbrace{\begin{bmatrix} -0.8944 & -0.3586 & -0.2673 \\ -0.4472 & -0.7171 & -0.5345 \\ 0 & 0.5976 & -0.8018 \end{bmatrix}}_{\mathsf{Q}_1} \underbrace{\begin{bmatrix} -2.2361 & 1.7889 & -0.4472 \\ 0 & -1.6733 & 1.9124 \\ 0 & 0 & -1.0690 \end{bmatrix}}_{\mathsf{R}_1} \\
\mathsf{A}_2 &= \mathsf{R}_1\mathsf{Q}_1 = \begin{bmatrix} 2.8000 & -0.7483 & 0 \\ -0.7483 & 2.3429 & -0.6389 \\ 0 & -0.6389 & 0.8571 \end{bmatrix} \\
&= \underbrace{\begin{bmatrix} -0.9661 & -0.2467 & -0.0761 \\ 0.2582 & -0.9231 & 0.2949 \\ 0 & 0.2949 & -0.9555 \end{bmatrix}}_{\mathsf{Q}_2} \underbrace{\begin{bmatrix} -2.8983 & 1.3279 & -0.1650 \\ 0 & -2.1665 & 0.8425 \\ 0 & 0 & -0.6370 \end{bmatrix}}_{\mathsf{R}_2} \\
\mathsf{A}_3 &= \mathsf{R}_2\mathsf{Q}_2 = \begin{bmatrix} 3.1429 & -0.5594 & 0 \\ -0.5594 & 2.2485 & -0.1878 \\ 0 & -0.1878 & 0.6087 \end{bmatrix} \\
&= \underbrace{\begin{bmatrix} -0.9845 & -0.1745 & -0.0155 \\ 0.1752 & -0.9807 & -0.0871 \\ 0 & 0.0884 & -0.9961 \end{bmatrix}}_{\mathsf{Q}_3} \underbrace{\begin{bmatrix} -3.1923 & 0.9448 & -0.0329 \\ 0 & -2.1240 & 0.2381 \\ 0 & 0 & -0.5900 \end{bmatrix}}_{\mathsf{R}_3} \\
\mathsf{A}_4 &= \mathsf{R}_3\mathsf{Q}_3 = \begin{bmatrix} 3.3084 & -0.3722 & 0 \\ -0.3722 & 2.1040 & -0.0522 \\ 0 & -0.0522 & 0.5876 \end{bmatrix} \\
&= \underbrace{\begin{bmatrix} -0.9937 & -0.1118 & -0.0028 \\ 0.1118 & -0.9934 & -0.0253 \\ 0 & 0.0254 & -1.000 \end{bmatrix}}_{\mathsf{Q}_4} \underbrace{\begin{bmatrix} -3.3292 & 0.6051 & -0.0058 \\ 0 & -2.0498 & 0.0668 \\ 0 & 0 & -0.5861 \end{bmatrix}}_{\mathsf{R}_4}
\end{aligned}$$

$$\mathsf{A}_5 = \mathsf{R}_4\mathsf{Q}_4 = \begin{bmatrix} 3.3761 & -0.2292 & 0 \\ -0.2292 & 2.0380 & -0.0149 \\ 0 & -0.0149 & 0.5859 \end{bmatrix}$$

$$= \underbrace{\begin{bmatrix} -0.9977 & -0.0766 & -0.0005 \\ 0.0677 & -0.9977 & -0.0094 \\ 0 & 0.0074 & -1.000 \end{bmatrix}}_{\mathsf{Q}_5} \underbrace{\begin{bmatrix} -3.3838 & 0.3666 & -0.0010 \\ 0 & -2.0179 & 0.0192 \\ 0 & 0 & -0.5858 \end{bmatrix}}_{\mathsf{R}_5}$$

$$\mathsf{A}_6 = \mathsf{R}_5\mathsf{Q}_5 = \begin{bmatrix} 3.4009 & -0.1367 & 0 \\ -0.1367 & 2.0133 & -0.0043 \\ 0 & -0.0043 & 0.5858 \end{bmatrix}$$

$$= \underbrace{\begin{bmatrix} -0.9991 & -0.0401 & -0.0001 \\ 0.0401 & -0.9992 & -0.0022 \\ 0 & 0.0022 & -1.0000 \end{bmatrix}}_{\mathsf{Q}_6} \underbrace{\begin{bmatrix} -3.4036 & 0.2174 & -0.0002 \\ 0 & -2.0062 & 0.0056 \\ 0 & 0 & -0.5858 \end{bmatrix}}_{\mathsf{R}_6}$$

$$\mathsf{A}_7 = \mathsf{R}_6\mathsf{Q}_6 = \begin{bmatrix} 3.4100 & -0.0805 & 0 \\ -0.0805 & 2.0046 & -0.0013 \\ 0 & -0.0013 & 0.5858 \end{bmatrix}$$

$$= \underbrace{\begin{bmatrix} -0.9997 & -0.0236 & 0 \\ 0.0236 & -0.9997 & -0.0006 \\ 0 & 0.0006 & -1.0000 \end{bmatrix}}_{\mathsf{Q}_7} \underbrace{\begin{bmatrix} -3.4106 & 0.1279 & 0 \\ 0 & -2.0021 & 0.0016 \\ 0 & 0 & -0.5858 \end{bmatrix}}_{\mathsf{R}_7}$$

$$\mathsf{A}_8 = \mathsf{R}_7\mathsf{Q}_7 = \begin{bmatrix} 3.4126 & -0.0473 & 0 \\ -0.0473 & 2.0016 & -0.0004 \\ 0 & -0.0004 & 0.5858 \end{bmatrix}$$

$$= \underbrace{\begin{bmatrix} -0.9990 & -0.0139 & 0 \\ 0.0139 & -0.9999 & -0.0002 \\ 0 & -0.0002 & -1.0000 \end{bmatrix}}_{\mathsf{Q}_8} \underbrace{\begin{bmatrix} -3.4130 & 0.0750 & 0 \\ 0 & -2.0007 & 0.0005 \\ 0 & 0 & -0.5858 \end{bmatrix}}_{\mathsf{R}_8}$$

$$\mathsf{A}_9 = \mathsf{R}_8\mathsf{Q}_8 = \begin{bmatrix} 3.4137 & -0.0277 & 0 \\ -0.0277 & 2.0005 & -0.0001 \\ 0 & -0.0001 & 0.5858 \end{bmatrix}.$$

In this case, 9 iterations yield a matrix A_9 whose diagonal entries are approximations to the eigenvalues of A, accurate to three decimal places.

Practical Considerations

Without further adornment, the QR method as outlined requires too much arithmetic to be practical. For a general matrix $\mathsf{A} \in \mathbb{R}^{n\times n}$, each QR decomposition $\mathsf{A}_m = \mathsf{Q}_m\mathsf{R}_m$ requires $\mathcal{O}(\frac{4}{3}n^3)$ operations. One can overcome this obstacle by performing an initial similarity transformation on A that reduces it to a form for which subsequent QR decompositions are much cheaper.

In particular, we initiate the QR method by converting A to a matrix $\tilde{\mathsf{A}}$

that is in **Hessenberg form**:

$$\tilde{\mathsf{A}} = \begin{bmatrix} \star & \star & \cdots & \star & \star \\ \star & \star & \cdots & \star & \star \\ & \star & \ddots & \vdots & \vdots \\ & & \ddots & \star & \star \\ & & & \star & \star \end{bmatrix}.$$

After this reduction, it is possible to compute the QR decomposition of the initial matrix $\mathsf{A}_0 := \tilde{\mathsf{A}}$ in $\mathcal{O}(n^2)$ operations, using an approach that we discuss shortly. We also show below that each of the subsequent iterates A_m remains in Hessenberg form. Consequently, one can assess the convergence of the QR method by monitoring the decay of the subdiagonal entries $a_{j,j-1}$ in the iterates A_m.

To reduce A to Hessenberg form, we use Householder transformations, as introduced in Section 5.3, to convert the columns of A one at a time. Let us examine in detail the operations needed for the first column. We seek a Householder transformation $\mathsf{H}_1 = \mathsf{I} - 2\mathbf{w}_1\mathbf{w}_1^\top$ such that

$$\mathsf{H}_1\mathsf{A} = \begin{bmatrix} \star & \star & \cdots & \star & \star \\ \star & \star & \cdots & \star & \star \\ 0 & \star & \cdots & \star & \star \\ \vdots & \vdots & \ddots & \vdots & \vdots \\ 0 & \star & \cdots & \star & \star \end{bmatrix}. \tag{5.4-1}$$

For this task, pick

$$\mathbf{w}_1 := \mu_1 \left(0, a_{2,1} - \alpha_2, a_{3,1}, \ldots, a_{n,n}\right)^\top,$$

where

$$\alpha_1 \quad := \quad -\mathrm{sgn}\,(a_{2,1}) \left(\sum_{j=2}^{n} a_{j,1}^2 \right)^{1/2},$$

$$\mu_1 \quad := \quad \frac{1}{\sqrt{2\alpha_1^2 - 2a_{2,1}\alpha_1}}.$$

(This construction recalls a similar use of Householder transformations in Section 5.3.) With this choice, $\mathsf{H}_1\mathsf{A}$ has the desired zero structure (5.4-1), and A is similar to the matrix

$$\mathsf{H}_1\mathsf{A}\mathsf{H}_1^\top = \mathsf{A} - 2\mathbf{w}_1\mathbf{w}_1^\top\mathsf{A} - 2\mathsf{A}\mathbf{w}_1\mathbf{w}_1^\top + 4\mathbf{w}_1^\top\mathsf{A}\mathbf{w}_1\mathbf{w}_1\mathbf{w}_1^\top.$$

One readily checks that H_1 has the zero structure

$$\mathsf{H}_1 = \begin{bmatrix} 1 & 0 & \cdots & 0 \\ 0 & \star & \cdots & \star \\ \vdots & \vdots & & \vdots \\ 0 & \star & \cdots & \star \end{bmatrix}.$$

Therefore, $\mathsf{H}_1\mathsf{A}\mathsf{H}_1^\top$ has the same zero structure (5.4-1) as $\mathsf{H}_1\mathsf{A}$.

Knowing in detail how to accomplish one step of the reduction to Hessenberg form, one can develop Householder transformations $\mathsf{H}_1, \mathsf{H}_2, \ldots, \mathsf{H}_{n-2}$ such that

$$\begin{aligned}\tilde{\mathsf{A}} &= \mathsf{H}_{n-2}\cdots\mathsf{H}_2\mathsf{H}_1\mathsf{A}\mathsf{H}_1^\top\mathsf{H}_2^\top\cdots\mathsf{H}_{n-2}^\top \\ &= \begin{bmatrix} \star & \star & \cdots & \star & \star \\ \star & \star & \cdots & \star & \star \\ & \star & & \vdots & \vdots \\ & & \ddots & \star & \star \\ & & & \star & \star \end{bmatrix}.\end{aligned} \tag{5.4-2}$$

Problem 8 asks for details and for a demonstration that this procedure costs $\mathcal{O}(n^3)$ arithmetic operations.

When the original matrix A is symmetric, the reduction is even nicer. It is an easy exercise to show that $\mathsf{A} = \mathsf{A}^\top$ implies that $\tilde{\mathsf{A}} = \tilde{\mathsf{A}}^\top$. This is good news: For symmetric matrices, the Hessenberg form is tridiagonal, and there are attractive opportunities for savings in both storage and arithmetic.

Now we apply the QR method to the initial matrix $\mathsf{A}_0 := \tilde{\mathsf{A}}$, producing a sequence $\{\mathsf{A}_0, \mathsf{A}_1, \mathsf{A}_2, \ldots\}$. We verify shortly that each of the matrices A_m remains in Hessenberg form, and we examine later in this section the circumstances under which

$$\mathsf{A}_m \to \begin{bmatrix} \lambda_1 & \star & \cdots & \star \\ & \lambda_2 & & \vdots \\ & & \ddots & \star \\ & & & \lambda_n \end{bmatrix},$$

as $m \to \infty$.

For now, let us discuss implementation. Algorithm 5.2 turns out not to be the best choice for the decompositions needed in the QR method, even though it is a reasonable way to factor general matrices. In the present context, the fact that the matrices A_m are in Hessenberg form makes it more efficient to use Givens transformations, instead of Householder transformations, to triangularize A_m at each iteration.

To see how this scheme works, suppose that we have completed stage k of the triangularization. Thus, we have applied k Givens transformations G_j to arrive at a partially triangularized matrix:

$$\mathsf{G}_k\cdots\mathsf{G}_2\mathsf{G}_1\mathsf{A}_m = \begin{matrix} & k & n-k \\ k & \mathsf{R}_k & \mathsf{B}_k \\ n-k & 0 & \mathsf{H}_k \end{matrix}.$$

Here, the block R_k is upper triangular and H_k is still in Hessenberg form.

For stage $k+1$, we choose another Givens transformation:

$$\mathsf{G}_{k+1} := \begin{bmatrix} 1 & & & & & & \\ & \ddots & & & & & \\ & & 1 & & & & \\ & & & g_{k+1,k+1} & g_{k+1,k+2} & & \\ & & & g_{k+2,k+1} & g_{k+2,k+2} & & \\ & & & & & 1 & \\ & & & & & & \ddots \\ & & & & & & & 1 \end{bmatrix},$$

where

$$g_{k+1,k+1} = g_{k+2,k+2} = \cos\theta_{k+1},$$

$$g_{k+1,k+2} = -g_{k+2,k+1} = \sin\theta_{k+1}.$$

Nominally, the goal is to pick θ_{k+1} such that

$$\mathsf{G}_{k+1}\mathsf{G}_k\cdots\mathsf{G}_1\mathsf{A}_m = \begin{matrix} \\ k+1 \\ n-k-1 \end{matrix} \begin{matrix} k+1 & n-k-1 \\ \left(\begin{matrix} \mathsf{R}_{k+1} \\ 0 \end{matrix}\right. & \left.\begin{matrix} \mathsf{B}_{k+1} \\ \mathsf{H}_{k+1} \end{matrix}\right) \end{matrix},$$

where again R_{k+1} is upper triangular and H_{k+1} is still in Hessenberg form. If we denote the (i,j)th entry of the matrix $\mathsf{G}_{k+1}\mathsf{G}_k\cdots\mathsf{G}_1\mathsf{A}_m$ by $a_{i,j}$, then we accomplish this goal by setting

$$g_{k+2,k+1}a_{k+1,k+1} + g_{k+2,k+2}a_{k+2,k+1} = 0,$$

that is,

$$-a_{k+1,k+1}\sin\theta_{k+1} + a_{k+2,k+1}\cos\theta_{k+1} = 0.$$

It follows that

$$g_{k+2,k+1} = \frac{-a_{k+2,k+1}}{\sqrt{a_{k+1,k+1}^2 + a_{k+2,k+1}^2}}, \qquad g_{k+2,k+2} = \frac{a_{k+1,k+1}}{\sqrt{a_{k+1,k+1}^2 + a_{k+2,k+1}^2}}.$$

In short, we compute G_{k+1} using a small number of arithmetic operations, never explicitly solving for θ_{k+1}.

Problem 9 shows that this use of Givens transformations to compute a QR decomposition of a matrix in Hessenberg form requires $\mathcal{O}(n^2)$ operations. This operation count stands in contrast with the $\mathcal{O}(n^3)$ operations required in Algorithm 5.2.

The $\mathcal{O}(n^3)$ operations required initially to reduce A to Hessenberg form A_0 would be largely for naught if the next iterate A_1 in the QR method were not in Hessenberg form. Fortunately, this does not happen. To see why A_1 and subsequent iterates remain in Hessenberg form, suppose that A_m is in Hessenberg form, and consider the QR decomposition of A_m via Givens transformations. We have

$$\underbrace{\mathsf{G}_{n-1}\cdots\mathsf{G}_2\mathsf{G}_1}_{\mathsf{Q}_m^\top}\mathsf{A}_m = \mathsf{R}_m,$$

where R_m is upper triangular. According to the QR method,

$$\mathsf{A}_{m+1} = \mathsf{R}_m\mathsf{Q}_m = \mathsf{R}_m\mathsf{G}_1^\top\mathsf{G}_2^\top\cdots\mathsf{G}_{n-1}^\top.$$

Since each of the Givens transformations $\mathsf{G}_k^\top$ has only two nonzero off-diagonal entries $g_{k+1,k}$ and $g_{k,k+1}$, the product $\mathsf{Q}_m := \mathsf{G}_1^\top\mathsf{G}_2^\top\cdots\mathsf{G}_{n-1}^\top$ is in Hessenberg form. This observation and the fact that R_m is upper triangular imply that the matrix $\mathsf{A}_{m+1} := \mathsf{R}_m\mathsf{Q}_m$ is also in Hessenberg form. Consequently, once we have performed the initial $\mathcal{O}(n^3)$ operations required to reduce A to Hessenberg form, subsequent iterations of the QR method require only $\mathcal{O}(n^2)$ operations apiece.

The following algorithm summarizes the method just outlined.

ALGORITHM 5.3 (QR METHOD). *Given a matrix* $\mathsf{A} \in \mathbb{R}^{n\times n}$ *and a tolerance* $\tau > 0$, *the following steps implement the* QR *method for determining the eigenvalues of* A. *In step 2, the notation "*$\mathsf{A}_0 \leftarrow$ Hessenberg (A)*" means, "reduce* A *to Hessenberg form." In step 6,* $a_{j,j-1}$ *denotes a subdiagonal entry of the iterate* A_{m+1}. *After numerical convergence, the approximate eigenvalues of* A *lie on the diagonal of* A_{m+1}.

1. $m \leftarrow 0$.
2. $\mathsf{A}_0 \leftarrow$ Hessenberg (A) (via Householder transformations).
3. $\mathsf{A}_m = \mathsf{Q}_m\mathsf{R}_m$ (via Givens transformations).
4. $\mathsf{A}_{m+1} \leftarrow \mathsf{R}_m\mathsf{Q}_m$.
5. $m \leftarrow m+1$.
6. If $\max_j |a_{j,j+1}| \geq \tau$ go to 3.
7. End.

Mathematical Details

We now turn to a convergence proof for the QR method. We give a detailed argument only for the case when the matrix A has eigenvalues that stand in the relationship $|\lambda_1| > |\lambda_2| > \cdots > |\lambda_n| > 0$. Convergence in other circumstances is a more complicated matter, discussed in a less rigorous fashion afterward. The analysis follows that given by Wilkinson [3, Chapter 8].

Much of the argument hinges on the behavior of powers of A, which the following lemma examines.

LEMMA 5.13 *If $\mathsf{A}_m = \mathsf{Q}_m\mathsf{R}_m$ denotes the* QR *decomposition of $\mathsf{A} \in \mathbb{R}^{n\times n}$ generated at stage m of the* QR *method, then the matrix A^m has* QR *decomposition*

$$\mathsf{A}^m = \mathsf{P}_m\mathsf{U}_m,$$

where $\mathsf{P}_m := \mathsf{Q}_1\mathsf{Q}_2\cdots\mathsf{Q}_m$ is orthogonal and $\mathsf{U}_m := \mathsf{R}_m\cdots\mathsf{R}_2\mathsf{R}_1$ is upper triangular.

PROOF: Observe that

$$\mathsf{A}_{m+1} = \mathsf{R}_m\mathsf{Q}_m = \mathsf{Q}_m^\top\mathsf{Q}_m\mathsf{R}_m\mathsf{Q}_m = \mathsf{Q}_m^\top\mathsf{A}_m\mathsf{Q}_m.$$

Repeated application of this identity gives

$$\mathsf{A}_{m+1} = \mathsf{Q}_m^\top\cdots\mathsf{Q}_2^\top\mathsf{Q}_1^\top\mathsf{A}\mathsf{Q}_1\mathsf{Q}_2\cdots\mathsf{Q}_m,$$

which implies that

$$\underbrace{\mathsf{Q}_1\mathsf{Q}_2\cdots\mathsf{Q}_m}_{\mathsf{P}_m}\mathsf{A}_{m+1} = \mathsf{A}\mathsf{Q}_1\mathsf{Q}_2\cdots\mathsf{Q}_m. \tag{5.4-3}$$

Now consider the product $\mathsf{P}_m\mathsf{U}_m$:

$$\begin{aligned}\mathsf{P}_m\mathsf{U}_m &= \mathsf{Q}_1\cdots\mathsf{Q}_{m-1}(\mathsf{Q}_m\mathsf{R}_m)\mathsf{R}_{m-1}\cdots\mathsf{R}_1\\ &= \mathsf{Q}_1\mathsf{Q}_2\cdots\mathsf{Q}_{m-1}\mathsf{A}_m\mathsf{R}_{m-1}\cdots\mathsf{R}_2\mathsf{R}_1.\end{aligned}$$

By virtue of Equation (5.4-3),

$$\begin{aligned}\mathsf{P}_m\mathsf{U}_m &= \mathsf{A}\mathsf{Q}_1\mathsf{Q}_2\cdots\mathsf{Q}_{m-1}\mathsf{R}_{m-1}\cdots\mathsf{R}_2\mathsf{R}_1\\ &= \mathsf{A}\mathsf{P}_{m-1}\mathsf{U}_{m-1}\\ &= \mathsf{A}^2\mathsf{P}_{m-2}\mathsf{U}_{m-2} = \cdots = \mathsf{A}^{m-1}\mathsf{P}_1\mathsf{Q}_1 = \mathsf{A}^m,\end{aligned}$$

as claimed. ∎

COROLLARY 5.14. *With A as in the previous lemma, there exists an orthogonal, diagonal matrix $\mathsf{S}_m \in \mathbb{R}^{n\times n}$ such that $\mathsf{A}^m = \mathsf{P}_m\mathsf{S}_m\mathsf{U}_m$, where $\mathsf{S}_m\mathsf{U}_m$ is upper triangular with positive diagonal entries.*

PROOF: This is an exercise. ∎

According to remarks on uniqueness of QR decompositions at the end of Section 5.3, the factoring $\mathsf{A}^m = \mathsf{P}_m(\mathsf{S}_m\mathsf{U}_m)$ is the unique QR decomposition of

A^m having positive diagonal entries in the upper triangular factor. We make further use of those uniqueness remarks in the arguments given below.

Lemma 5.13 and its corollary play key roles in the main convergence theorem:

THEOREM 5.15. *Suppose that the eigenvalues of* $\mathsf{A} \in \mathbb{R}^{n\times n}$ *stand in the relationship* $|\lambda_1| > |\lambda_2| > \cdots > |\lambda_n| > 0$. *Then the matrices* A_m *generated by the* QR *method converge to an upper triangular matrix with* $\lambda_1, \lambda_2, \ldots, \lambda_n$ *on the diagonal. If* A *is similar to a diagonal matrix* $\mathsf{D} := \mathrm{diag}(\lambda_1, \lambda_2, \ldots, \lambda_n)$ *via a similarity transformation* $\mathsf{A} = \mathsf{XDX}^{-1}$, *where* X^{-1} *has an* LU *decomposition with unit lower triangular factor* L, *then*

$$\mathsf{A}_m \to \begin{bmatrix} \lambda_1 & \star & \cdots & \star \\ & \lambda_2 & & \vdots \\ & & \ddots & \star \\ & & & \lambda_n \end{bmatrix}, \quad \text{as} \quad m \to \infty.$$

When A is symmetric, the iterates A_m tend to $\mathrm{diag}(\lambda_1, \lambda_2, \ldots, \lambda_n)$, as in the example given earlier in this section.

Some remarks about this theorem are in order. The hypothesis that the eigenvalues $\lambda_1, \lambda_2, \ldots, \lambda_n$ are separated in magnitude guarantees that $\mathsf{A} = \mathsf{XDX}^{-1}$ for *some* diagonal matrix D, which automatically has the eigenvalues as its diagonal entries. This fact is an immediate consequence of Corollary 5.4. In general, though, the matrix X^{-1} may not have an LU decomposition. Instead, pivoting may be necessary in the row reduction of X^{-1}. We may have to settle for a decomposition of the form $\mathsf{PX}^{-1} = \mathsf{LU}$, where P is a permutation matrix, as discussed in Section 2.2. In this case, the QR method still converges but not precisely in the fashion indicated in Theorem 5.15. We discuss this case and other exceptions shortly.

PROOF: Since each matrix A_m is similar to A, it suffices to show that A_m tends to an upper triangular matrix. For this task, we need only establish that Q_m tends to an orthogonal, diagonal matrix. We start with the observation that $\mathsf{A}^m = \mathsf{XD}^m\mathsf{X}^{-1}$. Decompose $\mathsf{X} = \mathsf{QR}$, where R has positive diagonal entries. (This decomposition is possible because $|\det \mathsf{R}| = |\det \mathsf{X}| \neq 0$.) Also, decompose $\mathsf{X}^{-1} = \mathsf{LU}$ as guaranteed by the hypotheses. Thus,

$$\mathsf{A}^m = \mathsf{QRD}^m\mathsf{LU} = \mathsf{QR}(\mathsf{D}^m\mathsf{LD}^{-m})\mathsf{D}^m\mathsf{U}, \tag{5.4-4}$$

where $\mathsf{D}^m\mathsf{LD}^{-m}$ is unit lower triangular with entries of the form $l_{j,k}(\lambda_j/\lambda_k)^m$ below the diagonal ($j > k$).

We conclude from the last observation that $\mathsf{D}^m\mathsf{LD}^{-m} = \mathsf{I} + \mathsf{E}_m$, where $\mathsf{E}_m \to 0$ as $m \to \infty$. This convergence lies at the heart of the convergence of

$\{\mathsf{Q}_m\}$. Let us rewrite Equation (5.4-4) as follows:

$$\begin{aligned} \mathsf{A}^m &= \mathsf{QR}(\mathsf{I}+\mathsf{E}_m)\mathsf{D}^m\mathsf{U} \\ &= \mathsf{Q}(\mathsf{I}+\mathsf{RE}_m\mathsf{R}^{-1})\mathsf{RD}^m\mathsf{U} \\ &= \mathsf{Q}(\mathsf{I}+\mathsf{F}_m)\mathsf{RD}^m\mathsf{U}, \end{aligned}$$

where $\mathsf{F}_m := \mathsf{RE}_m\mathsf{R}^{-1}$. Notice that $\mathsf{F}_m \to 0$ as $m \to \infty$, a consequence of the fact that F_m and E_m have the same spectral radius. Now decompose $\mathsf{I}+\mathsf{F}_m = \tilde{\mathsf{Q}}_m\tilde{\mathsf{R}}_m$, choosing the unique decomposition for which $\tilde{\mathsf{R}}_m$ has positive diagonal entries. Since $\mathsf{F}_m \to 0$ as $m \to \infty$, $\tilde{\mathsf{Q}}_m\tilde{\mathsf{R}}_m \to \mathsf{I}$. Using this fact it is a simple exercise to show that $\tilde{\mathsf{Q}}_m \to \mathsf{I}$ and $\tilde{\mathsf{R}}_m \to \mathsf{I}$.

We now have

$$\mathsf{A}^m = \underbrace{(\mathsf{Q}\tilde{\mathsf{Q}}_m)}_{\text{orthogonal}}\ \underbrace{(\tilde{\mathsf{R}}_m\mathsf{RD}^m\mathsf{U})}_{\substack{\text{upper}\\ \text{triangular}}}. \tag{5.4-5}$$

Our next goal is to relate this QR decomposition to the decomposition $\mathsf{A}^m = \mathsf{P}_m\mathsf{S}_m\mathsf{U}_m$ guaranteed in Corollary 5.14. Denote by $|\mathsf{D}|$ the diagonal matrix $\operatorname{diag}(|\lambda_1|, |\lambda_2|, \ldots, |\lambda_n|)$, so that $\mathsf{D} = \mathsf{D}_1|\mathsf{D}|$, where D_1 is diagonal with diagonal entries equal to ± 1. Also, write $\mathsf{U} = \mathsf{D}_2(\mathsf{D}_2^{-1}\mathsf{U})$, where D_2 is orthogonal and diagonal, its entries chosen so that $\mathsf{D}_2^{-1}\mathsf{U}$ has positive diagonal entries. Equation (5.4-5) becomes

$$\mathsf{A}^m = \mathsf{Q}\tilde{\mathsf{Q}}_m\mathsf{D}_2\mathsf{D}_1^m \underbrace{\left[(\mathsf{D}_2\mathsf{D}_1^m)^{-1}\tilde{\mathsf{R}}_m\mathsf{R}(\mathsf{D}_2\mathsf{D}_1^m)|\mathsf{D}|^m\mathsf{D}_2^{-1}\mathsf{U}\right]}_{\mathsf{T}}. \tag{5.4-6}$$

The matrix T in this equation is upper triangular and has positive diagonal entries, so T must be identical to the matrix $\mathsf{S}_m\mathsf{U}_m$ introduced in Corollary 5.14. It follows that the orthogonal factor P_m of that corollary is identical to the orthogonal factor on the right side of Equation (5.4-6): $\mathsf{P}_m = \mathsf{Q}\tilde{\mathsf{Q}}_m\mathsf{D}_2\mathsf{D}_1^m$. Since $\tilde{\mathsf{Q}}_m \to \mathsf{I}$ as $m \to \infty$,

$$\mathsf{Q}_1\mathsf{Q}_2\cdots\mathsf{Q}_m = \mathsf{P}_m \to \mathsf{QD}_2\mathsf{D}_1^m \qquad \text{as} \quad m \to \infty.$$

Therefore, $\mathsf{Q}_m \to \mathsf{D}_1$ as $m \to \infty$, and since D_1 is orthogonal and diagonal, the proof is complete. ∎

This proof indicates that the subdiagonal entries of A_m tend to zero at a rate limited by $\mathcal{O}(|\lambda_j/\lambda_k|)$, where $j > k$. This rate is slow if $\lambda_j \simeq \lambda_k$. As with the power methods of Section 5.2, we can often speed convergence by effecting shifts of origin. At iteration m we pick a value $\mu_m \in \mathbb{C} \setminus \sigma(\mathsf{A})$ that is close to λ_n, then perform the next step:

$$\mathsf{A}_m - \mu_m\mathsf{I} = \mathsf{Q}_m\mathsf{R}_m,$$

$$\mathsf{A}_{m+1} \leftarrow \mathsf{R}_m\mathsf{Q}_m + \mu_m\mathsf{I}.$$

One simple strategy is to select μ_m to be the (n, n)th entry of A_m, which we expect to approach λ_n as $m \to \infty$.

Further Remarks

We can relax several hypotheses in Theorem 5.14 without completely sacrificing the utility of the QR method. For example, the requirement that the matrix X^{-1} have an LU decomposition is not necessary for convergence of the sequence A_m. Suppose that $\mathsf{PX}^{-1} = \mathsf{LU}$, where P is a permutation matrix. Instead of Equation (5.4-4), we get

$$\mathsf{A}^m = \mathsf{QR}(\tilde{\mathsf{D}}^m \mathsf{L} \tilde{\mathsf{D}}^{-m})\tilde{\mathsf{D}}^m \mathsf{U},$$

where $\tilde{\mathsf{D}}^m := \mathsf{PD}^m\mathsf{P}^\top$ and $\mathsf{QR} = \mathsf{XP}^\top$. The convergence proof now proceeds as before, but the sequence $\{\mathsf{A}_m\}$ converges in this case to an upper triangular matrix with the eigenvalues $\lambda_1, \lambda_2, \ldots, \lambda_n$ ordered differently on the diagonal.

When several eigenvalues have equal magnitude, say $|\lambda_r| = |\lambda_{r+1}| = \cdots = |\lambda_{r+s-1}|$, the convergence of the QR method is a more complicated issue. When eigenvalues occur as complex conjugate pairs, the method cannot converge in the sense of Theorem 5.15, since the iterations produce only matrices having real entries. The QR method nevertheless yields useful information about $\sigma(\mathsf{A})$.

An illustrative case occurs when A has s eigenvalues of equal magnitude and all of its other eigenvalues are separated in magnitude. In this circumstance,

$$\mathsf{A}_m \to \begin{bmatrix} \lambda_1 & \star & & \cdots & & \star \\ & \ddots & & & & \\ & & \lambda_{r-1} & & & \\ & & & \mathsf{M} & & \vdots \\ & & & & \lambda_{r+s} & \\ & & & & \ddots & \star \\ & & & & & \lambda_n \end{bmatrix}, \qquad \text{as} \quad m \to \infty,$$

where $\mathsf{M} \in \mathbb{R}^{s \times s}$ has eigenvalues $\lambda_r, \lambda_{r+1}, \ldots, \lambda_{r+s-1}$. Writing the symbol "$\to$" in this case is slightly abusive: The sequence $\{\mathsf{A}_m\}$ does not converge. Instead, the entries in the $s \times s$ block eventually occupied by M settle down to a class of $s \times s$ arrays, the spectra of which converge to $\{\lambda_r, \lambda_{r+1}, \ldots, \lambda_{r+s-1}\}$. For details about this and related cases, we refer to Wilkinson [3, Chapter 8].

Geometrically compelling connections exist between the QR method and the power method. Imagine applying the power method simultaneously to k linearly independent vectors

$$\mathbf{z}_1^{(0)}, \mathbf{z}_2^{(0)}, \ldots, \mathbf{z}_k^{(0)},$$

in other words, to the subspace of $\mathbb{C}^n$ spanned by these vectors. The mth iteration of this procedure produces a subspace spanned by the vectors

$$\mathbf{z}_j^{(m)} := \mathsf{A}^m \mathbf{z}_j^{(0)}, \quad j = 1, 2, \ldots, k.$$

With appropriate rescaling and orthogonalization of the vectors at each stage, one can fashion from this **subspace iteration** a reasonable class of algorithms called **simultaneous iteration**. The QR algorithm amounts to a highly efficient scheme for simultaneous iteration. This view has a stronger geometric flavor than the more standard one presented in this chapter, and it may appeal to mathematically inclined readers. Watkins [2] gives an especially clear explanation.

5.5 Problems

PROBLEM 1. Suppose that $\mathsf{A} \in \mathbb{R}^{n \times n}$, that $\mathsf{A}\boldsymbol{v} = \lambda \boldsymbol{v}$, and that $\mathsf{A}^\top \boldsymbol{\omega} = \mu \boldsymbol{\omega}$, with $\lambda \neq \mu$. Prove that $\boldsymbol{\omega}^\top \boldsymbol{v} = 0$.

PROBLEM 2. The $n \times n$ matrix

$$\mathsf{A} = \begin{bmatrix} 0 & 1 & 0 & \cdots & 0 \\ 0 & 0 & \ddots & \ddots & \vdots \\ \vdots & & \ddots & 1 & 0 \\ 0 & & & 0 & 1 \\ -a_0 & -a_1 & \cdots & -a_{n-2} & -a_{n-1} \end{bmatrix}$$

has characteristic polynomial $p(\lambda) = \lambda^n + a_{n-1}\lambda^{n-1} + \cdots + a_1\lambda + a_0$. For this reason, we call A the **companion matrix** for p. Apply the Gerschgorin theorem to A and $\mathsf{A}^\top$ to estimate the zeros of p.

PROBLEM 3. Suppose that $\mathsf{A} \in \mathbb{R}^{n \times n}$ has a dominant eigenvalue associated with several linearly independent eigenvectors $\boldsymbol{v}_1, \boldsymbol{v}_2, \ldots, \boldsymbol{v}_s$. In the notation of Section 5.2, we therefore have $\lambda_1 = \lambda_2 = \cdots = \lambda_s$, with $|\lambda_1| > |\lambda_{s+1}| \geq \cdots \geq |\lambda_n|$. Prove that the power method still converges to λ_1, with $\mathbf{z}^{(m)}$ tending to a vector in span $\{\boldsymbol{v}_1, \boldsymbol{v}_2, \ldots, \boldsymbol{v}_s\}$.

PROBLEM 4. Show that the Givens transformation (5.3-1) is orthogonal and that its action on $\mathbf{x} \in \mathbb{R}^n$ is to rotate the vector through the angle θ in the (x_i, x_j)-plane, leaving the other coordinates of $\mathbf{x}$ unchanged.

PROBLEM 5. Show that the Householder transformation (5.3-2) is symmetric and orthogonal and that its action on $\mathbf{x} \in \mathbb{R}^n$ is to reflect the vector across the hyperplane perpendicular to $\mathbf{w}$.

PROBLEM 6. Compute, by hand, the QR decomposition of the matrix

$$\mathsf{A} := \begin{bmatrix} 1 & \frac{1}{2} & 3 \\ 2 & 0 & 1 \\ -2 & 1 & 0 \end{bmatrix}.$$

At each stage, describe geometrically the plane reflection used to advance the triangularization.

PROBLEM 7. Show that the product of two upper triangular matrices is upper triangular. Consider an upper triangular matrix $\mathsf{R} \in \mathbb{R}^{n\times n}$ whose diagonal entries are nonzero, and show that R^{-1} is upper triangular. (Hint: Use mathematical induction on the order of R.)

PROBLEM 8. Derive Householder transformations $\mathsf{H}_2, \mathsf{H}_3, \ldots, \mathsf{H}_{n-2}$ that complete the reduction of A to Hessenberg form in Equation (5.4-2). Show that the reduction costs $\mathcal{O}(n^3)$ arithmetic operations.

PROBLEM 9. Suppose that $\mathsf{A} \in \mathbb{R}^{n\times n}$ is in Hessenberg form. Show that QR decomposition of A by using Givens transformations requires $\mathcal{O}(n^2)$ arithmetic operations.

PROBLEM 10. Discuss the convergence properties of the QR method for the matrix

$$\mathsf{A} := \begin{bmatrix} 0 & 0 & 0 & 1 \\ 1 & 0 & 0 & 0 \\ 0 & 1 & 0 & 0 \\ 0 & 0 & 1 & 0 \end{bmatrix}.$$

PROBLEM 11. Let $\mathsf{A} \in \mathbb{R}^{n\times n}$ be symmetric and strictly (row) diagonally dominant (see Section 2.2). Use the Gerschgorin theorem to prove that, if the diagonal entries of A are positive, then A is positive definite.

PROBLEM 12. Show that the set $\mathcal{Q}$ of all orthogonal matrices $\mathsf{Q} \in \mathbb{R}^{n\times n}$ forms a group under multiplication. That is, (*i*) $\mathcal{Q}$ is closed under multiplication; (*ii*) multiplication in $\mathcal{Q}$ is associative; (*iii*) $\mathcal{Q}$ contains a multiplicative identity, and (*iv*) every element in $\mathcal{Q}$ has a multiplicative inverse in $\mathcal{Q}$.

5.6 References

1. G.H. Golub and J.M. Ortega, *Scientific Computing and Differential Equations,* Academic Press, San Diego, 1992.

2. D.S. Watkins, "Understanding the QR algorithm," *SIAM Review 24:4* (October, 1982), 427–440.

3. J.H. Wilkinson, *The Algebraic Eigenvalue Problem*, Oxford University Press, Oxford, U.K., 1965.

Chapter 6

Numerical Integration

6.1 Introduction

Elementary techniques for computing a definite integral $\int_a^b f(x)\,dx$ use the fundamental theorem of calculus: First find an antiderivative F, then compute the integral as $F(b) - F(a)$. For most functions f, finding an antiderivative is difficult at best, and it is necessary to abandon exact methods in favor of approximations.

One of the most fruitful ideas for approximating $\int_a^b f(x)\,dx$ is to replace f by an approximating function $\hat{f}$ whose antiderivatives are easier to find. Then $\int_a^b \hat{f}(x)\,dx$ serves as the approximation. We call such approximations **quadratures**, after the practice of the ancient Greeks of using inscribed and circumscribed rectangles to approximate areas of oddly shaped regions. Using ideas from Chapter 1, we estimate how well $\hat{f}$ approximates f on $[a,b]$. One purpose of the present chapter is to use such estimates to investigate how well $\int_a^b \hat{f}(x)\,dx$ approximates $\int_a^b f(x)\,dx$.

Methods for numerical integration typically lead to formulas having the form

$$\int_a^b f(x)\,dx \simeq \sum_j w_j f(x_j),$$

that is, a weighted sum of values of f at certain points $x_j \in [a,b]$. This form harks back to the definition of the Riemann integral as the limit of Riemann sums $\sum_j f(\overline{x}_j)h_j$. Here the coefficients h_j stand for the lengths of subintervals $[x_{j-1}, x_j]$ formed by a grid on $[a,b]$, and each point $\overline{x}_j$ lies in $[x_{j-1}, x_j]$. Indeed, Riemann sums furnish one approach to numerical integration. This chapter discusses more sophisticated techniques that yield accurate approximations for less computational effort.

6.2 Newton-Cotes Formulas

Motivation and Construction

One simple way to approximate $\int_a^b f(x)\,dx$ is to replace f by an interpolating polynomial $\hat{f}$. Let $h := (b-a)/N$ for some positive integer N, and define $x_j := a + jh$ for $j = 0, 1, \ldots, N$. These points define a uniform grid $\Delta = \{x_0, x_1, \ldots, x_N\}$, over which the Lagrange interpolating polynomial for f is

$$\hat{f}(x) = \sum_{j=0}^{N} f(x_j) L_j(x).$$

The functions $L_0, L_1, \ldots, L_N$ are the Lagrange interpolating basis functions of degree N. As discussed in Section 1.2, these have the form

$$L_i(x) = \prod_{j \neq i} \frac{x - x_j}{x_i - x_j},$$

where the product ranges over $j = 1, 2, \ldots, N$, with $j \neq i$. Making the change of variables $t := (x-a)/h$, we rewrite these basis functions as follows:

$$L_i(x) = \varphi_i(t) := \prod_{j \neq i} \frac{t - j}{i - j}.$$

Thus,

$$\int_a^b \hat{f}(x)\,dx = \sum_{j=0}^{N} f(x_j) \int_a^b L_j(x)\,dx = \sum_{j=0}^{N} f(x_j)\, h \int_0^N \varphi_j(t)\,dt.$$

Writing $\alpha_j := \int_0^N \varphi_j(t)\,dt$ reduces the approximation to the form

$$\int_a^b f(x)\,dx \simeq \int_a^b \hat{f}(x)\,dx = h \sum_{j=0}^{N} \alpha_j f(x_j). \tag{6.2-1}$$

The coefficients α_j are independent of the function f and of the interval $[a, b]$. Also, they are rational numbers having the property that

$$\sum_{j=0}^{N} \alpha_j = \frac{1}{h} \int_a^b \sum_{j=0}^{N} L_j(x)\,dx = \frac{1}{h} \int_a^b 1\,dx = \frac{b-a}{h} = N.$$

(The middle identity follows from the fact that polynomial interpolation is exact for constant functions.) These rational numbers have a common denominator d, so that each $\alpha_j = \sigma_j / d$ for some integer σ_j. We can therefore express the approximation to $\int_a^b f(x)\,dx$ in the following form:

$$\int_a^b f(x)\,dx \simeq \int_a^b \hat{f}(x)\,dx = \frac{b-a}{Nd} \sum_{j=0}^{N} \sigma_j f(x_j). \tag{6.2-2}$$

This representation is the Nth **Newton-Cotes formula**.

Table 6.1: The first four Newton-Cotes formulas

N	σ_j	Nd	Name	Degree of exactness
1	1,1	2	Trapezoid rule	1
2	1,4,1	6	Simpson rule	3
3	1,3,3,1	8	3/8 rule	3
4	7,32,12,32,7	90	Milne rule	5

Table 6.1 displays the first four Newton-Cotes formulas. The most familiar of these are the **trapezoid rule**,

$$\int_a^b f(x)\,dx \simeq \frac{b-a}{2}\left[f(x_0)+f(x_1)\right],$$

which corresponds to the case $N=1$, and the **Simpson rule**,

$$\int_a^b f(x)\,dx \simeq \frac{b-a}{6}\left[f(x_0)+4f(x_1)+f(x_2)\right],$$

corresponding to the case $N=2$. Figure 1 shows how the trapezoid rule approximates f by a line segment over the interval $[a,b]$.

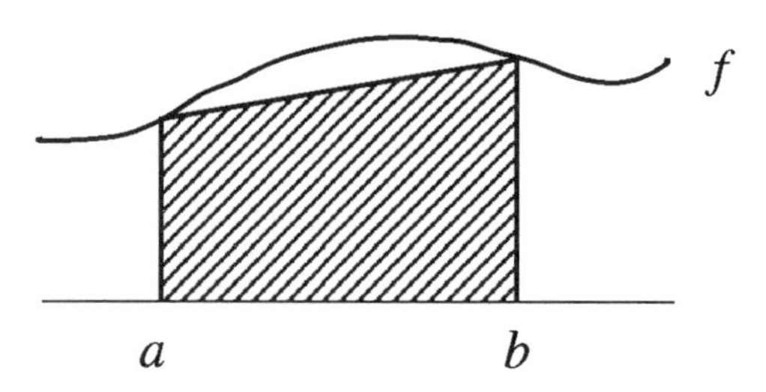

FIGURE 1. *Illustration of the trapezoid rule over an interval* $[a,b]$.

One can clearly integrate a polynomial f of degree N exactly by integrating its polynomial interpolant $\hat{f}$ having degree N, since $f=\hat{f}$ in this case. We restate this trivial fact as a formal proposition:

PROPOSITION 6.1. *If f is a polynomial of degree at most N, then the Nth Newton-Cotes formula for f yields $\int_a^b f(x)\,dx$ exactly.*

More generally, we say that a quadrature rule has **degree of exactness** k if it yields an exact value of $\int_a^b f(x)\,dx$ whenever f is a polynomial of degree at most k but there is a polynomial of degree $k+1$ for which the quadrature rule is inexact.

Certain Newton-Cotes formulas have a degree of exactness that is larger than expected. The proof of the following proposition is Problem 1:

PROPOSITION 6.2. *If N is an even positive integer and f is a polynomial of degree at most $N+1$, then the Nth Newton-Cotes formula yields $\int_a^b f(x)\,dx$ exactly.*

The last column of Table 6.1 lists the degree of exactness of the first four Newton-Cotes formulas.

A more important issue is how well the Newton-Cotes formulas approximate $\int_a^b f(x)\,dx$ when f is not a polynomial. In this case, a given quadrature approximation $\mathcal{I}(f)$ to $\int_a^b f(x)\,dx$ yields an error

$$\mathcal{E}(f) := \int_a^b f(x)\,dx - \mathcal{I}(f).$$

In general, $\mathcal{E}(f)$ increases in magnitude with both the interval length $b-a$ and the "roughness" of f, as measured by values of its derivative of some order. Estimates of $\mathcal{E}(f)$ for the Newton-Cotes formulas have the generic form

$$\mathcal{E}(f) = \text{const.}\,(b-a)^{n+1} f^{(n)}(\zeta),$$

where the constant and the integer n depend upon the method and ζ is some point in the interval (a,b). Later in this section we prove that estimates of this form hold provided that f and its derivatives $f', f'', \ldots, f^{(n)}$ are continuous on $[a,b]$. Table 6.2 summarizes the results for the first four Newton-Cotes formulas.

Practical Considerations: Composite Formulas

One rarely uses the Newton-Cotes formulas in the form just presented. The situation is analogous to that encountered in our discussion of polynomial interpolation versus piecewise polynomial interpolation in Chapter 1. According to the error estimates in Table 6.2, the error $\mathcal{E}(f)$ is proportional to a power of the interval length $b-a$. Hence we can expect the error to grow in magnitude as the interval length $b-a$ becomes large. However, if we apply Newton-Cotes formulas over small subintervals of $[a,b]$, then we can force the error to shrink by dividing $[a,b]$ into smaller and smaller pieces. In effect, this tactic approximates $\int_a^b f(x)\,dx$ by $\int_a^b \hat{f}(x)\,dx$, where $\hat{f}$ is a *piecewise* polynomial interpolant of f. The result is a **composite formula**.

Table 6.2: Error estimates for the first four Newton-Cotes formulas.

Formula	Error $\mathcal{E}(f)$
Trapezoid rule	$-\frac{1}{12}(b-a)^3 f''(\zeta)$
Simpson rule	$-\frac{1}{90}\left(\frac{b-a}{2}\right)^5 f^{(4)}(\zeta)$
3/8 rule	$-\frac{3}{80}\left(\frac{b-a}{3}\right)^5 f^{(4)}(\zeta)$
Milne rule	$-\frac{8}{945}\left(\frac{b-a}{4}\right)^7 f^{(6)}(\zeta)$

Consider, for example, the trapezoid rule. Divide the interval $[a, b]$ by constructing a grid $\Delta := \{x_0, x_1, \ldots, x_N\}$ with $a = x_0 < x_1 < \cdots < x_N = b$, denoting by h the maximum subinterval length $x_j - x_{j-1}$. Decompose $\int_a^b f(x)\, dx$ as follows:

$$\int_a^b f(x)\, dx = \sum_{j=1}^{N} \int_{x_{j-1}}^{x_j} f(x)\, dx.$$

By approximating each integral in the sum on the right by the trapezoid rule, we obtain an approximation of the form

$$\int_a^b f(x)\, dx \simeq \sum_{j=1}^{N} \frac{x_j - x_{j-1}}{2} \left[f(x_{j-1}) + f(x_j)\right].$$

In the special case when the grid is uniform, each subinterval has length $h = (b-a)/N$, and the approximation collapses to

$$\begin{aligned} \int_a^b f(x)\, dx &\simeq \sum_{j=1}^{N} \frac{h}{2} \left[f(x_{j-1}) + f(x_j)\right] \\ &= h\left[\tfrac{1}{2} f(x_0) + f(x_1) + f(x_2) + \cdots + f(x_{N-1}) + \tfrac{1}{2} f(N)\right] \end{aligned} \quad (6.2\text{-}3).$$

To estimate the error in the approximation (6.2-3), simply apply the appropriate error estimate from Table 6.2 to the quadrature over each subinterval $[x_{j-1}, x_j]$. The result is

$$\mathcal{E}(f) = -\sum_{j=1}^{N} \frac{h^3}{12} f''(\zeta_j) = -\frac{h^2}{12}\frac{b-a}{N}\sum_{j=1}^{N} f''(\zeta_j),$$

where each number ζ_j lies in the subinterval (x_{j-1}, x_j). We simplify this result by observing that f'' must be continuous on $[a,b]$ for the error estimates over each subinterval $[x_{j-1}, x_j]$ to hold. Therefore, f'' attains its minimum value m and its maximum value M over $[a,b]$, and

$$m \leq \frac{1}{N}\sum_{j=1}^{N} f''(\zeta_j) \leq M.$$

By the intermediate value theorem, there is a point $\zeta \in (a,b)$ such that

$$f''(\zeta) = \frac{1}{N}\sum_{j=1}^{N} f''(\zeta_j),$$

and there follows the error estimate

$$\mathcal{E}(f) = -\frac{b-a}{12} h^2 f''(\zeta).$$

For a given function f and a fixed interval $[a,b]$ of integration, we interpret this estimate by rewriting it in the form

$$|\mathcal{E}(f)| \leq \frac{b-a}{12} \|f''\|_\infty h^2. \tag{6.2-4}$$

This inequality shows that the error in the composite trapezoid rule shrinks at least as fast as h^2.

Composite formulas for the Simpson rule arise similarly: First, subdivide $[a,b]$ to construct a piecewise quadratic interpolant $\hat{f}$ to f. Thus we start with a grid $\Delta := \{x_0, x_1, \ldots, x_N\}$, where $x_0 = a$, $x_N = b$, $x_0 < x_1 < \cdots < x_N$, and N is an even integer, as Figure 2 illustrates. For uniform grids, we obtain

$$\int_a^b f(x)\,dx \simeq \frac{h}{3}[f(x_0)+4f(x_1)+2f(x_2)+\cdots+2f(x_{N-2})+4f(x_{N-1})+f(x_N)].$$

The error in this approximation is

$$\mathcal{E}(f) = -\sum_{j=1}^{N/2}\left(\frac{2h}{2}\right)^5 \frac{1}{90} f^{(4)}(\zeta_j) = -\frac{h^4}{90}\frac{b-a}{2}\frac{1}{N/2}\sum_{j=1}^{N/2} f^{(4)}(\zeta_j),$$

where $\zeta_j \in [x_{2j-2}, x_{2j-1}]$. Therefore,

$$|\mathcal{E}(f)| \leq \frac{b-a}{180}\|f^{(4)}(\zeta)\|_\infty h^4 = \mathcal{O}(h^4),$$

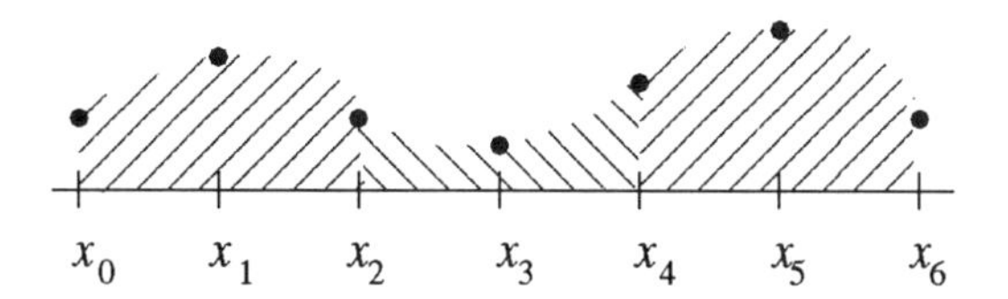

FIGURE 2. *Schematic illustration of a composite formula for the Simpson rule.*

where we have reasoned as for the estimate (6.2-4).

The idea of using composite formulas extends to other quadrature rules, some of which we discuss later in this chapter. Composite formulas also form the conceptual foundation for adaptive quadrature techniques, introduced in the next section.

Mathematical Details

We turn now to an analysis of the error $\mathcal{E}(f) := \int_a^b f(x)\,dx - \mathcal{I}(f)$ associated with Newton-Cotes approximations. The error functional $\mathcal{E}$ is linear, that is, $\mathcal{E}(af) = a\mathcal{E}(f)$ for any constant a, and $\mathcal{E}(f+g) = \mathcal{E}(f)+\mathcal{E}(g)$ for any pair f, g of integrable functions. Also, recall that $\Pi_n([a,b])$ denotes the vector space of all polynomials on the interval $[a,b]$ that have degree at most n.

At the heart of the analysis lies the Peano kernel theorem. The idea is to estimate the error $\mathcal{E}(f)$ for fairly general functions f by relating it to quadrature errors associated with a restricted class of functions, namely polynomials. We begin with a definition.

DEFINITION. *For a quadrature rule $\int_a^b f(x)\,dx \simeq \mathcal{I}(f)$ having error $\mathcal{E}(f) = \int_a^b f(x)\,dx - \mathcal{I}(f)$, the* **Peano kernel** *of degree n is*

$$K_n(t) := -\frac{1}{n!}\mathcal{E}\left((x-t)_+^n\right). \tag{6.2-5}$$

Here the expression $(x-t)_+^n$, considered as a function of x in computing the right side of Equation (6.2-5), has the form

$$(x-t)_+^n := \begin{cases} (x-t)^n, & \text{if} \quad x \geq t, \\ 0, & \text{if} \quad x < t. \end{cases}$$

For example, consider the Simpson rule for an integral $\int_{-1}^{1} f(x)\,dx$. The

Peano kernel of degree three in this case is

$$\begin{aligned} K_3(t) &= -\tfrac{1}{6}\mathcal{E}\left((x-t)_+^3\right) \\ &= \tfrac{1}{6}\Bigg[\tfrac{1}{3}\underbrace{(-1-t)_+^3}_{(\text{I})}+\tfrac{4}{3}\underbrace{(0-t)_+^3}_{(\text{II})}+\tfrac{1}{3}\underbrace{(1-t)_+^3}_{(\text{III})}-\int_{-1}^{1}(x-t)_+^3\,dx\Bigg]. \end{aligned}$$

By the definition of the Peano kernel, the factor labeled (I) in this expression vanishes for $t \in [-1,1]$; the factor (II) vanishes if $t \geq 0$ but equals $-t^3$ if $t < 0$, and the factor (III) equals $(1-t)^3$. Also, the integral on the right reduces to $\int_t^1 (x-t)^3\,dx$. Therefore, for the Simpson rule,

$$K_3(t) = \begin{cases} \frac{1}{72}(1-t)^3(1+3t), & \text{if } \; 0 \leq t \leq 1, \\ K_3(-t), & \text{if } \; -1 \leq t \leq 0. \end{cases}$$

This quantity is continuous on $[-1,1]$ and never changes sign there.

The central theorem is the following:

THEOREM 6.3 (PEANO KERNEL THEOREM). *Suppose that $\mathcal{I}(f) \simeq \int_a^b f(x)\,dx$ is a quadrature rule for which $\mathcal{E}(p) = 0$ whenever $p \in \Pi_n([a,b])$. If $f \in C^{n+1}([a,b])$, then*

$$\mathcal{E}(f) = -\int_a^b f^{(n+1)}(t)K_n(t)\,dt.$$

PROOF: The hypothesis that $f \in C^{n+1}([a,b])$ permits us to apply the Taylor theorem with integral remainder: For $x \in [a,b]$,

$$f(x) = f(a) + f'(a)(x-a) + \cdots + \frac{f^{(n)}(a)}{n!}(x-a)^n + R_n(x),$$

where

$$R_n(x) = \frac{1}{n!}\int_a^x f^{(n+1)}(t)(x-t)^n\,dt = \frac{1}{n!}\int_a^b f^{(n+1)}(t)(x-t)_+^n\,dt.$$

(One can derive this form of the Taylor theorem from the identity $f(x) - f(a) = \int_a^x f'(t)\,dt$, starting with the observation that $\int_a^x f''(t)(x-t)\,dt = -f'(a)(x-a) + \int_a^x f'(t)\,dt$ and repeatedly integrating by parts.) The linearity of the error functional $\mathcal{E}$ implies that

$$\mathcal{E}(f) = \mathcal{E}(f(a)) + \cdots + \mathcal{E}\left(\frac{f^{(n)}(a)}{n!}(x-a)^n\right) + \mathcal{E}(R_n(x)).$$

Since $\mathcal{E}(p) = 0$ for every polynomial $p \in \Pi_n([a,b])$, every term on the right except the last vanishes. What remains is

$$\mathcal{E}(f) = \frac{1}{n!}\mathcal{E}\left(\int_a^b f^{(n+1)}(t)(x-t)_+^n \, dt\right).$$

Recall that $\mathcal{E}$ operates on functions of x. Therefore, we expand this expression for $\mathcal{E}(f)$ as follows:

$$\mathcal{E}(f) = \underbrace{\int_a^b \left[\int_a^b f^{(n+1)}(t)(x-t)_+^n \, dt\right] dx}_{(\text{IV})} - \underbrace{\mathcal{I}\left(\int_a^b f^{(n+1)}(t)(x-t)_+^n \, dt\right)}_{(\text{V})}.$$

By recognizing the term labeled (V) as a quadrature approximation having the form (6.2-1), we find that

$$(\text{V}) = \sum_{j=0}^{K} \alpha_j \int_a^b f^{(n+1)}(t)(x_j - t)_+^n \, dt = \int_a^b f^{(n+1)}(t)\mathcal{I}\left((x-t)_+^n\right) dt.$$

Also, interchanging the order of integration in the term labeled (IV) gives

$$(\text{IV}) = \int_a^b f^{(n+1)}(t) \left[\int_a^b (x-t)_+^n \, dx\right] dt.$$

Substituting these representations into Equation (6.2-5) yields

$$\mathcal{E}(f) = \int_a^b f^{(n+1)}(t) \frac{1}{n!}\left[\int_a^b (x-t)_+^n \, dx - \mathcal{I}\left((x-t)_+^n\right)\right] dt.$$

The expression inside the square brackets is precisely $\mathcal{E}\left((x-t)_+^n\right)$; the conclusion of the theorem follows. ∎

The Peano kernel theorem has an immediate corollary that furnishes error estimates for quadrature rules:

COROLLARY 6.4. *Under the hypotheses of Theorem 6.3,*

$$|\mathcal{E}(f)| \le \|f^{(n+1)}\|_\infty \int_a^b |K_n(t)| \, dt. \tag{6.2-6}$$

The estimate (6.2-6) is crude. More refined estimates are available under an additional hypothesis:

COROLLARY 6.5. *Suppose that the hypotheses of Theorem 6.3 hold. If the Peano kernel $K_n(t)$ does not change sign on $[a, b]$, then there exists a point $\zeta \in (a, b)$ such that*

$$\mathcal{E}(f) = \frac{f^{(n+1)}(\zeta)}{(n+1)!}\mathcal{E}(x^{n+1}) \tag{6.2-7}$$

Several examples, discussed below, illustrate this corollary.

PROOF: The argument rests on the mean value theorem for integrals: Suppose that the functions f, g are both integrable on $[a, b]$, f is continuous on $[a, b]$, and g does not change sign there. Then there exists a point $\zeta \in (a, b)$ such that $\int_a^b f(x)g(x)\,dx = f(\zeta)\int_a^b g(x)\,dx$. This theorem, together with the Peano kernel theorem and the hypothesis that $K_n(t)$ does not change sign on $[a, b]$, implies that

$$\mathcal{E}(f) = -f^{(n+1)}(\zeta)\int_a^b K_n(t)\,dt,$$

for some point $\zeta \in (a, b)$. Also, an easy calculation shows that

$$\mathcal{E}(x^{n+1}) = -(n+1)!\int_a^b K_n(t)\,dt.$$

Equation (6.2-7) follows from these last two identities. ∎

As an example, consider the Simpson rule applied to a function $f \in C^4([-1, 1])$. In this case,

$$\mathcal{E}(f) = \int_{-1}^{1} f(x)\,dx - \tfrac{1}{3}f(-1) - \tfrac{4}{3}f(0) - \tfrac{1}{3}f(1).$$

Since $\mathcal{E}(f)$ vanishes if f is a polynomial of degree at most 3, we apply the Peano kernel theorem with $n = 3$. We have already seen that $K_3(t) \geq 0$ on $[-1, 1]$, so Corollary 6.5 implies that

$$\mathcal{E}(f) = \frac{f^{(4)}(\zeta)}{4!}\mathcal{E}(x^4) = -\frac{f^{(4)}(\zeta)}{24}\frac{4}{15} = -\frac{f^{(4)}(\zeta)}{90}.$$

Similar reasoning produces a representation for the error in the trapezoid rule:

$$\mathcal{E}(f) = \int_a^b f(x)\,dx - \frac{b-a}{2}[f(a) + f(b)] = -\frac{1}{12}(b-a)^3 f''(\zeta). \tag{6.2-8}$$

Problem 3 asks for details.

Further Remarks

The Peano kernel for any Newton-Cotes formula has constant sign. We refer readers to Steffensen [2] for proof. This fact allows one to estimate the error for the Nth Newton-Cotes formula using Corollary 6.5 and in particular to confirm the estimates of Table 6.2. We obtain

$$\mathcal{E}(f) = \begin{cases} \dfrac{\mathcal{E}(x^{N+1})}{(N+1)!} f^{(N+1)}(\zeta), & \text{if } N \text{ is even;} \\[2ex] \dfrac{\mathcal{E}(x^{N+2})}{(N+2)!} f^{(N+2)}(\zeta), & \text{if } N \text{ is odd.} \end{cases}$$

6.3 Romberg and Adaptive Quadrature

This section surveys two methods for enhancing the accuracy of composite quadrature rules. The first method, Romberg quadrature, allows one to use approximations to $\int_a^b f(x)\,dx$ that have low-order accuracy to compute high-order approximations. The second approach, adaptive quadrature, encompasses a class of strategies for tailoring composite rules to local, idiosyncratic behavior in the integrand f. Much of the following discussion is heuristic, delving briefly into theory at several junctures.

Romberg Quadrature

At the core of Romberg quadrature lies the method of **Richardson extrapolation**. Suppose that $I(h)$ is any numerical approximation whose value depends continuously on the mesh size h of a uniform grid. Presumably, $I(0) := \lim_{h\to 0} I(h)$ is the exact value of the integral. Suppose further that we can represent the error associated with $I(h)$ asymptotically, as $h \to 0$, in the form of a power series in h about the point $h = 0$:

$$I(h) = I(0) + a_1 h^{p_1} + \mathcal{O}(h^{p_2}). \tag{6.3-1}$$

Here, a_1 denotes some constant, and $p_2 > p_1 > 0$. In this case, for any $q > 0$, we also have

$$I(qh) = I(0) + a_1 (qh)^{p_1} + \mathcal{O}(h^{p_2}). \tag{6.3-2}$$

By multiplying Equation (6.3-1) by q^{p_1} and subtracting Equation (6.3-2), we eliminate the error terms that are $\mathcal{O}(h^{p_1})$, obtaining

$$I(0) = I(h) + \frac{I(h) - I(qh)}{q^{p_1} - 1} + \mathcal{O}(h^{p_2}).$$

In other words, by evaluating the $\mathcal{O}(h^{p_1})$-accurate approximation $I(h)$ on two different grids, we derive an $\mathcal{O}(h^{p_2})$-accurate approximation by a few additional arithmetic operations.

This line of reasoning generalizes:

THEOREM 6.6 (RICHARDSON EXTRAPOLATION). *Suppose that the approximation $I(h)$, associated with uniform grids of mesh size h, has the asymptotic expansion*

$$I(h) = a_0 + a_1 h^{p_1} + a_2 h^{p_2} + \cdots, \qquad \text{as} \quad h \to 0,$$

where $0 < p_1 < p_2 < \cdots$. Let $q > 0$, and define

$$\begin{aligned} I_1(h) &:= I(h), \\ I_{k+1}(h) &:= I_k(h) + \frac{I_k(h) - I_k(qh)}{q^{p_k} - 1}, \quad k > 1. \end{aligned}$$

Then $I_n(h)$ has an asymptotic expansion of the form

$$I_n(h) = a_0 + a_n^{(n)} h^{p_n} + a_{n+1}^{(n+1)} h^{p_{n+1}} + \cdots, \qquad \text{as} \quad h \to 0.$$

PROOF: This is Problem 5. ∎

Romberg quadrature is the application of Richardson extrapolation to composite Newton-Cotes formulas. As an example, consider the composite trapezoid rule, which we denote as $I(h) \simeq \int_a^b f(x)\,dx$. We demonstrate below that this approximation has an asymptotic expansion of the form

$$I(h) = \underbrace{\int_a^b f(x)\,dx}_{I(0)} + a_1 h^2 + a_2 h^4 + a_3 h^6 + \cdots. \tag{6.3-3}$$

(We already know from Section 6.2 that the lowest-order error term has the form $a_1 h^2$; what we show later is that the higher-order error terms involve only even powers of the mesh size h.) Let $q = 2$, and proceed in stages, which we index as $k = 1, 2, 3, \ldots$. At stage 1, compute $I_{m,1} := I(2^m h)$ for $m = 1, 2, \ldots, M$, where M signifies some prescribed positive integer. Thus stage 1 yields composite trapezoid approximations to $\int_a^b f(x)\,dx$ computed on grids having mesh size $2^m h$. At subsequent stages $k = 2, 3, \ldots$, compute

$$I_{m,k} = I_{m,k-1} + \frac{I_{m,k-1} - I_{m-1,k-1}}{2^{2k} - 1}, \tag{6.3-4}$$

for $m = k, k+1, \ldots, M$. This procedure terminates when we have computed $I_{M,M}$.

Suppose, for concreteness, that we have computed $I_{1,1}$, $I_{2,1}$, $I_{3,1}$, and $I_{4,1}$. We use Equation (6.3-4) to construct the following tableau:

$$\begin{array}{ccccccccc}
k=1 & & k=2 & & k=3 & & k=4 & & \\
I_{1,1} & & & & & & & & \\
 & \searrow & & & & & & & \\
I_{2,1} & \rightarrow & I_{2,2} & & & & & & \\
 & \searrow & & \searrow & & & & & \\
I_{3,1} & \rightarrow & I_{3,2} & \rightarrow & I_{3,3} & & & & \\
 & \searrow & & \searrow & & \searrow & & & \\
I_{4,1} & \rightarrow & I_{4,2} & \rightarrow & I_{4,3} & \rightarrow & I_{4,4} & = & \int_a^b f(x)\,dx + \mathcal{O}(h^8).
\end{array}$$

This procedure yields an $\mathcal{O}(h^8)$-accurate approximation to $\int_a^b f(x)\,dx$ using four $\mathcal{O}(h^2)$-accurate trapezoid approximations. More generally, one computes an $\mathcal{O}(h^{2M})$-accurate approximation to $\int_a^b f(x)\,dx$ by applying Richardson extrapolation to a set of M trapezoid approximations.

The asymptotic expansion (6.3-3) forms the theoretical underpinning of this method. Following Davis and Rabinowitz ([1], Section 2.9), we now justify this expansion.

THEOREM 6.7 (EULER-MACLAURIN FORMULA). *Let $f \in C^{2k+1}([a,b])$, and let N be a positive integer. Define $h := (b-a)/N$, and construct a uniform grid $\Delta = \{x_0, x_1, \ldots, x_N\}$ on $[a,b]$, where $x_j := a + jh$. Then there exist real numbers $B_2, B_4, B_6, \ldots, B_{2k}$ such that, as $h \to 0$,*

$$\begin{aligned}
& h\left[\tfrac{1}{2}f(x_0) + f(x_1) + \cdots + f(x_{N-1}) + \tfrac{1}{2}f(x_N)\right] \\
= & \int_a^b f(x)\,dx + \frac{B_2}{2!}h^2\left[f'(b) - f'(a)\right] + \frac{B_4}{4!}h^4\left[f^{(3)}(b) - f^{(3)}(a)\right] \qquad (6.3\text{-}5) \\
& + \cdots + \frac{B_{2k}}{(2k)!}h^{2k}\left[f^{(2k-1)}(b) - f^{(2k-1)}(a)\right] + \mathcal{O}(h^{2k+1}).
\end{aligned}$$

Before proving this theorem, we make two remarks. First, the numbers $B_2, B_4, B_6, \ldots$ that appear in Equation (6.3-5) are **Bernoulli numbers**. One way to define them is as the coefficients in the expansion

$$\frac{t}{e^t - 1} = \sum_{k=0}^{\infty} \frac{B_k t^k}{k!}.$$

It happens that $B_0 = 1$, $B_1 = -\frac{1}{2}$, $B_{2j+1} = 0$ for $j = 1, 2, 3, \ldots$, and

$$B_{2j} = (-1)^{j-1}(2j)!\sum_{n=1}^{\infty} \frac{2}{(2\pi n)^{2j}}, \qquad j = 1, 2, 3, \ldots.$$

The first few of these even-indexed Bernoulli numbers are as follows:

$$B_2 = \tfrac{1}{6}, \quad B_4 = -\tfrac{1}{30}, \quad B_6 = \tfrac{1}{42}, \quad B_8 = -\tfrac{1}{30}, \quad B_{10} = \tfrac{5}{66}.$$

The second remark is more practical: Equation (6.3-5) implies that the trapezoid rule is an excellent choice for functions that are periodic on $[a, b]$ and for functions whose derivatives near the endpoints of the interval decay rapidly as $k \to \infty$.

PROOF: We establish the following identity: For any $g \in C^{2k+1}([0, N])$,

$$\begin{aligned}
&\tfrac{1}{2}g(0) + g(1) + \cdots + g(N-1) + \tfrac{1}{2}g(N) \\
&= \int_0^N g(x)\,dx + \frac{B_2}{2!}\left[g'(N) - g'(0)\right] + \frac{B_4}{4!}\left[g^{(3)}(N) - g^{(3)}(0)\right] \\
&\quad + \cdots + \frac{B_{2k}}{(2k)!}\left[g^{(2k+1)}(N) - g^{(2k+1)}(0)\right] + \int_0^N p_{2k+1}(x) g^{(2k+1)}(x)\,dx.
\end{aligned} \tag{6.3-6}$$

where p_{2k+1} is a function identified below. The Euler-MacLaurin formula (6.3-5) for $f \in C^{2k+1}([a, b])$ follows when we apply Equation (6.3-6) to $g(x) := f(a + hx)$.

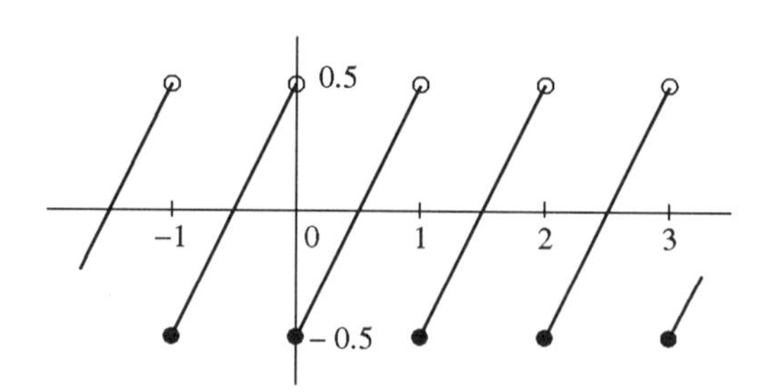

FIGURE 1. *Graph of the function $p_1(x)$.*

The proof begins with the observation that, for $k = 0, 1, 2, \ldots, N-1$,

$$\frac{1}{2}[g(k) + g(k+1)] = \int_k^{k+1} g(x)\,dx + \int_k^{k+1} p_1(x) g'(x)\,dx. \tag{6.3-7}$$

Here, $p_1(x) := x - \operatorname{int}(x) - \frac{1}{2}$, where $\operatorname{int}(x)$ denotes the largest integer that is less than or equal to x. Problem 6 asks for verification of this identity. As Figure 1 shows, p_1 is periodic with period 1.

The function p_1 has Fourier series

$$-\sum_{n=1}^{\infty} \frac{2\sin(2\pi n x)}{2\pi n}.$$

Since p_1 is piecewise continuously differentiable with discontinuities that are at worst jumps, this series converges uniformly to p_1 for all x except near the loci of the discontinuities, and we find successive antiderivatives of p_1 by formally integrating the series term-by-term. For $0 \le x \le 1$, define

$$p_2(x) := \sum_{n=1}^{\infty} \frac{2\cos(2\pi nx)}{(2\pi n)^2},$$

$$p_3(x) := \sum_{n=1}^{\infty} \frac{2\sin(2\pi nx)}{(2\pi n)^3},$$

and so forth, extending these functions to $[0, N]$ by setting $p_k(x+n) = p_k(x)$ for $n = 1, 2, \ldots, N-1$. In general,

$$p_{2j}(x) := (-1)^{j-1}\sum_{n=1}^{\infty} \frac{2\cos(2\pi nx)}{(2\pi n)^{2j}},$$

$$p_{2j+1}(x) := (-1)^{j-1}\sum_{n=1}^{\infty} \frac{2\sin(2\pi nx)}{(2\pi n)^{2j+1}}.$$

These functions have the following properties:

(i) Each function p_k is piecewise polynomial with degree k.

(ii) Each function p_k is periodic with period 1.

(iii) If p_k is continuous at x, then $p'_{k+1}(x) = p_k(x)$.

(iv) If k is an odd integer, then $p_k(0) = p_k(1) = 0$.

(v) If k is an even integer, then

$$p_k(0) = p_k(1) = (-1)^{(k/2)-1}\sum_{n=1}^{\infty} \frac{2}{(2\pi n)^k} = \frac{B_k}{k!}.$$

Now sum the identity (6.3-7) from $k = 0$ to $k = N-1$ to get

$$\tfrac{1}{2}g(0) + g(1) + \cdots + g(N-1) + \tfrac{1}{2}g(N)$$

$$= \int_0^N g(x)\,dx + \int_0^N p_1(x)g'(x)\,dx.$$

Integrating by parts gives

$$\tfrac{1}{2}g(0) + g(1) + \cdots + g(N-1) + \tfrac{1}{2}g(N)$$

$$= \int_0^N g(x)\,dx + p_2(x)g'(x)\Big|_0^N - \int_0^N p_2(x)g''(x)\,dx.$$

Evaluating the second term on the right side and integrating again by parts gives

$$\tfrac{1}{2}g(0) + g(1) + \cdots + g(N-1) + \tfrac{1}{2}g(N)$$

$$= \int_0^N g(x)\,dx + \frac{B_2}{2!}[g'(N) - g'(0)] - p_3(x)g''(x)\Big|_0^N + \int_0^N p_3(x)g^{(3)}(x)\,dx.$$

The third term on the right vanishes owing to properties (*ii*) and (*iv*) listed above. Repeatedly integrating by parts in this way yields the desired identity (6.3-6). ∎

Adaptive Quadrature

Frequently the function to be integrated has highly localized behavior that makes it difficult to approximate on a uniform grid. Figure 2 shows an example of such a function. Over most of the region, it seems reasonable to approximate f using a low-order piecewise polynomial $\hat{f}$. By "reasonable" in this application we mean that we expect the local approximations $\int_{x_{j-1}}^{x_j} \hat{f}(x)\,dx$ to produce small contributions to the overall error, which we write as

$$\mathcal{E} = \int_a^b f(x)\,dx - \int_a^b \hat{f}(x)\,dx.$$

(For the rest of this section, we abbreviate notation by suppressing the dependence of quadrature approximations $\mathcal{I}(f)$ and their errors $\mathcal{E}(f)$ on the integrand f.)

In the middle of the region, however, f oscillates rapidly. Here, good resolution requires much finer grids. In this region, we expect the local approximations

$$\mathcal{I}_{[x_j-1,x_j]} := \int_{x_{j-1}}^{x_j} \hat{f}(x)\,dx$$

to contribute significantly to $\mathcal{E}$. One way to think of this difficulty is as an imbalance in the magnitudes of the local errors

$$\mathcal{E}_{[x_{j-1},x_j]} := \int_{x_{j-1}}^{x_j} f(x)\,dx - \int_{x_{j-1}}^{x_j} \hat{f}(x)\,dx$$

associated with the subintervals $[x_{j-1}, x_j]$. In some regions, $\mathcal{E}_{[x_{j-1},x_j]}$ is small, while in others it may be significant.

Using a *uniform* grid, fine enough to make every local error $\mathcal{E}_{[x_{j-1},x_j]}$ small, certainly reduces the overall error. However, this strategy is inefficient. It resolves local behavior in f, but it also requires unnecessary computation by demanding fine-grid local approximations $\mathcal{I}_{[x_{j-1},x_j]}$ even where coarse-grid approximations are adequate. Again, the problem is one of imbalance: In some regions, $\mathcal{E}_{[x_{j-1},x_j]}$ is appropriately small in magnitude, while in others

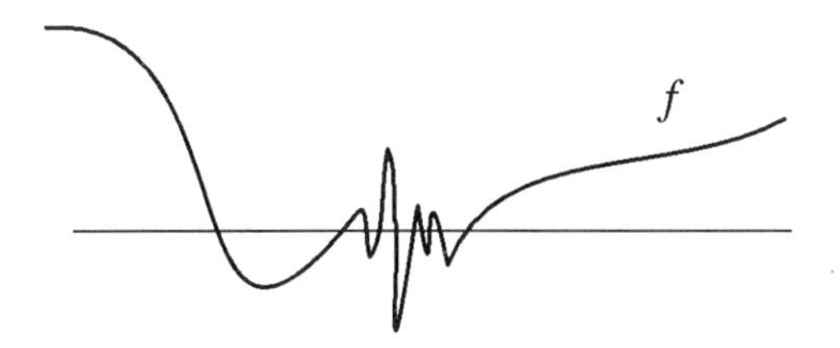

FIGURE 2. *Graph of a function f that exhibits highly localized oscillatory behavior calling for adaptive quadrature.*

we could accept larger local error contributions (requiring less arithmetic) without significantly affecting the overall error.

The idea behind **adaptive quadrature** is to balance the local error contributions $\mathcal{E}_{[x_{j-1},x_j]}$. Typically, we do this by approximating the integrand f with piecewise polynomial interpolants on *non*uniform grids. We construct these grids so that each subinterval $[x_{j-1}, x_j]$ contributes roughly the same amount to the overall error, which we force to be smaller in magnitude than some prescribed tolerance. The ideal adaptive quadrature algorithm takes as input the integrand f, the interval $[a, b]$, and an error tolerance $\tau > 0$ and automatically constructs a quadrature approximation that satisfies the condition $|\mathcal{E}| < \tau$, at least approximately, while balancing the local error contributions.

There are myriad such algorithms. Many existing adaptive quadrature routines use gridding strategies and error indicators that are quite sophisticated, and a study of the state of the art lies beyond the scope of this book. (See [1], Chapter 6, for an introduction.) Instead, we review one simple algorithm, based on the Simpson rule, that has practical utility and conveys the flavor of the field.

The aim of the algorithm is to construct an approximation of the form

$$\int_a^b f(x)\,dx \simeq \mathcal{I} = \sum_{j=1}^{N} \mathcal{I}_{[x_{j-1},x_j]}.$$

To balance the local error contributions, we try to choose the grid $\Delta = \{x_0, x_1, \ldots, x_N\}$ on $[a, b]$ so that

$$\left|\mathcal{E}_{[x_{j-1},x_j]}\right| < \frac{x_j - x_{j-1}}{b-a}\tau. \tag{6.3-8}$$

This condition ensures that

$$\left|\int_a^b f(x)\,dx - \mathcal{I}\right| \le \sum_{j=1}^{N}\left|\int_{x_{j-1}}^{x_j} f(x)\,dx - \mathcal{I}_{[x_{j-1},x_j]}\right|$$

$$= \sum_{j=1}^{N}\left|\mathcal{E}_{[x_{j-1},x_j]}\right| < \tau.$$

To do this, start with a coarse grid, say the two-point grid $\{a,b\}$, and use the Simpson rule to compute

$$\mathcal{I}_{[a,b]} := \frac{b-a}{6}\left[f(a) + 4f(\tfrac{1}{2}(a+b)) + f(b)\right].$$

Then estimate the error $\mathcal{E}_{[a,b]}$. If magnitude of the estimate equals or exceeds the tolerance τ, split $[a,b]$ into two subintervals $[a,(a+b)/2]$ and $[(a+b)/2,b]$, apply the Simpson rule to each subinterval, and estimate the errors $\mathcal{E}_{[a,(a+b)/2]}$ and $\mathcal{E}_{[(a+b)/2,b]}$. If the estimate for the left half satisfies the criterion (6.3-8), then accept $\mathcal{I}_{[a,(a+b)/2]}$ as the contribution to the overall quadrature from the interval $[a,(a+b)/2]$. Otherwise, split $[a,(a+b)/2]$ into two subintervals $[a,(a+b)/4]$ and $[(a+b)/4,(a+b)/2]$, apply the Simpson rule to each, and estimate errors. Proceed similarly for the right half, either accepting $I_{[(a+b)/2,b]}$ or splitting $[(a+b)/2,b]$ and applying the Simpson rule to each half. By repeatedly splitting subintervals in this way, we hope to arrive at a partitioning of $[a,b]$ into subintervals $[x_{j-1},x_j]$ such that each of the contributions $\mathcal{I}_{[x_{j-1},x_j]}$ is accurate enough to satisfy the error criterion (6.3-8).

It is possible to construct functions f that defeat this strategy. In implementing the algorithm one should restrict the number of interval splittings that a computer code can effect before halting.

So far we have left unspecified how to estimate the errors $\mathcal{E}_{[x_{j-1},x_j]}$. This part of the algorithm relies on heuristics. From Table 6.2,

$$\int_{x_{j-1}}^{x_j} f(x)\,dx = \mathcal{I}_{[x_{j-1},x_j]} - \frac{1}{90}(x_j - x_{j-1})^5 f^{(4)}(\zeta), \qquad (6.3\text{-}9)$$

for some point $\zeta \in (x_{j-1},x_j)$. In the absence of detailed information about $f^{(4)}$, this estimate alone offers little help. However, a simple observation together with an additional assumption allow us to estimate the quadrature error without explicitly estimating $f^{(4)}$. Consider the error associated with the Simpson rule applied to the split interval:

$$\begin{aligned}
\int_{x_{j-1}}^{x_j} f(x)\,dx &= \int_{x_{j-1}}^{x_{j-1/2}} f(x)\,dx + \int_{x_{j-1/2}}^{x_j} f(x)\,dx \\
&= \mathcal{I}_{[x_{j-1},x_{j-1/2}]} - \frac{1}{90}(x_{j-1/2} - x_{j-1})^5 f^{(4)}(\zeta_1) \\
&\quad + \mathcal{I}_{[x_{j-1/2},x_j]} - \frac{1}{90}(x_j - x_{j-1/2})^5 f^{(4)}(\zeta_2).
\end{aligned}$$

Since $f^{(4)}$ is continuous, the intermediate value theorem guarantees the existence of a point $\overline{\zeta} \in (x_{j-1}, x_j)$ such that $f^{(4)}(\overline{\zeta}) = \frac{1}{2}\left[f^{(4)}(\zeta_1) + f^{(4)}(\zeta_2)\right]$. Therefore, since $x_{j-1/2} - x_{j-1} = x_j - x_{j-1/2} = (x_j - x_{j-1})/2$,

$$\begin{aligned} \int_{x_{j-1}}^{x_j} f(x)\,dx &= \mathcal{I}_{[x_{j-1},x_{j-1/2}]} + \mathcal{I}_{[x_{j-1/2},x_j]} \\ &\quad - \frac{1}{16}\cdot\frac{1}{90}(x_j - x_{j-1})^5 f^{(4)}(\overline{\zeta}). \end{aligned} \tag{6.3-10}$$

Now assume that $f^{(4)}$ is roughly constant over intervals having length $\max_j\{x_j - x_{j-1}\}$. This assumption is ad hoc, but it allows us to eliminate the terms involving $f^{(4)}(\zeta)$ and $f^{(4)}(\overline{\zeta})$ in Equations (6.3-9) and (6.3-10). Specifically, if we multiply Equation (6.3-9) by 4/15, multiply Equation (6.3-10) by 16/15, and subtract the results, we obtain

$$\begin{aligned} \int_{x_{j-1}}^{x_j} f(x)\,dx &- \left(\mathcal{I}_{[x_{j-1},x_{j-1/2}]} + \mathcal{I}_{[x_{j-1/2},x_j]}\right) \\ &\simeq \frac{1}{15}\left(\mathcal{I}_{[x_{j-1},x_{j-1/2}]} + \mathcal{I}_{[x_{j-1/2},x_j]} - \mathcal{I}_{[x_{j-1},x_j]}\right). \end{aligned}$$

In other words, the composite quadrature approximation on the split interval is roughly 15 times as good an approximation of the unknown quantity $\int_{x_{j-1}}^{x_j} f(x)\,dx$ as it is of the known quantity $\mathcal{I}_{[x_{j-1},x_j]}$. Thus we can test the error criterion (6.3-8) by checking whether

$$\frac{1}{15}\left(\mathcal{I}_{[x_{j-1},x_{j-1/2}]} + \mathcal{I}_{[x_{j-1/2},x_j]} - \mathcal{I}_{[x_{j-1},x_j]}\right) < \frac{x_j - x_{j-1}}{b-a}\tau. \tag{6.3-11}$$

As soon as the quadrature approximation $\mathcal{I}_{[x_{j-1},x_{j-1/2}]} + \mathcal{I}_{[x_{j-1/2},x_j]}$ satisfies this condition, we stop splitting the interval $[x_{j-1}, x_j]$ and move on to other subintervals of $[a, b]$.

Coding these ideas requires some care to avoid unnecessary storage and an infinite algorithm. It is useful to associate with each interval $[x_{j-1}, x_j]$ a work vector

$$\mathbf{w} := \left(x_{j-1},\, h,\, f(x_{j-1}),\, f(x_{j-1/2}),\, f(x_j),\, \mathcal{I}_{[x_{j-1},x_j]}\right)^\top,$$

where

$$h := \frac{x_j - x_{j-1}}{2}, \quad \mathcal{I}_{[x_{j-1},x_j]} := \frac{h}{3}\left[f(x_{j-1}) + 4f(x_{j-1/2}) + f(x_j)\right].$$

At a typical stage in the algorithm, we have a stack $\left(\mathbf{w}^{(1)}, \mathbf{w}^{(2)}, \ldots, \mathbf{w}^{(l)}\right)$ of such vectors, along with a partially accumulated sum of acceptable subinterval contributions to the approximation $\mathcal{I}$ of $\int_a^b f(x)\,dx$.

We work on the vector $\mathbf{w}^{(l)} = (w_1^{(l)}, \ldots, w_6^{(l)})^\top$, associated with a subinterval $[x_{j-1}, x_j]$, that currently occupies the last position in the stack. If the quadrature approximation associated with the split interval $[x_{j-1}, x_{j-1/2}] \cup [x_{j-1/2}, x_j]$ fails to satisfy the convergence criterion (6.3-11), then we remove $\mathbf{w}_l$ from the stack and replace it with two new vectors $\mathbf{w}_l$ and $\mathbf{w}_{l+1}$ associated with the subintervals $[x_{j-1}, x_{j-1/2}]$ and $[x_{j-1/2}, x_j]$. Otherwise, we accept the approximation

$$\int_{x_{j-1}}^{x_j} f(x)\,dx \simeq \mathcal{I}_{[x_{j-1},x_{j-1/2}]} + \mathcal{I}_{[x_{j-1/2},x_j]} + \frac{1}{15}\left(\mathcal{I}_{[x_{j-1},x_{j-1/2}]} + \mathcal{I}_{[x_{j-1/2},x_j]} - \mathcal{I}_{[x_{j-1},x_j]}\right),$$

add it to the partial sum for $\mathcal{I}$, remove $\mathbf{w}_l$ from the stack, and work on the new last entry in the stack. To avoid an infinite algorithm, the algorithm aborts when the length l of the stack exceeds a prescribed integer L.

Algorithm 6.1 implements this approach. The notation used in the following description indicates how to minimize the number of evaluations of the integrand $f(x)$ by addressing information stored in previously computed work vectors $\mathbf{w}^{(k)}$.

ALGORITHM 6.1 (ADAPTIVE QUADRATURE BASED ON THE SIMPSON RULE). *The following algorithm computes an adaptive quadrature approximation I to $\int_a^b f(x)\,dx$, given an error tolerance $\tau > 0$. The algorithm employs a stack $(\mathbf{w}^{(1)}, \mathbf{w}^{(2)}, \ldots, \mathbf{w}^{(l)})$ of work vectors; the algorithm fails when the length of this stack exceeds some prescribed limit $L > 0$.*

1. $k \leftarrow 1$.
2. $I \leftarrow 0$.
3. $h \leftarrow (b-a)/2$.
4. $f_1 \leftarrow f(a), \quad f_2 \leftarrow f(a+h), \quad f_3 \leftarrow f(b)$.
5. $I_L \leftarrow (f_1 + 4f_2 + f_3)h/3$.
6. $\mathbf{w}^{(1)} \leftarrow (a, h, f_1, f_2, f_3, I_L)^\top$.
7. If $l \geq 1$ then:
8. $\quad h \leftarrow \frac{1}{2} w_1^{(l)}$.
9. $\quad f_1 \leftarrow f(w_1^{(l)} + h), \quad f_2 \leftarrow f(w_1^{(l)} + 3h)$.
10. $\quad I_L \leftarrow (w_3^{(l)} + 4f_1 + w_4^{(l)})h/3$.
11. $\quad I_R \leftarrow (w_4^{(l)} + 4f_2 + w_5^{(l)})h/3$.

12. $\delta \leftarrow I_L + I_R - w_6^{(l)}$.

13. If $|\delta| > 30\tau h/(b-a)$ then:

14. If $l \geq L$ then stop; algorithm fails.

15. $f_3 \leftarrow w_5^{(l)}$.

16. $\mathbf{w}^{(l)} \leftarrow (w_1^{(l)}, h, w_3^{(l)}, f_1, w_4^{(l)}, I_L)^\top$.

17. $\mathbf{w}^{(l+1)} \leftarrow (w_1^{(l)} + 2h, h, w_5^{(l)}, f_2, f_3, I_R)^\top$.

18. $l \leftarrow l + 1$.

19. Go to 7.

20. End if.

21. $I \leftarrow I + \delta$.

22. $l \leftarrow l - 1$.

23. Go to 7.

24. End if.

25. End.

6.4 Gauss Quadrature

Motivation and Construction

So far the discussion of quadrature rules

$$\int_a^b f(x)\,dx \simeq \sum_{j=0}^{n} w_j f(x_j)$$

gives scant consideration to the possibility of choosing different evaluation points x_j. In this section we demonstrate that one can achieve surprising gains in accuracy by choosing these points carefully.

To frame the discussion, consider quadrature rules that replace f by a polynomial interpolant $\hat{f}$ having degree n on a nonuniform grid $\{a \leq x_0, x_1, \ldots, x_n \leq b\}$. From Section 1.2, $f(x) = \hat{f}(x) + R_n(x)$, where

$$\hat{f}(x) = \sum_{j=0}^{n} f(x_j) L_j(x),$$

the functions L_j being the standard Lagrange basis functions. Also, by the estimate (1.2-4), the remainder term $R_n(x)$ obeys the bound

$$\|R_n\|_\infty \leq \frac{\|f^{(n+1)}\|_\infty (b-a)^{n+1}}{4(n+1)}.$$

Therefore,

$$\int_a^b f(x)\,dx = \underbrace{\int_a^b \hat{f}(x)\,dx}_{(\text{I})} + \underbrace{\int_a^b R_n(x)\,dx}_{(\text{II})}.$$

The term labeled (I) serves as the quadrature approximation:

$$\int_a^b \hat{f}(x)\,dx = \sum_{j=0}^{n} f(x_j) \int_a^b L_j(x)\,dx = \sum_{j=0}^{n} w_j f(x_j), \tag{6.4-1}$$

where $w_j := \int_a^b L_j(x)\,dx$.

By estimating the term labeled (II), we arrive at a bound for the quadrature error:

$$\begin{aligned} |E| = \left| \sum_{j=0}^{n} w_j f(x_j) - \int_a^b f(x)\,dx \right| &\leq \|R_n\|_\infty \int_a^b dx \\ &\leq \frac{\|f^{(n+1)}\|_\infty (b-a)^{n+2}}{4(n+1)}. \end{aligned}$$

This error estimate holds for an arbitrary choice of distinct evaluation points $x_0, x_1, \ldots, x_n$. In particular, the quadrature approximation is exact when $f^{(n+1)}$ vanishes identically, that is, when the integrand f is a polynomial having degree at most n. The idea behind Gauss quadrature methods is that, by clever choice of $x_0, x_1, \ldots, x_n$, one can concoct quadrature formulas of the form (6.4-1) that are exact for polynomials of even higher degree. These formulas also yield astonishingly favorable error estimates for more general integrands.

The concept that makes this additional accuracy possible is that of *orthogonal systems of polynomials.* For concreteness, let us examine one such system:

DEFINITION. *The* **Legendre polynomials** *constitute a set* $\{P_0, P_1, P_2, \ldots\}$ *of polynomials satisfying the following conditions:*

(i) *Each polynomial* P_n *has degree exactly* n.

(ii) *The function* P_n *is orthogonal to* $P_0, P_1, \ldots, P_{n-1}$ *with respect to the inner product defined by* $\langle f, g\rangle := \int_{-1}^{1} f(x)g(x)\,dx$.

(iii) *Each polynomial P_n is* **monic**, *that is, the coefficient multiplying x^n is 1.*

The first few Legendre polynomials after P_0 are as follows:

$$P_1(x) = x, \qquad P_2(x) = x^2 - \tfrac{1}{3}, \qquad P_3(x) = x^3 - \tfrac{3}{5}x.$$

Each Legendre polynomial P_n has n distinct, real zeros, all lying in the interval $(-1, 1)$. Moreover, any polynomial having degree n is a linear combination $a_0 P_0(x) + a_1 P_1(x) + \cdots + a_n P_n(x)$ of the first $n + 1$ Legendre polynomials. From this observation and property (ii) it follows that P_n is orthogonal to all polynomials having degree less than n.

For the moment, consider the task of approximating $\int_{-1}^{1} f(x)\,dx$; we show later how to scale to the more general case $\int_a^b f(x)\,dx$. A simple and elegant theorem elucidates the connection between orthogonal systems of polynomials and the exactness of quadrature methods:

THEOREM 6.8. *Let $\overline{x}_0, \overline{x}_1, \ldots, \overline{x}_n \in [0, 1]$ be the zeros of P_{n+1}. Then the quadrature approximation*

$$\int_{-1}^{1} f(x)\,dx \simeq \sum_{j=0}^{n} w_j f(\overline{x}_j),$$

with w_j defined as in Equation (6.4-1), yields $\int_{-1}^{1} f(x)\,dx$ exactly when f is a polynomial having degree at most $2n + 1$.

PROOF: If f is a polynomial having degree at most $2n + 1$, then $f(x) = q(x)P_{n+1}(x) + r(x)$, where $q(x)$ is a polynomial having degree at most n and the remainder $r(x)$ is also a polynomial having degree at most n. Integrating this expression yields

$$\begin{aligned} \int_{-1}^{1} f(x)\,dx &= \int_{-1}^{1} q(x)P_{n+1}(x)\,dx + \int_{-1}^{1} r(x)\,dx \\ &= \int_{-1}^{1} r(x)\,dx, \end{aligned} \tag{6.4-2}$$

the integral involving $q(x)$ vanishing by the fact that P_{n+1} is orthogonal to all polynomials having lower degree. On the other hand, the quadrature approximation reduces to the following:

$$\sum_{j=0}^{n} w_j f(\overline{x}_j) = \sum_{j=0}^{n} w_j [q(\overline{x}_j)P_{n+1}(\overline{x}_j) + r(\overline{x}_j)] = \sum_{j=0}^{n} w_j r(\overline{x}_j),$$

the sum involving $q(\overline{x}_j)$ vanishing since each point $\overline{x}_j$ is a root of P_{n+1}. But, as we have already observed,

$$\sum_{j=0}^{n} w_j r(\overline{x}_j) = \int_{-1}^{1} r(x)\,dx,$$

by virtue of the fact that such $n+1$-point quadrature formulas are exact for polynomials having degree n or less. Combining this observation with Equation (6.4-2), we obtain

$$\sum_{j=0}^{n} w_j f(\overline{x}_j) = \int_{-1}^{1} f(x)\,dx,$$

as desired. ∎

We call quadrature approximation using zeros of the Legendre polynomials as evaluation points **Gauss-Legendre quadrature**.

As an example, let us approximate $\int_{-1}^{1} f(x)\,dx$ using the zeros of the cubic Legendre polynomial P_3. These zeros are $\overline{x}_1 \simeq -0.774600$, $\overline{x}_2 = 0$, and $\overline{x}_3 \simeq 0.774600$, accurate to six decimal digits. The corresponding weights are, respectively, $w_1 \simeq 0.555556$, $w_2 \simeq 0.888889$, and $w_3 \simeq 0.555556$, again accurate to six digits. The approximation is therefore

$$\int_{-1}^{1} f(x)\,dx \simeq 0.555556 f(-0.774600) + 0.888889 f(0) + 0.555556 f(0.774600).$$

Table 6.3 describes the first four Gauss-Legendre rules. The table lists the number $n+1$ of evaluation points $\overline{x}_j$, often called **Gauss-Legendre points** or just **Gauss points**; the coordinates of the Gauss points and the weights w_j associated with them, to six decimal digits; and the degree $2n+1$ of polynomial that the quadrature rule integrates exactly according to Theorem 6.8. For higher-degree Gauss-Legendre rules, readers should consult Stroud and Secrest [3], Chapter 6.

Practical Considerations

It is essential to extend Gauss quadrature to intervals more general than $[-1,1]$. We scale the results for $\int_{-1}^{1} f(x)\,dx$ to the computation of $\int_a^b f(x)\,dx$ by a change of variables. Let $\xi := (a+b-2x)/(a-b)$. Then ξ ranges over the interval $[-1,1]$ as x ranges over $[a,b]$, and

$$\begin{aligned}\int_a^b f(x)\,dx = \frac{b-a}{2}\int_{-1}^{1} f(x(\xi))\,d\xi &\simeq \frac{b-a}{2}\sum_{j=0}^{n} w_j f(x(\overline{\xi}_j)) \\ &= \sum_{j=0}^{n} W_j f(x(\overline{\xi}_j)).\end{aligned}$$

Table 6.3: Gauss-Legendre quadrature rules for $n = 0, 1, 2, 3$, with weights w_j and Gauss points $\overline{x}_j$ accurate to six decimal digits.

n	w_j	$\overline{x}_j$	$2n+1$
0	2	0	1
1	$1, 1$	$-0.577350, 0.577350$	3
2	$0.555556, 0.888889, 0.555556$	$-0.774600, 0, 0.744600$	5
3	$0.347854, 0.652145,$	$-0.861136, -0.339981,$	7
	$0.347854, 0.652145,$	$0.339981, 0.861136$	

In this last expression, the new weights are $W_j := \frac{1}{2}(b-a)w_j$, and the new evaluation points are $\overline{x}_j = x(\overline{\xi}_j) := \frac{1}{2}[a + b + (b-a)\overline{\xi}_j]$.

Intuition suggests that, unless a Gauss rule yields $\int_a^b f(x)\,dx$ exactly, its error increases with the interval length $b - a$. (Theorem 6.8 says nothing about the case when f is not a polynomial having degree $2n+1$ or less.) In the last part of this section, we show that this expectation is correct. Suppose that $f \in C^{2n+2}([a,b])$ and

$$\mathcal{I}(f) := \sum_{j=0}^{n} W_j f(\overline{x}_j).$$

Let $\mathcal{E}(f) := \int_a^b f(x)\,dx - \mathcal{I}(f)$ denote the error associated with the $n+1$-point Gauss-Legendre rule. We prove at the end of this section that

$$|\mathcal{E}(f)| \leq \frac{\|f^{(2n+2)}\|_\infty}{2(2n+2)!}(b-a)^{2n+3}. \tag{6.4-3}$$

The estimate (6.4-3) has significant implications for composite rules. We start by constructing a grid $\Delta = \{a = x_0, x_1, \ldots, x_N = b\}$ on $[a,b]$. This step can involve a variety of special considerations, including perhaps the use of adaptive quadrature ideas such as those discussed in Section 6.3. For now, assume that Δ is uniform with mesh size $h = (b-a)/N$. We select n and apply the $n+1$-point Gauss-Legendre quadrature on each subinterval $[x_{i-1}, x_i]$ formed by the grid Δ. The result is

$$\int_a^b f(x)\,dx = \sum_{i=1}^{N} \mathcal{I}_{[x_{i-1},x_i]}(f),$$

where each contribution $\mathcal{I}_{[x_{i-1},x_i]}(f)$ has the form

$$\mathcal{I}_{[x_{i-1},x_i]}(f) = \sum_{j=0}^{n} W_j f(\overline{x}_j).$$

In this last expression, we understand that the $n+1$ points $\overline{x}_j$ depend upon the interval $[x_{i-1}, x_i]$ under consideration. Figure 1 shows a typical uniform grid and the locations of the evaluation points for a composite two-point Gauss-Legendre rule.

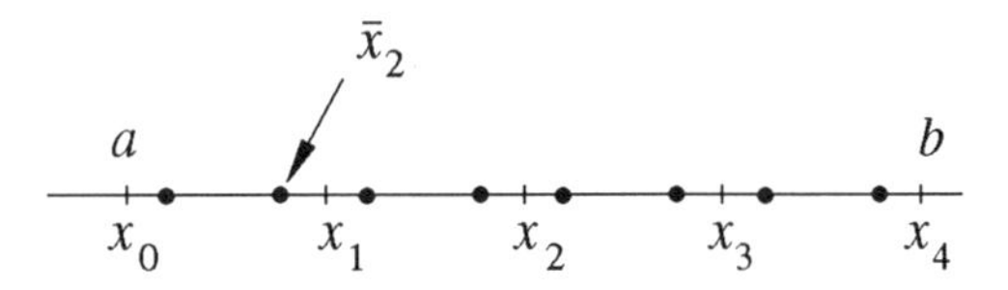

FIGURE 1. *A uniform grid on an interval $[a, b]$, with locations of the evaluation points for two-point composite Gauss-Legendre quadrature shown by the symbol* •.

The error $\mathcal{E}(f)$ associated with such a composite scheme is simply the sum of the errors $\mathcal{E}_{[x_{i-1},x_i]}(f) := \int_{x_{i-1}}^{x_i} f(x)\,dx - I_{[x_{i-1},x_i]}(f)$ over the subintervals. We have

$$\begin{aligned} |\mathcal{E}(f)| \leq \sum_{i=1}^{N} |\mathcal{E}_{[x_{i-1},x_i]}(f)| &\leq N \frac{\|f^{(2n+2)}\|_\infty}{2(2n+2)!} h^{2n+3} \\ &= \frac{b-a}{2(2n+2)!} \|f^{(2n+2)}\|_\infty h^{2n+2}. \end{aligned} \tag{6.4-4}$$

Thus, the error associated with a composite two-point Gauss-Legendre rule on a uniform grid is $\mathcal{O}(h^4)$; that associated with a composite three-point Gauss-Legendre rule is $\mathcal{O}(h^6)$, and so forth.

Mathematical Details

We now explore the theory of Gauss quadrature. In doing so, we generalize the framework established above, based on the Legendre polynomials, to include other Gauss quadrature schemes based on different orthogonal systems of polynomials. In each case, there is a specific interval of integration associated with the basic quadrature scheme. However, each scheme readily extends to more general intervals via the change-of-variables tactic described earlier.

Denote by J an interval, which may be finite, as when $J = [a, b]$, or infinite, as when $J = [a, \infty)$ or $J = (-\infty, \infty)$. As usual, $L^2(J)$ signifies the vector space of functions $f: J \to \mathbb{R}$ that satisfy the condition $\int_J |f(x)|^2 \, dx < \infty$. Also, recall that Π_n is the vector space consisting of all polynmials having degree at most n. Denote by Π_n^1 the subset of Π_n containing only monic polynomials of degree n, that is, only the polynomials

$$p(x) = x^n + a_{n-1}x^{n-1} + \cdots + a_1 x + a_0$$

whose leading coefficients are 1.

We begin with a discussion of orthogonal systems of polynomials.

DEFINITION. *A function $\omega: J \to \mathbb{R}$ is a* **weight function** *if*

(i) *For all $x \in J$, $\omega(x) > 0$.*

(ii) *For $k = 0, 1, 2, \ldots$, $\left|\int_J x^k \omega(x) \, dx\right| < \infty$.*

The goal is to analyze quadrature rules having the form

$$\int_J f(x)\omega(x) \, dx \simeq \sum_{j=0}^{n} w_j f(x_j).$$

This form generalizes the Gauss-Legendre quadrature rule developed earlier.

The first task is to show that there is an orthogonal system of polynomials associated with any weight function. The following proposition is an easy exercise.

PROPOSITION 6.9. *If $\omega: J \to \mathbb{R}$ is a weight function, then the expression*

$$\langle f, g \rangle_\omega := \int_J f(x)g(x)\omega(x) \, dx$$

determines an inner product on $L^2(J)$.

(See Section 0.3 for the axioms defining inner products.) Two functions f and g are orthogonal with respect to the inner product $\langle \cdot, \cdot \rangle_\omega$ if $\langle f, g \rangle_\omega = 0$. The next theorem gives the correspondence between orthogonal systems of polynomials and weight functions.

THEOREM 6.10. *For any weight function ω, there is a set $\{p_0, p_1, p_2, \ldots\}$ of polynomials, with each $p_j \in \Pi_j^1$, such that $\langle p_i, p_j \rangle_\omega = 0$ whenever $i \neq j$.*

PROOF: Define $\{p_0, p_1, p_2, \ldots\}$ inductively as follows:

$$\begin{aligned} p_0(x) &:= 1 \\ p_{j+1}(x) &:= (x - \alpha_{j+1})p_j(x) - \beta_j p_{j-1}(x), \end{aligned} \tag{6.4-5}$$

where $\alpha_{j+1} := \langle xp_j, p_j\rangle_\omega / \langle p_j, p_j\rangle_\omega$, and

$$\beta_j := \begin{cases} 0, & \text{if} \quad j = 0, \\ \langle p_j, p_j\rangle_\omega / \langle p_{j-1}, p_{j-1}\rangle_\omega, & \text{if} \quad j = 1, 2, 3, \ldots. \end{cases}$$

We prove by induction that each $p_j \in \Pi_j^1$ and that each p_j is orthogonal to $p_0, p_1, \ldots, p_{j-1}$ with respect to the inner product $\langle \cdot, \cdot \rangle_\omega$.

For the case $j = 0$ there is little to prove: $p_0 \in \Pi_0^1 = \{p_0\}$ when $p_0(x) = 1$ identically. Suppose that we have constructed $p_0, p_1, \ldots, p_n$ according to the relationships (6.4-5) and that these functions have the properties that $p_j \in \Pi_j^1$ and $\langle p_i, p_j\rangle_\omega = 0$ whenever $j \neq i$. We argue that the polynomial

$$p_{n+1}(x) = (x - \alpha_{n+1})p_n(x) + c_{n-1}p_{n-1}(x) + \cdots + c_0 p_0(x)$$

is monic and that the orthogonality conditions force p_{n+1} to have the form specified in Equations (6.4-5). Monicity is trivial, since any polynomial having this form has 1 as its leading coefficient. By the induction hypothesis, $\langle p_i, p_j\rangle_\omega = 0$ whenever i, j are distinct indices in the set $\{0, 1, \ldots, n\}$, so the orthogonality constraints on p_{n+1} are equivalent to the following equations:

$$0 = \langle p_{n+1}, p_n\rangle_\omega = \langle xp_n, p_n\rangle_\omega - \alpha_{n+1}\langle p_n, p_n\rangle_\omega, \qquad (6.4\text{-}6)$$

and

$$0 = \langle p_{n+1}, p_{j-1}\rangle_\omega = \langle (x - \alpha_{n+1})p_n, p_{j-1}\rangle_\omega + c_{j-1}\langle p_{j-1}, p_{j-1}\rangle_\omega, \qquad (6.4\text{-}7)$$

for $j = 1, 2, \ldots, n$.

Since none of the polynomials $p_0, p_1, \ldots, p_n$ is identically zero and $\langle \cdot, \cdot \rangle_\omega$ is an inner product, the $n + 1$ numbers $\langle p_0, p_0\rangle_\omega, \langle p_1, p_1\rangle_\omega, \ldots, \langle p_n, p_n\rangle_\omega$ are all positive. Thus, for example, Equation (6.4-6) holds precisely when

$$\alpha_{n+1} = \frac{\langle xp_n, p_n\rangle_\omega}{\langle p_n, p_n\rangle_\omega}.$$

Similarly, since $\langle \alpha p_n, p_{j-1}\rangle_\omega = 0$ for $j \leq n$, Equations (6.4-7) hold if and only if

$$c_{j-1} = -\frac{\langle xp_n, p_{j-1}\rangle_\omega}{\langle p_{j-1}, p_{j-1}\rangle_\omega}, \qquad j = 1, 2, \ldots, n.$$

But the defining relationships (6.4-5) for $p_0, p_1, \ldots, p_n$ imply that

$$\begin{aligned} \langle xp_n, p_{j-1}\rangle_\omega &= \langle xp_{j-1}, p_n\rangle_\omega \\ &= \langle p_j, p_n\rangle_\omega + \alpha_j\langle p_{j-1}, p_n\rangle_\omega + \beta_{j-1}\langle p_{j-2}, p_n\rangle_\omega. \end{aligned}$$

The last two terms on the right vanish by the inductive hypothesis, so Equation (6.4-7) holds precisely when

$$c_{j-1} = \begin{cases} 0, & \text{if} \quad j = 1, 2, \ldots, n, \\ -\beta_{j+1}, & \text{if} \quad j = n. \end{cases}$$

The induction is now complete. ∎

In particular, for a given weight function ω and interval J, the corresponding polynomial $p_n \in \Pi_n^1$ is orthogonal, with respect to the inner product $\langle \cdot, \cdot \rangle_\omega$, to all polynomials having degree less than n.

Several specific weight functions $\omega(x)$ and intervals J give rise to classical orthogonal systems of polynomials. Perhaps the simplest choice is $\omega(x) = 1$ and $J = [-1, 1]$, which, as discussed earlier, yields the **Legendre polynomials**. Another choice, $\omega(x) = (1 - x^2)^{-1/2}$ and $J = [-1, 1]$, produces the **Chebyshev polynomials**, the first few of which are as follows:

$$p_1(x) = x, \qquad p_2(x) = x^2 - \tfrac{1}{2}, \qquad p_3(x) = x^3 - \tfrac{3}{4}x.$$

The choice $\omega(x) = e^{-x}$, $J = [0, \infty)$ yields the **Laguerre polynomials**, the first few of which are

$$p_1(x) = x - 1, \qquad p_2(x) = x^2 + 4x - 2, \qquad p_3(x) = x^3 - 9x^2 + 18x - 6.$$

Finally, the choice $\omega = \exp(-x^2)$, $J = (-\infty, \infty)$ gives the **Hermite polynomials**,

$$p_1(x) = x, \qquad p_2(x) = x^2 - 1, \qquad p_3(x) = x^3 - 3x,$$

and so forth.

The orthogonality properties of the Legendre polynomials serve as the foundation for Gauss-Legendre quadrature, which is applicable to integrals of the form $\int_{-1}^{1} f(x)\,dx$. The orthogonality properties of other orthogonal systems of polynomials lead, in a similar fashion, to Gauss quadrature rules applicable to different types of integrands $f(x)\omega(x)$ and intervals J. We specify the forms of such rules shortly. To generalize Gauss quadrature to these other orthogonal systems, however, we show first that each polynomial p_n in a given system has n real zeros in the interval J.

THEOREM 6.11. *Let $\{p_0, p_1, p_2, \ldots\}$ be an orthogonal system of polynomials associated with a weight function ω and an interval J. All of the zeros $\overline{x}_0, \overline{x}_1, \ldots, \overline{x}_{n-1}$ of each polynomial p_n in the system are real and simple, and these zeros all lie in the interior of the interval J.*

PROOF: Denote by $\overline{x}_0, \overline{x}_1, \ldots, \overline{x}_m$ the collection of distinct zeros of p_n that lie in J and have odd multiplicity. Obviously $m \le n - 1$; to prove the theorem, we show that $m = n - 1$. If $m < n - 1$, define a polynomial $q \in \Pi_m^1$ as follows:

$$q(x) := (x - \overline{x}_0)(x - \overline{x}_1) \cdots (x - \overline{x}_m).$$

The polynomial $p_n q$ does not change sign in the interval J, since each of its zeros has even multiplicity. It follows that $\int_J p_n(x)q(x)\omega(x)\,dx = \langle p_n, q \rangle_\omega \neq 0$. (See Problem 8.) This observation implies that $q \notin \Pi_j$ for $j = 0, 1, \ldots, n-$

1. Therefore, q must have at least n zeros, that is, $m \geq n-1$. Hence $m = n-1$. ∎

We now possess the background needed to examine Gauss quadrature in a general setting. The central task is to extend Theorem 6.8 to rules associated with general orthogonal systems of polynomials. We begin with a definition.

DEFINITION. *A set $\{f_0, f_1, \ldots, f_n\}$ of real-valued functions defined on an interval $J \subset \mathbb{R}$ is a* **Chebyshev system** *if, whenever $t_0, t_1, \ldots, t_n \in J$ are distinct, the matrix*

$$\mathsf{T} := \begin{bmatrix} f_0(t_0) & \cdots & f_0(t_n) \\ \vdots & & \vdots \\ f_n(t_0) & \cdots & f_n(t_n) \end{bmatrix} \tag{6.4-8}$$

is nonsingular.

Orthogonal systems of polynomials give rise to Chebyshev systems, as the following lemma demonstrates.

LEMMA 6.12. *The first $n+1$ polynomials in any system of orthogonal polynomials form a Chebyshev system.*

PROOF: We argue by contradiction. Assume that the matrix T defined in Equation (6.4-8) is singular for the orthogonal polynomials $p_0, p_1, \ldots, p_n$. Then $\mathsf{T}^\top$ is singular, and there exists a nonzero vector $\mathbf{c} = (c_0, c_1, \ldots, c_n)^\top \in \mathbb{R}^n$ such that

$$\mathbf{0} = \mathsf{T}^\top \mathbf{c} = \left(\sum_{j=0}^{n} c_j p_j(t_0), \ldots, \sum_{j=0}^{n} c_j p_j(t_n) \right)^\top .$$

In other words, the polynomial

$$p(x) := \sum_{j=0}^{n} c_j p_j(x) \in \Pi_n$$

has $n+1$ zeros. This is possible only if p is the zero polynomial, that is, only if $c_0 = c_1 = \cdots = c_n = 0$, contradicting our assumption that $\mathbf{c} \neq \mathbf{0}$. ∎

The next theorem generalizes Theorem 6.8.

THEOREM 6.13. *Given an interval $J \subset \mathbb{R}$ and a weight function ω, let $\overline{x}_0, \overline{x}_1, \ldots, \overline{x}_n$ be the zeros of the function p_{n+1} in the associated orthogonal* *system of polynomials. Denote by* $\mathbf{w} = (w_0, w_1, \ldots, w_n)^\top$ *the solution to the linear system*

$$\begin{bmatrix} p_0(\overline{x}_0) & \cdots & p_0(\overline{x}_n) \\ p_1(\overline{x}_0) & \cdots & p_1(\overline{x}_n) \\ \vdots & & \vdots \\ p_n(\overline{x}_0) & \cdots & p_n(\overline{x}_n) \end{bmatrix} \begin{bmatrix} w_0 \\ w_1 \\ \vdots \\ w_n \end{bmatrix} = \begin{bmatrix} \langle p_0, p_0 \rangle_\omega \\ 0 \\ \vdots \\ 0 \end{bmatrix}. \tag{6.4-9}$$

Then the following conclusions hold:

(i) *Whenever the polynomial* $p \in \Pi_{2n+1}$,

$$\int_J p(x)\omega(x)\,dx = \sum_{j=0}^{n} w_j p(\overline{x}_j). \tag{6.4-10}$$

(ii) *Each of the numbers* $w_0, w_1, \ldots, w_n$ *is positive.*

Three remarks are in order before the proof. First, Equation (6.4-10) generalizes the fact that Gauss-Legendre quadrature using $n+1$ Gauss-Legendre points $\overline{x}_0, \overline{x}_1, \ldots, \overline{x}_n$ yields $\int_{-1}^{1} p(x)\,dx$ exactly when $p \in \Pi_{2n+1}$. Second, Equation (6.4-9) indeed has exactly one solution $\mathbf{w}$: According to Theorem 6.11 the points $\overline{x}_0, \overline{x}_1, \ldots, \overline{x}_n$ are distinct, and Lemma 6.12 ensures that $\{p_0, p_1, \ldots, p_n\}$ is a Chebyshev system. Third, the entries $w_0, w_1, \ldots, w_n$ of the solution vector $\mathbf{w}$ are the **weights** associated with the quadrature scheme

$$\int_J f(x)\omega(x)\,dx \simeq \sum_{j=0}^{n} w_j f(\overline{x}_j). \tag{6.4-11}$$

This approximation is the $n+1$-point **Gauss rule** associated with the orthogonal system $\{p_0, p_1, \ldots, p_n\}$ of polynomials. For example, if the system $\{p_0, p_1, p_2, \ldots\}$ is the set of Chebyshev polynomials, then the approximation (6.4-11) is the $n+1$-point **Gauss-Chebyshev** rule. Stroud and Secrest [3] tabulate 14 classes of such rules, for n ranging from 1 up to values as large as 512.

PROOF: Start with assertion (i). One can express any polynomial $p \in \Pi_{2n+1}$ as $p(x) = p_{n+1}(x)q(x) + r(x)$, where $q, r \in \Pi_n$. Also, the polynomials q and r are expressible as linear combinations of the polynomials $p_0, p_1, \ldots, p_n$ in the orthogonal system, say

$$q(x) = \sum_{k=0}^{n} a_k p_k(x), \qquad r(x) = \sum_{k=0}^{n} b_k p_k(x).$$

Therefore,

$$\int_J p(x)\omega(x)\,dx = \underbrace{\int_J p_{n+1}(x)q(x)\omega(x)\,dx}_{(\text{I})} + \underbrace{\int_J r(x)\cdot 1\omega(x)\,dx}_{(\text{II})}.$$

The term labeled (I) is $\langle p_{n+1}, q\rangle_\omega$, which vanishes since p_{n+1} is orthogonal to all functions in Π_n. For the same reason,

$$(\text{II}) = \left\langle \sum_{k=0}^{n} b_k p_k, p_0 \right\rangle_\omega = b_0\langle p_0, p_0\rangle_\omega.$$

Hence

$$\int_J p(x)\omega(x)\,dx = b_0\langle p_0, p_0\rangle_\omega.$$

On the other hand, since $p_{n+1}(\overline{x}_j) = 0$ for $j = 0, 1, \ldots, n$,

$$\begin{aligned}\sum_{j=0}^{n} w_j p(\overline{x}_j) &= \sum_{j=0}^{n} w_j p_{n+1}(\overline{x}_j)q(\overline{x}_j) + \sum_{j=0}^{n} w_j r(\overline{x}_j)\\ &= \sum_{k=0}^{n} b_k \left[\sum_{j=0}^{n} w_j p_k(\overline{x}_j)\right].\end{aligned}$$

The quantity in square brackets is the kth row in the matrix-vector product on the left side of Equation (6.4-9). It follows that

$$\sum_{j=0}^{n} w_j p(\overline{x}_j) = b_0\langle p_0, p_0\rangle_\omega,$$

and we have established part (*i*).

To prove part (*ii*), consider a particular weight w_k and the special choice

$$p(x) := (x-\overline{x}_0)^2(x-\overline{x}_1)^2\cdots(x-\overline{x}_{k-1})^2(x-\overline{x}_{k+1})^2\cdots(x-\overline{x}_n)^2,$$

which is not identically zero. Clearly, $p \in \Pi_{2n} \subset \Pi_{2n+1}$, so by part (*i*) the Gauss rule (6.4-11) yields $\int_J p(x)\omega(x)\,dx$ exactly:

$$\int_J p(x)\omega(x)\,dx = \sum_{j=0}^{n} w_j p(\overline{x}_j). \tag{6.4-12}$$

But $p(x) \geq 0$, since it is the product of squared monomial factors, so

$$\int_J p(x)\omega(x)\,dx > 0.$$

(See Problem 8.) Moreover, since $p(\overline{x}_j) = 0$ except when $j = k$, the right side of Equation (6.4-12) collapses to $w_k p(\overline{x}_k)$. Since the left side of Equation (6.4-12) is positive, $w_k p(\overline{x}_k) > 0$. The fact that $p(\overline{x}_k) > 0$ now implies that $w_k > 0$, completing the proof. ∎

Remarkable though it seems that one can integrate a polynomial having degree $2n+1$ by sampling at only $n+1$ points, one might ask whether it is possible to do even better. The answer is no:

THEOREM 6.14. *There is no quadrature formula*

$$\int_J f(x)\omega(x)\,dx \simeq \sum_{j=0}^{n} w_j f(\overline{x}_j)$$

that yields $\int_J p(x)\omega(x)\,dx$ *exactly for all* $p \in \Pi_{2n+2}$.

PROOF: We argue by contradiction. Suppose that such a rule exists. Then it must be exact for the polynomial

$$p(x) := (x-\overline{x}_0)^2(x-\overline{x}_1)^2\cdots(x-\overline{x}_n)^2.$$

But this choice of $p \in \Pi_{2n+2}$ is nonnegative and not identically zero, so

$$0 < \int_J p(x)\omega(x)\,dx = \sum_{j=0}^{n} w_j p(\overline{x}_j).$$

Each factor $p(\overline{x}_j)$ vanishes, leading to the absurd conclusion that $0 < 0$. ∎

We close by indicating how to derive error estimates for Gauss quadrature formulas when the integrand f is not a polynomial. The next theorem facilitates such estimates.

THEOREM 6.15. *If* $f \in C^{2n+2}(J)$, *then there exists a point* η *in the interior of* J *such that*

$$\mathcal{E}(f) := \int_J f(x)\omega(x)\,dx - \sum_{j=0}^{n} w_j f(\overline{x}_j) = \frac{f^{(2n+2)}(\eta)}{(2n+2)!}\langle p_n, p_n\rangle_\omega.$$

PROOF: Let $\hat{f} \in \Pi_{2n+1}$ be the Hermite interpolant of f satisfying the constraints

$$\hat{f}(\overline{x}_j) = f(\overline{x}_j), \qquad \hat{f}'(\overline{x}_j) = f'(\overline{x}_j) \qquad j = 0, 1, \ldots, n.$$

(See Section 1.4.) Since the $(n+1)$-point Gauss rule yields $\int_J \hat{f}(x)\omega(x)\,dx$ exactly,

$$\int_J \hat{f}(x)\omega(x)\,dx = \sum_{j=0}^{n} w_j \hat{f}(\overline{x}_j) = \sum_{j=0}^{n} w_j f(\overline{x}_j).$$

Therefore,

$$\int_J f(x)\omega(x)\,dx - \sum_{j=0}^{n} w_j f(\overline{x}_j) = \int_J \left[f(x) - \hat{f}(x)\right]\omega(x)\,dx.$$

In Section 1.4 we establish the following expression for the error in Hermite interpolation:

$$f(x) - \hat{f}(x) = \frac{f^{(2n+2)}(\zeta(x))}{(2n+2)!} \underbrace{(x-\overline{x}_0)^2(x-\overline{x}_1)^2\cdots(x-\overline{x}_n)^2}_{p_{n+1}^2(x)}, \tag{6.4-13}$$

for some point $\zeta(x)$ in the interval spanned by the points $x, \overline{x}_0, \overline{x}_1, \ldots, \overline{x}_n$. (We write $\zeta(x)$ to indicate that the value of ζ generally depends on the choice of x.) Since f and $\hat{f}$ are both continuous, the right side of Equation (6.4-13) is, too. Therefore, the mean value theorem for integrals holds for this integrand: For some number $\eta \in J$,

$$\begin{aligned}\int_J f(x)\omega(x)\,dx \quad - \quad & \sum_{j=0}^{n} w_j f(\overline{x}_j) = \int_J \left[f(x) - \hat{f}(x)\right]\omega(x)\,dx \\ = \quad & \frac{1}{(2n+2)!}\int_J f^{(2n+2)}(\zeta(x))p_{n+1}^2(x)\omega(x)\,dx \\ = \quad & \frac{f^{(2n+2)}(\eta)}{(2n+2)!}\langle p_{n+1}, p_{n+1}\rangle_\omega.\end{aligned}$$

The proof is now complete. ∎

Using this result, we prove the error estimate (6.4-3) for Gauss-Legendre quadrature over arbitrary intervals $[a,b]$.

COROLLARY 6.16. *For a function $f \in C^{2n+2}([a,b])$, the error associated with $n+1$-point Gauss-Legendre quadrature obeys the bound*

$$|\mathcal{E}(f)| \leq \frac{\|f^{(2n+2)}\|_\infty}{2(2n+2)!}(b-a)^{2n+3}.$$

PROOF: Employing the change of variables $\xi = (a+b-2x)/(b-a)$, we write the Gauss-Legendre approximation for $\int_a^b f(x)\,dx$ as follows:

$$\int_a^b f(x)\,dx \simeq \sum_{j=0}^{n} \frac{b-a}{2} w_j f(x(\overline{\xi}_j)),$$

where the points $\overline{\xi}_j$ are the usual Gauss-Legendre points in $[-1,1]$ and the numbers w_j are the associated weights. Under the change of variables, $x(\xi) = \frac{1}{2}[a+b+(b-a)\xi]$. Theorem 6.15 now yields the following estimate for the quadrature error over the interval $[a,b]$:

$$\begin{aligned}
\mathcal{E}(f) &= \frac{b-a}{2}\left[\int_{-1}^{1} f(x(\xi))\,d\xi - \sum_{j=0}^{n} w_j f(x(\overline{\xi}_j))\right] \\
&= \frac{b-a}{2}\left(\frac{b-a}{2}\right)^{2n+2} \frac{1}{(2n+2)!}\frac{d^{2n+2}f}{d\xi^{2n+2}}(x(\eta))\langle P_{n+1}, P_{n+1}\rangle_w .
\end{aligned} \tag{6.4-14}$$

But

$$\langle P_{n+1}, P_{n+1}\rangle_w = \int_{-1}^{1} (x-\overline{x}_0)^2(x-\overline{x}_2)^2\cdots(x-\overline{x}_n)^2\,d\xi .$$

Furthermore, the integrand in this integral is nonnegative and has 2^{2n+2} as an upper bound. Substituting this bound in the estimate (6.4-14) and simplifying finishes the proof. ∎

6.5 Problems

PROBLEM 1. Prove Propositon 6.2. (Hint: It suffices to prove that there is some polynomial of degree $N+1$ that the rule integrates exactly. Consider $f(x) := [x - \frac{1}{2}(b+a)]^{N+1}$.)

PROBLEM 2. The **midpoint** or **rectangle rule** approximates $\int_a^b f(x)\,dx$ by $(b-a)f((a+b)/2)$. Prove that, if $f \in C^2([a,b])$, then the error for this rule is $E(f) = f''(\zeta)(b-a)^3/24$, for some $\zeta \in (a,b)$.

PROBLEM 3. Prove the error representation (6.2-8) for the trapezoid rule.

PROBLEM 4. Derive the following quadrature rule based on approximating f by its Hermite cubic interpolant:

$$\int_a^b f(x)\,dx \simeq \frac{b-a}{2}[f(a)+f(b)] + \frac{(b-a)^2}{12}[f'(a)-f'(b)].$$

Use the Peano kernel theorem to show that the error in this approximation is

$$E(f) = -\frac{(b-a)^5}{720} f^{(4)}(\zeta),$$

for some $\zeta \in (a,b)$. Develop a corresponding composite rule.

PROBLEM 5. Prove Theorem 6.6.

PROBLEM 6. Let $g \in C^1([k, k+1])$, where k is a positive integer, and denote by $\text{int}(x)$ the largest integer that is less than or equal to x. Prove that

$$\frac{1}{2}[g(k) + g(k+1)] = \int_k^{k+1} g(x)\,dx + \int_k^{k+1} \left[x - \text{int}(x) - \frac{1}{2}\right] g'(x)\,dx.$$

PROBLEM 7. Suppose that f is a function that is not identically zero and does not change sign on an interval $J \subset \mathbb{R}$ and that satisfies $\int_I f(x)\,dx \neq 0$ for every subinterval $I \subset J$. Let ω be a weight function on J. Prove that $\int_J f(x)\omega(x)\,dx \neq 0$. Prove that if, in addition, $f(x) \geq 0$ for $x \in J$, then $\int_J f(x)\omega(x)\,dx > 0$.

PROBLEM 8. Estimate $\int_0^2 \int_1^3 |x-2| \sin y\,dx\,dy$ using four-point Gauss quadrature in each coordinate direction. Compare this result with the answer you get by dividing each of the intervals $[1,3]$ and $[0,2]$ into two equal-length subintervals and then using a composite two-point Gauss rule in each coordinate direction. (The exact answer is $1 - \cos 2$.)

PROBLEM 9. Apply Romberg quadrature with the trapezoid rule to compute an approximation to $\ln 100 = \int_1^{100} x^{-1} dx$ that is accurate to six significant digits. (The correct answer, to eight significant digits, is 4.6051702.)

PROBLEM 10. One can approximate integrals of the form

$$\int_a^b \int_{c(x)}^{d(x)} f(x,y)\,dy\,dx$$

by using Newton-Cotes formulas in each coordinate direction, allowing the step size in the y-direction to depend on x. Use this idea to compute

$$\int_{-1}^{1} \int_x^{2x} \exp(x+y)\,dy\,dx$$

using the Simpson rule in each coordinate direction.

PROBLEM 11. Show that the change of variables $y = 1/x$ converts the integral $\int_1^\infty f(x)\,dx$ to $\int_0^1 y^{-2} f(y^{-1})\,dy$. Discuss the application of this observation to the approximation of improper integrals having the form $\int_a^\infty f(x)\,dx$.

6.6 References

1. P.J. Davis and P. Rabinowitz, *Methods of Numerical Integration*, Academic Press, New York, 1975.

2. J.F. Steffensen, *Interpolation*, Chelsea, New York, 1950.

3. A.H. Stroud and D. Secrest, *Gaussian Quadrature Formulas*, Prentice-Hall, Englewood Cliffs, NJ, 1966.

Chapter 7

Ordinary Differential Equations

7.1 Introduction

An astonishing variety of natural phenomena give rise to mathematical models involving rates of change. For example, the decay of radionuclides, the kinetics of simple chemical reactions, and the dynamics of certain populations obey equations having the form $u'(t) = ku(t)$, where u is an unknown function of time t and k is a constant. This **ordinary differential equation** (ODE) has solutions of the form $C\exp(kt)$, where C is an arbitrary constant that one can determine from knowledge of $u(t)$ at some particular value of t.

In most realistic applications, the ODEs of interest are so complicated that we cannot determine their solutions exactly. Often we must settle for numerical solutions and whatever qualitative knowledge we can glean from theory. This chapter explores basic numerical methods for solving ODEs approximately. We consider two important classes of schemes: **one-step methods**, in which one computes each new value of the solution using one previous value, and **multistep methods**, which utilize several previously computed values of the solution. The methods have in common the tactic of replacing differential operators by algebraic analogs that are more amenable to digital arithmetic.

Several practical questions arise. How accurate are the algebraic analogs as approximations to the original differential operators? Do the approximations amplify small errors as the calculations progress, or do small errors undergo numerical damping? Do the approximate solutions improve as the algebraic analogs approach the exact derivatives? One purpose of the theory in this chapter is to elucidate the connections among these questions.

Before discussing numerics, it is useful to review some elementary facts about ODEs. In the simplest case, we consider first-order ODEs having the form

$$u'(t) = f(t, u(t)),$$

where the independent variable t ranges over an interval $[a, b]$; $u: [a, b] \to \mathbb{R}$ is an unknown, differentiable function of t; and f is a function defined on some subset of $\mathbb{R}^2$. More generally, we consider first-order systems,

$$\mathbf{u}'(t) = \begin{bmatrix} u_1'(t) \\ u_2'(t) \\ \vdots \\ u_k'(t) \end{bmatrix} = \begin{bmatrix} f_1(t, \mathbf{u}(t)) \\ f_2(t, \mathbf{u}(t)) \\ \vdots \\ f_k(t, \mathbf{u}(t)) \end{bmatrix} = \mathbf{f}(t, \mathbf{u}(t)), \qquad (7.1\text{-}1)$$

where again $t \in [a, b]$, and the unknown is now a differentiable function $\mathbf{u}: [a, b] \to \mathbb{R}^k$. We emphasize the vector case (7.1-1). Throughout this chapter, the notation $\|\cdot\|$ signifies an arbitrary norm on $\mathbb{R}^k$.

Equation (7.1-1) typically has infinitely many solutions if it has any at all. To guarantee unique solutions, one must impose extra conditions. In this chapter we focus on **initial-value problems** (IVPs) [5], in which the extra conditions take the form

$$\mathbf{u}(t_0) = \mathbf{u}_I, \quad \text{some} \quad t_0 \in [a, b).$$

Here, $\mathbf{u}_I$ is a known **initial value**. Other types of problems occur in some applications. Noteworthy among these are **boundary-value problems**, in which one knows information about $\mathbf{u}$ or its derivatives at the points a and b. We treat problems of this type briefly in Chapter 9.

The first-order form (7.1-1) is not as special as it may appear. A standard procedure called **reduction in order** allows one to rewrite any kth-order ODE

$$u^{(k)}(t) = f(t, u(t), u'(t), \ldots, u^{(k-1)}(t))$$

as a first-order system,

$$\mathbf{v}' = \begin{bmatrix} v_1' \\ \vdots \\ v_{k-1}' \\ v_k' \end{bmatrix} = \begin{bmatrix} v_2 \\ \vdots \\ v_k \\ f(t, v_1, \ldots, v_k) \end{bmatrix},$$

if we identify $v_j(t) = u^{(j-1)}(t)$ for $j = 1, 2, \ldots, k$. In what follows, therefore, we discuss only IVPs involving Equation (7.1-1).

The performance of numerical methods for IVPs hinges in part on whether the IVPs themselves are well behaved. The following definition makes the notion of "good behavior" more specific.

DEFINITION. *The* IVP

$$\begin{aligned} \mathbf{u}'(t) &= \mathbf{f}(t, \mathbf{u}(t)), \quad t \in [a, b], \\ \mathbf{u}(t_0) &= \mathbf{u}_I, \quad \text{some} \quad t_0 \in [a, b], \end{aligned} \qquad (7.1\text{-}2)$$

is **well posed** *if both of the following conditions hold:*

(i) *There exists a unique, differentiable function* $\mathbf{u}: [a,b] \to \mathbb{R}^k$ *such that* $\mathbf{u}(t_0) = \mathbf{u}_I$ *and* $\mathbf{u}'(t) = \mathbf{f}(t, \mathbf{u}(t))$ *for all* $t \in [a,b]$.

(ii) *The function* $\mathbf{u}$ *depends continuously upon the data* $\mathbf{f}$ *and* $\mathbf{u}_I$. *In other words, there exist constants* $\epsilon > 0$ *and* $K > 0$ *such that the perturbed* IVP

$$\begin{aligned} \tilde{\mathbf{u}}'(t) &= \tilde{\mathbf{f}}(t, \tilde{\mathbf{u}}(t)), \\ \tilde{\mathbf{u}}(t_0) &= \tilde{\mathbf{u}}_I \end{aligned}$$

has a unique, differentiable solution $\tilde{\mathbf{u}}: [a,b] \to \mathbb{R}^k$ *satisfying* $\|\tilde{\mathbf{u}}(t) - \mathbf{u}(t)\| < K\epsilon$ *for all* $t \in [a,b]$ *whenever both* $\|\tilde{\mathbf{u}}_I - \mathbf{u}_I\| < \epsilon$ *and* $\|\tilde{\mathbf{f}}(s,\mathbf{v}) - \mathbf{f}(s,\mathbf{v})\| < \epsilon$ *for all points* $(s,\mathbf{v}) \in [a,b] \times \mathbb{R}^k$.

Part (*ii*) essentially requires that small changes in the data yield small changes in the solution. The following fundamental theorem furnishes conditions under which IVPs are well posed.

THEOREM 7.1. *The initial-value problem (7.1-2) is well posed provided that* $\mathbf{f}$ *is continuous on the strip* $[a,b] \times \mathbb{R}^k \subset \mathbb{R}^{k+1}$ *and there exists a constant* $L > 0$ *such that*

$$\|\mathbf{f}(s,\mathbf{v}) - \mathbf{f}(s,\mathbf{w})\| \le L\|\mathbf{v} - \mathbf{w}\| \tag{7.1-3}$$

for every pair of points $(s,\mathbf{v}), (s,\mathbf{w}) \in [a,b] \times \mathbb{R}^k$.

The inequality (7.1-3) is a **Lipschitz condition**, and the constant L is the **Lipschitz constant** for f (see Section 3.6). For proof, see Brauer and Nohel ([1], Chapter 3).

7.2 One-Step Methods

Motivation and Construction

The exact solution $\mathbf{u}$ to the IVP (7.1-2) in some sense has an infinite number of degrees of freedom, namely the values of $\mathbf{u}(t)$ throughout $[a,b]$. Solving for $\mathbf{u}$ digitally is therefore out of the question. Instead, we settle for an approximate solution that requires computing only finitely many degrees of freedom.

Toward this end, consider a grid $\Delta = \{a = t_0, t_1, \ldots, t_N \le b\}$. For simplicity, assume that $t_n = t_0 + nh$, where h is the mesh size or **stepsize**. (Later in this section we relax the assumption that Δ is uniform.) We seek an approximate solution in the form of a **grid function** $\mathbf{U}: \Delta \to \mathbb{R}^k$, where $\mathbf{u}_n := \mathbf{U}(t_n)$ serves as an approximation of the unknown exact value $\mathbf{u}(t_n)$. As the discussion in Section 7.1 suggests, we typically devise schemes for determining $\mathbf{U}$ by replacing derivative operators in the original ODE with algebraic analogs.

One of the simplest methods for doing this exploits the definition of the derivative as a limit of difference quotients:

$$\mathbf{u}'(t) = \lim_{h \to 0} \frac{\mathbf{u}(t+h) - \mathbf{u}(t)}{h}.$$

We approximate $\mathbf{u}'(t_n)$ by the difference quotient:

$$\mathbf{u}'(t_n) \simeq \frac{\mathbf{U}(t_n + h) - \mathbf{U}(t_n)}{h} = \frac{\mathbf{u}_{n+1} - \mathbf{u}_n}{h}.$$

Our algebraic analog of the ODE (7.1-1) thus becomes

$$\frac{\mathbf{u}_{n+1} - \mathbf{u}_n}{h} = \mathbf{f}(t_n, \mathbf{u}_n), \quad n = 0, 1, \ldots, N-1,$$

or

$$\mathbf{u}_{n+1} = \mathbf{u}_n + h\mathbf{f}(t_n, \mathbf{u}_n), \quad n = 0, 1, \ldots, N-1. \tag{7.2-1}$$

Figure 1 shows how this scheme uses the slope $f(t_n, u_n)$ to extrapolate from u_n to u_{n+1} in the case of a scalar ODE ($k = 1$).

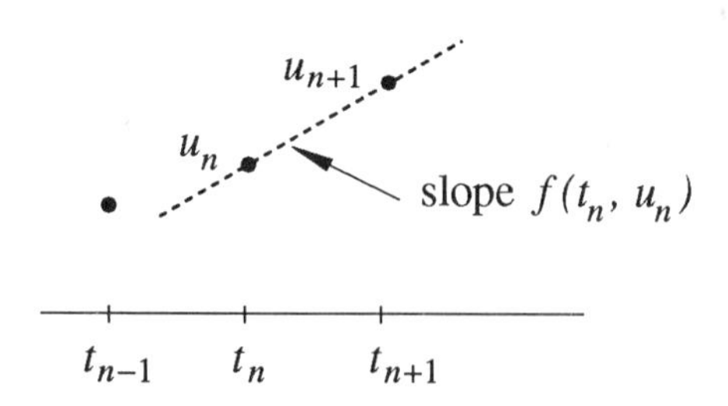

FIGURE 1. *Extrapolation from u_n to u_{n+1} in the explicit Euler method.*

Algorithmically, we regard Equation (7.2-1) as a prescription for computing the unknown $\mathbf{u}_{n+1}$ in terms of known values associated with the previous grid point t_n. To start the calculations we need a value for $\mathbf{u}_0$, which is available from the initial condition in the IVP (7.1-2). Thus $\mathbf{u}_0 = \mathbf{u}_I$. Since Equation (7.2-1) determines each successive unknown $\mathbf{u}_{n+1}$ explicitly in terms of previously computed information, we say that the scheme is **explicit**. Equation (7.2-1) is the **explicit Euler method**.

The explicit Euler method is the simplest of a family of discrete approximations having the general form

$$\mathbf{u}_{n+1} = \mathbf{u}_n + h\mathbf{\Phi}_h(t_n, \mathbf{u}_n). \tag{7.2-2}$$

For each fixed choice of mesh size $h > 0$, $\mathbf{\Phi}_h : [a,b] \times \mathbb{R}^k \to \mathbb{R}^k$. Discrete schemes having the form (7.2-2) are called **explicit one-step methods**, since they give a new value $\mathbf{u}_{n+1}$ in terms of one previously computed step. In the explicit Euler method, $\mathbf{\Phi}_h(s, \mathbf{v}) := \mathbf{f}(s, \mathbf{v})$. (We use the symbols s and $\mathbf{v}$ instead of t and $\mathbf{u}$ here to indicate that these are generic arguments, not specific values of the independent variable t and the unknown solution $\mathbf{u}$.)

Other explicit one-step schemes are possible [4]. For example, consider the **Heun method**,

$$\mathbf{u}_{n+1} = \mathbf{u}_n + \frac{h}{2}\left[\mathbf{f}(t_n, \mathbf{u}_n) + \mathbf{f}(t_{n+1}, \mathbf{u}_n + h\mathbf{f}(t_n, \mathbf{u}_n))\right], \tag{7.2-3}$$

where $\mathbf{\Phi}_h(s, \mathbf{v}) := \frac{1}{2}[\mathbf{f}(s, \mathbf{v}) + \mathbf{f}(s+h, \mathbf{v} + h\mathbf{f}(s, \mathbf{v}))]$. The heuristic underlying this scheme is most apparent when the original ODE (7.1-1) has the simple form $\mathbf{u}' = \mathbf{f}(t)$. In this case, $\mathbf{u}(t) = \mathbf{u}_0 + \int_a^t \mathbf{f}(s)\, ds$, and the Heun method amounts to using the trapezoid rule to approximate the definite integral.

Another example of an explicit one-step method is the **modified Euler method**,

$$\mathbf{u}_{n+1} = \mathbf{u}_n + h\mathbf{f}\left(t_n + \frac{h}{2}, \mathbf{u}_n + \frac{h}{2}\mathbf{f}(t_n, \mathbf{u}_n)\right), \tag{7.2-4}$$

where $\mathbf{\Phi}_h(s, \mathbf{v}) := \mathbf{f}(s + \frac{1}{2}h, \mathbf{v} + \frac{1}{2}h\mathbf{f}(s, \mathbf{v}))$. When the ODE has the form $\mathbf{u}'(t) = \mathbf{f}(t)$, this scheme corresponds to the rectangle rule for computing $\int_a^t \mathbf{f}(s)\, ds$.

Among the most popular explicit one-step methods is the **Runge-Kutta method**. Here,

$$\mathbf{\Phi}_h(s, \mathbf{v}) := \tfrac{1}{6}(\mathbf{f}_1 + 2\mathbf{f}_2 + 2\mathbf{f}_3 + \mathbf{f}_4), \tag{7.2-5}$$

where

$$\begin{aligned}
\mathbf{f}_1 &:= \mathbf{f}(s, \mathbf{v}), \\
\mathbf{f}_2 &:= \mathbf{f}(s + \tfrac{1}{2}h, \mathbf{v} + \tfrac{1}{2}h\mathbf{f}_1), \\
\mathbf{f}_3 &:= \mathbf{f}(s + \tfrac{1}{2}h, \mathbf{v} + \tfrac{1}{2}h\mathbf{f}_2), \\
\mathbf{f}_4 &:= \mathbf{f}(s + h, \mathbf{v} + h\mathbf{f}_3).
\end{aligned}$$

This scheme corresponds to the Simpson rule for $\int_a^t \mathbf{f}(s)\, ds$ when $\mathbf{u}'(t) = \mathbf{f}(t)$.

The general form (7.2-2) by no means exhausts the possibilities. More generally, one-step schemes have the form

$$\mathbf{u}_{n+1} = \mathbf{u}_n + h\mathbf{\Phi}_h(t_n, \mathbf{u}_n, t_{n+1}, \mathbf{u}_{n+1}). \tag{7.2-6}$$

In such schemes, it is typically impossible to compute the right side explicitly using known information, since $\mathbf{u}_{n+1}$ remains unknown. We call discrete methods having the form (7.2-6) **implicit one-step methods**, since they determine $\mathbf{u}_{n+1}$ as an implicit function of t_n, t_{n+1}, and $\mathbf{u}_n$.

The simplest implicit one-step method is the **implicit Euler method**,

$$\mathbf{u}_{n+1} = \mathbf{u}_n + h\mathbf{f}(t_{n+1}, \mathbf{u}_{n+1}). \tag{7.2-7}$$

For typical functions $\mathbf{f}$, Equation (7.2-7) is nonlinear in the unknown $\mathbf{u}_{n+1}$. To solve for $\mathbf{u}_{n+1}$ we use methods like those discussed in Chapter 3. For instance, successive substitution yields the equation

$$\mathbf{u}_{n+1}^{(m+1)} = \mathbf{u}_n + h\mathbf{f}(t_{n+1}, \mathbf{u}_{n+1}^{(m)}).$$

The idea is to iterate until $\mathbf{u}_{n+1}^{(m+1)}$ provides an acceptable approximation to $\mathbf{u}_{n+1}$, then to proceed to the next level in the independent variable t. It is also a worthwhile exercise to apply Newton's method to Equation (7.2-7).

Another implicit one-step scheme is the **trapezoid method**,

$$\mathbf{u}_{n+1} = \mathbf{u}_n + \frac{h}{2}\left[\mathbf{f}(t_n, \mathbf{u}_n) + \mathbf{f}(t_{n+1}, \mathbf{u}_{n+1})\right]. \tag{7.2-8}$$

Problem 7 explores this method in more detail.

The analysis of implicit one-step methods falls more naturally into the framework for multistep methods. We examine these in Sections 7.3 and 7.4. For the remainder of this section, we focus on explicit one-step methods.

Practical Considerations

As the analogies between various one-step methods and quadrature approximations suggest, some schemes are more accurate than others. The main issue is how fast the error in the approximation $\mathbf{u}_n$ shrinks as we decrease the stepsize h. As we show later, the Euler explicit scheme (7.2-1), for all its simplicity, is not very accurate. By comparison, the Runge-Kutta method (7.2-5) is quite accurate. The accuracies of the other methods mentioned above lie between these two extremes.

One can gain insight into the Euler explicit scheme by examining a Taylor expansion of $\mathbf{u}$ about a typical grid point t_n. For some point $\zeta \in (t_n, t_{n+1})$,

$$\mathbf{u}(t_{n+1}) = \mathbf{u}(t_n) + h\mathbf{u}'(t_n) + \frac{h^2}{2}\mathbf{u}''(\zeta).$$

Rearranging this expansion gives

$$\frac{\mathbf{u}(t_{n+1}) - \mathbf{u}(t_n)}{h} = \mathbf{u}'(t) + \mathcal{O}(h).$$

The expression on the left is precisely the derivative approximation that motivates the Euler approximation. Thus the Euler explicit scheme amounts to replacing $\mathbf{u}'(t)$ by an $\mathcal{O}(h)$-accurate approximation, then applying this approximation to the grid function $\mathbf{u}_n$.

There is a logical distinction between the accuracy of the approximation to $\mathbf{u}'$ and the accuracy of the resulting numerical solution $\mathbf{u}_n$. Nevertheless, the two are connected, as we explore later in this section. For now, we state a rule of thumb: Under reasonable conditions, a discrete method arising from an $\mathcal{O}(h^p)$-accurate approximation to $\mathbf{u}'(t)$ yields $\mathcal{O}(h^p)$-accurate numerical solutions $\mathbf{u}_n$. For example, the Euler explicit scheme yields numerical solutions $\mathbf{u}_n$ that obey an error estimate of the form $\|\mathbf{u}(t_n) - \mathbf{u}_n\| = \mathcal{O}(h)$. In particular, as $h \to 0$, halving h in the Euler explicit scheme reduces the maximum error $E_h := \max_{0 \leq n \leq N} \|\mathbf{u}(t_n) - \mathbf{u}_n\|$ by a factor of 2.

Table 7.1 illustrates this phenomenon, to four decimal digits, for the simple IVP

$$\mathbf{u}'(t) = -\mathbf{u}(t), \quad \mathbf{u}(0) = 1.$$

This problem has solution $\mathbf{u}(t) = e^{-t}$, and the Euler explicit scheme gives $\mathbf{u}_{n+1} = (1-h)\mathbf{u}_n$, with $\mathbf{u}_0 = 1$. To verify that $E_h = \mathcal{O}(h)$ numerically, one can construct a convergence plot by graphing $\log E_h$ versus $\log h$. The points on the graph should lie roughly on a line having unit slope.

Table 7.1: The explicit Euler method for $\mathbf{u}' = -\mathbf{u}$ with $\mathbf{u}(0) = 1$, using three different values of stepsize h.

t	$\mathbf{U}(t)$ $(h = 0.2)$	$\mathbf{U}(t)$ $(h = 0.1)$	$\mathbf{U}(t)$ (h=0.05)	$\mathbf{u}(t)$
0.00	1.0000	1.0000	1.0000	1.0000
0.05			0.9500	0.9512
0.10		0.9000	0.9025	0.9048
0.15			0.8574	0.8607
0.20	0.8000	0.8100	0.8145	0.8187
0.25			0.7738	0.7788
0.30		0.7290	0.7351	0.7408
0.35			0.6983	0.7047
0.40	0.6400	0.6561	0.6634	0.6703

Similar numerical experiments illustrate the accuracy of other methods. For example, the Runge-Kutta method is based on an $\mathcal{O}(h^4)$-accurate approximation to $\mathbf{u}'(t)$. Thus we expect a convergence plot of $\log E_h$ versus h for this scheme to produce points lying close to a line having slope 4. Problem 1 calls for numerical experiments of this type.

Mathematical Details

We distinguish between two logically different questions. First, how well does a given discrete analog approximate the original ODE? Second, how well does the solution to the discrete analog approximate the solution to the ODE? The first question is typically easier to answer, while the answer to the second question is more important from a practical point of view. Fortunately, there are connections between the questions, which we now explore for explicit one-step methods.

To address the first question — whether the discrete scheme is a reasonable approximation to the original ODE — we adopt some definitions. Consider a point $(s, \mathbf{v})$ belonging to the region $[a, b] \times \mathbb{R}^k$, as drawn in Figure 2 for the case $k = 1$. Denote by $\mathbf{w}(t)$ the solution to the IVP

$$\mathbf{w}'(t) = \mathbf{f}(t, \mathbf{w}(t)), \quad \mathbf{w}(s) = \mathbf{v},$$

that is, the unique solution to the original ODE that passes through the given point $(s, \mathbf{v})$. Difference quotients involving $\mathbf{w}(t)$ have the form

$$\mathbf{\Delta}_h(s, \mathbf{v}) := \begin{cases} \dfrac{\mathbf{w}(s+h) - \mathbf{w}(s)}{h} = \dfrac{\mathbf{w}(s+h) - \mathbf{v}}{h}, & \text{if } h \neq 0, \\ \mathbf{f}(s, \mathbf{v}), & \text{if } h = 0. \end{cases}$$

Of particular interest are comparisons between $\mathbf{\Delta}_h(s, \mathbf{v})$ and the quantity $\mathbf{\Phi}_h(s, \mathbf{v})$, which serves as an approximation to $\mathbf{u}'(t)$ in the explicit one-step scheme (7.2-2).

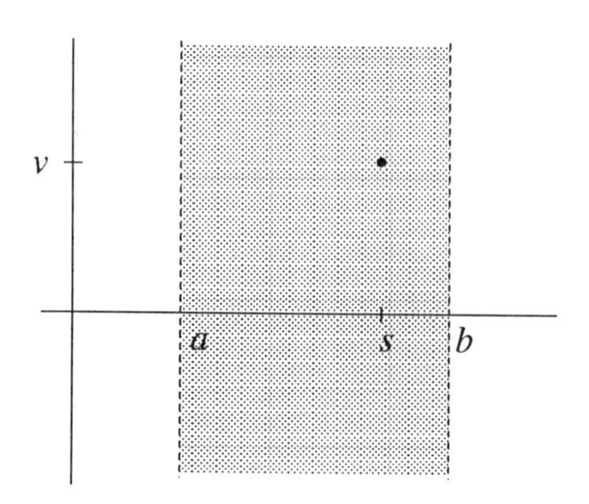

FIGURE 2. *The strip $[a, b] \times \mathbb{R}$, showing a typical point (s, v).*

DEFINITION. *The* **truncation error** *(or* **local discretization error***) associated with the explicit one-step method (7.2-2) is*

$$\tau_h(s, \mathbf{v}) := \mathbf{\Delta}_h(s, \mathbf{v}) - \mathbf{\Phi}_h(s, \mathbf{v}), \tag{7.2-9}$$

defined for all points $(s, \mathbf{v}) \in [a, b] \times \mathbb{R}^k$*. The scheme (7.2-2) is* **consistent** *if, whenever* $\mathbf{f}$ *is smooth enough,*

$$\lim_{h \to 0} \tau_h(s, \mathbf{v}) = 0,$$

for all $(s, \mathbf{v}) \in [a, b] \times \mathbb{R}^k$.

Since we are interested for the moment in the behavior of the numerical scheme (characterized by $\mathbf{\Phi}_h$) and not in the particular ODE, we refrain from fretting over the precise properties of $\mathbf{f}$.

This definition recasts the question whether a given discrete scheme reasonably approximates the original ODE: We simply ask whether the scheme is consistent.

Taylor series provide a useful vehicle for assessing consistency. Consider, for example, the Euler explicit scheme. When the $\mathbf{f}$ is sufficiently smooth,

$$\mathbf{w}(s + h) = \mathbf{w}(s) + h\mathbf{w}'(s) + \frac{h^2}{2}\mathbf{w}''(s) + \mathcal{O}(h^3). \tag{7.2-10}$$

But $\mathbf{w}'(s) = \mathbf{f}(s, \mathbf{v})$. Also, if we denote by $\partial_j \varphi$ the partial derivative of a function φ with respect to its jth argument and by $\mathsf{J}_w(s, \mathbf{w}(s))$ the $k \times k$ matrix-valued function whose entries are the derivatives $\partial_j f_i(s, \mathbf{w}(s))$, we have by the chain rule

$$\begin{aligned} \mathbf{w}''(s) &= \partial_1 \mathbf{f}(s, \mathbf{v}) + \mathsf{J}_w(s, \mathbf{v})\mathbf{w}'(s) \\ &= \partial_1 \mathbf{f}(s, \mathbf{v}) + \mathsf{J}_w(s, \mathbf{v})\mathbf{f}(s, \mathbf{v}). \end{aligned}$$

Substituting these expressions into Equation (7.2-10) yields

$$\begin{aligned} \tau_h(s, \mathbf{v}) &= \frac{\mathbf{w}(s + h) - \mathbf{w}(s)}{h} - \mathbf{\Phi}_h(s, \mathbf{v}) \\ &= \mathbf{f}(s, \mathbf{v}) + \frac{h}{2}\left[\partial_1 \mathbf{f}(s, \mathbf{v}) + \mathsf{J}_w(s, \mathbf{v})\mathbf{f}(s, \mathbf{v})\right] + \mathcal{O}(h^2) - \mathbf{f}(s, \mathbf{v}). \end{aligned} \tag{7.2-10}$$

Since the quantity inside square brackets is independent of h, $\tau_h(s, \mathbf{v}) \to 0$ as $h \to 0$.

Equation (7.2-11) not only implies that the truncation error τ_h vanishes as $h \to 0$; it also furnishes quantitative information about the *rate* at which $\tau_h \to 0$. The following terminology formalizes this concept:

DEFINITION. *The one-step scheme (7.2-2) for the* IVP *(7.1-2) is* **consistent with order** p *if* $\tau_h(s, \mathbf{v}) = \mathcal{O}(h^p)$ *as* $h \to 0$*, that is, if there are positive constants* $C, \overline{h}$*, independent of* h*, such that* $\|\tau_h(s, \mathbf{v})\| < Ch^p$ *whenever* $h < \overline{h}$ *and* $\mathbf{f} \in C^{p+1}([a, b] \times \mathbb{R}^k)$.

Equation (7.2-11) shows that the Euler explicit scheme is consistent with order $p = 1$. To construct methods having higher-order consistency, one can adopt any of several approaches. One appears in Problem 2, which asks for verification that the Heun method and the modified Euler method are both consistent with order $p = 2$. Problem 3 suggests another approach, based on the direct use of Taylor expansions. Still other, more involved methods exist. For example, it is an unreasonably tedious exercise to show that the Runge-Kutta method (7.2-5) is consistent with order $p = 4$ (see [2], Section 2.4).

We turn now to the second question, whether the solution to the discrete analog reasonably approximates to the solution of the original IVP.

DEFINITION. *The* **solution error** *(or* **global discretization error***) associated with the one-step method (7.2-2) with stepsize h is*

$$\varepsilon_h(t_n) := \mathbf{u}(t_n) - \mathbf{u}_n.$$

The critical question is whether $\varepsilon_h(t) \to 0$ for all $t \in [a, b]$, as the stepsize $h \to 0$. Here we encounter a slight technicality. We wish to treat $\varepsilon_h(t)$ as a function of the continuous variable t. However, for a specified value of t, only a discrete set of choices, namely, $t - t_0, (t - t_0)/2, (t - t_0)/3, \ldots$, exist for the stepsize h. Hence we interpret assertions regarding $\lim_{h\to 0} \varepsilon_h(t)$ as statements about sequences of values of $\varepsilon_h(t)$, where h ranges over the discrete set of values that make sense for the given value of t. With this understanding, we adopt the following definition:

DEFINITION. *The explicit one-step method (7.2-2) for the* IVP *(7.1-2)* **converges** *(or is* **convergent***) if, for every initial point $(t_0, \mathbf{u}_I) \in [a, b] \times \mathbb{R}^k$ and every function* $\mathbf{f}$ *satisfying the hypotheses of Theorem 7.1,*

$$\lim_{h \to 0} \varepsilon_h(t) = 0$$

for every $t \in [a, b]$.

Proving convergence directly can be difficult. A more convenient approach is to establish a connection between consistency and convergence. We show that $\varepsilon_h(t) = \mathcal{O}(h^p)$ whenever the one-step scheme (7.2-2) is consistent with order p and satisfies certain additional "tameness" conditions.

We begin with a lemma about the growth of sequences.

LEMMA 7.2. *Suppose that there are constants $\delta > 0$ and $\beta \geq 0$ such that the nonnegative sequence $\{\xi_j\}$ satisfies the inequality*

$$\xi_{j+1} \leq (1 + \delta)\xi_j + \beta, \quad j = 0, 1, 2, \ldots.$$

Then

$$\xi_n \le \xi_0 e^{n\delta} + \frac{e^{n\delta}-1}{\delta}\beta.$$

PROOF: We have

$$\begin{aligned}
\xi_1 &\le (1+\delta)\xi_0 + \beta \\
\xi_2 &\le (1+\delta)\xi_1 + \beta \le (1+\delta)^2\xi_0 + \beta(1+\delta) + \beta \\
&\vdots \\
\xi_n &\le (1+\delta)^n\xi_0 + \beta\sum_{j=0}^{n-1}(1+\delta)^j \\
&= (1+\delta)^n\xi_0 + \beta\frac{(1+\delta)^n - 1}{\delta}.
\end{aligned}$$

(Problem 4 asks for proof of the last step.) But $1+\delta < e^{\delta}$, so

$$\xi_n \le e^{n\delta}\xi_0 + \frac{e^{n\delta}-1}{\delta}\beta,$$

strict inequality holding when $\xi_0 \neq 0$. ∎

The main convergence theorem is as follows:

THEOREM 7.3. *Consider the one-step scheme (7.2-2) for the IVP (7.1-2). Suppose that there exists a stepsize $\overline{h} > 0$ such that, whenever $0 < h \le \overline{h}$, the following conditions hold:*

(i) *There exists a positive constant γ such that $\mathbf{\Phi}_h$ is continuous on the region*

$$R := \left\{(s,\mathbf{v}) \in [a,b] \times \mathbb{R}^k : \|\mathbf{v} - \mathbf{u}(s)\| < \gamma\right\}$$

(see Figure 3).

(ii) *$\mathbf{\Phi}_h$ satisfies a Lipschitz condition with Lipschitz constant L on R.*

(iii) *The one-step method (7.2-2) is consistent with order p.*

Then the one-step method (7.2-2) converges. Moreover, $\varepsilon_h(t) = \mathcal{O}(h^p)$ as $h \to 0$.

Before we embark on the proof, two remarks are appropriate. First, in the present context the conclusion $\varepsilon_h(t) = \mathcal{O}(h^p)$ means that there exist constants

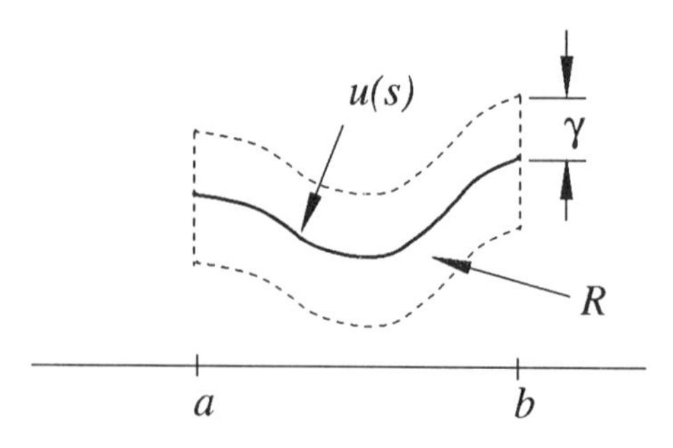

FIGURE 3. *The region* $R = \Big\{(s,v) \in [a,b] \times \mathbb{R} : |v - u(s)| < \gamma\Big\}$.

$h_0 \in (0, \bar{h}]$ and $C \geq 0$, independent of t, such that $\|\boldsymbol{\varepsilon}_h(t)\| \leq C|h^p|$ whenever $h \leq h_0$. In the special case when we can take $R = [a,b] \times \mathbb{R}^k$ (that is, when we can take $\gamma = \infty$), $h_0 = \bar{h}$.

Second, by analogy with terminology used in Theorem 7.1, the condition (*ii*) means that $\|\boldsymbol{\Phi}_h(s, \mathbf{v}_1) - \boldsymbol{\Phi}_h(s, \mathbf{v}_2)\| \leq L\|\mathbf{v}_1 - \mathbf{v}_2\|$ for every pair of points $(s, \mathbf{v}_1), (s, \mathbf{v}_2) \in R$. This requirement hardly seems onerous if we regard it as a natural analog of the condition on $\mathbf{f}$ needed to guarantee well-posedness of the original IVP. However, (*ii*) has an interesting interpretation: It guarantees that the discrete scheme is **stable**, in a sense that Problem 8 makes clearer. Viewed in this way, Theorem 7.3 establishes a connection between consistency and stability — properties that one can readily test — and convergence, the desirable property that may be difficult to check directly. This motif reappears later in the book.

PROOF: The first task is to extend $\boldsymbol{\Phi}_h : R \to \mathbb{R}^k$ continuously to a function $\tilde{\boldsymbol{\Phi}}_h : [a,b] \times \mathbb{R}^k \to \mathbb{R}^k$. We do this as follows: For any point $(s, \mathbf{v}) \in [a,b] \times \mathbb{R}^k$, denote by $\boldsymbol{v}(s) := [\mathbf{v} - \mathbf{u}(s)]/\|\mathbf{v} - \mathbf{u}(s)\|$ the unit vector in the direction of $\mathbf{v} - \mathbf{u}(s)$. Then define

$$\tilde{\boldsymbol{\Phi}}_h(s, \mathbf{v}) := \begin{cases} \boldsymbol{\Phi}_h(s, \mathbf{v}), & \text{if } (s, \mathbf{v}) \in R, \\ \boldsymbol{\Phi}_h(s, \mathbf{u}(s) + \gamma \boldsymbol{v}(s)), & \text{if } s \in [a,b] \text{ and } \|\mathbf{v} - \mathbf{u}(s)\| > \gamma. \end{cases}$$

One easily verifies that $\tilde{\boldsymbol{\Phi}}_h$ satisfies a Lipschitz condition with Lipschitz constant L on all of the region $[a,b] \times \mathbb{R}^k$. Moreover, since $\tilde{\boldsymbol{\Phi}}_h$ agrees with $\boldsymbol{\Phi}_h$ on R, the hypothesis (*iii*) implies that

$$\|\boldsymbol{\Delta}_h(t, \mathbf{u}(t)) - \tilde{\boldsymbol{\Phi}}_h(t, \mathbf{u}(t))\| \leq Ch^p$$

for all $t \in [a,b]$.

The extended function $\tilde{\boldsymbol{\Phi}}_h$ generates a one-step scheme that, given an initial value $\tilde{\mathbf{u}}_0 := \mathbf{u}_I$, produces a sequence

$$\tilde{\mathbf{u}}_{n+1} = \tilde{\mathbf{u}}_n + h\tilde{\boldsymbol{\Phi}}_h(t_n, \tilde{\mathbf{u}}_n).$$

This sequence does not necessarily yield points $(t_n, \tilde{\mathbf{u}}_n)$ lying in R, an inconvenience that we address later in the proof.

The solution to the IVP (7.1-2) has corresponding values

$$\mathbf{u}(t_{n+1}) = \mathbf{u}(t_n) + h\mathbf{\Delta}_h(t_n, \mathbf{u}(t_n)).$$

Therefore, the error $\tilde{\varepsilon}_h(t_{n+1}) := \mathbf{u}(t_{n+1}) - \tilde{\mathbf{u}}_{n+1}$ obeys the estimate

$$\begin{aligned} \|\tilde{\varepsilon}_h(t_{n+1})\| &\leq \|\tilde{\varepsilon}_h(t_n)\| + h\|\mathbf{\Delta}_h(t_n, \mathbf{u}(t_n)) - \mathbf{\Phi}_h(t_n, \mathbf{u}(t_n))\| \\ &\qquad + h\|\tilde{\mathbf{\Phi}}_h(t_n, \mathbf{u}(t_n)) - \tilde{\mathbf{\Phi}}_h(t_n, \tilde{\mathbf{u}}_n)\| \\ &\leq \|\tilde{\varepsilon}_h(t_n)\| + hL\|\tilde{\varepsilon}_h(t_n)\| + Ch^{p+1} \\ &= (1 + hL)\|\tilde{\varepsilon}_h(t_n)\| + Ch^{p+1}. \end{aligned}$$

Applying Lemma 7.2 with $\xi_n = \|\tilde{\varepsilon}_h(t_n)\|$ (and hence with $\xi_0 = 0$) yields

$$\|\tilde{\varepsilon}_h(t_n)\| \leq Ch^p \frac{e^{nhL} - 1}{L}. \tag{7.2-12}$$

It remains to show that we can restrict the stepsize h so that each point $(t_n, \tilde{\mathbf{u}}_n)$, generated by the extended one-step method associated with $\tilde{\mathbf{\Phi}}_h$, lies in the region R. Such a restriction guarantees that the numerical solution $\tilde{\mathbf{u}}_n$ coincides with a solution $\mathbf{u}_n$ generated by the original one-step method associated with $\mathbf{\Phi}_h$. We begin by fixing a value $t > t_0$ in $[a, b]$. If n is any positive integer and $h := (t - t_0)/n$, then the inequality (7.2-12) ensures that

$$\|\tilde{\varepsilon}_h(t)\| \leq Ch^p \frac{\exp[(t - t_0)L] - 1}{L} \leq Ch^p \frac{\exp[(b - a)L] - 1}{L}.$$

Since $\gamma > 0$, we can pick a number $h_0 \in (0, \overline{h}]$ such that $\|\tilde{\varepsilon}_h(t)\| \leq \gamma$ for every $t \in [a, b]$, whenever $0 \leq h \leq h_0$. For any such h, all points $(t_n, \tilde{\mathbf{u}}_n)$ generated by the one-step method associated with $\tilde{\mathbf{\Phi}}_h$ lie in R, where $\tilde{\mathbf{\Phi}}_h$ coincides with $\mathbf{\Phi}_h$. Thus we have

$$\|\varepsilon_h(t)\| \leq Ch^p \frac{\exp[(t - t_0)L] - 1}{L} = \mathcal{O}(h^p).$$

This establishes convergence at the desired rate. ∎

Further Remarks: The Runge-Kutta-Fehlberg Algorithm

One-step methods tend to produce errors that increase with the stepsize h. Many applications involve ODEs whose solutions exhibit some regions requiring a small stepsize for reasonable accuracy, while in other regions a larger, less costly stepsize may suffice. A key aim of modern software design is to develop *adaptive* algorithms, that is, algorithms that automatically vary step-sizes to accommodate local peculiarities in the solution. Section 6.3 presents ideas along this line in the context of numerical integration. We devote the rest of this section to a sketch of an adaptive method for controlling truncation error in the numerical solution of ODEs.

Problem 2 contains the germ of the idea. By considering schemes having the general form

$$\mathbf{\Phi}_h(s,\mathbf{v}) = \alpha_1\mathbf{f}(s,\mathbf{v}) + \alpha_2\mathbf{f}(s+\alpha_3 h, \mathbf{v}+\alpha_4 h\mathbf{f}(s,\mathbf{v})),$$

one can choose the coefficients $\alpha_1, \alpha_2, \alpha_3, \alpha_4$ to force cancellation of low powers of h in the Taylor expansion of the truncation error. In this way, one generates the $\mathcal{O}(h^2)$-accurate Heun and modified Euler methods. Similarly, by considering schemes of the form

$$\mathbf{\Phi}_h(s,\mathbf{v}) = \alpha_1\mathbf{f}(s,\mathbf{v}) + \alpha_2\mathbf{f}(s+\alpha_3 h, \mathbf{v}+\alpha_4 h\mathbf{f}(s+\alpha_5 h, \mathbf{v}+\alpha_6 h\mathbf{f}(s,\mathbf{v}))),$$

one can select constants $\alpha_1, \alpha_2, \ldots, \alpha_6$ to yield a truncation error having magnitude $\mathcal{O}(h^3)$.

One can produce successively higher-order one-step schemes by extending this procedure. The calculations are straightforward, but they require increasingly arduous (and unenlightening) algebra to determine suitable constants α_j. This procedure underlies the $\mathcal{O}(h^4)$ Runge-Kutta scheme (7.2-5).

For the present, we consider two higher-order schemes of the Runge-Kutta type. The first scheme has $\mathcal{O}(h^4)$ truncation error:

$$\tilde{\mathbf{\Phi}}_h(s,\mathbf{v}) := \tfrac{25}{216}\mathbf{k}_1 + \tfrac{1408}{2565}\mathbf{k}_3 + \tfrac{2197}{4104}\mathbf{k}_4 - \tfrac{1}{5}\mathbf{k}_5, \tag{7.2-13}$$

with the vectors $\mathbf{k}_j$ defined below. The second has $\mathcal{O}(h^5)$ truncation error:

$$\overline{\mathbf{\Phi}}_h(s,\mathbf{v}) := \tfrac{16}{135}\mathbf{k}_1 + \tfrac{6656}{12825}\mathbf{k}_3 + \tfrac{28561}{56430}\mathbf{k}_4 - \tfrac{9}{50}\mathbf{k}_5 + \tfrac{2}{55}\mathbf{k}_6. \tag{7.2-14}$$

These two schemes share several evaluations of the function $\mathbf{f}$:

$$\begin{aligned}
\mathbf{k}_1 &:= \mathbf{f}(s,\mathbf{v}),\\
\mathbf{k}_2 &:= \mathbf{f}(s+\tfrac{1}{4}h, \mathbf{v}+\tfrac{1}{4}\mathbf{k}_1),\\
\mathbf{k}_3 &:= \mathbf{f}(s+\tfrac{3}{8}h, \mathbf{v}+\tfrac{3}{32}\mathbf{k}_1+\tfrac{9}{32}\mathbf{k}_2),\\
\mathbf{k}_4 &:= \mathbf{f}(s+\tfrac{12}{13}h, \mathbf{v}+\tfrac{1932}{2197}\mathbf{k}_1-\tfrac{7200}{2197}\mathbf{k}_2+\tfrac{7296}{2197}\mathbf{k}_3),\\
\mathbf{k}_5 &:= \mathbf{f}(s+h, \mathbf{v}+\tfrac{439}{216}\mathbf{k}_1-8\mathbf{k}_2+\tfrac{3680}{513}\mathbf{k}_3-\tfrac{845}{4104}\mathbf{k}_4),\\
\mathbf{k}_6 &:= \mathbf{f}(s+\tfrac{1}{2}h, \mathbf{v}-\tfrac{8}{27}\mathbf{k}_1+2\mathbf{k}_2-\tfrac{3544}{2565}\mathbf{k}_3+\tfrac{1859}{4104}\mathbf{k}_4-\tfrac{11}{40}\mathbf{k}_5).
\end{aligned}$$

Thus, with six distinct evaluations of $\mathbf{f}$ we compute two Runge-Kutta steps, each having a different order truncation error.

This difference in the truncation errors is the key to adaptivity. The objective is as follows: Given an initial stepsize h, we wish at any stage to choose a scaling factor $\sigma > 0$ such that using the stepsize σh in the $\mathcal{O}(h^5)$ scheme (7.2-14) yields a truncation error $\overline{\boldsymbol{\tau}}_h$ smaller than some prescribed tolerance $\delta > 0$. The idea is to use information generated by the $\mathcal{O}(h^4)$ scheme (7.2-13) to limit the size of $\overline{\boldsymbol{\tau}}_h$. We proceed via a sequence of approximations.

Using the definition (7.2-9), we have the following approximate expressions for the truncation errors at level t_n:

$$\tilde{\boldsymbol{\tau}}_h(t_n, \tilde{\mathbf{u}}_n) \simeq h^{-1}[\mathbf{u}(t_{n+1}) - \tilde{\mathbf{u}}_n] - \tilde{\boldsymbol{\Phi}}_h(t_n, \tilde{\mathbf{u}}_n) \qquad (\mathcal{O}(h^4)),$$

$$\overline{\boldsymbol{\tau}}_h(t_n, \overline{\mathbf{u}}_n) \simeq h^{-1}[\mathbf{u}(t_{n+1}) - \overline{\mathbf{u}}_n] - \overline{\boldsymbol{\Phi}}_h(t_n, \overline{\mathbf{u}}_n) \qquad (\mathcal{O}(h^5)).$$

Here, $\tilde{\mathbf{u}}_n$ and $\overline{\mathbf{u}}_n$ denote the numerical solutions generated by the schemes (7.2-13) and (7.2-14), respectively. It follows that

$$\begin{aligned} \mathbf{u}(t_{n+1}) - \tilde{\mathbf{u}}_{n+1} &= \mathbf{u}(t_{n+1}) - \tilde{\mathbf{u}}_n - h\tilde{\boldsymbol{\Phi}}_h(t_n, \tilde{\mathbf{u}}_n) \\ &\simeq h\tilde{\boldsymbol{\tau}}_h(t_n, \tilde{\mathbf{u}}_n). \end{aligned}$$

Thus

$$\begin{aligned} \tilde{\boldsymbol{\tau}}_h(t_n, \tilde{\mathbf{u}}_n) &\simeq h^{-1}[\mathbf{u}(t_{n+1}) - \tilde{\mathbf{u}}_{n+1}] \\ &= h^{-1}[\mathbf{u}(t_{n+1}) - \overline{\mathbf{u}}_{n+1}] + h^{-1}(\overline{\mathbf{u}}_{n+1} - \tilde{\mathbf{u}}_{n+1}) \\ &= h^{-1}\left[\mathbf{u}(t_{n+1}) - \overline{\mathbf{u}}_n - h\overline{\boldsymbol{\Phi}}_h(t_n, \overline{\mathbf{u}}_n)\right] + h^{-1}(\overline{\mathbf{u}}_{n+1} - \tilde{\mathbf{u}}_{n+1}) \\ &\simeq \overline{\boldsymbol{\tau}}_h(t_n, \overline{\mathbf{u}}_n) + h^{-1}(\overline{\mathbf{u}}_{n+1} - \tilde{\mathbf{u}}_{n+1}). \end{aligned}$$

But $\tilde{\boldsymbol{\tau}}_h(t_n, \tilde{\mathbf{u}}_n) = \mathcal{O}(h^4)$, while $\overline{\boldsymbol{\tau}}_h(t_n, \overline{\mathbf{u}}_n) = \mathcal{O}(h^5)$. By reasoning that the latter is much smaller than the former, we arrive at the approximation

$$\tilde{\boldsymbol{\tau}}_h(t_n, \tilde{\mathbf{u}}_n) \simeq \frac{\overline{\mathbf{u}}_{n+1} - \tilde{\mathbf{u}}_{n+1}}{h}. \tag{7.2-15}$$

The fact that $\tilde{\boldsymbol{\tau}}_h(t_n, \tilde{\mathbf{u}}_n) = \mathcal{O}(h^4)$ implies that

$$\tilde{\boldsymbol{\tau}}_h(t_n, \tilde{\mathbf{u}}_n) \simeq Ch^4, \tag{7.2-16}$$

where C is some positive constant. Therefore,

$$\frac{\overline{\mathbf{u}}_{n+1} - \tilde{\mathbf{u}}_{n+1}}{h} \simeq Ch^4. \tag{7.2-17}$$

Using the approximations (7.2-16) and (7.2-17), we can now estimate the truncation error in the $\mathcal{O}(h^4)$ scheme (7.2-13) that results when we use a different stepsize σh:

$$\tilde{\boldsymbol{\tau}}_{\sigma h}(t_n, \tilde{\mathbf{u}}_n) \simeq C(\sigma h)^4 \simeq \sigma^4 \frac{\overline{\mathbf{u}}_{n+1} - \tilde{\mathbf{u}}_{n+1}}{h}.$$

This approximation suggests that $||\tilde{\tau}_{\sigma h}|| < \delta$ (and hence $||\overline{\tau}_{\sigma h}|| < \delta$) provided that

$$0 < \sigma < \left(\frac{\delta h}{||\overline{\mathbf{u}}_{n+1} - \tilde{\mathbf{u}}_{n+1}||}\right)^{1/4}. \qquad (7.2\text{-}18).$$

These observations allow us to control the stepsize h adaptively. To compute the new value $\mathbf{u}_{n+1}$ from $\mathbf{u}_n$ for a given stepsize h, we first compute $\tilde{\mathbf{u}}_n$ and $\overline{\mathbf{u}}_{n+1}$ using the fourth- and fifth-order Runge-Kutta schemes (7.2-13) and (7.2-14). We then estimate the truncation error using the approximation (7.2-15). If the estimate is at least as large in magnitude as the prescribed tolerance δ, we reject the current stepsize, make the "midstep correction" $h \leftarrow \sigma h$, and try again. Otherwise, we accept $\overline{\mathbf{u}}_{n+1}$ as the new value of $\mathbf{u}_{n+1}$, adjust the stepsize if appropriate, and proceed to the next step.

If done correctly, adjusting the stepsize for the next step can greatly enhance the efficiency of the algorithm. One should steer a moderate course between excessive caution (using a stepsize much smaller than is really needed for accuracy) and excessive boldness (using a stepsize likely to require midstep adjustments to h and hence more evaluations of $\mathbf{f}$ during the step). Conservatism suggests a strategy that modifies Equation (7.2-18): After computing the quantity

$$\overline{\sigma} := \left(\frac{\delta h}{2||\overline{\mathbf{u}}_{n+1} - \tilde{\mathbf{u}}_{n+1}||}\right)^{1/4},$$

set

$$\sigma = \begin{cases} 1/10, & \text{if } \overline{\sigma} \leq 1/10, \\ \overline{\sigma}, & \text{if } 1/20 < \overline{\sigma} < 4, \\ 4, & \text{if } \overline{\sigma} \geq 4. \end{cases} \qquad (7.2\text{-}19)$$

This strategy avoids extreme modifications in stepsize, possibly at the expense of some efficiency or even of the algorithm's success.

We can now state the adaptive algorithm succinctly. Denote by $\tilde{\mathbf{u}}_{n+1} \leftarrow \text{RK}_4(t_n, \mathbf{u}_n, h)$ the fourth-order Runge-Kutta scheme (7.2-13) and by $\overline{\mathbf{u}}_{n+1} \leftarrow \text{RK}_5(t_n, \mathbf{u}_n, h)$ the fifth-order scheme (7.2-14).

ALGORITHM 7.1 (RUNGE-KUTTA-FEHLBERG ADAPTIVE STEPSIZING). *This algorithm generates a numerical solution $\mathbf{u}_n$ to the* IVP *(7.1-2) given an initial stepsize h, bounds $h_{\min}, h_{\max}$ on the stepsize, and a tolerance $\delta > 0$ on the stepsize h.*

1. $n \leftarrow 0$.

2. $t_n \leftarrow a$.

3. $\mathbf{u}_n \leftarrow \mathbf{u}_I$.

4. $\tilde{\mathbf{u}} \leftarrow \text{RK}_4(t_n, \mathbf{u}_n, h)$.

5. $\overline{\mathbf{u}} \leftarrow \text{RK}_5(t_n, \mathbf{u}_n, h)$.

6. If $\|\bar{\mathbf{u}} - \tilde{\mathbf{u}}\|/h < \delta$ then:

7. $\quad t_{n+1} \leftarrow t_n + h.$

8. $\quad \mathbf{u}_{n+1} \leftarrow \bar{\mathbf{u}}.$

9. $\quad$ If $t_{n+1} \geq b$ stop.

10. $\quad n \leftarrow n + 1.$

11. End if.

12. Compute σ using Equation (7.2-19).

13. $h \leftarrow \sigma h.$

14. If $h > h_{\max}$ set $h \leftarrow h_{\max}$.

15. If $h < h_{\min}$ stop (algorithm fails).

16. Go to 4.

17. End.

7.3 Multistep Methods: Consistency and Stability

Motivation

Multistep methods extend one-step methods by utilizing computed values $\mathbf{u}_n$ from several previous steps. The following definition captures this extension:

DEFINITION. *A* **multistep method** *for the* ODE *(7.1-1) is a scheme of the form*

$$\mathbf{u}_{n+r} + \underbrace{a_{r-1}\mathbf{u}_{n+r-1} + \cdots + a_0\mathbf{u}_n}_{\text{known}} = h\mathbf{\Phi}_h(t_n, \mathbf{u}_{n+r}, \underbrace{\mathbf{u}_{n+r-1}, \ldots, \mathbf{u}_n}_{r \text{ steps}}). \quad (7.3\text{-}1)$$

Here, $h > 0$ is the stepsize. A **solution** *to the scheme (7.3-1) is a sequence $\{\mathbf{u}_n\}$ in $\mathbb{C}^k$ for which Equation (7.3-1) holds for $n = r, r+1, r+2, \ldots$.*

We call Equation (7.3-1) an r-step method, since it determines the unknown value $\mathbf{u}_{n+r}$ in terms of the r values $\mathbf{u}_{n+r-1}, \mathbf{u}_{n+r-2}, \ldots, \mathbf{u}_n$ computed previously. When the function $\mathbf{\Phi}_h$ does not depend upon the unknown $\mathbf{u}_{n+r}$, we call the scheme **explicit**. Otherwise, it is typically necessary to solve nonlinear equations for $\mathbf{u}_{n+r}$, and we call the method **implicit**. Later in this chapter we explore uses of explicit and implicit multistep methods.

Equation (7.3-1) admits an important special case in which $\mathbf{\Phi}_h$ is a linear combination of values of the function $\mathbf{f}$:

$$\mathbf{\Phi}_h(t_n, \mathbf{u}_{n+r}, \ldots, \mathbf{u}_n) = b_r \mathbf{f}(t_{n+r}, \mathbf{u}_{n+r}) + \cdots + b_0 \mathbf{f}(t_n, \mathbf{u}_n).$$

Here, $b_0, b_1, \ldots, b_r$ are scalars. In this case, the multistep method is **linear**. A linear multistep method is explicit when $b_r = 0$ and implicit otherwise. This section and the next focus on linear multistep methods.

There are at least two good reasons for studying multistep methods. The first involves a practical observation: In the most commonly used multistep methods, every evaluation of the function $\mathbf{f}$ involves an argument $\mathbf{u}_k$ that is an approximate value of the solution $\mathbf{u}(t)$. Thus multistep methods stand in contrast to such one-step methods as the Heun method (7.2-3) or the Runge-Kutta method (7.2-5), which require evaluations of $\mathbf{f}$ at several "intermediate" arguments that are not useful approximations to $\mathbf{u}(t)$. In this sense, multistep methods are more economical than higher-order one-step schemes.

The second reason for studying multistep methods is more theoretical: The analysis of multistep methods is both highly developed and fairly general. The form (7.3-1) includes both explicit and implicit one-step methods as the special case $r = 1$. Thus our theoretical treatment of multistep methods casts one-step methods in a framework more systematic than that captured in Theorem 7.3. Moreover, the convergence theory for multistep methods has analogies in the analysis of discrete methods for partial differential equations, as we demonstrate in Chapter 8.

In this section we examine the construction of two popular classes of linear multistep methods: the Adams-Bashforth methods, which are explicit, and the Adams-Moulton methods, which are implicit. We also examine what it means for linear multistep methods to be consistent and stable. We use these notions to outline some of the practical aspects of the methods, including the construction of predictor-corrector methods. At the end of the section, we examine the notion of stability.

As with one-step methods, there are noteworthy connections between the concepts of consistency and stability and the convergence of multistep methods. We explore these connections in Section 7.4.

Adams-Bashforth and Adams-Moulton Methods

Several important multistep methods arise from the following observation: If $\mathbf{u}(t)$ is the solution to the IVP (7.1-2), then by the fundamental theorem of calculus

$$\mathbf{u}(t_{n+k}) - \mathbf{u}(t_{n-j}) = \int_{t_{n-j}}^{t_{n+k}} \mathbf{u}'(t)\, dt = \int_{t_{n-j}}^{t_{n+k}} \mathbf{f}(t, \mathbf{u}(t))\, dt.$$

Pursuing an idea from the quadrature methods of Section 6.2, let us replace the integrand $\mathbf{f}(t, \mathbf{u}(t))$ by an approximating function $\mathbf{p}(t)$. Choose $\mathbf{p}$ so that

$\mathbf{p}(t_l, \mathbf{u}_l)$ agrees with the approximate value of $\mathbf{u}'$ computed during several previous steps, say

$$\mathbf{p}(t_l) = \mathbf{f}(t_l, \mathbf{u}_l), \quad l = n, n-1, \ldots, n-q.$$

A simple way to do this is to use Lagrange interpolating polynomials:

$$\mathbf{p}(t) := \sum_{l=0}^{q} \mathbf{f}(t_{n-l}, \mathbf{u}_{t-l}) L_{n-l}(t),$$

where

$$L_{n-l}(t) = \prod_{\substack{m=0 \\ m \neq l}}^{q} \frac{t - t_{n-m}}{t_{n-l} - t_{n-m}}.$$

This approximation suggests schemes having the form

$$\mathbf{u}_{n+k} - \mathbf{u}_{n-j} = \sum_{l=0}^{q} \mathbf{f}(t_{n-l}, \mathbf{u}_{n-l}) \underbrace{\int_{t_{n-l}}^{t_{n+k}} L_{n-l}(t)\, dt}_{b_l}.$$

Specifying values for the integers k, j, and q yields particular linear multistep schemes. For example, the choices $k = 1$, $j = 0$ yield a class of explicit methods called **Adams-Bashforth methods**. The first few of these schemes are as follows:

$$\begin{aligned}
q = 1: \quad \mathbf{u}_{n+1} &= \mathbf{u}_n + \tfrac{1}{2}h(3\mathbf{f}_n - \mathbf{f}_{n-1}) && \text{2-step,} \\
q = 2: \quad \mathbf{u}_{n+1} &= \mathbf{u}_n + \tfrac{1}{12}h(23\mathbf{f}_n - 16\mathbf{f}_{n-1} + 5\mathbf{f}_{n-2}) && \text{3-step,} \\
q = 3: \quad \mathbf{u}_{n+1} &= \mathbf{u}_n + \tfrac{1}{24}h(55\mathbf{f}_n - 59\mathbf{f}_{n-1} + 37\mathbf{f}_{n-2} - 9\mathbf{f}_{n-3}) && \text{4-step.}
\end{aligned} \tag{7.3-2}$$

Here, $\mathbf{f}_n := \mathbf{f}(t_n, \mathbf{u}_n)$. The choices $k = 0$, $j = 1$ yield a sequence of implicit methods called **Adams-Moulton methods**. Here are the first few:

$$\begin{aligned}
q = 2: \quad \mathbf{u}_{n+1} &= \mathbf{u}_n + \tfrac{1}{112}h(\mathbf{f}_{n+1} + 8\mathbf{f}_n - \mathbf{f}_{n-1}) && \text{2-step,} \\
q = 3: \quad \mathbf{u}_{n+1} &= \mathbf{u}_n + \tfrac{1}{24}h(9\mathbf{f}_{n+1} + 19\mathbf{f}_n - 5\mathbf{f}_{n-1} + \mathbf{f}_{n-2}) && \text{3-step,} \\
q = 4: \quad \mathbf{u}_{n+1} &= \mathbf{u}_n + \tfrac{1}{720}h(251\mathbf{f}_{n+1} + 646\mathbf{f}_n - 264\mathbf{f}_{n-1} \\
&\quad + 106\mathbf{f}_{n-2} - 19\mathbf{f}_{n-3}) && \text{4-step.}
\end{aligned} \tag{7.3-3}$$

Each of these schemes requires values of the computed solution from at least two previous levels. To get started, we need values $\mathbf{u}_1, \mathbf{u}_2, \ldots, \mathbf{u}_{r-1}$ in addition to the initial value $\mathbf{u}_0$. The typical procedure for starting an r-step method is to use the initial value $\mathbf{u}_0 = u_I$ together with values $\mathbf{u}_1, \mathbf{u}_2, \ldots, \mathbf{u}_{r-1}$ computed via a one-step method having accuracy comparable to the multistep method.

Consistency of Multistep Methods

We assess consistency of multistep methods in a manner analogous to that employed for one-step methods. To keep the discussion simple, we restrict attention to linear schemes,

$$\mathbf{u}_{n+r} + \sum_{j=0}^{r-1} a_j \mathbf{u}_{n+j} = h \sum_{j=0}^{r} b_j \mathbf{f}_{n+j}. \tag{7.3-4}$$

DEFINITION. *Given a point* $(s, \mathbf{v}) \in [a,b] \times \mathbb{R}^k$, *let* $\mathbf{w}\colon [a,b] \to \mathbb{R}^k$ *denote the solution to the* ODE *(7.1-1) corresponding to the initial condition* $\mathbf{w}(s) = \mathbf{v}$. *The* **truncation error** *(or* **local discretization error***) of the multistep scheme (7.3-4) is*

$$\boldsymbol{\tau}_h(s, \mathbf{v}) := \frac{1}{h}\left[\mathbf{w}(s+rh) + \sum_{j=0}^{r-1} a_j \mathbf{w}(s+jh) - h\sum_{j=0}^{r} b_j \mathbf{f}(s+jh, \mathbf{w}(s+jh))\right].$$

Think of $\boldsymbol{\tau}_h$ as a difference quotient that gauges how far the numerical solution can stray from the exact solution during r steps. Heuristically,

$$\boldsymbol{\tau}_h(s, \mathbf{v}) = \frac{1}{h}\left[\begin{pmatrix}\text{exact solution} \\ \text{at } s+rh\end{pmatrix} - \begin{pmatrix}\text{numerical solution at } s+rh \\ \text{using starting values} \\ \mathbf{w}(s), \ldots, \mathbf{w}(s+(r-1)h)\end{pmatrix}\right].$$

For a multistep scheme (7.3-4) to be a reasonable approximation to the original ODE, the truncation error must vanish as $h \to 0$. Moreover, we cannot expect good approximations unless the starting values $\mathbf{u}_0, \mathbf{u}_1, \ldots, \mathbf{u}_{r-1}$ also approach the exact values $\mathbf{u}(t_0), \mathbf{u}(t_1), \ldots, \mathbf{u}(t_{r-1})$ as $h \to 0$. The following definition formalizes these requirements:

DEFINITION. *Suppose that the r-step method (7.3-4) uses the starting values* $\mathbf{u}_0, \mathbf{u}_1, \ldots, \mathbf{u}_{r-1}$. *The method is* **consistent** *if both of the following conditions hold:*

(i) $\lim_{h \to 0} \boldsymbol{\tau}_h(s, \mathbf{v}) = \mathbf{0}$ *for all* $(s, \mathbf{v}) \in [a,b] \times \mathbb{R}^k$,

(ii) $\lim_{h \to 0} [\mathbf{u}_n - \mathbf{u}(t_n)] = \mathbf{0}$ *for* $n = 0, 1, \ldots, r-1$.

As the next section explores, consistency is necessary for convergence.

For now, consider an example in the one-dimensional case $u\colon [a, b] \to \mathbb{R}$. The truncation error for the three-step Adams-Bashforth method is

$$\begin{aligned}\tau_h(s, v) &= \frac{1}{h}\left\{w(s+h) - w(s) - \frac{h}{12}\Big[23f(s, w(s))\right.\\ &\left.\quad -16f(s-h, w(s-h)) + 5f(s-2h, w(s-2h))\Big]\right\}\\ &= \frac{1}{h}\left[\int_s^{s+h} f(t, w(t))\, dt - \int_s^{s+h} p(t)\, dt\right].\end{aligned}$$

Here, p is the quadratic polynomial that interpolates f over the interval $[s-2h, s]$, as drawn in Figure 1. If $f \in C^3([a, b])$, then we can use the interpolation error estimates of Section 1.2 to rewrite this equation as follows:

$$\begin{aligned}\tau_h(s, v) &= \frac{1}{h}\int_s^{s+h} \frac{f^{(3)}(\zeta(t), w(\zeta(t)))}{3!}(t-s)(t-s+h)(t-s+2h)\, dt\\ &= \frac{1}{3!h}h^4 \int_0^1 f^{(3)}(\zeta, w(\zeta))\alpha(\alpha+1)(\alpha+2)\, d\alpha\\ &= \frac{h^3}{3!}\underbrace{f^{(3)}(\xi, w(\xi))}_{w^{(4)}(\xi)}\int_0^1 \alpha(\alpha+1)(\alpha_2)\, d\alpha\\ &= \frac{3}{8}h^3 w^{(4)}(\xi),\end{aligned}$$

where ξ is some point belonging to the interval $(s, s+h)$. The third line in this sequence of identities follows from the mean value theorem for integrals. Problem 1 asks for the details in this calculation. We conclude that the three-step Adams-Bashforth method has truncation error $\tau_h = \mathcal{O}(h^3)$.

This calculation rests on a seemingly ad hoc assumption about the smoothness of f. In evaluating truncation error, we do not fret over such assumptions, the focus being on the *method* and not on the particular ODE being solved. In this sense truncation error estimates come with the unspoken proviso that the right side of Equation (7.1-1) is as smooth as needed.

Table 7.2 lists the truncation errors for the first few Adams-Bashforth and Adams-Moulton methods.

As with one-step schemes, we distinguish among various orders of consistency:

DEFINITION. *The multistep scheme (7.3-4) is* **consistent with order** p *if* $\tau_h(s, \mathbf{v}) = \mathcal{O}(h^p)$.

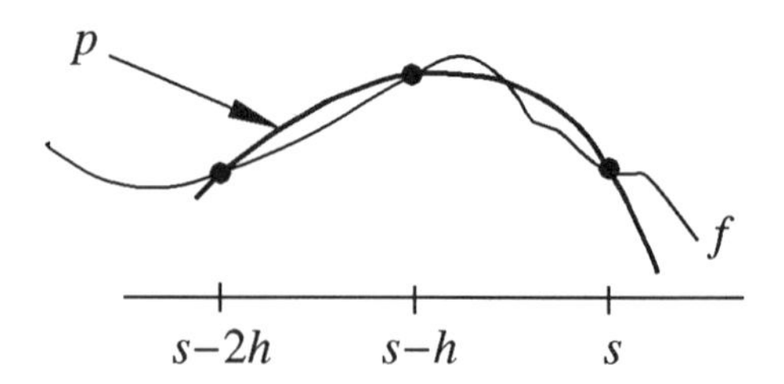

FIGURE 1. *Quadratic polynomial p interpolating f over the interval $[s-2h, s]$.*

Table 7.2: Truncation errors for r-step Adams-Bashforth and Adams-Moulton methods.

r	Adams-Bashforth	Adams-Moulton
2	$\frac{5}{12}\mathbf{w}^{(3)}(\xi)h^2$	$-\frac{1}{24}\mathbf{w}^{(4)}(\xi)h^3$
3	$\frac{3}{8}\mathbf{w}^{(4)}(\xi)h^3$	$-\frac{19}{720}\mathbf{w}^{(5)}(\xi)h^4$
4	$\frac{251}{720}\mathbf{w}^{(5)}(\xi)h^4$	$-\frac{3}{160}\mathbf{w}^{(6)}(\xi)h^5$

The following proposition leads to a convenient way to check the consistency of linear multistep schemes.

PROPOSITION 7.4. *If the function $\mathbf{f}$ in the ODE (7.1-1) is sufficiently differentiable, then the truncation error for the multistep scheme (7.3-4) has the expansion*

$$\tau_h(s, \mathbf{v}) = \frac{1}{h}\left[C_0\mathbf{v}(s) + C_1 h\mathbf{v}'(s) + \cdots + C_p h^p \mathbf{v}^{(p)}(s) + \mathcal{O}(h^{p+1})\right],$$

where

$$C_0 := 1 + a_{r-1} + a_{r-2} + \cdots + a_0,$$

$$C_1 := r + (r-1)a_{r-1} + (r-2)a_{r-2} + \cdots + a_1 - \sum_{j=0}^{r} b_j,$$

$$\vdots$$

$$C_p := \frac{1}{p!}\left[r^p + (r-1)^p a_{r-1} + \cdots + a_1\right] - \frac{1}{(p-1)!}\sum_{j=0}^{r} j^{p-1} b_j.$$

PROOF: The expansion follows directly (but somewhat tediously) if one substitutes the Taylor series

$$\begin{aligned}\mathbf{v}(s+jh) &= \mathbf{v}(s)+jh\mathbf{u}'(s)+\tfrac{1}{2}j^2h^2\mathbf{u}''(s)+\cdots,\\ h\mathbf{f}(s+jh,\mathbf{v}(s+jh)) &= h\mathbf{v}'(s+jh)=h\mathbf{v}'(s)+jh^2\mathbf{v}''(s)+\cdots\end{aligned}$$

into the definition for $\boldsymbol{\tau}_h$. ∎

COROLLARY 7.5. *Assume that the starting values for the linear multistep scheme (7.3-4) satisfy the condition* $\lim_{h\to 0}[\mathbf{u}_n-\mathbf{u}(t_n)]=\mathbf{0}$, $n=0,1,\ldots,r-1$. *Then the scheme is consistent if and only if* $C_0=C_1=0$. *It is consistent with order* p *if and only if* $C_0=C_1=\cdots=C_p=0$.

Stability of Multistep Methods

Theorem 7.3 asserts that, for "reasonable" one-step methods, consistency suffices for convergence. For multistep methods, consistency is not enough, even if the function $\boldsymbol{\Phi}_h$ on the right side of Equation (7.3-1) obeys a Lipschitz condition. One can devise consistent schemes of the form (7.3-4) that yield wildly divergent sequences $\{\mathbf{u}_n\}$ in response to tiny perturbations in the starting values $\mathbf{u}_0,\mathbf{u}_1,\ldots,\mathbf{u}_{r-1}$. Such schemes are computationally useless, since essentially all digital calculations have finite precision and hence introduce perturbations automatically. To avoid this pathology, we impose an additional constraint on multistep schemes that ensures their stability against small arithmetic errors.

The following example gives some hint of the difficulties that can arise. Consider the IVP

$$u'(t)=0,\quad t\in[0,T];\qquad u(0)=u_I. \tag{7.3-5}$$

The solution to this problem is $u(t)=u_I$. The two-step scheme

$$u_{n+2}-4u_{n+1}+3u_n=-2hf_n$$

for this equation is consistent with order 2, since

$$\begin{aligned}C_0 &= 1-4+3=0,\\ C_1 &= 2-4-2=0,\\ C_2 &= \frac{1}{2!}(2^2-4)-0=0,\\ C_3 &= \frac{1}{3!}(2^3-4)-0\neq 0.\end{aligned}$$

For the IVP in question, the scheme reduces to the **homogeneous linear difference equation**

$$u_{n+2} - 4u_{n+1} + 3u_n = 0. \tag{7.3-6}$$

To initialize this scheme, we use starting values u_0, u_1 that are good approximations to $u(0) = u_I$ and $u(h) = u_I$, respectively.

To find nontrivial solutions $\{u_n\}$ to this problem, let us try expressions of the form $u_n = \zeta^n$, where $\zeta \neq 0$. Substituting this trial solution into Equation (7.3-6), we find that $\zeta^2 - 4\zeta + 3 = 0$, so $\zeta = 1$ or $\zeta = 3$ will work. In fact, any linear combination $u_n := c_1 \cdot 1^n + c_2 \cdot 3^n$, where c_1 and c_2 are constants, is a solution to the linear difference equation (7.3-6). To determine c_1 and c_2, we use the starting values u_0 and u_1. With exact starting values, we find that

$$\begin{aligned} c_1 \cdot 1^0 + c_2 \cdot 3^0 &= u_I, \\ c_1 \cdot 1^1 + c_2 \cdot 3^1 &= u_I, \end{aligned}$$

so $c_1 = 1$, $c_2 = 0$, and $u_n = u_I$, as expected.

Suppose, though, that we compute another solution $\tilde{u}_n$ by using the inexact (but consistent) starting values $\tilde{u}_0 = u_I$ and $\tilde{u}_1 = u_I + \delta$, where $\delta \to 0$ as $h \to 0$. For any positive stepsize h, these starting values imply that $c_1 = u_I - \delta/2$ and $c_2 = \delta/2$. The numerical solution in this case is $\tilde{u}_n = u_I - \frac{1}{2}\delta + \frac{1}{2}3^n\delta$. The factor 3^n in this solution causes $|u_n - \tilde{u}_n|$ to grow without bound with n, and reducing the stepsize causes the difference to increase even more rapidly with t. Since small errors in the starting values yield enormous errors in the computed solution, this scheme has no practical utility. The difficulty in this case stems from the zero $\zeta = 3$ of the polynomial $\psi(\zeta) := \zeta^2 - 4\zeta + 3$.

This example serves as a paradigm for the stability analysis of more general schemes. We begin with a definition:

DEFINITION. *The linear multistep scheme (7.3-4) is* **stable** *if, when applied to the* IVP *(7.3-5), there exist positive constants $\bar{h}$ and M such that, whenever $0 \leq h < \bar{h}$, the following condition holds: For every $\delta > 0$, if the two sets*

$$\left\{u_0, u_1, \ldots, u_{r-1}\right\},$$

$$\left\{\tilde{u}_0, \tilde{u}_1, \ldots, \tilde{u}_{r-1}\right\}$$

satisfy $|u_n - \tilde{u}_n| < \delta$ for $n = 0, 1, \ldots, r-1$, then $|u_n - \tilde{u}_n| < M\delta$ for every index $n = r, r+1, \ldots, T/h$.

In other words, nearby starting values yield nearby solutions — precisely the condition needed to guard against disasters arising from inexact arithmetic.

The definition does not immediately furnish a convenient way to check stability a priori. A more practical approach, as the example above suggests, is to analyze the polynomial that we get when we substitute the trial solution ζ^n into the scheme, assuming that $f(t, u(t)) = 0$ for all t:

DEFINITION. *The* **characteristic polynomial** *for the multistep scheme (7.3-4) is*

$$\psi(\zeta) := \zeta^r + \sum_{j=0}^{r-1} a_j \zeta^{r-j-1}.$$

Corollary 7.5 guarantees that $\zeta = 1$ is a zero of $\psi(\zeta)$ for any consistent scheme. Indeed, in the example above this zero corresponds to the component of the numerical solution that is a good approximation of the exact solution $u(t)$. We call $\zeta = 1$ the **principal root** of ψ. Other zeros of ψ are **parasitic roots**. Parasitic roots correspond to solutions of the multistep scheme that do not give good approximations to the exact solution.

One way to characterize stable schemes of the form (7.3-4) is by stipulating that the components of the numerical solution corresponding to parasitic roots of ψ do not grow. In the example above, the parasitic root $\zeta = 3$ corresponds to a solution component $c_2 \cdot 3^n$, which grows rapidly as n increases. This reasoning leads to the following definition:

DEFINITION. *A linear multistep scheme (7.3-4) satisfies the* **root condition** *if, whenever $\zeta \in \mathbb{C}$ is a zero of the characteristic polynomial ψ, the following statements hold:*

(i) $|\zeta| \leq 1$;

(ii) *If $|\zeta| = 1$, then ζ is a simple root of ψ.*

We show later in this section that the root condition is equivalent to stability. The role of condition (i) in preventing growth in ζ^n should be clear. The motivation for condition (ii) may be less transparent. We show below that, if ζ is a zero of ψ having multiplicity q, then the functions

$$\zeta^n, n\zeta^n, n^2\zeta^n, \ldots, n^{q-1}\zeta^n$$

are all solutions to the linear difference equation (7.3-6). Therefore, zeros having unit magnitude and multiplicity greater than 1 also lead to growth in the parasitic components of $\{u_n\}$.

Using the root condition, we check the stability of any linear multistep scheme simply by examining the zeros of its characteristic polynomial. For example, the characteristic polynomial for any one-step method (7.2-2) is $\psi(\zeta) = \zeta - 1$, which has only the simple zero $\zeta = 1$. Therefore, all one-step methods are stable in our sense.

The Adams-Bashforth and Adams-Moulton methods all have characteristic polynomials of the form $\psi(\zeta) = \zeta^r - \zeta^{r-1} = \zeta^{r-1}(\zeta - 1)$. According to the root condition, then, these methods are also all stable.

Parasitic roots ζ of ψ correspond to errors that grow at least as fast as $|\zeta|^n$. This observation suggests that one should avoid methods having parasitic roots of unit magnitude. These roots give rise to errors that fail to decay as n increases. We call schemes for which ψ has two or more roots of unit magnitude **weakly stable**. The following four-step method, while consistent, is weakly stable; its characteristic polynomial $\psi(\zeta) = \zeta^4 - 1$ has zeros $1, -1, i, -i$:

$$\mathbf{u}_{n+4} - \mathbf{u}_n = \frac{h}{3}\left(8\mathbf{f}_{n+3} - 4\mathbf{f}_{n+2} + 8\mathbf{f}_{n+1}\right).$$

Weakly stable schemes allow certain components of arithmetic errors to persist in magnitude from step to step, so in practice the effects of such errors tend to accumulate slowly as the imprecise digital arithmetic progresses.

If $\zeta = 1$ is the only zero of ψ with $|\zeta| = 1$, then the multistep scheme is **strongly stable**. The Adams-Bashforth and Adams-Moulton methods are all strongly stable, since their parasitic roots all vanish.

Predictor-Corrector Methods

Although consistent and stable, the Adams-Bashforth and Adams-Moulton methods have drawbacks. For example, while the explicitness of the Adams-Bashforth methods makes them convenient from a programmer's viewpoint, it also implies that the methods rely only on "old" information to advance the numerical solution from t_n to t_{n+1}. Extrapolation of this sort can be risky when the ODE being solved imposes rapid changes on the solution over intervals that are short compared with the stepsize h.

The Adams-Moulton schemes, being implicit, avoid the risks associated with extrapolation. However, they also require one to solve nonlinear algebraic equations. Consequently, straightforward application of the Adams-Moulton methods involves iteration *within* steps, usually accompanied, as Problem 9 indicates, by stepsize restrictions. The need to iterate poses a strategic dilemma: Should we use a fairly large stepsize and invest computational effort in iterations to solve the nonlinear equations accurately? Or, should we take a larger number of small steps, in which previous values of $\mathbf{u}_n$ serve as good initial guesses and lead to numerical convergence in fewer iterations?

The answer is that we should invest in smaller stepsizes. Smaller stepsizes lead to more accurate approximations to the original ODE. Ideally, we should start each small step with a good initial guess, then compute an accurate answer in one iteration before proceeding to the next step. This strategy admits roles for both explicit and implicit schemes. We use an explicit scheme as a **predictor**, to extrapolate from the known level to a good initial guess for the next level. Then we use one iteration of an implicit method as a **corrector**, to advance from the initial guess to a final approximation for that level.

Most implementations of this **predictor-corrector** strategy use predictors and correctors that have comparable truncation errors. For example, one might use the fourth-order (four-step) Adams-Bashforth method as a predictor and the fourth-order (three-step) Adams-Moulton method as a corrector. To generate the four starting values needed to initiate the algorithm, one might use the initial value together with three steps of a fourth-order one-step method, such as the Runge-Kutta scheme (7.2-5). The following algorithm employs this approach.

ALGORITHM 7.2 (FOURTH-ORDER PREDICTOR-CORRECTOR SCHEME). *This algorithm generates approximate solutions* $\{\mathbf{u}_n\}$ *to (7.1-1) using the fourth-order Runge-Kutta scheme* RK4 *to calculate starting values, the fourth-order Adams-Bashforth scheme* AB4 *as a predictor, and the fourth-order Adams-Moulton scheme* AM4 *as a corrector. Here,* h *denotes the stepsize,* T *is the maximum value of the independent variable* t, *and* $\mathbf{u}_n^*$ *signifies the initial guess for* $\mathbf{u}_n$ *furnished by the predictor.*

1. $n \leftarrow 0$.

2. $t_n \leftarrow t_0$.

3. $\mathbf{u}_n \leftarrow \mathbf{u}_I$.

4. For $n = 0, 1, 2$:

5. $\quad t_n \leftarrow t_0 + nh$.

6. $\quad \mathbf{u}_{n+1} \leftarrow \mathrm{RK}_4(t_n, \mathbf{u}_n)$.

7. Next n.

8. $\mathbf{u}_{n+1}^* \leftarrow \mathrm{AB}_4(t_n, \mathbf{u}_n, \ldots, t_{n-3}, \mathbf{u}_{n-3})$.

9. $\mathbf{u}_{n+1} \leftarrow \mathrm{AM}_4(t_{n+1}, \mathbf{u}_{n+1}^*, t_n, \mathbf{u}_n, t_{n-1}, \mathbf{u}_{n-1}, t_{n-2}, \mathbf{u}_{n-2})$.

10. $n \leftarrow n + 1$.

11. If $n \leq T/h$ go to 8.

12. End.

Mathematical Details: Stability and the Root Condition

We devote the rest of this section to the equivalence between the root condition and stability of multistep methods. The exposition has the following plan: First, we show that the difference between two numerical solutions to the IVP (7.3-5), generated by the scheme (7.3-4) with nearby starting values, obeys a homogeneous linear difference equation. We then examine the properties of solutions to such equations and their connections with the characteristic polynomial $\psi(\zeta)$. Finally, we characterize the growth of errors in terms of the zeros of ψ.

Consider two solutions to the multistep method (7.3-4) for the IVP (7.3-5): one, $\{u_n\}$, generated using the starting values

$$u_0, u_1, \ldots, u_{r-1},$$

and the other, $\{\tilde{u}_n\}$, generated using starting values

$$\tilde{u}_0, \tilde{u}_1, \ldots, \tilde{u}_{r-1}.$$

Let $\varepsilon_n := u_n - \tilde{u}_n$. The scheme (7.3-4) is stable if one can bound $|\varepsilon_n|$ in terms of the starting "errors" $|\varepsilon_0|, |\varepsilon_1|, \ldots, |\varepsilon_{r-1}|$, at least for some range $0 \le h < \overline{h}$ of stepsizes. By the linearity of the multistep scheme, the sequence $\{\varepsilon_n\}$ obeys the homogeneous linear difference equation

$$\varepsilon_{n+r} + \sum_{j=0}^{r-1} a_j \varepsilon_{n+j} = 0. \tag{7.3-7}$$

We are interested only in the nonzero sequences $\{\varepsilon_n\}$ that satisfy this equation, since the trivial sequence $\{0\}$ cannot contribute to growth in the error. If $a_0 = a_1 = \cdots = a_{r-\sigma-1} = 0$, then any nonzero solution $\{\varepsilon_n\}$ to Equation (7.3-7) obeys the slightly simpler equation

$$\varepsilon_{n+\sigma} + \sum_{j=0}^{\sigma-1} a_{j+r-\sigma} \varepsilon_{n+j} = 0. \tag{7.3-8}$$

In what follows, therefore, we focus on these nonzero error sequences by assuming that $a_0 = a_1 = \cdots = a_{r-\sigma-1} = 0$ but $a_{r-\sigma} \neq 0$.

To analyze the growth of ε_n, we examine all possible solutions to the difference equation (7.3-8). Several facts are fundamental in this regard.

PROPOSITION 7.6. *The homogeneous linear difference equation (7.3-8) has a unique solution $\{\varepsilon_n\}$ for any set $\varepsilon_0, \varepsilon_1, \ldots, \varepsilon_{\sigma-1} \in \mathbb{C}$ of starting values.*

PROOF: Existence is easy to establish inductively: Just solve Equation (7.3-8) for ε_{n+r}. Uniqueness follows if we note that, for any two solutions $\{\varepsilon_n\}$ and $\{\tilde{\varepsilon}_n\}$ that satisfy Equation (7.3-8) and have the given starting values, the difference $\{\varepsilon_n - \tilde{\varepsilon}_n\}$ also satisfies the homogeneous linear difference equation (7.3-8). Moreover, this sequence has starting values $\varepsilon_n - \tilde{\varepsilon}_n = 0$ for $n = 0, 1, \ldots, \sigma - 1$. Induction shows that $\varepsilon_n - \tilde{\varepsilon}_n = 0$ for $n = \sigma, \sigma + 1, \ldots$. ∎

The trivial solution $\varepsilon_n = 0$ is always a solution to Equation (7.3-8). More is true:

PROPOSITION 7.7. *The set $\mathcal{V}$ of all complex-valued sequences $\{\varepsilon_n\}$ that are solutions to the homogeneous linear difference equation (7.3-8) is a vector space.*

PROOF: This is a simple exercise. ∎

Instrumental in our characterization of stability is the fact that $\mathcal{V}$ has dimension σ. To establish this fact, we construct a basis

$$\Big\{\{e_{1_n}\}, \{e_{2_n}\}, \ldots, \{e_{\sigma_n}\}\Big\}$$

of solutions to Equation (7.3-8). We must show that this set of sequences spans $\mathcal{V}$ and is linearly independent, in the sense that the only constants $c_1, c_2, \ldots, c_\sigma \in \mathbb{C}$ for which

$$c_1 e_{1_n} + c_2 e_{2_n} + \cdots + c_\sigma e_{\sigma_n} = 0 \quad \text{for} \quad n = 0, 1, 2, \ldots$$

are $c_1 = c_2 = \cdots = c_\sigma = 0$. The discussion of linear independence is a bit involved; Theorem 7.8, proved below, settles the issue of spanning.

Knowing the zeros $\zeta_1, \zeta_2, \ldots, \zeta_r$ of the characteristic polynomial ψ allows us to identify the desired basis for $\mathcal{V}$. We begin by noting that, when $a_0 = a_1 = \cdots = a_{r-\sigma-1} = 0$, the polynomial ψ has $\zeta = 0$ as a zero of multiplicity $r - \sigma$. Therefore, $\psi(\zeta) = \zeta^{r-\sigma}\psi_\sigma(\zeta)$ for some polynomial ψ_σ having degree σ. Our assumption that $a_{r-\sigma} \neq 0$ implies that the polynomial $\psi_\sigma(\zeta)$ does not have $\zeta = 0$ as a zero.

Now return to the observation, made earlier, that for any zero ζ_k of ψ the sequence $\{\zeta_k^n\}$ is a solution to Equation (7.3-7), since

$$\zeta_k^{n+r} + \sum_{j=0}^{r-1} a_j \zeta_k^{n+j} = \zeta_k^n \psi(\zeta_k) = 0.$$

However, the trivial zero $\zeta = 0$ is of no interest, since it yields only the trivial solution to the difference equation and does not contribute to growth in the error ε_n. The nontrivial zeros, which are of interest, are precisely the zeros of ψ_σ. Henceforth, denote these zeros by $\zeta_1, \zeta_2, \ldots, \zeta_\sigma$. (The existence of at least one such zero, namely $\zeta = 1$, follows from the consistency of the original multistep scheme.) The nature of the basis for $\mathcal{V}$ depends on whether or not any of the zeros $\zeta_1, \zeta_2, \ldots, \zeta_\sigma$ has multiplicity greater than 1.

In the first case, each zero ζ_k of ψ_σ has multiplicity exactly one. Choose as basis elements the sequences

$$\left\{\zeta_1^n\right\}, \left\{\zeta_2^n\right\}, \ldots, \left\{\zeta_\sigma^n\right\}$$

(that is, $\{e_{k_n}\} := \{\zeta_k^n\}$). To prove that these are linearly independent, it suffices by Proposition 7.6 to show that the σ vectors $(\zeta_k^0, \zeta_k^1, \ldots, \zeta_k^{\sigma-1})$, $k = 1, 2, \ldots, \sigma$, form a linearly independent set in $\mathbb{C}^\sigma$. Equivalently, we show that the matrix

$$\begin{bmatrix} 1 & \zeta_1 & \cdots & \zeta_1^{\sigma-1} \\ 1 & \zeta_2 & \cdots & \zeta_2^{\sigma-1} \\ \vdots & \vdots & & \vdots \\ 1 & \zeta_\sigma & \cdots & \zeta_\sigma^{\sigma-1} \end{bmatrix} \in \mathbb{C}^{\sigma\times\sigma}$$

has σ linearly independent rows. This last assertion follows from the fact that the following determinant, known as the **Vandermonde determinant**, is nonzero:

$$\det \begin{bmatrix} 1 & \zeta_1 & \cdots & \zeta_1^{\sigma-1} \\ 1 & \zeta_2 & \cdots & \zeta_2^{\sigma-1} \\ \vdots & \vdots & & \vdots \\ 1 & \zeta_\sigma & \cdots & \zeta_\sigma^{\sigma-1} \end{bmatrix} = \prod_{j=0}^{\sigma-2} \prod_{k=j+1}^{\sigma-1} (\zeta_k - \zeta_j) \neq 0.$$

In the second, more general case, ψ_σ has nontrivial zeros $\zeta_1, \zeta_2, \ldots, \zeta_m$ with multiplicities $q_1, q_2, \ldots, q_m$, respectively. In this case, each of the σ sequences

$$\left\{\zeta_k^n\right\}, \left\{n\zeta_k^n\right\}, \ldots, \left\{n(n-1)\cdots(n-q_k+2)\zeta_k^n\right\}, \qquad k = 1, 2, \ldots, m, \quad (7.3\text{-}9)$$

is a solution to Equation (7.3-8). To see this, observe that each of the numbers ζ_k is a zero of multiplicity q_k of the function $\varphi(\zeta) := \zeta^n \psi_\sigma(\zeta)$ for any nonnegative integer n. Thus, for $n \geq \sigma$,

$$\begin{aligned}
0 &= \varphi(\zeta_k) = \zeta_k^{n+r} + \sum_{j=r-\sigma}^{r-1} a_j \zeta_k^{n+j}, \\
0 &= \varphi'(\zeta_k) = (n+\sigma)\zeta_k^{n+\sigma-1} + \sum_{j=r-\sigma}^{r-1} a_j (n+j)\zeta_k^{n+j-1}, \\
&\vdots \\
0 &= \varphi^{(q_k-1)}(\zeta_k) = (n+\sigma)(n+\sigma-1)\cdots(n+\sigma-q_k+2)\zeta_k^{n+\sigma-q_k+1} \\
&\quad + \sum_{j=r-\sigma}^{r-1} a_j (n+j)(n+j-1)\cdots(n+j-q_k+2)\zeta_k^{n+j-q_k+1}.
\end{aligned}$$

Dividing through by ζ_k^{n-p} in the equation for $\varphi^{(p)}(\zeta_k)$ establishes that the σ sequences in the list (7.3-9) are solutions to Equation (7.3-8).

It remains to show that the sequences (7.3-9) are linearly independent. The argument here is analogous to that used for the case when all zeros ζ_k are simple, except that now we encounter a matrix in $\mathbb{C}^{\sigma\times\sigma}$ whose determinant is

$$\prod_{j=1}^{m}\prod_{k=1}^{j-1}(\zeta_k-\zeta_j)^{q_k+q_j}\prod_{j=1}^{m}(q_k-1)!!\neq 0.$$

Here, $0!!:=1$, and $k!!:=k!(k-1)!\cdots 2!1!$ for $k=1,2,\ldots$.

We can reduce the sequences (7.3-9) to a set of slightly more streamlined sequences. Specifically, we take linear combinations to arrive at the following σ linearly independent solutions to Equation (7.3-8):

$$\left\{\zeta_k^n\right\},\left\{n\zeta_k^n\right\},\ldots,\left\{n^{q_k-1}\zeta_k^n\right\},\qquad k=1,2,\ldots,m.$$

This listing of proposed basis sequences covers both the special case when each zero ζ_k has unit multiplicity (so $m=\sigma$) and the more general case in which some zeros have larger multiplicity.

The following theorem settles the spanning issue, confirming that we have identified a basis for the space $\mathcal{V}$ of solutions to the homogeneous linear difference equation (7.3-8).

THEOREM 7.8. *If the set*

$$\left\{\{e_{1_n}\},\{e_{2_n}\},\ldots,\{e_{\sigma_n}\}\right\}$$

of σ solutions to Equation (7.3-8) is linearly independent, then we can express any solution $\{\varepsilon_n\}$ of Equation (7.3-8) as a linear combination

$$\varepsilon_n=\sum_{k=1}^{\sigma}c_j e_{k_n},\qquad n=0,1,2,\ldots.$$

PROOF: By Proposition 7.6, it suffices to show that we can solve the linear system

$$\varepsilon_n=\sum_{j=1}^{\sigma}c_j e_{k_n},\qquad n=0,1,\ldots,\sigma-1,$$

for the coefficients $c_1,c_2,\ldots,c_\sigma$. This system has the matrix form

$$\begin{bmatrix} e_{1_0} & \cdots & e_{\sigma_0}\\ \vdots & & \vdots\\ e_{1_{\sigma-1}} & \cdots & e_{\sigma_{\sigma-1}}\end{bmatrix}\begin{bmatrix} c_1\\ \vdots\\ c_\sigma\end{bmatrix}=\begin{bmatrix}\varepsilon_0\\ \vdots\\ \varepsilon_{\sigma-1}\end{bmatrix}. \qquad (7.3\text{-}10)$$

To show that the matrix of this system is nonsingular, we demonstrate by contradiction that the columns $\boldsymbol{\kappa}_k := (e_{k_0}, e_{k_2}, \ldots, e_{k_{\sigma-1}})^{\mathsf{T}}$ are linearly independent. Assume that they are not, so that $\alpha_1\boldsymbol{\kappa}_1 + \alpha_2\boldsymbol{\kappa}_2 + \cdots + \alpha_\sigma\boldsymbol{\kappa}_\sigma = \mathbf{0}$ with not all of the coefficients α_k vanishing. Using Equation (7.3-8) together with a simple induction argument, we deduce that

$$\alpha_1 e_{1_n} + \alpha_2 e_{2_n} + \cdots + \alpha_\sigma e_{\sigma_n} = 0, \quad n = \sigma, \sigma+1, \sigma+2, \ldots.$$

This conclusion contradicts the hypothesis of linear independence. Therefore, the columns of the matrix in Equation (7.3-10) are linearly independent, and we can indeed solve for the coefficients $c_1, c_2, \ldots, c_\sigma$. ∎

Finally, we confirm the equivalence between stability and the root condition.

THEOREM 7.9. *A consistent linear multistep scheme (7.3-4) is stable if and only if it satisfies the root condition.*

PROOF: We first show that stability implies the root condition. We prove the contrapositive: If the root condition fails, then there are nearby starting values that generate sequences whose difference grows without bound. Assume that the characteristic polynomial ψ has a zero ζ with $|\zeta| > 1$. In this case, sequences of the form $c_1 + c_2\zeta^n$ are among the solutions to the scheme (7.3-4) applied to the IVP (7.3-5). Pick $\delta > 0$, and examine two sets of starting values:

$$\begin{bmatrix} u_0 \\ u_1 \\ \vdots \\ u_{r-1} \end{bmatrix} := \begin{bmatrix} u_I \\ u_I \\ \vdots \\ u_I \end{bmatrix}, \qquad \begin{bmatrix} \tilde{u}_0 \\ \tilde{u}_1 \\ \vdots \\ \tilde{u}_{r-1} \end{bmatrix} := \begin{bmatrix} u_I + \beta\delta \\ u_I + \beta\zeta\delta \\ \vdots \\ u_I + \beta\zeta^{r-1}\delta \end{bmatrix},$$

where $0 < \beta < |\zeta^{r-1}|^{-1}$. By construction, $|u_n - \tilde{u}_n| < \delta$ for $n = 0, 1, \ldots, r-1$. However, the first set of starting values yields the solution $u_n = u_I$ for $n = 0, 1, 2, \ldots$, while the second set yields $\tilde{u}_n = u_I + \beta\zeta^n\delta$. We make the difference $|u_n - \tilde{u}_n| = |\beta\zeta^n\delta|$ as large as desired for the step $n = T/h$ by taking the stepsize h small enough. Therefore, in this case the multistep scheme is unstable.

We leave the argument for the case when ψ has a multiple zero of unit magnitude for Problem 10.

Now we show that the root condition implies stability. Assume that the scheme (7.3-4) satisfies the root condition, and denote by

$$\Big\{\{e_{1_n}\}, \{e_{2_n}\}, \ldots, \{e_{\sigma_n}\}\Big\}$$

the basis identified in Theorem 7.8. Consider two solutions $\{u_n\}$ and $\{\tilde{u}_n\}$, expressible as

$$\begin{aligned} u_n &:= c_1 e_{1_n} + c_2 e_{2_n} + \cdots + c_\sigma e_{\sigma_n}, \\ \tilde{u}_n &:= \tilde{c}_1 e_{1_n} + \tilde{c}_2 e_{2_n} + \cdots + \tilde{c}_\sigma e_{\sigma_n}, \end{aligned}$$

such that $|u_n - \tilde{u}_n| < \delta$ for $n = 0, 1, \ldots, r-1$. By the linearity of the multistep scheme,

$$\varepsilon_n := u_n - \tilde{u}_n = d_1 e_{1_n} + d_2 e_{2_n} + \cdots + d_\sigma e_{\sigma_n},$$

where $d_k := c_k - \tilde{c}_k$.

To bound the growth of ε_n, we need an estimate for the vector $\mathbf{d} := (d_1, d_2, \ldots, d_\sigma)^\top$. The estimate arises from the fact that $\mathbf{d}$ is the solution to the following linear system:

$$\underbrace{\begin{bmatrix} e_{1_0} & \cdots & e_{\sigma_0} \\ \vdots & & \vdots \\ e_{1_{\sigma-1}} & \cdots & e_{\sigma_{\sigma-1}} \end{bmatrix}}_{\mathsf{E}} \underbrace{\begin{bmatrix} d_1 \\ \vdots \\ d_\sigma \end{bmatrix}}_{\mathbf{d}} = \underbrace{\begin{bmatrix} \varepsilon_0 \\ \vdots \\ \varepsilon_{\sigma-1} \end{bmatrix}}_{\boldsymbol{\varepsilon}}.$$

Theorem 7.8 ensures that E is nonsingular, so $\mathbf{d} = \mathsf{E}^{-1}\boldsymbol{\varepsilon}$. Therefore,

$$\|\mathbf{d}\|_1 \leq \|\mathsf{E}^{-1}\|_1 \, \|\boldsymbol{\varepsilon}\|_1 \leq \sigma \, \|\mathsf{E}^{-1}\|_1 \, \|\boldsymbol{\varepsilon}\|_\infty.$$

Since $|\varepsilon_n| < \delta$ for $n = 0, 1, \ldots, r-1$,

$$\|\mathbf{d}\|_1 < \sigma \, \|\mathsf{E}^{-1}\|_1 \, \delta. \tag{7.3-11}$$

Next we estimate ε_n for arbitrary step number $n \geq 0$. According to the root condition, each of the basis sequences $\{e_{k_n}\}$ has the form $e_{k_n} = n^\alpha \zeta^n$, where ζ is a nontrivial zero of the characteristic polynomial ψ. The possible values for α depend upon the nature of ζ. If $|\zeta| = 1$, then ζ must have multiplicity 1, and $\alpha = 0$. If $|\zeta| < 1$, then $\alpha \leq \sigma - 1$. Therefore, $|e_{k_n}| = 1$ when $|\zeta| = 1$ and $|e_{k_n}| \leq n^{\sigma-1}|\zeta|^n$ when $0 < |\zeta| < 1$. For each nontrivial zero ζ_k of ψ, the function $\gamma_k \colon [0, \infty) \to \mathbb{R}$ defined by $\gamma_k(n) := n^{\sigma-1}|\zeta_k|^n$ is nonnegative and has a maximum value $\gamma_k(n_{\max}(k))$. Let

$$\Gamma := \max_{|\zeta_k|<1} \left\{1, \gamma_k(n_{\max}(k))\right\},$$

so that each basis sequence $\{e_{k_n}\}$ obeys the bound $|e_{k_n}| \leq \Gamma$. By the triangle inequality,

$$|\varepsilon_n| \leq \Gamma \left(|d_1| + |d_2| + \cdots + |d_\sigma|\right) = \Gamma \, \|\mathbf{d}\|_1,$$

for $n = 0, 1, 2, \ldots$. From the estimate (7.3-11), it follows that

$$|\varepsilon_n| \leq \Gamma \, \|\mathbf{d}\|_1 < \Gamma \, \sigma \, \|\mathsf{E}^{-1}\|_1 \, \delta.$$

Taking $M = \Gamma \, \sigma \, \|\mathsf{E}^{-1}\|_1$, we have $|\varepsilon_n| < M\delta$, as desired. ∎

7.4 Convergence of Multistep Methods

This section concludes the discussion of multistep methods by establishing conditions under which they converge. The development is more theoretical than that given in the first part of Section 7.3. In broad terms, the theory yields a result that relates convergence to stability and consistency — two properties that we can often check with relative ease. The following schematic summarizes the connection:

$$\begin{array}{c}\text{CONSISTENCY}\\ + \\ \text{STABILITY}\end{array} \iff \text{CONVERGENCE.}$$

This idea appears again in the context of numerical schemes for partial diff-ferential equations, which we explore in Chapter 8. (However, that discussion is accessible without the development in this section.)

Our approach follows a line of reasoning presented in Henrici ([3], Chapter 5) and has an analytic flavor. It is also possible to develop the proofs from linear-algebraic points of view, as Ortega ([6], Chapter 4) and Gear ([2], Chapter 10) illustrate.

To keep the arguments manageable, we restrict attention to the linear multistep scheme

$$u_{n+r} + \sum_{j=0}^{r-1} a_j u_{n+j} = h \sum_{j=0}^{r} b_j f_{n+j}, \tag{7.4-1}$$

applied to the scalar IVP

$$u'(t) = f(t,u), \quad t \in [a,b]; \qquad u(t_0) = u_I. \tag{7.4-2}$$

We assume that this IVP is well posed.

We begin with a formal definition of convergence.

DEFINITION. *The linear multistep method (7.4-1) is* **convergent** *when applied to the* IVP *(7.4-2) if, whenever the starting values satisfy the condition*

$$\lim_{h\to 0} u_n = u_I, \qquad n = 0, 1, \ldots, r-1,$$

the sequence $\{u_n\}$ generated by Equation (7.4-1) satisfies the condition

$$\lim_{\substack{h \to 0 \\ t_n = t}} u_n = u(t) \qquad \text{for all} \quad t \in [a,b]. \tag{7.4-3}$$

This definition does not require exact starting values $u_0, u_1, \ldots, u_{r-1}$. Instead, it calls for good approximations, such as those generated by a convergent one-step method. Also, the unusual notation in Equation (7.4-3) is a concession to the fact that, for a given value of $t \in [a,b]$, only a discrete set of stepsizes h can yield approximations u_n to $u(t)$. Therefore,

$$\lim_{\substack{h \to 0 \\ t_n = t}} u_n := \lim_{n\to\infty} \{u_n : nh = t - t_0\}.$$

Convergence turns out to be a comparatively "strong" property, in the sense that the proofs that convergence implies stability and consistency are relatively straightforward.

THEOREM 7.10. *A linear multistep scheme is stable whenever it is convergent.*

PROOF: The fact that the scheme converges for the general IVP (7.4-2) implies that it converges for the specific IVP

$$u'(t) = 0, \qquad u(0) = 0,$$

for which the solution is $u(t) = 0$ identically. Therefore,

$$\lim_{\substack{h \to 0 \\ t_n = t}} u_n = 0$$

whenever the starting values obey the condition $\lim_{h\to 0} u_n = 0$ for $n = 0, 1, \ldots, r-1$. We use this fact to show that the scheme obeys the root condition, which by Theorem 7.9 is equivalent to stability.

Suppose that $\zeta = \rho \exp(i\theta)$ is a zero of the characteristic polynomial ψ. We show first that $|\zeta| \leq 1$, then examine the case when $|\zeta| = 1$. The sequence $\{h\zeta^n\} = \{h\rho^n[\cos n\theta + i\sin(n\theta)]\}$ is a solution to the homogeneous linear difference equation

$$u_{n+r} + \sum_{j=0}^{r-1} a_j u_{n+j} = 0 \tag{7.4-4}$$

for any $\theta \in [0, 2\pi)$. By linearity, the sequence $\{u_n\} := \{h\rho^n \cos(n\theta)\}$ is also a solution to Equation (7.4-4). If $\theta = 0$ or $\theta = \pi$, then the convergence of the scheme implies that $|u^n| = |(t/n)\rho^n| \to 0$ as $n \to \infty$, which can be true only if $\rho = |\zeta| \leq 1$. For other values of $\theta \in [0, \pi)$, one uses the identity $\cos(\alpha+\beta)\cos(\alpha-\beta) = \cos^2\alpha - \sin^2\beta$ to show that

$$\frac{t^2}{n^2}\rho^{2n} = h^2\rho^{2n} = \frac{u_n^2 - u_{n+1}u_{n-1}}{\sin^2\theta}.$$

Convergence implies that the expression on the right tends to zero as $n \to \infty$, so we conclude again that $\rho = |\zeta| \leq 1$.

It remains to prove that ζ is a simple zero of ψ when $|\zeta| = 1$. We prove the contrapositive: Assume that $\zeta = \rho\exp(i\theta)$ has multiplicity greater than 1. Since the sequence $\{\sqrt{h}n\zeta^n\}$ satisfies the difference equation (7.4-4), the sequence $\{u_n\} := \{\sqrt{h}n\rho^n \cos(n\theta)\}$ does, too. When $\theta = 0$ or $\theta = \pi$, we have $|u_n| = \sqrt{h}n\rho^n = \sqrt{t/n}\,n\rho^n = \rho^n\sqrt{nt}$. Convergence requires that $u_n \to 0$ as $n \to \infty$, which occurs only if $\rho = |\zeta| < 1$. For other values of $\theta \in [0, 2\pi)$, define a new sequence $\{v_n\} := \{u_n/(n\sqrt{h})\}$. Convergence requires that $v_n \to 0$ as $n \to \infty$, so employing the trigonometric identity used above yields

$$\rho^{2n} = \frac{v_n^2 - v_{n+1}v_{n-1}}{\sin^2\theta} \to 0 \quad \text{as} \quad n \to 0.$$

Again we conclude that $\rho = |\zeta| < 1$. Therefore the multistep scheme (7.4-1) satisfies the root condition. ∎

THEOREM 7.11. *A linear multistep scheme is consistent whenever it is convergent.*

PROOF: Recall from Proposition 7.4 that the truncation error for the linear multistep scheme (7.4-1) has the form

$$\tau_h(s, v) = \frac{1}{h}[C_0 v(s) + C_1 h v'(s) + \mathcal{O}(h^2)],$$

where

$$\begin{aligned} C_0 &:= 1 + a_r + a_{r-1} + \cdots + a_0, \\ C_1 &:= r + (r-1)a_{r-1} + (r-2)a_{r-2} + \cdots + a_1 - \sum_{j=0}^{r} b_j. \end{aligned}$$

By Corollary 7.5, it suffices to show that, if the scheme is convergent, then $C_0 = C_1 = 0$. We do this by examining two special initial-value problems.

To prove that $C_0 = 0$, consider the IVP

$$u'(t) = 0, \qquad u(0) = 1,$$

for which the solution is $u(t) = 1$ identically. Applying the multistep scheme to this IVP yields the homogeneous difference equation (7.4-4). According to the hypothesis of convergence, this equation must generate a sequence $\{u_n\}$ such that $u_n \to 1$ as $h = t/n \to 0$, whenever we use the exact starting values $u_0 = u_1 = \cdots = u_{r-1} = 1$. Therefore, taking the limit of Equation (7.4-4) as $h \to 0$, we get

$$1 + \sum_{j=0}^{r-1} a_j = C_0 = 0.$$

To prove that $C_1 = 0$, consider the IVP

$$u'(t) = 1, \qquad u(0) = 0,$$

for which the solution is $u(t) = t$. For this IVP, the multistep scheme reduces to

$$u_{n+r} + \sum_{j=0}^{r-1} a_j u_{n+j} = h \sum_{j=0}^{r} b_j. \tag{7.4-5}$$

Since the scheme is convergent, Theorem 7.10 ensures that it is stable, so its characteristic polynomial ψ cannot have $\zeta = 1$ as a multiple root. Therefore,

$\psi\prime(1) = r + (r-1)a_{r-1} + \cdots + a_0 \neq 0$. Given this fact, one can easily check that the sequence $\{u_n\} := \{nhR\}$, where

$$R := \frac{1}{r + (r-1)a_{r-1} + \cdots + a_1} \sum_{j=0}^{r} b_j,$$

is a solution to the difference equation (7.4-5). The starting values in this sequence also satisfy the condition

$$\lim_{h \to 0} u_n = 0, \qquad n = 0, 1, \ldots, r-1.$$

The convergence of the scheme therefore implies that $u_n \to t_n$ as $h \to 0$. Hence,

$$\lim_{\substack{h \to 0 \\ t_n = t}} u_n = \lim_{\substack{h \to 0 \\ t_n = t}} nhR = \lim_{\substack{h \to 0 \\ t_n = t}} t_n R = t_n.$$

It follows that $R = 1$, that is, that $C_1 = 0$. Therefore the scheme is consistent. ∎

Since consistency and stability are usually easier to check than convergence, the proofs that consistency and stability are necessary for convergence seem less useful than the fact that these properties are sufficient. Unfortunately, the proof of sufficiency is more involved. We begin with three lemmas, the first two of which concern analytic properties of the characteristic polynomial $\psi(\zeta)$.

LEMMA 7.12. *Suppose that the linear multistep scheme (7.4-1) with characteristic polynomial $\psi(\zeta) = \zeta^r + a_{r-1}\zeta^{r-1} + \cdots + a_0$ satisfies the root condition. Define $\hat{\psi}(\zeta)$ as follows:*

$$\hat{\psi}(\zeta) := 1 + a_{r-1}\zeta + \cdots + a_0\zeta^r.$$

If $\gamma_0, \gamma_1, \gamma_2, \ldots \in \mathbb{C}$ denote the coefficients in the expansion

$$\frac{1}{\hat{\psi}(\zeta)} = \gamma_0 + \gamma_1\zeta + \gamma_2\zeta^2 + \cdots, \qquad |\zeta| < 1,$$

then $\{\gamma_k\}$ is a bounded sequence in $\mathbb{C}$.

PROOF: Observe first that ζ is a zero of ψ if and only if ζ^{-1} is a root of $\hat{\psi}$. Therefore, by hypothesis $\hat{\psi}$ has no zeros in the unit disk $|\zeta| < 1$, and hence the function $1/\hat{\psi}(\zeta)$ has no singularities there. Moreover, $\hat{\psi}$ has the following form:

$$\hat{\psi}(\zeta) = a_0(\zeta - \zeta_1^{-1}) \cdots (\zeta - \zeta_m^{-1})(\zeta - \xi_1^{-1})^{q_1} \cdots (\zeta - \xi_n^{-1})^{q_n},$$

where $\zeta_1, \zeta_2, \ldots, \zeta_m$ denote the simple zeros of $\hat{\psi}$ with unit magnitude and $\xi_1, \xi_2, \ldots, \xi_n$ denote the zeros of $\hat{\psi}$, having multiplicities $q_1, q_2, \ldots, q_n$ respectively, that lie inside the unit disk. As a consequence, $1/\hat{\psi}$ has a partial fraction expansion of the form

$$\frac{1}{\hat{\psi}(\zeta)} = \underbrace{\sum_{j=1}^{m} \frac{A_j}{\zeta - \zeta_j^{-1}}}_{(\text{I})} + \underbrace{\sum_{j=1}^{n} \sum_{l=1}^{q_j} \frac{B_{j,l}}{(\zeta - \xi_j^{-1})^l}}_{(\text{II})}, \tag{7.4-6}$$

for some finite set of constants $A_j, B_{j,l}$.

To establish the lemma, it suffices to show that each term in Equation (7.4-6) has an expansion in powers of ζ in which the coefficients are bounded. Consider first the terms in the sum labeled (I). We write each of these terms as follows for $|\zeta| < 1$:

$$\frac{A_j}{\zeta - \zeta_j^{-1}} = -\frac{A_j \zeta_j}{1 - \zeta \zeta_j} = -A_j \zeta_j \sum_{k=0}^{\infty} \zeta_j^k \zeta^k.$$

The coefficients $A_j \zeta_j^{k+1}$ in this expansion are certainly bounded, since each zero ζ_j has unit magnitude.

For the terms in the sum labeled (II), we have

$$\frac{B_{j,ll}}{(\zeta - \xi_j)^l} = B_{j,l} \frac{(-1)^{l-1}}{(l-1)!} \frac{d^{l-1}}{d\zeta^{l-1}} \left(\frac{1}{\zeta - \xi_j^{-1}} \right) = \frac{d^{l-1}}{d\zeta^{l-1}} \left[-\xi_j \left(\frac{1}{1 - \zeta \xi_j} \right) \right].$$

Problem 11 asks for proof that this expression has an expansion in powers of ζ whose coefficients are bounded. ∎

LEMMA 7.13. *With the coefficients $\gamma_0, \gamma_1, \gamma_2, \ldots$ defined as in Lemma 7.12 and $\gamma_n := 0$ for $n = -1, -2, -3, \ldots$,*

$$\gamma_n + a_{r-1}\gamma_{n-1} + \cdots + a_0 \gamma_{n-r} = \begin{cases} 1, & \text{if } n = 0, \\ 0, & \text{if } n > 0. \end{cases}$$

PROOF: This is a simple exercise: Multiply the expansion

$$\frac{1}{\hat{\psi}(\zeta)} = \gamma_0 + \gamma_1 \zeta + \gamma_2 \zeta^2 + \cdots$$

through by $\hat{\psi}(\zeta)$. ∎

Now consider a sequence $\{\xi_n\}$ that satisfies a linear difference equation

$$\xi_{n+r} + \sum_{j=0}^{r-1} a_j \xi_{n+j} = h \sum_{j=0}^{r} b_{j,n} \xi_{n+j} + \lambda_n, \tag{7.4-7}$$

having characteristic polynomial $\psi(\zeta) = \zeta^r + a_{r-1}\zeta^{r-1} + \cdots + a_0$. The next lemma establishes bounds on the growth rate of $\{\xi_n\}$. In the final theorem, a sequence $\{\varepsilon_n\}$ will play the role of $\{\xi_n\}$. This "growth-rate" lemma furnishes a multistep analog of Lemma 7.2, proved for one-step methods. Unfortunately, both the statement and the proof of the current lemma are more complicated.

LEMMA 7.14. *Suppose that the sequence $\{\xi_n\}$ obeys the linear difference equation (7.4-7) and that the characteristic polynomial ψ satisfies the root condition. Denote by $\{\gamma_n\}$ the coefficients defined in Lemma 7.12. Furthermore, let B, β, and Λ be positive constants such that, for $n = 0, 1, \ldots, N$,*

$$\sum_{j=0}^{r} |b_{j,n}| \leq B, \tag{7.4-8}$$

$$|b_{j,n}| \leq \beta, \tag{7.4-9}$$

$$|\lambda_n| \leq \Lambda. \tag{7.4-10}$$

Let $h \in [0, 1/\beta)$, and suppose that the starting values for $\{\xi_n\}$ have the bound

$$\Xi := \max_{0 \leq n \leq r-q} |\xi_n|. \tag{7.4-11}$$

Then the sequence $\{\xi_n\}$ obeys the bound

$$|\xi_n| \leq K \exp(nhB\Gamma), \qquad n = 0, 1, \ldots, N.$$

Here,

$$\begin{aligned} K &:= \Gamma(N\Lambda + A\Xi r), \\ \Gamma &:= \frac{\sup|\gamma_n|}{1 - h\beta}, \\ A &:= 1 + |a_{r-1}| + |a_{r-2}| + \cdots + |a_0|. \end{aligned}$$

(Notice that $\sup|\gamma_n|$ exists by Lemma 7.12.)

PROOF: The proof involves tedious algebra, details of which we merely sketch. Start by summing the following $n - r + 1$ equations, obtained from Equation (7.4-7):

$$\begin{aligned} \gamma_0 \left(\xi_r + \sum_{j=0}^{r-1} a_j \xi_j \right) &= \gamma_0 \left(h \sum_{j=0}^{r} b_{j,0} \xi_j + \lambda_0 \right), \\ &\vdots \\ \gamma_{n-r} \left(\xi_n + \sum_{j=0}^{r-1} a_j \xi_{j+n-r} \right) &= \gamma_{n-r} \left(h \sum_{j=0}^{r} b_{j,n-r} \xi_{j+n-r} + \lambda_{n-r} \right). \end{aligned}$$

One can use Lemma (7.13) to reduce the left side of the resulting equation to the expression $\xi_n + (\text{I})$, where

$$(\text{I}) := (a_{r-1}\gamma_{n-r} + \cdots + a_0\gamma_{n-2r+1})\xi_{n-1} + \cdots + a_0\gamma_{n-r}\xi_0.$$

The right side of the resulting equation has the form $(\text{II}) + (\text{III}) + (\text{IV})$, where

$$\begin{aligned}
(\text{II}) &:= hb_{r,n-r}\gamma_0\xi_n, \\
(\text{III}) &:= h[(b_{r-1,n-r}\gamma_0 + b_{r,n-r-1}\gamma_1)\xi_{n-1} \\
&\quad + \cdots + (b_{0,n-r}\gamma_0 + \cdots + b_{r,n-2r}\gamma_r)\xi_{n-r} + b_{0,0}\gamma_{n-r}\xi_0], \\
(\text{IV}) &:= \lambda_{n-r}\gamma_0 + \lambda_{n-r-1}\gamma_1 + \cdots + \lambda_0\gamma_{n-r}.
\end{aligned}$$

Now set $\xi_n + (\text{I}) = (\text{II}) + (\text{III}) + (\text{IV})$, rearrange, and use the bounds (7.4-8) through (7.4-11) to obtain the inequality

$$|\xi_n| \leq \underbrace{h\beta|\xi_n|}_{(\text{II})} + \underbrace{hB\sup|\gamma_k|\sum_{j=0}^{n-1}|\xi_j|}_{(\text{III})} + \underbrace{N\Lambda\sup|\gamma_k|}_{(\text{IV})} + \underbrace{A\Xi r\sup|\gamma_k|}_{(\text{I})},$$

where the underbraces indicate the origins of the terms on the right. Since $1 - h\beta > 0$ by hypothesis,

$$(1 - h\beta)|\xi_n| \leq hB\sup|\gamma_k|\sum_{j=0}^{n-1}|\xi_j| + (N\Lambda + A\Xi r)\sup|\gamma_k|,$$

which implies that

$$|\xi_n| \leq hB\Gamma\sum_{j=0}^{n-1}|\xi_j| + \underbrace{\Gamma(N\Lambda + A\Xi r)}_{K}.$$

We now claim that $|\xi_n| \leq K(1+hB\Gamma)^n$, which suffices to show that $|\xi_n| \leq K[\exp(hB\Gamma)]^n$ and finish the proof. To establish this claim, observe that it holds for the starting values $\xi_0, \xi_1, \ldots, \xi_{r-1}$, since Lemma 7.13 implies that $A\sup|\gamma_k| \geq 1$ and hence that $K \geq \Xi$. We now use mathematical induction, showing that the claim is true for ξ_n whenever it holds for $\xi_0, \xi_1, \ldots, \xi_{n-1}$. Problem 15 asks for details of this induction. ∎

Finally, we arrive at the main theorem: Consistency and stability imply convergence.

THEOREM 7.15. *Whenever the function f in the* IVP *(7.4-2) is continuous and satisfies a Lipschitz condition with Lipschitz constant $L > 0$, a consistent, stable multistep scheme of the form (7.4-1) is convergent.*

PROOF: Denote by $\{u_n\}$ the sequence generated by the multistep scheme

$$u_{n+r} + \sum_{j=0}^{r-1} a_j u_{n+j} - h \sum_{j=0}^{r} b_j f_{n+j}, \tag{7.4-12}$$

with starting values $u_0, u_1, \ldots, u_{r-1}$. Also, define

$$\delta_h := \max_{0 \le n \le r-1} |u(t_n) - u_n|,$$

where $u(t)$ signifies the exact solution to the IVP (7.1-1), and assume that $\lim_{h\to 0} \delta_h = 0$. We must show that

$$\lim_{\substack{h \to 0 \\ t_n = t}} [u(t_n) - u_n] = 0,$$

for all $t \in [a, b]$. The argument proceeds in three stages, the first of which involves the most work.

In the first stage, we estimate the truncation error $\tau_h(t_n, u_n)$. For this task, let $w(t)$ be the solution to the IVP

$$u'(t) = f(t, u), \qquad u(t_n) = u_n.$$

By definition,

$$\tau_h(t_n, u_n) = \frac{1}{h}\left[w(t_{n+r}) + \sum_{j=0}^{r-1} a_j w(t_{n+j}) - h \sum_{j=0}^{r} b_j \underbrace{f(t_{n+j}, w(t_{n+j}))}_{w'(t_{n+j})}\right]. \tag{7.4-13}$$

Now for any $\epsilon \ge 0$ define

$$d(\epsilon) := \sup_{t\in[a,b]} \left\{ |w'(\tilde{t}) - w'(t)| \,:\, \tilde{t} \in [t - \epsilon, t + \epsilon] \cap [a, b] \right\}.$$

(See Figure 1.) If $\epsilon_1 > \epsilon_2$, then $d(\epsilon_1) \ge d(\epsilon_2)$, since $d(\epsilon_1)$ is the supremum of a possibly larger set. Thus d is a nondecreasing function. Also, since $w'(t) = f(t, w(t))$ is a continuous function of t on the closed interval $[a, b]$, it is uniformly continuous there, and consequently $d(\epsilon) \to 0$ as $\epsilon \to 0$. From these observations, we deduce that

$$w'(t_{n+j}) = w'(t_n) + \theta_j d(jh), \tag{7.4-14}$$

for some number $\theta_j \in [-1, 1]$. Furthermore, by the mean value theorem, there exists a number $\zeta_j \in (t_n, t_{n+j})$ such that $w(t_{n+j}) = w(t_n) + jhw'(\zeta_j)$, a relationship that we rewrite in the following form:

$$w(t_{n+j}) = w(t_n) + jh\left[w'(t_n) + \eta_j d(jh)\right], \tag{7.4-15}$$

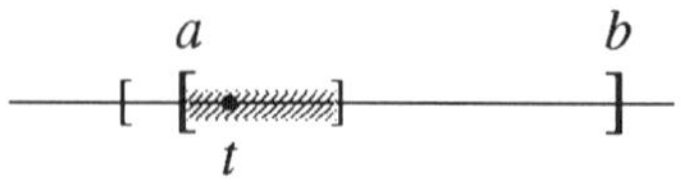

FIGURE 1. *The interval* $[t-\epsilon, t+\epsilon] \cap [a,b]$ *for a particular value of* $t \in [a,b]$.

for some $\eta_j \in [-1,1]$.

Substituting the expressions (7.4-14) and (7.4-15) into the truncation error (7.4-13) and rearranging gives

$$\begin{aligned}\tau_h(t_n, u_n) &= \frac{1}{h}\Bigg\{\underbrace{\left(1+\sum_{j=0}^{r-1} a_j\right)}_{C_0} w(t_n) + \underbrace{\left(r+\sum_{j=0}^{r-1} ja_j - \sum_{j=0}^{r} b_j\right)}_{C_1} hw'(t_n) \\ &+ h\left[r\eta_r d(rh) + \sum_{j=0}^{r-1} ja_j\eta_j d(jh)\right] - h\sum_{j=0}^{r} b_j\theta_j d(jh)\Bigg\}.\end{aligned}$$

Consistency implies that $C_0 = C_1 = 0$. From this observation and the fact that d is nondecreasing, we obtain the desired estimate,

$$|\tau_h(t_n, u_n)| \leq \left(r + \sum_{j=0}^{r-1} j|a_j| + \sum_{j=0}^{r} |b_j|\right) d(rh) = Cd(rh),$$

where C is a positive constant. We rewrite this estimate in the slightly more useful form

$$\frac{1}{h}\left[w(t_{n+r}) + \sum_{j=0}^{r-1} a_j w(t_{n+j}) - h\sum_{j=0}^{r} b_j f(t_{n+j}, w(t_{n+j}))\right] = \kappa_n Cd(rh), \quad (7.4\text{-}16)$$

for some number $\kappa_n \in [-1,1]$.

In the second stage of the proof, we derive a recursion relation for the local error $\varepsilon_n := w(t_n) - u_n$. Multiplying Equation (7.4-16) by h and subtracting Equation (7.4-12) from the result yields

$$\varepsilon_{n+r} + \sum_{j=0}^{r-1} a_j\varepsilon_{n+j} - h\sum_{j=0}^{r} \underbrace{b_j g_{n+j}}_{b_{j,n}} \varepsilon_{n+j} = \underbrace{h\kappa_n Cd(rh)}_{\lambda_n},$$

where

$$g_m := \begin{cases} \varepsilon_m^{-1}[f(t_m, w(t_m)) - f(t_m, u_m)], & \text{if } \varepsilon_m \neq 0, \\ 0 & \text{if } \varepsilon_m = 0. \end{cases}$$

The final stage of the proof uses Lemma 7.14 to estimate the error ε_n for all levels $n = 0, 1, 2, \ldots$. The hypothesis that f obeys a Lipschitz condition implies that $|f(t_m, w(t_m)) - f(t_m, u_m)| \leq L|w(t_m) - u_m| = L|\varepsilon_m|$. Hence $|g_m| \leq L$. Now we use the hypothesis of stability, together with Lemma 7.14 and the identifications

$$\{\xi_n\} = \{\varepsilon_n\}, \qquad \Xi = \delta_h, \qquad \Lambda = hCd(rh),$$

$$N = (t_n - t_0)/h, \qquad B = L \sum_{j=0}^{r} |b_j|,$$

to deduce that the sequence $\{\varepsilon_h\}$ of errors obeys an estimate of the form

$$|\varepsilon_n| \leq \Gamma \left[\frac{t - t_0}{h} \cdot hCd(rh) + A\delta_h \right] \exp[(t_n - t_0)L].$$

But both $d(rh)$ and δ_h tend to zero as $h \to 0$, so $|\varepsilon_n| \to 0$ as $h \to 0$. ∎

7.5 Problems

PROBLEM 1. Using a convergence plot (log(||error||) versus log(h)), graphically estimate the order of convergence for the Euler explicit method, the Heun method, and the Runge-Kutta method (7.2-5) applied to the initial-value problem $y' = xy^{1/3}$, $y(1) = 1$. (The exact solution is $y = [(x^2 + 2)/3]^{3/2}$.)

PROBLEM 2. Consider schemes of the form (7.2-2), with

$$\Phi_h(s, v) = \alpha_1 f(s, v) + \alpha_2 f\left(s + \alpha_3 h, v + \alpha_4 h f(s, v)\right).$$

Here, $\alpha_1, \alpha_2, \alpha_3, \alpha_4$ are parameters that one can choose to force cancellation in the low powers of h occurring in the Taylor expansion of τ_h. Show that, for smooth enough f,

$$\begin{aligned} \Phi_h(s, v) &= \alpha_1 f(s, v) + \alpha_2 \big[f(s, v) + \alpha_3 h \partial_1 f(s, v) \\ &\quad + \alpha_4 h f(s, v) \partial_2 f(s, v) + \mathcal{O}(h^2) \big]. \end{aligned}$$

Show that any choices of $\alpha_1, \alpha_2, \alpha_3, \alpha_4$ for which $\alpha_1 + \alpha_2 = 1$, $\alpha_2 \alpha_3 = \frac{1}{2}$, and $\alpha_2 \alpha_4 = \frac{1}{2}$ yield an explicit one-step scheme that is consistent with order 2. Show that the Heun method and the modified Euler method have this form.

PROBLEM 3. **Taylor methods** for $u' = f(t, u)$ have the following form:

$$u_{n+1} = u_n + h u_n' + \cdots + \frac{1}{k!} h^k u_n^{(k)}.$$

Here, $u'_n, u''_n, \ldots$ are computable from the differential equation as

$$u'_n = f(t_n, u_n), \quad u''_n = f'(t_n, u_n), \ldots.$$

Estimate $u(5)$ for the initial-value problem given in Problem 1, using a Taylor method with $k = 3$. Use a convergence plot to estimate the order of convergence of the method.

PROBLEM 4. Prove that, whenever $\delta > 0$,

$$\frac{(1+\delta)^n - 1}{\delta} = \sum_{i=0}^{n-1} (1+\delta)^i.$$

PROBLEM 5. Consider the initial-value problem $u' = t + u$, $u(0) = 0$. Compute $u(1)$ to four significant digits using Euler's explicit scheme in conjunction with repeated Richardson extrapolation (see Theorem 6.6).

PROBLEM 6. Consider a one-step method $u_{n+1} = u_n + h\Phi_h(t_n, t_n)$ having stepsize h. The **absolute stability region** R_A of the method is the set of all values of qh in the complex plane for which numerical solutions to the test problem $u' = qu$, $u(0) = 1$, remain bounded as $n \to \infty$. Find R_A for the Euler explicit scheme.

PROBLEM 7. The **trapezoid method** for $u' = f(t, u)$ is as follows:

$$u_{n+1} = u_n + \frac{1}{2}h[f(t_n, u_n) + f(t_{n+1}, u_{n+1})].$$

(A) Explain the name of the method.

(B) The method is implicit. Explain the heuristic behind the iterative procedure in which one uses

$$u_{n+1}^{(m+1)} = u_n + \frac{1}{2}h\left[f(t_n, u_n) + f(t_{n+1}, u_{n+1}^{(m)})\right]$$

to solve for $f(t_{n+1}, u_{n+1})$. When will this iterative procedure converge?

(C) What is the method's absolute stability region? (See Problem 6.)

PROBLEM 8. Consider the explicit one-step method (7.2-2), applied to any ODE (7.1-1) for which f satisfies the hypotheses of Theorem 7.1. Imagine starting the scheme with different initial values u_0, v_0, thereby computing numerical solutions $\{u_n\}_{n=0}^N$ and $\{v_n\}_{n=0}^N$. We call the method **stable** if there exist positive constants K, h_0, independent of step number n, such that, whenever $0 < h \leq h_0$, $|u_n - v_n| \leq K|u_0 - v_0|$ for $n = 0, 1, 2, \ldots, N$. Prove that a one-step method is stable whenever Φ_h satisfies a Lipschitz condition.

PROBLEM 9. One can write an implicit multistep method (7.3-4) for a single ODE in the simplified form

$$u_{n+r} = hb_r f(t_{n+r}, u_{n+r}) + g,$$

where g is computable from previously calculated values of u_n. Consider solving this nonlinear equation for u_{n+r} by successive substitution. If f has Lipschitz constant L, show that this iterative scheme converges when $h < (|b_r|L)^{-1}$.

PROBLEM 10. Complete the proof of Theorem 7.9 by showing that the multistep scheme (7.3-4) is unstable when its characteristic polynomial has a zero ζ whose magnitude is unity and whose multiplicity is greater than 1.

PROBLEM 11. Complete the proof of Lemma 7.11.

PROBLEM 12. Compute truncation errors $\tau_h(s, v)$ for the Adams-Bashforth 2-step and 3-step methods.

PROBLEM 13. Write a computer code to implement a fourth-order predictor-corrector scheme using a Runge-Kutta method for starting values, an Adams-Bashforth method as the predictor, and one iteration of an Adams-Moulton method as the corrector. Generate numerical solutions to the initial-value problem $y' = y^2$, $y(0) = 1$. Graphically verify the convergence rate of the scheme.

PROBLEM 14. Determine constants a and b so that the multistep scheme

$$u_{n+3} + au_{n+2} - au_{n+1} - u_n = hb(f_{n+2} + f_{n+1})$$

is consistent to order at least 3. Comment on the stability of the scheme.

PROBLEM 15. Suppose that a sequence $\{\xi_n\}$ obeys an inequality of the form

$$|\xi_n| \leq M \sum_{j=0}^{n-1} |\xi_j| + K,$$

where $M, K > 0$, and that the estimate $|\xi_j| \leq K(1 + M)^j$ holds for $j = 0, 1, \ldots, n-1$. Complete the induction in Lemma 7.14 by showing that $|\xi_n| \leq K(1 + M)^n$ for $n = 0, 1, 2, \ldots$.

7.6 References

1. F. Brauer and J.A. Nohel, *The Qualitative Theory of Ordinary Differential Equations: An Introduction*, Dover, Mineola, NY, 1969.

2. C.W. Gear, *Numerical Initial Value Problems in Ordinary Differential Equations*, Prentice-Hall, Englewood Cliffs, NJ, 1971.

3. P. Henrici, *Discrete Variable Methods in Ordinary Differential Equations*, Wiley, New York, 1962.

4. E. Isaacson and H.B. Keller, *Analysis of Numerical Methods*, John Wiley and Sons, New York, 1966.

5. J.D. Lambert, *Numerical Methods for Ordinary Differential Systems: The Initial Value Problem*, Wiley, New York, 1991.

6. J.M. Ortega, *Numerical Analysis: A Second Course*, Academic Press, New York, 1972.

Chapter 8

Difference Methods for PDEs

8.1 Introduction

Partial differential equations (PDEs) occur so frequently in mathematics, natural science, and engineering that they constitute a lingua franca for the sciences. PDEs arise in problems involving rates of change of functions of several independent variables. The following examples involve two independent variables:

$$-\nabla^2 u := -\frac{\partial^2 u}{\partial x^2} - \frac{\partial^2 u}{\partial y^2} = f(x,y), \qquad \textbf{Poisson equation;}$$

$$\frac{\partial u}{\partial t} + v\frac{\partial u}{\partial x} = 0, \qquad \textbf{advection equation;}$$

$$\frac{\partial u}{\partial t} - D\frac{\partial^2 u}{\partial x^2} = 0, \qquad \textbf{heat equation;}$$

$$\frac{\partial^2 u}{\partial t^2} - c^2\frac{\partial^2 u}{\partial x^2} = 0, \qquad \textbf{wave equation.}$$

Here, v, D, and c are real, positive constants. In the Poisson equation, the unknown function u depends upon the independent variables x and y, often regarded as space coordinates. In the advection, heat, and wave equations, u depends on t and x, often viewed as time and space coordinates, respectively. For a discussion of how these equations arise from fundamental physical principles, see Allen et al. ([2], Chapter 1).

These four examples serve as prototypes, but they by no means exhaust the possibilities. Realistic applications often involve three space dimensions instead of just one or two. Therefore, in practice one typically encounters

PDEs in three or four independent variables. Also, most applications involve coefficients that vary in space and time. The coefficients also may vary as functions of the unknown function u and possibly its derivatives. Finally, there are important applications in which the relevant PDEs involve derivatives of u having order higher than 2.

This chapter focuses on finite-difference schemes for the four simple examples listed above. We occasionally refer to simple generalizations but make no attempt at exhaustive treatment. The numerical approximation of PDEs occupies an immense and thinly settled territory. Solving realistic problems in this realm often calls for inspired combinations of theory, physically motivated heuristics, and computational experiment. Several excellent texts introduce the main ideas in more depth than we can provide in one chapter; interested readers should consult Ames [3], Lapidus and Pinder [6], and Strikwerda [9].

We devote the remainder of this section to the classification of PDEs; to a brief discussion of characteristic curves, which furnish insight into the geometry of PDEs; and to the rudiments of grid functions, which serve as the foundation for finite-difference approximations. Throughout, we restrict attention to equations in two independent variables.

Classification

PDEs in two independent variables x and y have the form

$$\Phi\left(x, y, u, \frac{\partial u}{\partial x}, \frac{\partial u}{\partial y}, \frac{\partial^2 u}{\partial x^2}, \frac{\partial^2 u}{\partial x \partial y}, \frac{\partial^2 u}{\partial y^2}, \ldots\right) = 0, \tag{8.1-1}$$

where the symbol Φ stands for some functional relationship. Given such a general form, one can say little about such fundamental properties of the solution $u(x, y)$ as its existence, uniqueness, or smoothness. For this reason, it is essential to adopt classifications that help distinguish the few classes of equations that have practical utility from the infinity of cases that are of no interest.

The **order** of a PDE (8.1-1) is the order of the highest derivative that appears. The advection equation, for example, is a first-order PDE, while the Poisson, heat, and wave equations are second-order PDEs. The PDE (8.1-1) is **linear** if Φ is linear in each of the quantities

$$u, \frac{\partial u}{\partial x}, \frac{\partial u}{\partial y}, \frac{\partial^2 u}{\partial x^2}, \ldots.$$

Thus the Poisson, advection, heat, and wave equations are all linear, as is the following generalization of the heat equation:

$$\frac{\partial u}{\partial t} - \frac{\partial}{\partial x}\left[D(t, x)\frac{\partial u}{\partial x}\right] = f(t, x).$$

(Notice that Φ need not be linear in the independent variables, in this case t and x.) The following equations, however, are nonlinear:

$$\frac{\partial u}{\partial t} + u\frac{\partial u}{\partial x} = 0, \qquad \text{Burgers's equation;}$$

$$\frac{\partial u}{\partial t} - \frac{\partial}{\partial x}\left[D(u)\frac{\partial u}{\partial x}\right] = 0, \qquad \text{nonlinear heat equation.}$$

Most of the mathematical theory of PDEs concerns linear equations of first or second order; equations of order higher than four seldom appear in applications.

After order and linearity, the most important classification scheme for PDEs involves geometry. We introduce the ideas in the simple context of a first-order, linear equation,

$$\alpha(t,x)\frac{\partial u}{\partial t} + \beta(t,x)\frac{\partial u}{\partial x} = \gamma(t,x). \tag{8.1-2}$$

A solution $u(t,x)$ to this PDE defines a surface $\{(t,x,u(t,x))\}$ lying over some region of the (t,x)-plane, as shown in Figure 1.

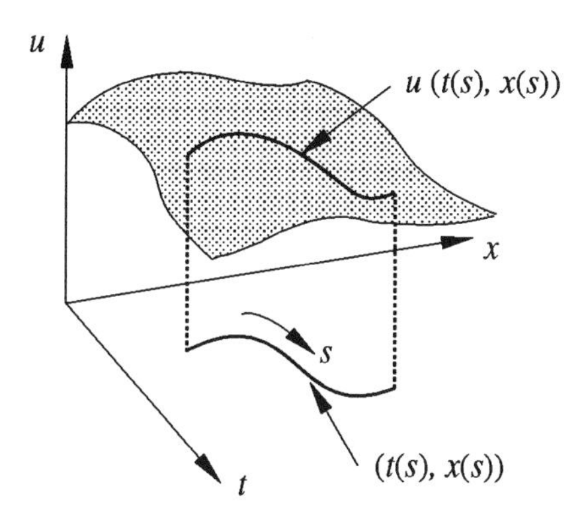

FIGURE 1. *Solution surface $(t,x,u(t,x))$ associated with Equation (8.1-2), along with a smooth path $(t(s),x(s))$ in the (t,x)-plane.*

Consider any smooth path in the (t,x)-plane, lying below the solution surface $(t,x,u(t,x))$ as depicted in Figure 1. Such a path has a parametrization $(t(s),x(s))$, where the parameter s measures progress along the path. What is the rate of change du/ds of the solution as we travel along the path $(t(s),x(s))$? The chain rule provides the answer:

$$\frac{dt}{ds}\frac{\partial u}{\partial t} + \frac{dx}{ds}\frac{\partial u}{\partial x} = \frac{du}{ds}. \tag{8.1-3}$$

Equation (8.1-3) holds for an arbitrary smooth path in the (t, x)-plane. Restricting attention to a specific family of paths leads to a useful observation: When

$$\frac{dt}{ds} = \alpha(t, x) \qquad \text{and} \qquad \frac{dx}{ds} = \beta(t, x), \tag{8.1-4}$$

the simultaneous validity of Equations (8.1-2) and (8.1-3) requires that

$$\frac{du}{ds} = \gamma(t, x). \tag{8.1-5}$$

Equations (8.1-4) define a family of curves $(t(s), x(s))$, called **characteristic curves**, in the (t, x)-plane. Equation (8.1-5) is an ODE, called the **characteristic equation**, that the solution $u(t, x)$ must satisfy along any characteristic curve. Thus the original PDE collapses to an ODE along the characteristic curves.

This geometry furnishes a useful solution technique for many first-order PDEs. Consider, for example, the advection equation,

$$\frac{\partial u}{\partial t} + v\frac{\partial u}{\partial x} = 0, \tag{8.1-6}$$

where $v > 0$. This equation serves as a simple model of the transport of contaminant, having concentration $u(t, x)$, in a one-dimensional stream flowing with constant velocity. Let us examine the contaminant transport given the auxiliary conditions

$$u(0, x) = \begin{cases} 1, & \text{if} \quad 0 < x < 1, \\ 0, & \text{if} \quad x \geq 1, \end{cases} \tag{8.1-7}$$

and

$$u(t, 0) = 1, \quad t > 0. \tag{8.1-8}$$

Equation (8.1-7) is an **initial condition**, specifying the concentration at $t = 0$, and Equation (8.1-8) is a **boundary condition**, giving the concentration at $x = 0$. These conditions imply that the initial concentration distribution has the form of a square wave and that the concentration at $x = 0$ remains constant at 1. Given the PDE (8.1-6) and these auxiliary conditions, we seek the concentration $u(t, x)$ for all $t \geq 0$ and all $x \geq 0$.

According to Equations (8.1-4), the characteristic curves for this problem satisfy the ODEs

$$\frac{dx}{ds} = v, \qquad \frac{dt}{ds} = 1.$$

It follows that $dx/dt = v$. Therefore, the characteristic curves are rays of the form $x = vt + \text{constant}$. Figure 2 shows these rays. They coincide with the paths of contaminant particles in the river.

Along the characteristic curves, the PDE (8.1-6) reduces to the ODE $du/ds = du/dt = 0$. That is, the contaminant concentration remains constant along

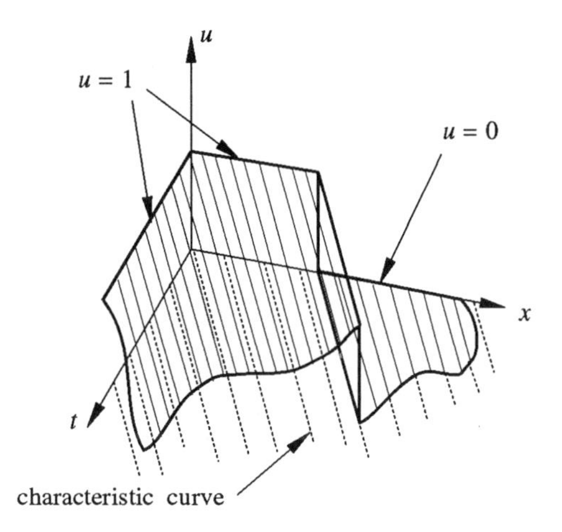

FIGURE 2. *Geometry of the advection problem, showing the characteristic curves and the propagation of solution values from the curves $x = 0$ and $t = 0$.*

the curves $x = vt +$constant. To compute the value of u at any point (t_1, x_1), we simply trace the characteristic curve that passes through (t_1, x_1) back to its intersection with one of the curves $x = 0$ or $t = 0$, where we have prescribed the auxiliary conditions. The value $u(t_1, x_1)$ is the value of u at this intersection. Figure 2 illustrates this construction. Explicitly,

$$u(t,x) = \begin{cases} u(0, x - vt), & \text{if} \quad x - vt > 0, \\ u(t - x/v, 0), & \text{if} \quad x - vt \leq 0, \end{cases}$$

or

$$u(t,x) = \begin{cases} 0, & \text{if} \quad x - vt \geq 1, \\ 1, & \text{if} \quad x - vt < 1. \end{cases}$$

Therefore, the initial square wave propagates to the right (downstream) with time, as Figure 3 shows.

Characteristic curves are paths along which information about the solution to the PDE propagates from points where initial or boundary values are known. This idea extends to second-order PDEs, but it does not always lead to an effective solution procedure. Instead, it leads to a scheme for classifying PDEs according to the nature of their characteristic curves.

Consider a second-order PDE having the form

$$\alpha(x,y)\frac{\partial^2 u}{\partial x^2} + \beta(x,y)\frac{\partial^2 u}{\partial x \partial y} + \gamma(x,y)\frac{\partial^2 u}{\partial y^2} = \Psi\left(x, y, u, \frac{\partial u}{\partial x}, \frac{\partial u}{\partial y}\right). \qquad (8.1\text{-}9)$$

Along an arbitrary smooth curve $(x(s), y(s))$ in the (x, y)-plane, the gradient

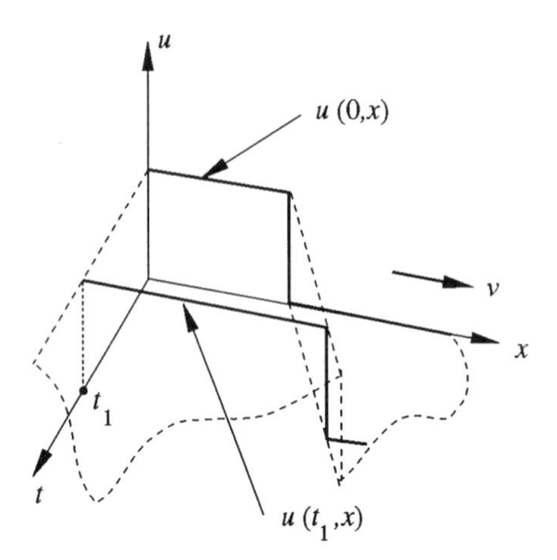

FIGURE 3. *Downstream propagation of the initial square wave of contaminant concentration.*

$(\partial u/\partial x, \partial u/\partial y)$ of the solution varies according to the chain rule:

$$\begin{aligned}\frac{dx}{ds}\frac{\partial^2 u}{\partial x^2}+\frac{dy}{ds}\frac{\partial^2 u}{\partial y \partial x} &= \frac{d}{ds}\left(\frac{\partial u}{\partial x}\right),\\ \frac{dx}{ds}\frac{\partial^2 u}{\partial x \partial y}+\frac{dy}{ds}\frac{\partial^2 u}{\partial y^2} &= \frac{d}{ds}\left(\frac{\partial u}{\partial y}\right).\end{aligned}$$

If the solution $u(x, y)$ is continuously differentiable, then these relationships together with the original PDE (8.1-9) yield the following system:

$$\begin{bmatrix} \alpha & \beta & \gamma \\ dx/ds & dy/ds & 0 \\ 0 & dx/ds & dy/ds \end{bmatrix}\begin{bmatrix} \partial^2 u/\partial x^2 \\ \partial^2 u/\partial x \partial y \\ \partial^2 u/\partial y^2 \end{bmatrix} = \begin{bmatrix} \Psi \\ d(\partial u/\partial x)/ds \\ d(\partial u/\partial y)/ds \end{bmatrix}. \tag{8.1-10}$$

By analogy with the first-order case, we determine the characteristic curves by asking where the PDE (8.1-9) is redundant with the chain rule. This occurs when the determinant of the matrix in Equation (8.1-10) vanishes, that is, when

$$\alpha\left(\frac{dy}{ds}\right)^2-\beta\frac{dx}{ds}\frac{dy}{ds}+\gamma\left(\frac{dx}{ds}\right)^2=0.$$

Eliminating the parameter s reduces this equation to the equivalent condition

$$\alpha\left(\frac{dy}{dx}\right)^2-\beta\frac{dy}{dx}+\gamma=0.$$

Formally solving this quadratic equation for dy/dx, we find that

$$\frac{dy}{dx} = \frac{\beta \pm \sqrt{\beta^2 - 4\alpha\gamma}}{2\alpha}. \tag{8.1-11}$$

This pair of ODEs determines the characteristic curves.

The nature of the characteristic curves in any region of the (x, y)-plane depends on the value of the function $\beta^2 - 4\alpha\gamma$. In regions where this function is positive, Equation (8.1-11) determines two families of real characteristic curves. In this case, the PDE (8.1-9) is **hyperbolic**. For example, the wave equation is hyperbolic: Replacing y by t gives $\beta^2 - 4\alpha\gamma = 0 - 4 \cdot (-c^2) \cdot 1$. For second-order hyperbolic equations the characteristic curves are paths along which information about the value of the solution propagates, and in principle one can use this idea to devise solution techniques.

In regions where $\beta^2 - 4\alpha\gamma = 0$, Equation (8.1-11) determines exactly one family of real characteristic curves. In this case, the PDE is **parabolic**. The heat equation, for example, is parabolic: Replacing y by t gives $\beta^2 - 4\alpha\gamma = 0 - 4 \cdot D \cdot 0$. Finally, in regions where $\beta^2 - 4\alpha\gamma < 0$, there are no real characteristic curves, and the PDE is **elliptic**. The Poisson equation, with $\beta^2 - 4\alpha\gamma = 0 - 4 \cdot 1 \cdot 1$, is one of the simplest elliptic equations.

There are rough correspondences between these categories, based on characteristic curves, and certain qualitative aspects of the physical systems being modeled. Broadly speaking, hyperbolic equations model nondissipative or weakly dissipative physical systems. Examples include the propagation of light and sound and the behavior of waves in elastic solids. We often associate parabolic equations with transient, dissipative phenomena, such as unsteady heat conduction, diffusion, and viscous fluid flows. Finally, elliptic equations commonly arise in modeling the steady states of systems whose transient behavior obeys parabolic equations. While there are exceptions, these correspondences can be useful as heuristic guides to the development of numerical approximations.

Grid Functions and Difference Operators

As in Chapter 7, we construct difference approximations to PDEs by replacing derivatives with algebraic analogs. These analogs involve functions defined only at discrete sets of points in the domain of the original problem. Before embarking on the construction of difference schemes for specific problems, let us review some of the associated terminology and notation.

If $\Delta = \{x_0, x_1, \ldots, x_M\}$ is a grid on an interval $[a, b]$, then a **grid function** on Δ is a function $U: \Delta \to \mathbb{R}$ (or $U: \Delta \to \mathbb{C}$). As shorthand, denote the value $U(x_j)$ by U_j. Any function $u: [a, b] \to \mathbb{R}$ gives rise to an associated grid function by the identity $u_j := u(x_j)$; we denote this obvious restriction of u by $u: \Delta \to \mathbb{R}$. Also, define $h_j := x_j - x_{j-1}$, and denote by

$$h := \max_{1 \le j \le M} h_j$$

the mesh size of Δ.

The extension of this notation to two dimensions is straightforward if we consider rectangular domains. Let $\Delta_x = \{x_0, x_1, \ldots, x_M\}$ be a grid on $[a, b]$ and $\Delta_y = \{y_0, y_1, \ldots, y_N\}$ be a grid on $[c, d]$. Then $\Delta := \Delta_x \times \Delta_y$ is a rectangular grid on the two-dimensional domain $[a, b] \times [c, d]$, as discussed in Section 1.5 and illustrated in Figure 4. For grid functions $U: \Delta \to \mathbb{R}$ (or $U: \Delta \to \mathbb{C}$), abbreviate $U(x_j, y_l)$ as $U_{j,l}$. Also, define $h_j := x_j - x_{j-1}$ and $k_l := y_l - y_{l-1}$. We use the symbols h and k to denote the mesh sizes of Δ_x and Δ_y, respectively. When one of the two dimensions is time, we typically alter this notation slightly, writing U_j^n for $U(t_n, x_j)$. The treatment of non-rectangular domains is somewhat more complicated; Lapidus and Pinder [6] offer suggestions.

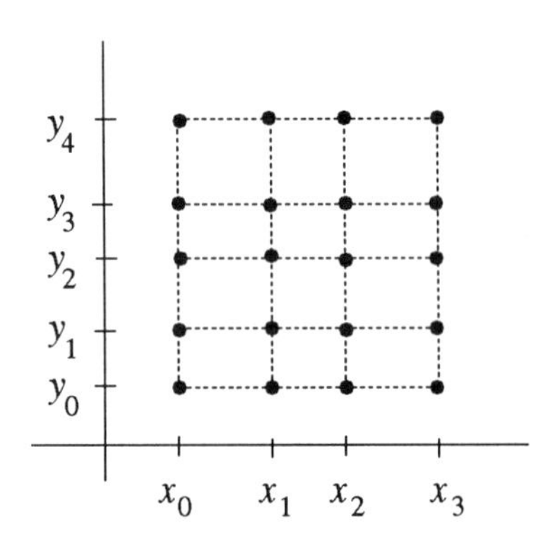

FIGURE 4. *A rectangular grid* $\Delta = \Delta_x \times \Delta_y$.

To construct algebraic analogs of derivative operators, we use the following **difference operators**:

$$\delta^+ U_j := \frac{1}{h_{j+1}}(U_{j+1} - U_j), \qquad \textbf{forward difference;}$$

$$\delta^- U_j := \frac{1}{h_j}(U_j - U_{j-1}), \qquad \textbf{backward difference;}$$

$$U_{j\pm 1/2} := \tfrac{1}{2}(U_j + U_{j\pm 1}), \qquad \textbf{average;}$$

$$\delta U_j := \frac{1}{h_{j+1/2}}\left(U_{j+1/2} - U_{j-1/2}\right), \qquad \textbf{centered difference.}$$

The expressions $\delta^+ U_j$, $\delta^- U_j$, and δU_j have obvious parallels with the difference quotients used to define derivatives.

For grid functions defined on rectangular grids, these difference operators act in one coordinate direction at a time. We specify which direction by attaching a subscript:

$$\begin{aligned}
\delta_x U_{j,l} &:= \frac{1}{h_{j+1/2}}\big(U_{j+1/2,l} - U_{j-1/2,l}\big),\\
\delta_y^+ U_{j,l} &:= \frac{1}{k_{l+1}}(U_{j,l+1} - U_{j,l}),\\
\delta_t^- U_j^n &:= \frac{1}{k_n}\big(U_j^n - U_j^{n-1}\big),
\end{aligned}$$

and so forth. Difference operators are linear, and they commute. For example, if κ is a constant, then

$$\delta_x(\kappa U_{j,l} + V_{j,l}) = \kappa\,\delta_x U_{j,l} + \delta_x V_{j,l},$$

and

$$\delta_x(\delta_y U_{j,l}) = \delta_y(\delta_x U_{j,l}).$$

The strategy behind finite-difference approximations is to convert the differential equations to equations that involve only arithmetic operations. We do this by replacing derivative expressions with difference operators. Examples include

$$\frac{\partial u}{\partial t}(x_j, t_n) \simeq \delta_t^- U_j^n$$

and

$$\begin{aligned}
\frac{\partial^2 u}{\partial x^2}(x_j, y_l) &\simeq \delta_x^2 U_{j,l}\\
&= \frac{1}{h_{j+1/2}}\left(\frac{U_{j+1,l} - U_{j,l}}{h_{j+1}} - \frac{U_{j,l} - U_{j-1,l}}{h_j}\right).
\end{aligned}$$

We then solve the resulting sets of algebraic equations for the unknown grid function U that approximates u at the nodes of Δ. We expect the approximations to become more accurate as the mesh sizes h and k shrink. Much of this chapter is devoted to an inquiry into when this expectation is valid.

8.2 The Poisson Equation

In this section we examine the most common finite-difference approximation to the Poisson equation. Consider the following model boundary-value problem (BVP):

$$\begin{aligned}
-\nabla^2 u(x,y) &= f(x,y), \qquad (x,y) \in \Omega := (0,1)\times(0,1),\\
u(x,y) &= g(x,y), \qquad (x,y) \in \partial\Omega.
\end{aligned} \tag{8.2-1}$$

The boundary condition in this problem, prescribing values of $u(x, y)$ along $\partial\Omega$, is a **Dirichlet condition**. After introducing the basic difference approximation and discussing its computational aspects, we sketch extensions of the method to generalizations of the problem (8.2-1) that involve variable coefficients and different boundary conditions. The section closes with a discussion of the scheme's consistency and convergence.

The Five-Point Scheme

The first step in constructing a finite-difference approximation to Equations (8.2-1) is to establish a grid on the domain. In th model problem, the domain is the unit square, $\overline{\Omega} := \Omega \cup \partial\Omega = [0,1] \times [0,1]$. For simplicity, consider a uniform grid. Partition the unit interval $[0,1]$ in each coordinate direction into N segments $[x_{j-1}, x_j]$ or $[y_{l-1}, y_l]$, each of which has length $h := 1/N$. This construction yields a grid

$$\Delta := \Big\{(x_j, y_l) \in [0,1] \times [0,1] \,:\, x_j = jh, \quad y_l = lh\Big\},$$

as shown in Figure 1. For later use, we identify two subsets of Δ. The first contains the interior nodes:

$$\Omega_h := \Big\{(x_j, y_l) \in \Delta \,:\, 1 \le j, l \le N-1\Big\}.$$

The second subset contains the boundary nodes, except for the corners of the square:

$$\partial\Omega_h := \Big\{(x_0, y_l), (x_N, y_l), (x_j, y_0), (x_j, y_N) \in \Delta \,:\, 1 \le j, l \le N-1\Big\}.$$

Figure 1 illustrates these sets.

The aim is to find a grid function $U_h \colon \Omega_h \cup \partial\Omega_h \to \mathbb{R}$ such that $U_{j,l} := U_h(x_j, y_l)$ approximates $u_{j,l} := u(x_j, y_l)$. We do this by demanding that U_h satisfy a discrete analog of the true BVP (8.2-1). Straightforward application of the centered difference operator yields the following discrete equations:

$$\begin{aligned} -\left(\delta_x^2 U_{j,l} + \delta_y^2 U_{j,l}\right) &= f_{j,l}, \qquad (x_j, y_l) \in \Omega_h, \\ U_{j,l} &= g_{j,l}, \qquad (x_j, y_l) \in \partial\Omega_h. \end{aligned} \tag{8.2-2}$$

None of the equations in this set involves the corner nodes (x_0, y_0), (x_0, y_N), (x_N, y_0), (x_N, y_N).

There are many useful ways to view the difference scheme (8.2-2). One is to expand the difference expressions to yield detailed algebraic equations. Since the grid Δ is uniform with mesh size h,

$$\delta_x^2 U_{j,l} = h^{-2}(U_{j+1,l} - 2U_{j,l} + U_{j-1,l})$$

and

$$\delta_y^2 U_{j,l} = h^{-2}(U_{j,l+1} - 2U_{j,l} + U_{j,l-1}).$$

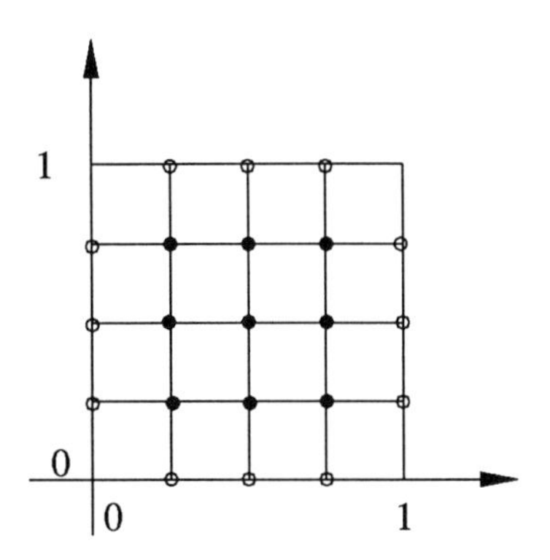

FIGURE 1. *Uniform grid Δ on the square $\Omega = (0,1) \times (0,1)$, showing points in the sets Ω_h ($\bullet$) and $\partial\Omega_h$ ($\circ$).*

Therefore, the difference equation reduces to the following:

$$-\left(U_{j-1,l} + U_{j,l-1} - 4U_{j,l} + U_{j,l+1} + U_{j+1,l}\right) = h^2 f_{j,l}. \tag{8.2-3}$$

A second way to view Equations (8.2-2) is to regard the difference equation as being centered at a typical node $(x_j, y_l) \in \Omega_h$ and to ask which nodal values the equation couples via nonzero coefficients. Equation (8.2-3) couples the five values $U_{j,l}$, $U_{j\pm1,l}$, and $U_{j,l\pm1}$. These values correspond to five points of the grid, namely (x_j, y_l) and its four nearest neighbors, so we call Equation (8.2-3) the **five-point scheme** for the Poisson equation. We depict this scheme by drawing the **stencil** of the difference operator

$$\nabla_h^2 := \delta_x^2 + \delta_y^2,$$

as shown in Figure 2.

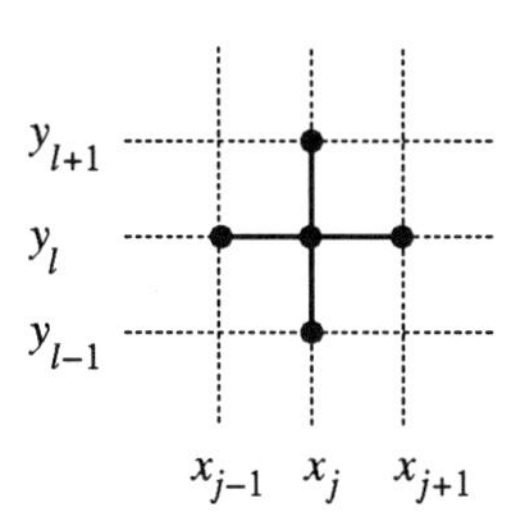

FIGURE 2. *Stencil of the five-point scheme ∇_h^2 for the differential operator ∇^2.*

A third way to view Equation (8.2-2) is to examine the corresponding linear system. There is one equation for each interior node in Ω_h and one for each boundary node in $\partial\Omega_h$. Since the equations $U_{j,l} = g_{j,l}$ associated with the boundary nodes simply assign certain values $U_{j,l}$ to the grid function U_h, it is possible to use these equations to eliminate some of the nodal values in the difference equations $-\nabla_h^2 U_{j,l} = f_{j,l}$. The resulting linear system has a matrix form in which the matrix has at most five nonzero entries in any row, a fact that the stencil in Figure 2 makes apparent.

If we order the unknowns $U_{j,l}$ lexicographically, then the matrix equation has the following block-tridiagonal structure:

$$\begin{matrix} & \begin{matrix} N-1 & N-1 & \cdots & & \end{matrix} \\ \begin{matrix} N-1 \\ N-1 \\ \vdots \\ \\ \\ \end{matrix} & \begin{pmatrix} \mathsf{T} & \mathsf{I} & & & \\ \mathsf{I} & \mathsf{T} & \mathsf{I} & & \\ & & \ddots & & \\ & & \mathsf{I} & \mathsf{T} & \mathsf{I} \\ & & & \mathsf{I} & \mathsf{T} \end{pmatrix} \end{matrix} \begin{pmatrix} \mathbf{u}_1 \\ \mathbf{u}_2 \\ \vdots \\ \mathbf{u}_{N-2} \\ \mathbf{u}_{N-1} \end{pmatrix} = -h^2 \begin{pmatrix} \mathbf{r}_1 \\ \mathbf{r}_2 \\ \vdots \\ \mathbf{r}_{N-2} \\ \mathbf{r}_{N-1} \end{pmatrix}.$$

Each block $\mathsf{T} \in \mathbb{R}^{(N-1)\times(N-1)}$ is tridiagonal:

$$\mathsf{T} := \begin{bmatrix} -4 & 1 & & & \\ 1 & -4 & 1 & & \\ & & \ddots & & \\ & & 1 & -4 & 1 \\ & & & 1 & -4 \end{bmatrix},$$

and I denotes the identity matrix in $\mathbb{R}^{(N-1)\times(N-1)}$. The vector of unknowns under this ordering has blocks of the form

$$\mathbf{u}_j := \begin{bmatrix} U_{j,1} \\ U_{j,2} \\ \vdots \\ U_{j,N-1} \end{bmatrix}.$$

The vector on the right side contains blocks that decompose as $\mathbf{r}_j = \mathbf{b}_j - \mathbf{f}_j$. The block vectors $\mathbf{f}_j$ contain values $f_{j,l}$, ordered lexicographically, while the block vectors $\mathbf{b}_j$ contain the known boundary values. For $j = 2, 3, \ldots, N-2$, these "boundary vectors" have the form

$$\mathbf{b}_j := -\begin{bmatrix} g_{j,0} \\ 0 \\ \vdots \\ 0 \\ g_{j,N} \end{bmatrix},$$

while

$$\mathbf{b}_1 := -\begin{bmatrix} g_{0,1} + g_{1,0} \\ g_{0,2} \\ \vdots \\ g_{0,N-2} \\ g_{0,N-1} + g_{1,N} \end{bmatrix}, \qquad \mathbf{b}_{N-1} := -\begin{bmatrix} g_{N,1} + g_{N-1,0} \\ g_{N,2} \\ \vdots \\ g_{N,N-2} \\ g_{N,N-1} + g_{N-1,N} \end{bmatrix}.$$

The matrix in this system is symmetric and positive definite, so the system has a unique solution vector. Common methods for solving systems like this include the successive overrelaxation and preconditioned conjugate gradient algorithms discussed in Chapter 4. Problem 1 calls for such an approach.

Generalizations of the Five-Point Scheme

One can extend the approximation (8.2-2) to accommodate variable coefficients and other boundary conditions. First consider variable coefficients. The Poisson equation generalizes to the following PDE:

$$-\nabla \cdot [a(x,y)\nabla u] = f(x,y),$$

that is,

$$-\frac{\partial}{\partial x}\left[a(x,y)\frac{\partial u}{\partial x}\right] - \frac{\partial}{\partial y}\left[a(x,y)\frac{\partial u}{\partial y}\right] = f(x,y).$$

This equation is elliptic if there exists a constant $a_0 > 0$ such that $a(x,y) \geq a_0$ for every point (x,y) in the domain of the problem.

Constructing a five-point difference approximation to this equation is straightforward if one uses the centered difference operators:

$$-\delta_x\left[a_{j,l}\delta_x U_{j,l}\right] - \delta_y\left[a_{j,l}\delta_y U_{j,l}\right] = f_{j,l}.$$

By definition of the operators δ_x and δ_y, these difference equations involve values of a having the form $a(x_{j-1/2}, y_l)$, $a(x_j, y_{l-1/2})$, and so forth. Problem 2 asks for numerical experiments involving a scheme of this form.

Handling boundary conditions other than Dirichlet conditions is slightly less straightforward. Let us investigate the typical treatment of the boundary condition

$$\frac{\partial u}{\partial x}(0,y) = g(0,y), \tag{8.2-4}$$

applied to the Poisson equation. Such boundary conditions, prescribing values of the directional derivative $\partial u/\partial n$ of u normal to the boundary $\partial\Omega$, are called **Neumann conditions**.

To approximate the condition (8.2-4), we establish a line of "fictitious" nodes (x_{-1}, y_l) along the left boundary of Ω, as shown in Figure 3. We then discretize Equation (8.2-4) by applying the centered difference operator:

$$\delta_x U_{0,l} = \frac{1}{2h}\left(U_{1,l} - U_{-1,l}\right) = g_{0,l}. \tag{8.2-5}$$

The problem with this equation is that it involves an unknown nodal value $U_{-1,l}$ that does not appear in the original list of unknowns for the numerical domain $\Omega_h \cup \partial\Omega_h$. We rectify this problem by assuming that the difference equation (8.2-3) holds at each of the boundary nodes (x_0, y_l). Thus,

$$U_{-1,l} + U_{0,l-1} - 4U_{0,l} + U_{0,l+1} + U_{1,l} = -h^2 f_{0,l}.$$

The discrete approximation (8.2-5) implies that $U_{-1,l} = U_{1,l} - 2hg_{0,l}$, so we replace the five-point approximation centered at (x_0, y_l) by the equation

$$U_{0,l-1} - 4U_{0,l} + U_{0,l+1} + 2U_{1,l} = -h^2 f_{0,l} + 2h\, g_{0,l}.$$

This equation incorporates centered differences into the boundary condition while referring only to nodal values $U_{j,l}$ associated with the original computational domain $\Omega_h \cup \partial\Omega_h$.

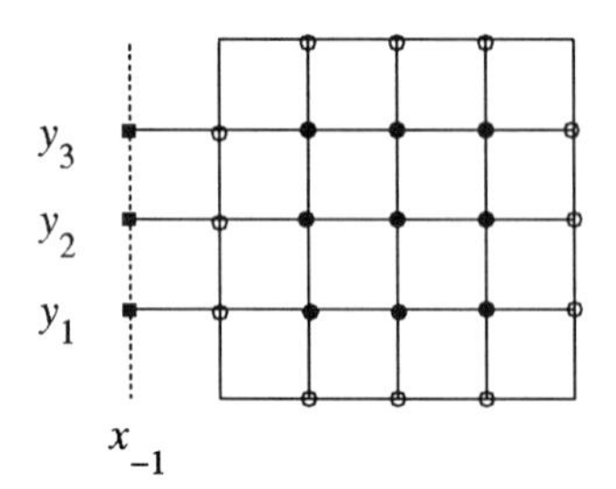

FIGURE 3. *Finite-difference grid on the unit square Ω, showing the line of fictitious nodes (x_{-1}, y_l) used to approximate Neumann conditions on the boundary segment $x = 0$.*

Problem 1 provides an opportunity to apply this approach. It is a worthwhile exercise to derive an analogous technique for approximating **Robin conditions**, which have the form

$$u(x, y) + b(x, y)\frac{\partial u}{\partial n}(x, y) = g(x, y), \quad (x, y) \in \Omega.$$

Consistency and Convergence

We now ask how well the grid function U_h, determined using the five-point scheme (8.2-2), approximates the exact solution u of the Poisson problem (8.2-1). As for the numerical solutions to ODEs, it is important to distinguish between two ways of assessing a difference approximation. The first measures how well a difference analog approximates a differential operator:

DEFINITION. *Let $\mathcal{L}_h$ denote a finite-difference approximation, associated with a grid Δ_h having mesh size h, to a partial differential operator $\mathcal{L}$ defined on* *a simply connected, open set $D \subset \mathbb{R}^n$. For a given function $\varphi \in C^\infty(D)$, the* **truncation error** *of $\mathcal{L}_h$ is*

$$\tau_h(\mathbf{x}) := (\mathcal{L} - \mathcal{L}_h)\varphi(\mathbf{x}).$$

The approximation $\mathcal{L}_h$ is **consistent** *with $\mathcal{L}$ if*

$$\lim_{h\to 0} \tau_h(\mathbf{x}) = 0, \tag{8.2-6}$$

for all $\mathbf{x} \in D$ and all $\varphi \in C^\infty(D)$. The approximation is consistent **to order** *p if $\tau_h(\mathbf{x}) = \mathcal{O}(h^p)$.*

The notation $\tau_h(\mathbf{x})$ masks the formal dependence of truncation error on the function φ. This convention suggests that h is the important parameter and that we should view φ as an arbitrary function, chosen simply as a "test bed" for the assessment of $\mathcal{L}_h$.

To illustrate the definition, we prove the following:

PROPOSITION 8.1. *The five-point difference analog $-\nabla_h^2$ is consistent to order 2 with $-\nabla^2$.*

PROOF: Pick $\varphi \in C^\infty(\Omega)$, and let $(x, y) \in \Omega$ be a point such that $(x \pm h, y), (x, y \pm h) \in \Omega \cup \partial\Omega$. By the Taylor theorem,

$$\begin{aligned} \varphi(x \pm h, y) &= \varphi(x, y) \pm h\frac{\partial\varphi}{\partial x}(x, y) + \frac{h^2}{2}\frac{\partial^2\varphi}{\partial x^2}(x, y) \\ &\quad \pm \frac{h^3}{3!}\frac{\partial^3\varphi}{\partial x^3}(x, y) + \frac{h^4}{4!}\frac{\partial^4\varphi}{\partial x^4}(\zeta^\pm, y), \end{aligned}$$

where $\zeta^\pm$ are points in the interval $(x-h, x+h)$. Adding this pair of equations and rearranging the result, we arrive at the following relationship:

$$\begin{aligned} \frac{1}{h^2}\left[\varphi(x+h, y) - 2\varphi(x, y) + \varphi(x-h, y)\right] &- \frac{\partial^2\varphi}{\partial x^2}(x, y) \\ = \frac{h^2}{4!}\left[\frac{\partial^4\varphi}{\partial x^4}(\zeta^+, y)\right. &+ \left.\frac{\partial^4\varphi}{\partial x^4}(\zeta^-, y)\right]. \end{aligned} \tag{8.2-7}$$

By the intermediate value theorem,

$$\left[\frac{\partial^4\varphi}{\partial x^4}(\zeta^+, y) + \frac{\partial^4\varphi}{\partial x^4}(\zeta^-, y)\right] = 2\frac{\partial^4\varphi}{\partial x^4}(\zeta, y),$$

for some $\zeta \in (x-h, x+h)$. Also, the expression in square brackets on the left side of Equation (8.2-7) is just $\delta_x^2\varphi(x, y)$. Therefore,

$$\delta_x^2\varphi(x, y) = \frac{\partial^2\varphi}{\partial x^2}(x, y) + \frac{h^2}{12}\frac{\partial^4\varphi}{\partial x^4}(\zeta, y),$$

for some $\zeta \in (x-h, x+h)$. Similar reasoning shows that

$$\delta_y^2 \varphi(x,y) = \frac{\partial^2 \varphi}{\partial y^2}(x,y) + \frac{h^2}{12}\frac{\partial^4 \varphi}{\partial y^4}(x,\eta),$$

for some point $\eta \in (y-h, y+h)$. We conclude that $\tau_h(x,y) = (\nabla_h^2 - \nabla^2)(x,y) = \mathcal{O}(h^2)$. ∎

Consistency does not guarantee that the solution to the difference equations approximates the exact solution to the PDE. To capture the latter concept, we make the following definition:

DEFINITION. *Let $\mathcal{L}_h U_h(\mathbf{x}_j) = f(\mathbf{x}_j)$ be a finite-difference approximation, defined on a grid having mesh size h, to a* PDE *$\mathcal{L}u(\mathbf{x}) = f(\mathbf{x})$ on a simply connected open set $D \subset \mathbb{R}^n$. Assume that $U_h(\mathbf{x}_j) = u(\mathbf{x}_j)$ at all grid points $\mathbf{x}_j$ that lie on the boundary ∂D. The finite-difference scheme* **converges** *(or is* **convergent***) if*

$$\max_j \Big| u(\mathbf{x}_j) - U_h(\mathbf{x}_j) \Big| \to 0 \quad \text{as} \quad h \to 0.$$

For the five-point scheme applied to the Poisson equation, there is a direct connection between consistency and convergence. We show that the estimate $\tau_h(x,y) = \mathcal{O}(h^2)$ implies that $\max_j |u_{j,l} - U_{j,l}| = \mathcal{O}(h^2)$ for the problem (8.2-1). Underlying this connection is an argument based on the following principle:

THEOREM 8.2 (DISCRETE MAXIMUM PRINCIPLE). *If $\nabla_h^2 V_{j,l} \geq 0$ for all points $(x_j, y_l) \in \Omega_h$, then*

$$\max_{(x_j,y_l)\in\Omega_h} V_{j,l} \leq \max_{(x_j,y_l)\in\partial\Omega_h} V_{j,l}.$$

If $\nabla_h^2 V_{j,l} \leq 0$ for all $(x_j, y_l) \in \Omega_h$, then

$$\min_{(x_j,y_l)\in\Omega_h} V_{j,l} \geq \min_{(x_j,y_l)\in\partial\Omega_h} V_{j,l}.$$

In other words, a grid function V for which $\nabla_h^2 V$ is nonnegative on Ω_h attains its maximum on the boundary $\partial\Omega_h$ of the grid. Similarly, if $\nabla_h^2 V$ is nonpositive on Ω_h, then V attains its minimum value on $\partial\Omega_h$.

PROOF: The proof is by contradiction. We argue for the case $\nabla_h^2 V_{j,l} \geq 0$, the reasoning for the case $\nabla_h^2 V_{j,l} \leq 0$ being similar. Assume that V

attains its maximum value M at an interior grid point (x_J, y_L) and that $\max_{(x_j,y_l)\in\partial\Omega_h} V_{j,l} < M$. The hypothesis $\nabla_h^2 V_{j,l} \geq 0$ implies that

$$V_{J,L} \leq \frac{1}{4}(V_{J+1,L} + V_{J-1,L} + V_{J,L+1} + V_{J,L-1}).$$

This relationship cannot hold unless $V_{J+1,L} = V_{J-1,L} = V_{J,L+1} = V_{J,L-1} = M$. If any of the corresponding grid points $(x_{J+1}, y_L), (x_{J-1}, y_L), (x_J, y_{L+1})$, or (x_J, y_{L-1}) lies in $\partial\Omega_h$, then we have reached the desired contradiction. Otherwise, we continue arguing in this way until we conclude that $V_{J+j,L+l} = M$ for some point $(x_{J+j}, y_{L+l}) \in \partial\Omega_h$, which again gives a contradiction. ∎

The discrete maximum principle leads to several interesting conclusions that have close parallels in the theory of partial differential equations:

COROLLARY 8.3.

(i) *The zero grid function (for which $U_{j,l} = 0$ for all $(x_j, y_l) \in \Omega_h \cup \partial\Omega_h$) is the only solution to the finite-difference problem*

$$\begin{aligned} \nabla_h^2 U_{j,l} &= 0 \qquad \textit{for} \quad (x_j, y_l) \in \Omega_h, \\ U_{j,l} &= 0 \qquad \textit{for} \quad (x_j, y_l) \in \partial\Omega_h. \end{aligned}$$

(ii) *For prescribed grid functions $f_{j,l}$ and $g_{j,l}$, there exists a unique solution to the problem*

$$\begin{aligned} \nabla_h^2 U_{j,l} &= f_{j,l} \qquad \textit{for} \quad (x_j, y_l) \in \Omega_h, \\ U_{j,l} &= g_{j,l} \qquad \textit{for} \quad (x_j, y_l) \in \partial\Omega_h. \end{aligned}$$

PROOF: Both statements are exercises. ∎

For our immediate purpose, the discrete maximum principle allows us to estimate a grid function V in terms of its discrete Laplacian $\nabla_h^2 V$. Two norms are useful in stating and proving the estimate:

DEFINITION. *For any grid function $V: \Omega_h \cup \partial\Omega_h \to \mathbb{R}$,*

$$\begin{aligned} \|V\|_\Omega &:= \max_{(x_j,y_l)\in\Omega_h} |V_{j,l}|, \\ \|V\|_{\partial\Omega} &:= \max_{(x_j,y_l)\in\partial\Omega_h} |V_{j,l}|. \end{aligned}$$

LEMMA 8.4. *If the grid function $V: \Omega_h \cup \partial\Omega_h \to \mathbb{R}$ satisfies the boundary condition $V_{j,l} = 0$ for $(x_j, y_l) \in \partial\Omega_h$, then*

$$\|V\|_\Omega \le \frac{1}{8}\|\nabla_h^2 V\|_\Omega.$$

PROOF: Let $\nu := \|\nabla_h^2 V\|_\Omega$. Clearly, for all points $(x_j, y_l) \in \Omega_h$,

$$-\nu \le \nabla_h^2 V_{j,l} \le \nu. \tag{8.2-8}$$

Now define $W: \Omega_h \cup \partial\Omega_h \to \mathbb{R}$ by setting $W_{j,l} := \frac{1}{4}[(x_j - \frac{1}{2})^2 + (y_l - \frac{1}{2})^2]$, which is nonnegative. One can check that $\nabla_h^2 W_{j,l} = 1$ and that $\|W\|_{\partial\Omega} = \frac{1}{8}$. The inequality (8.2-8) implies that, for all points $(x_j, y_l) \in \Omega_h$,

$$\begin{aligned} \nabla_h^2(V_{j,l} + \nu W_{j,l}) &\ge 0, \\ \nabla_h^2(V_{j,l} - \nu W_{j,l}) &\le 0. \end{aligned}$$

By the discrete maximum principle and the fact that V vanishes on $\partial\Omega_h$,

$$\begin{aligned} V_{j,l} \;\le\; V_{j,l} + \nu W_{j,l} &\le \nu\|W\|_{\partial\Omega}, \\ V_{j,l} \;\ge\; V_{j,l} - \nu W_{j,l} &\ge -\nu\|W\|_{\partial\Omega}. \end{aligned}$$

Since $\|W\|_{\partial\Omega} = \frac{1}{8}$, $\|V\|_\Omega \le \frac{1}{8}\nu = \frac{1}{8}\|\nabla_h^2 V\|_\Omega$. ■

Finally we prove that the five-point scheme for the Poisson equation is convergent.

THEOREM 8.5. *Let u be the solution to the BVP (8.2-1), and let U_h be the grid function that satisfies the discrete analog*

$$\begin{aligned} -\nabla_h^2 U_{j,l} &= f_{j,l}, \qquad \text{for} \quad (x_j, y_l) \in \Omega_h, \\ U_{j,l} &= g_{j,l}, \qquad \text{for} \quad (x_j, y_l) \in \partial\Omega_h. \end{aligned}$$

Then there exists a positive constant κ such that

$$\|u - U_h\|_\Omega \le \kappa M h^2,$$

where

$$M := \max\left\{ \left\|\frac{\partial^4 u}{\partial x^4}\right\|_\infty, \left\|\frac{\partial^4 u}{\partial^3 x \partial y}\right\|_\infty, \ldots, \left\|\frac{\partial^4 u}{\partial y^4}\right\|_\infty \right\}.$$

The statement of the theorem implicitly assumes that $u \in C^4(\overline{\Omega})$. This assumption holds if f and g are smooth enough.

PROOF: Following the proof of Proposition 8.1, we have

$$\left(\nabla_h^2 - \nabla^2\right) u_{j,l} = \frac{h^2}{12}\left[\frac{\partial^4 u}{\partial x^4}(\zeta_j, y_l) + \frac{\partial^4 u}{\partial y^4}(x_j, \eta_l)\right],$$

for some points $\zeta_j \in (x_{j-1}, x_{j+1})$ and $\eta_l \in (y_{l-1}, y_{l+1})$. Therefore,

$$-\nabla_h^2 u_{j,l} = f_{j,l} - \frac{h^2}{12}\left[\frac{\partial^4 u}{\partial x^4}(\zeta_j, y_l) + \frac{\partial^4 u}{\partial y^4}(x_j, \eta_l)\right].$$

If we subtract from this identity the equation $-\nabla_h^2 U_{j,l} = f_{j,l}$ and note that $u - U_h$ vanishes on $\partial\Omega_h$, we find that

$$\nabla_h^2 (u_{j,l} - U_{j,l}) = \frac{h^2}{12}\left[\frac{\partial^4 u}{\partial x^4}(\zeta_j, y_l) + \frac{\partial^4 u}{\partial y^4}(x_j, \eta_l)\right].$$

It follows by Lemma 8.4 that $\|u - U_h\|_\Omega \leq \frac{1}{8}\|\nabla_h^2(u - U_h)\|_\Omega \leq \kappa M h^2$. ∎

8.3 The Advection Equation

The numerical solution of the advection equation raises new issues. The differences arise from the nature of the equation's characteristic curves, which Section 8.1 describes. Consider the model initial-value problem (IVP) that consists of the PDE

$$\frac{\partial u}{\partial t} + v\frac{\partial u}{\partial x} = 0, \qquad \text{for} \quad -\infty < x < \infty, \quad t > 0, \tag{8.3-1}$$

with $v > 0$, together with the initial condition

$$u(0, x) = f(x), \qquad \text{for} \quad -\infty < x < \infty.$$

This problem has a solution whose value remains constant along the characteristic curves $x - vt =$ constant. In particular, $u(t, x) = f(x - vt)$. A more suggestive view of this fact is as follows: The value of the solution u at a point (t_0, x_0) depends only upon the values of u at points in the set

$$\left\{(t, x) \in \mathbb{R}^2 \,:\, x - vt = x_0 - vt_0,\ 0 \leq t < t_0\right\},$$

illustrated in Figure 1. We call this set the **exact domain of dependence** of the PDE at (t_0, x_0).

This geometry plays a crucial role in numerical solutions. In addition to being consistent, a finite-difference approximation to the advection equation must respect the equation's exact domains of dependence, in a sense that we clarify below. Otherwise, the resulting approximate solution fails to depend continuously upon the initial and boundary values imposed. Such approximate solutions are useless.

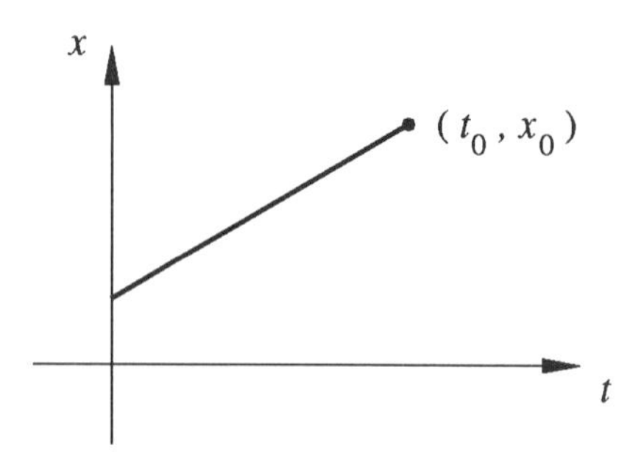

FIGURE 1. *Exact domain of dependence for the advection equation at a point* (t_0, x_0).

This consideration leads us to the notion of stability. Later in this section we examine a method, attributed to von Neumann, for testing the stability of finite-difference approximations to PDEs. We then establish a useful theoretical relationship among consistency, stability, and convergence, paralleling results for ODEs discussed in Section 7.4.

The Courant-Friedrichs-Lewy Condition

We begin by examining geometric constraints imposed by the exact domains of dependence. Define a grid in the (t, x)-plane as follows: Let

$$\Delta_t := \left\{nk \,:\, n = 0, 1, 2, \ldots\right\},$$

$$\Delta_x := \left\{jh \,:\, j = 0, \pm 1, \pm 2, \ldots\right\},$$

and set $\Delta := \Delta_t \times \Delta_x$, as drawn in Figure 2. Consider the following finite-difference approximation to Equation (8.3-1):

$$\delta_t^+ U_j^n + v\,\delta_x^- U_j^n = \frac{U_j^{n+1} - U_j^n}{k} + v\,\frac{U_j^n - U_{j-1}^n}{h} = 0. \tag{8.3-2}$$

We denote this scheme as FB, since it uses the forward difference operator in t and the backward difference operator in x.

Equation (8.3-2) reduces to the simpler form

$$U_j^{n+1} = U_j^n - C(U_j^n - U_{j-1}^n), \tag{8.3-3}$$

where $C := vk/h$ is the **Courant number**. Thus FB is an explicit scheme: One can solve for each value U_j^{n+1} at the new time level $n+1$ in terms of previously computed values associated with time level n.

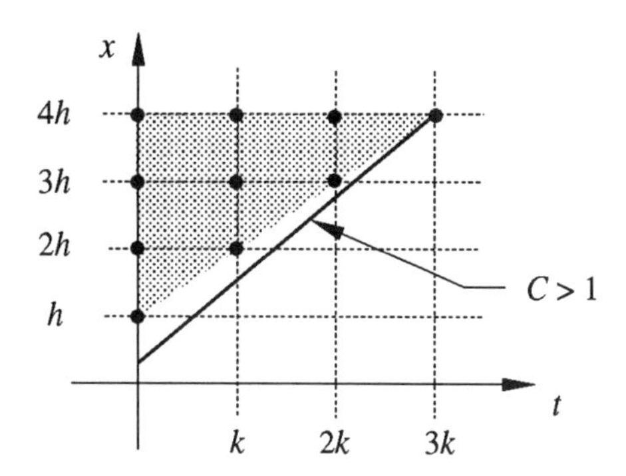

FIGURE 2. *A grid in the (t, x)-plane, showing the exact domain of dependence of the advection equation at the point $(3k, 4h)$ and the numerical domain of dependence associated with a forward-backward difference scheme.*

The simplified form (8.3-3) reveals a peculiar feature of the scheme FB. Consider the "ancestry" of a typical value U_j^n, say U_4^3. According to Equation (8.3-3), we compute this value using the two values U_4^2 and U_3^2. These values, in turn, depend upon the three values U_4^1, U_3^1, and U_2^1, which depend upon the prescribed initial values U_4^0, U_3^0, U_2^0, and U_1^0. In short, U_4^3 depends upon the values U_j^n associated with nodes (t_n, x_j) that lie in the region shaded in Figure 2. We call this region the **numerical domain of dependence** of the scheme FB at $(3k, 4h)$.

When $C > 1$, the characteristic curve $x - vt =$ constant passing through the point $(3k, 4h)$ lies outside the shaded region. In other words, the exact domain of dependence at $(3k, 4h)$ lies outside the numerical domain of dependence of FB at this point. The value of U_4^3 determined using FB with $C > 1$ cannot "see" values at the previous time levels that determine the exact solution $u(3k, 4h)$. We demonstrate below that this circumstance prevents the values generated by the numerical scheme from depending continuously upon the initial data.

One can generalize this observation:

COURANT-FRIEDRICHS-LEWY (CFL) CONDITION. *A necessary condition for the convergence of a finite-difference approximation to an initial-value problem is that the numerical domain of dependence at any grid point contain the exact domain of dependence at that point.*

It is easy to show that FB satisfies the CFL condition when $C \leq 1$. If we think of computations on a fixed spatial grid with a given value of $v > 0$, then the CFL condition $k \leq h/v$ imposes a restriction on the allowable time step k. Restrictions like this are typical of explicit schemes. (An interesting but less useful observation is that, when v is constant, FB yields exact solutions to the model problem when $C = 1$.)

We can interpret the CFL condition in terms of the growth of errors in the initial data. Define a **shift operator** S on grid functions by the equation $\mathsf{S}U_j^n := U_{j+1}^n$. This operator has an inverse defined by the relationship $\mathsf{S}^{-1}U_j^n = U_{j-1}^n$. In terms of S we can rewrite Equation (8.3-3) for the model IVP in the form

$$\begin{aligned} U_j^n = (1 - C + C\mathsf{S}^{-1})U_j^{n-1} = \quad \cdots \quad &= (1 - C + C\mathsf{S}^{-1})^n U_j^0 \\ &= (1 - C + C\mathsf{S}^{-1})^n f_j, \end{aligned}$$

where we have used the initial condition to rewrite U_j^0. The binomial theorem then yields

$$\begin{aligned} U_j^n &= \sum_{m=0}^{n} \binom{n}{m} (1 - C)^m (C\mathsf{S}^{-1})^{n-m} f_j \\ &= \sum_{m=0}^{n} \binom{n}{m} (1 - C)^m C^{n-m} f_{j-(n-m)}. \end{aligned}$$

If, instead of the exact initial data f_j, we use erroneous data $f_j \pm \epsilon$, where $\epsilon > 0$, then the computations yield an inexact solution

$$\hat{U}_j^n = \sum_{m=0}^{n} \binom{n}{m} (1 - C)^m C^{n-m} (f_{j-(m-n)} \pm \epsilon).$$

Subtracting the equations for U_j^n and $\hat{U}_j^n$ and using the triangle inequality then yields the estimate

$$\left| U_j^n - \hat{U}_j^n \right| \leq \epsilon \sum_{m=0}^{n} \binom{n}{m} |1 - C|^m C^{n-m}.$$

Now consider how different values of the Courant number C affect this estimate. When $0 < C \leq 1$,

$$\sum_{m=0}^{n} \binom{n}{m} |1 - C|^m C^{n-m} = (1 - C + C)^n = 1.$$

In this case, $|U_j^n - \hat{U}_j^n| \leq \epsilon$; small errors in the initial data yield small errors in the solution computed at later time levels. However, when $C > 1$,

$$\sum_{m=0}^{n} \binom{n}{m} |1 - C|^m C^{n-m} = (2C - 1)^n.$$

In this case, the error estimate reduces to

$$|U_j^n - \hat{U}_j^n| \leq (2C-1)^n \epsilon. \tag{8.3-4}$$

This inequality is sharp: Equality occurs when the erroneous initial data have the form $f_j + (-1)^j \epsilon$. We conclude that the scheme FB with $C > 1$ allows small initial errors to grow arbitrarily large as the calculations progress. Especially distressing is the fact that, as we reduce the time step k, the exponent n required to reach a fixed time level increases; hence smaller time steps can lead to more rapid error growth.

In light of this reasoning, we associate the CFL condition with stability. Before elaborating, let us establish directly that the scheme FB is convergent when the CFL condition holds.

THEOREM 8.6. *If the* CFL *condition holds and the solution $u(t,x)$ is twice continuously differentiable, then the approximation* FB *to the advection equation is convergent.*

PROOF: Begin by examining the truncation error $\tau_{k,h}(t,x)$ of the difference scheme. By the Taylor theorem,

$$u(t, x-h) = u(t,x) - h\frac{\partial u}{\partial x}(t,x) + \frac{h^2}{2}\frac{\partial^2 u}{\partial x^2}(t,\zeta).$$

Therefore,

$$\delta_x^- u(t,x) = \frac{\partial u}{\partial x}(t,x) - \frac{h}{2}\frac{\partial^2 u}{\partial x^2}(t,\zeta).$$

Similarly,

$$\delta_t^+ u(t,x) = \frac{\partial u}{\partial t}(t,x) + \frac{k}{2}\frac{\partial^2 u}{\partial t^2}(\eta,x).$$

We conclude that

$$\delta_t^+ u(t,x) + v\delta_x^- u(t,x) = -\tau_{k,h}(t,x),$$

where $\tau_{k,h}(t,x) = \mathcal{O}(k+h)$. Subtracting the difference scheme (8.3-2) from this equation yields

$$\delta_t^+ \varepsilon_j^n + v\delta_x^- \varepsilon_j^n = -\tau_{k,h}(t_n, x_j),$$

where $\varepsilon_j^n := u_j^n - U_j^n$. By rearranging and using the triangle inequality, we find that

$$|\varepsilon_j^{n+1}| \leq |1-C||\varepsilon_j^n| + C|\varepsilon_{j-1}^n| + k|\tau_{k,h}(t_n,x_j)|.$$

Since $0 < C \leq 1$, $|1-C| = 1-C$. Letting

$$E^n := \max_j |\varepsilon_j^n|, \qquad T^n := \max_j |\tau_{k,h}(t_n,x_j)|,$$

we obtain

$$\begin{aligned} E^{n+1} &\leq (1-C+C)E^n + kT^n \\ &\leq E^{n-1} + kT^{n-1} + kT^n \\ &\leq \cdots \leq E^0 + k\sum_{m=1}^{n} T^m. \end{aligned}$$

Using exact initial data for U_j^0 guarantees that $E^0 = 0$. Therefore,

$$E^{n+1} \leq nk \max_{1\leq m\leq n} T^n = t\mathcal{O}(k+h).$$

Since the right side tends to 0 as $h, k \to 0$, the scheme FB converges. ∎

Stability

While the CFL condition has a clear connection with the growth of errors in certain difference schemes, not all approximations to time-dependent PDEs admit such a compelling picture. We turn now to a more general view of stability and to a popular method for assessing it.

We treat the grid functions under consideration as being defined on a spatial grid $\Delta_x = \{x_j = jh : j = 0, \pm1, \pm2, \ldots\}$. The temporal grid is $\Delta_t = \{t_n = nk : n = 0, 1, 2, \ldots\}$. For any grid function $V: \Delta_x \to \mathbb{C}$, define the **discrete L^2 norm** $\|V\|_2$ as follows:

$$\|V\|_2^2 := h \sum_{j=-\infty}^{\infty} |V_j|^2.$$

A finite-difference scheme for a time-dependent PDE generates a sequence $\{U^n\}$ of spatial grid functions, starting with initial data in the form of a spatial grid function U^0. At issue is whether the data U^0 control the growth of $\{U^n\}$.

DEFINITION. *A finite-difference scheme for a time-dependent* PDE *is* **stable** *if, for any time $T > 0$, there exists a constant $K > 0$, independent of time step k and spatial mesh size h, such that, for any initial data U^0, the sequence $\{U^n\}$ generated by the scheme satisfies*

$$\|U^n\|_2 \leq K\|U^0\|_2,$$

whenever $0 \leq nk \leq T$.

(The constant K may depend upon T but not on the number of time steps taken to reach a fixed time $t \leq T$.)

To illustrate, let U_j^0 and $\hat{U}_j^0$ denote exact and erroneous initial data, respectively, with $\|U^0 - \hat{U}^0\|_2^2 = \epsilon$. Applying a stable, linear difference scheme to a well posed PDE, with $U_j^0 - \hat{U}_j^0$ as initial data, yields a sequence of spatial grid functions $U_j^n - \hat{U}_j^n$ that satisfy the inequality $\|U^n - \hat{U}^n\|_2^2 \leq K\epsilon$, for any time level $n \leq T/k$. In other words, the error depends continuously upon the initial error. This continuous dependence does not hold for the scheme FB with $C > 1$. On the contrary, the estimate (8.3-4) shows that the magnitude of the error at any fixed time level T may increase without bound as $k \to 0$, since $(2C-1)^n \to \infty$ as $n = T/k \to \infty$.

Testing for stability by applying the definition directly is difficult for most schemes. A technique proposed by von Neumann often makes the task easier. Roughly speaking, we decompose a hypothetical grid function error into harmonics and examine how the difference scheme propagates a typical harmonic as the timestepping progresses.

We use ideas from Fourier analysis. Consider first the case when $h = 1$ and $\Delta_x = \{\ldots, -2, -1, 0, 1, 2, \ldots\}$. Given a grid function $V: \Delta_x \to \mathbb{C}$, its **Fourier transform** is

$$(\mathcal{F}V)(\xi) := \frac{1}{\sqrt{2\pi}} \sum_{j=-\infty}^{\infty} e^{-ij\xi} V_j,$$

where $i = \sqrt{-1}$. Thus $\mathcal{F}V: [-\pi, \pi] \to \mathbb{C}$, and $(\mathcal{F}V)(-\pi) = (\mathcal{F}V)(\pi)$. The Fourier transform of a grid function V defined on Δ_x is the complex-valued function that has the nodal values V_j as its Fourier coefficients. Given the Fourier transform $\mathcal{F}V$, we recover the nodal values by computing its Fourier coefficients:

$$V_j = \frac{1}{\sqrt{2\pi}} \int_{-\pi}^{\pi} e^{ij\xi} (\mathcal{F}V)(\xi)\, d\xi.$$

This relationship is the **Fourier inversion formula**. By the Parseval identity and the fact that $h = 1$,

$$\|\mathcal{F}V\|_{L^2([-\pi,\pi])}^2 = \int_{-\pi}^{\pi} |(\mathcal{F}V)(\xi)|^2\, d\xi = \sum_{j=-\infty}^{\infty} |V_j|^2 = \|V\|_2^2. \tag{8.3-5}$$

(See Problem 4 for proof.)

When $h \neq 1$, we obtain the Fourier transform of a grid function $V: \Delta_x \to \mathbb{C}$ through a change of variables, from the interval $[-\pi, \pi]$ to $[-\pi/h, \pi/h]$. The nodal values

$$V_j = \frac{1}{\sqrt{2\pi}} \int_{-\pi/h}^{\pi/h} e^{ijh\xi} (\mathcal{F}V)(\xi)\, d\xi \tag{8.3-6}$$

are now Fourier coefficients of the function

$$(\mathcal{F}V)(\xi) = \frac{h}{\sqrt{2\pi}} \sum_{j=-\infty}^{\infty} e^{-ijh\xi} V_j. \tag{8.3-7}$$

In this more general setting, Equation (8.3-7) defines the Fourier transform, and Equation (8.3-6) serves as the corresponding Fourier inversion formula.

The Parseval identity now gives

$$\|\mathcal{F}V\|^2_{L^2([-\pi/h,\pi/h])} = \int_{-\pi/h}^{\pi/h} |(\mathcal{F}V)(\xi)|^2\,d\xi = h\sum_{j=-\infty}^{\infty} |V_j|^2 = \|V\|_2^2. \quad (8.3\text{-}8)$$

This identity is the key to von Neumann stability analysis. Suppose that a finite-difference scheme for a time-dependent PDE generates a sequence $\{U^n\}$ of grid functions defined on Δ_x. The scheme is stable if and only if, for any time $T > 0$, there exists a positive constant K, independent of time step k, such that, for all initial data U^0,

$$\|\mathcal{F}U^n\|_{L^2([-\pi/h,\pi/h])} \leq K\|\mathcal{F}U^0\|_{L^2([-\pi/h,\pi/h])},$$

whenever $0 \leq nk \leq T$.

This observation suggests that we translate stability analyses of finite-difference schemes to the Fourier transform domain. In this latter realm the analysis frequently reduces to algebra involving complex exponentials. To see how the procedure works, consider the scheme FB, which we write as $U_j^n = (1-C)U_j^{n-1} + CU_{j-1}^{n-1}$. This relationship and the Fourier inversion formula (8.3-6) allow us to rewrite the grid function U^n, yielding

$$U_j^n = \frac{1}{\sqrt{2\pi}}\int_{-\pi/h}^{\pi/h} e^{ijh\xi}(1 - C + Ce^{-ih\xi})(\mathcal{F}U^{n-1})(\xi)\,d\xi.$$

We also rewrite U^n by using the Fourier inversion formula directly:

$$U_j^n = \frac{1}{\sqrt{2\pi}}\int_{-\pi/h}^{\pi/h} e^{ijh\xi}(\mathcal{F}U^n)(\xi)\,d\xi.$$

It follows that

$$(\mathcal{F}U^n)(\xi) = \underbrace{(1 - C + Ce^{-ih\xi})}_{A(h\xi)}(\mathcal{F}U^{n-1})(\xi).$$

Repeated application of this relationship yields

$$(\mathcal{F}U^n)(\xi) = [A(h\xi)]^n(\mathcal{F}U^0)(\xi).$$

Hence,

$$\|\mathcal{F}U^n\|^2_{L^2([-\pi/h,\pi/h])} = \int_{-\pi/h}^{\pi/h} |A(h\xi)|^{2n}|\mathcal{F}U^0|^2\,d\xi.$$

We call the function $A(\theta) := 1 - C + Ce^{-i\theta}$ the **amplification factor** for FB, since the magnitude of this function determines the growth of the Fourier transform $\mathcal{F}U^n$ and hence of the grid function U^n as timestepping proceeds. To keep $\|U^n\|_2$ bounded, it suffices to demand that $|A(\theta)| \leq 1$ for all values of θ. For the scheme FB, the values of $A(\theta) = 1 - C + Ce^{-i\theta}$ lie on the circle of radius C centered at the point $1 - C$ in the complex plane, as drawn in Figure 3. We conclude that $|A(\theta)| \leq 1$ for all values of θ if and only if $|C| \leq 1$. This condition is identical to the constraint deduced from the CFL condition. Because the scheme FB is stable only for certain values of the Courant number $C = vk/h$, we say that the scheme is **conditionally stable**.

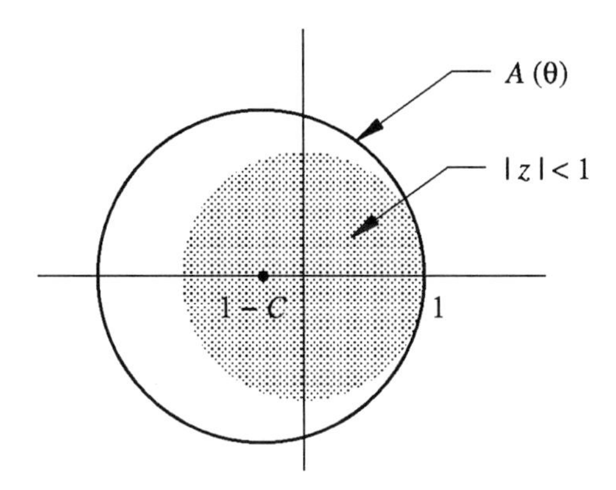

FIGURE 3. *Locus of the values of the amplification factor $A(\theta)$ for the difference scheme* FB.

To streamline the technique, observe that one obtains identical results simply by substituting a typical nonzero Fourier mode $\upsilon^n e^{ijh\xi}$ for U_j^n in the difference scheme. For example, consider the scheme BB ("backward in t, backward in x") for the advection equation:

$$\frac{U_j^n - U_j^{n-1}}{k} + v\frac{U_j^n - U_{j-1}^n}{h} = 0.$$

Replacing U_j^n by $\upsilon^n e^{ijh\xi}$ and rearranging, we obtain

$$\upsilon^n e^{ijh\xi} - \upsilon^{n-1} e^{ijh\xi} + \frac{vk}{h}\left(\upsilon^n e^{ijh\xi} - \upsilon^n e^{i(j-1)h\xi}\right) = 0.$$

Dividing through by $e^{ijh\xi}$, we find that

$$\left[1 + C\left(1 - e^{-ih\xi}\right)\right] \upsilon^n = \upsilon^{n-1}.$$

Therefore, $\upsilon^n = A(h\xi)\upsilon^{n-1}$, where the amplification factor is $A(\theta) = [1 + C(1 - e^{-i\theta})]^{-1}$. Since

$$\begin{aligned} |1 + C(1 - e^{-i\theta})| &\geq |1 + C| - |Ce^{-i\theta}| \\ &= 1 + C - C = 1, \end{aligned}$$

we conclude that $|A(\theta)| \leq 1$, for all values of θ, independent of the Courant number C. Therefore the scheme BB is **unconditionally stable**.

Problem 5 examines the issue of stability computationally, and Problems 6 and 7 call for von Neumann analyses of particular difference approximations. Table 8.1 summarizes properties of a variety of elementary difference approximations to the advection equation. In this table, the notation $C < \infty$ means that the scheme is unconditionally stable. The names of the schemes are mnemonic; for example, BC means "backward in t, centered in x."

Table 8.1: Properties of difference schemes for the advection equation with $v > 0$.

Scheme	Formula	Consistency	Stability
BB	$(\delta_t^- + v\delta_x^-)U_j^n = 0$	$\mathcal{O}(k+h)$	$C < \infty$
BC	$(\delta_t^- + v\delta_x)U_j^n = 0$	$\mathcal{O}(k+h^2)$	$C < \infty$
BF	$(\delta_t^- + v\delta_x^+)U_j^n = 0$	$\mathcal{O}(k+h)$	$C \geq 1$
CB	$(\delta_t + v\delta_x^-)U_j^n = 0$	$\mathcal{O}(k^2+h)$	unstable
CC	$(\delta_t + v\delta_x)U_j^n = 0$	$\mathcal{O}(k^2+h^2)$	$C \leq 1$
CF	$(\delta_t + v\delta_x^+)U_j^n = 0$	$\mathcal{O}(k^2+h)$	unstable
FB	$(\delta_t^+ + v\delta_x^-)U_j^n = 0$	$\mathcal{O}(k+h)$	$C \leq 1$
FC	$(\delta_t^+ + v\delta_x)U_j^n = 0$	$\mathcal{O}(k+h^2)$	unstable
FF	$(\delta_t^+ + v\delta_x^+)U_j^n = 0$	$\mathcal{O}(k+h)$	unstable

Sufficient Conditions for Convergence

For many finite-difference approximations it is difficult to establish convergence directly. By contrast, for consistency and stability we have techniques — Taylor series and Fourier analysis, respectively — that are often easy to apply. Fortunately, under rather general circumstances, one can deduce that a difference scheme is convergent by checking that it is both consistent and stable. This connection parallels results developed for ODEs in Section 7.4. We devote the remainder of this section to a brief introduction to the theory applicable to time-dependent PDEs. The discussion follows that presented by Thomée ([10], Section II.3).

We begin by establishing notation. Consider a pure IVP having the form

$$\begin{aligned} \frac{\partial u}{\partial t} &= \mathcal{L}u, \qquad (t,x) \in (0,T) \times \mathbb{R}, \\ u(0,x) &= f(x), \qquad x \in \mathbb{R}. \end{aligned} \tag{8.3-9}$$

Assume that the differential operator $\mathcal{L}$ is linear and that $\mathcal{L}$ and the initial data f have all the properties needed to ensure that a solution $u(t,x)$ exists, is unique, belongs to $C^\infty([0,T] \times \mathbb{R})$, and depends continuously upon the initial data.

The next step is to express finite-difference approximations to this problem in a somewhat abstract way that facilitates the analysis. Given grids $\Delta_t := \{0, k, 2k, \ldots\}$ and $\Delta_x := \{\ldots, -h, 0, h, 2h, \ldots\}$, denote by $\mathcal{V}$ the vector space containing all grid functions $V: \Delta_x \to \mathbb{C}$. A difference approximation to the problem (8.3-9) presumably generates a sequence $\{U^n\}$ of functions in $\mathcal{V}$ such that $U_j^n \simeq u_j^n := u(t_n, x_j)$. In assessing the accuracy of this approximation, we henceforth treat u as a grid function by restricting attention to arguments (t,x) that belong to the grid $\Delta_t \times \Delta_x$. With this convention, the issue at hand is whether the finite-difference scheme converges in the sense that $\|U^n - u^n\|_2 \to 0$ as $h, k \to 0$.

One can write a typical finite-difference approximation to the problem (8.3-9) as follows:

$$\mathsf{B}U^{n+1} = \mathsf{A}U^n, \qquad n = 0, 1, 2, \ldots. \tag{8.3-10}$$

Here, $\mathsf{A}, \mathsf{B}: \mathcal{V} \to \mathcal{V}$ are operators, depending upon the mesh sizes k and h, that assign to a given grid function another function in $\mathcal{V}$. For "reasonable" difference schemes, B is invertible, with $\|\mathsf{B}^{-1}V\|_2 \le B\|V\|_2$ for some constant $B > 0$ that is independent of k and h. This being the case, $U^{n+1} = \mathsf{E}U^n$, where $\mathsf{E} := \mathsf{B}^{-1}\mathsf{A}$. As a consequence, $U^n = \mathsf{E}^n U^0$. We assume in what follows that the numerical initial data are exact, that is, that $U_j^0 = f_j$ for $j = 0, \pm 1, \pm 2, \ldots$.

A couple of examples make this notation more concrete. Consider first the scheme FB for the advection equation (8.3-1):

$$U_j^{n+1} = (1 + C)U_j^n - CU_{j-1}^n,$$

where $C := vk/h$ is the Courant number. For this scheme,

$$\begin{aligned} (\mathsf{A}V)_j &= (1 + C)V_j - CV_{j-1}, \\ (\mathsf{B}V)_j &= V_j, \end{aligned}$$

for any $V \in \mathcal{V}$. Thus B is the identity operator on $\mathcal{V}$, and we may take $B = 1$. Now consider the scheme BB for the same equation:

$$(1 + C)U_j^{n+1} - CU_{j-1}^{n+1} = U_j^n.$$

Here,

$$\begin{aligned}(\mathsf{A}V)_j &= V_j,\\ (\mathsf{B}V)_j &= (1+C)V_j - CV_{j-1}.\end{aligned}$$

This operator B has an inverse defined by the equation

$$(\mathsf{B}^{-1}V)_j = \cdots + \frac{C^2}{(1+C)^3}V_{j-2} + \frac{C}{(1+C)^2}V_{j-1} + \frac{1}{1+C}V_j.$$

In this case, $\|\mathsf{B}^{-1}V\|_2 \le \|V\|_2$, and again we may take $B=1$.

One can construct the operators A and B so that the truncation error at time level n is the grid function

$$\tau_{k,h}^n := k^{-1}(\mathsf{B}u^n - \mathsf{A}u^{n-1}).$$

(By using Taylor expansions of the smooth solution $u^n(x) := u(t_n, x)$, check that the two difference approximations just mentioned have this form.) It follows that

$$u^n = \mathsf{E}u^{n-1} + u^n - \mathsf{E}u^{n-1} = \mathsf{E}u^{n-1} + k\mathsf{B}^{-1}\tau_{k,h}^n. \tag{8.3-11}$$

For positive numbers p, q, the finite-difference scheme (8.3-10) is consistent to order (p,q) if

$$\tau_{k,h}^n = \mathcal{O}(k^p + h^q) \quad \text{as} \quad k, h \to 0,$$

uniformly for $0 \le nk \le T$. As before, the scheme is stable if there exists a constant $K > 0$, independent of U^0, k and h, such that $\|U^n\|_2 \le K\|U^0\|_2$. We rewrite this condition as $\|\mathsf{E}^n U^0\|_2 \le K\|U^0\|_2$.

The following theorem establishes that consistency and stability together guarantee convergence:

THEOREM 8.7. *Suppose that the finite-difference scheme (8.3-10) is consistent to order (p,q) with the* PDE *(8.3-9), where $p, q > 0$, and that the scheme is stable. If we use exact initial data ($U^0 = f$ on Δ_x), then, as $h, k \to 0$,*

$$\|U^n - u^n\|_2 = \mathcal{O}(k^p + h^q), \tag{8.3-12}$$

for $0 \le nk \le T$.

PROOF: By subtracting the identity (8.3-11) from the difference equation $U^n = \mathsf{E}U^{n-1}$, we have

$$\begin{aligned}U^n - u^n &= \mathsf{E}(U^{n-1} - u^{n-1}) - k\mathsf{B}^{-1}\tau_{k,h}^n\\ &= \mathsf{E}^2(U^{n-2} - u^{n-2}) - k\mathsf{E}\mathsf{B}^{-1}\tau_{k,h}^{n-1} - k\mathsf{B}^{-1}\tau_{k,h}^n\\ &\;\;\vdots\\ &= \mathsf{E}^n(U^0 - u^0) - k\sum_{m=1}^{n} \mathsf{E}^{n-m}\mathsf{B}^{-1}\tau_{k,h}^m.\end{aligned}$$

Since $U^0 - u^0 = U^0 - f = 0$,

$$\begin{aligned} \|U^n - u^n\|_2 &\leq nk \max_{1\leq m\leq n} \|\mathsf{E}^{n-m}\mathsf{B}^{-1}\tau^m_{k,h}\|_2 \\ &\leq T \max_{1\leq m\leq n} K\|\mathsf{B}^{-1}\tau^m_{k,h}\|_2, \end{aligned}$$

the last step following from the hypothesis of stability. Using the boundedness of B^{-1} and the hypothesis of consistency, we conclude that

$$\begin{aligned} \|U^n - u^n\|_2 &\leq TKB \max_{1\leq m\leq n} \|\tau^m_{k,h}\|_2 \\ &= \mathcal{O}(k^p + h^q), \end{aligned}$$

completing the proof. ∎

Further Remarks

Regarding connections among consistency, stability, and convergence, Theorem 8.7 barely scratches the surface. To begin with, the theorem as proved treats only pure initial-value problems for homogeneous PDEs. Incorporating boundary-value approximations and inhomogeneous terms into the argument adds only minor complications; see Thomée ([10], Section II.3) and Isaacson and Keller ([5], Section 9.5) for details. Also, while the proof given above applies to equations that are first-order in time, the results extend to equations involving higher-order time derivatives by reductions in order. For example, the second-order wave equation

$$\frac{\partial^2 u}{\partial t^2} - c^2\frac{\partial^2 u}{\partial x^2} = 0$$

is equivalent to the first-order system

$$\frac{\partial}{\partial t}\begin{bmatrix} v \\ w \end{bmatrix} + c\frac{\partial}{\partial x}\begin{bmatrix} v \\ -w \end{bmatrix} = \begin{bmatrix} w \\ 0 \end{bmatrix}.$$

The extension of the theorem to PDEs in more than one spatial variable is straightforward.

Much deeper connections exist:

LAX EQUIVALENCE THEOREM: *Given a consistent finite-difference approximation to a well posed initial-value problem, stability is a necessary and sufficient condition for convergence.*

The proof that stability is necessary requires ideas from elementary functional analysis; we do not delve into the argument here. Richtmyer and Morton ([8], Section 3.5) give a detailed proof of this important fact.

8.4 Other Time-Dependent Equations

Using ideas developed in the previous section, one can construct and analyze finite-difference methods for other time-dependent PDEs. This section introduces the rudiments of such approximations for three important PDEs: the heat equation, an extension of the heat equation known as the advection-diffusion equation, and the wave equation. In discussing each equation, we start by reviewing some properties of the equation's solutions, deduced from the exact theory of PDEs. We then introduce common difference approximations and discuss the convergence of each, using Theorem 8.7.

Throughout this section, $\Delta_t := \{0, k, 2k, \ldots\}$ denotes the temporal grid, while $\Delta_x := \{\ldots, -h, 0, h, 2h, \ldots\}$ is the spatial grid.

The Heat Equation

For the heat equation, consider the following initial-value problem:

$$\begin{aligned} \frac{\partial u}{\partial t} - D\frac{\partial^2 u}{\partial x^2} &= 0, \qquad (t,x) \in (0,T] \times \mathbb{R}, \\ u(0,x) &= f(x), \qquad x \in \mathbb{R}. \end{aligned} \tag{8.4-1}$$

Here, D denotes a positive constant. This PDE serves as a simple analog of the energy balance for rigid bodies in which heat flows occur; in this context, $u(t,x)$ signifies the temperature of the body. The equation also arises in simple problems involving diffusion; in this context, $u(t,x)$ represents the concentration of some constituent in a mixture.

The problem (8.4-1) furnishes a convenient test for finite-difference approximations to more complicated heat-flow and diffusion problems, since it has a closed-form solution:

$$u(t,x) = \frac{1}{\sqrt{4\pi Dt}} \int_{-\infty}^{\infty} f(\xi) \exp\left[\frac{-(x-\xi)^2}{4Dt}\right] d\xi.$$

From this expression, one can deduce several qualitative properties of solutions to the heat equation. First, whenever the initial function f is square-integrable (that is, $f \in L^2(\mathbb{R})$), the solution is smooth in the sense that $u \in C^\infty((0,T] \times \mathbb{R})$. Thus rough initial data give rise to smooth solutions in infinitesimal time. Second, the solution obeys the following **maximum principle**:

$$\sup_{x\in\mathbb{R}} |u(t,x)| \leq \sup_{x\in\mathbb{R}} |f(x)|, \qquad \text{for} \quad t \in (0,T].$$

In particular, the initial data bound the behavior of the solution at later times, a property that we associate with the stability of the PDE. An analogous property figures prominently in convergence arguments for discrete approximations to the PDE.

Third, no finite speed characterizes the propagation of signals in the spatial domain. To see what this statement means, consider the response to a localized **point source** taking the form of a Dirac distribution, $f(x) = \delta(x)$. For this initial function,

$$u(t,x) = \frac{1}{\sqrt{4\pi Dt}} \exp\left(\frac{-x^2}{4Dt}\right),$$

for $t > 0$. For fixed $t > 0$, this function has the graph shown in Figure 1. The figure illustrates that information from highly localized initial data propagates throughout the spatial domain in infinitesimal time. In this sense, the heat equation is characterized by infinite propagation speed.

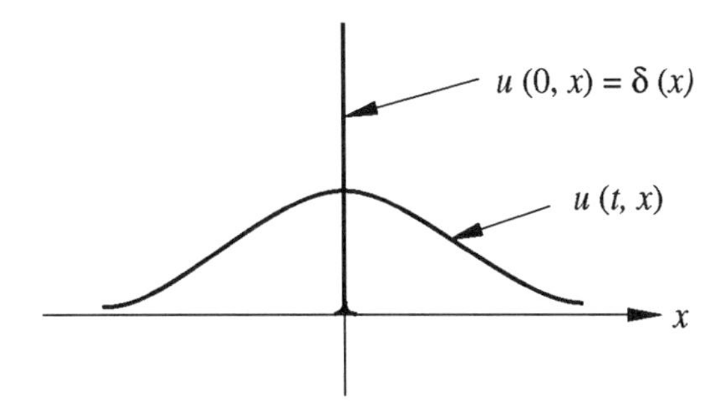

FIGURE 1. *Initial Dirac distribution and subsequent solution to the heat equation at $t > 0$.*

We examine five finite-difference schemes for the problem (8.4-1). The first is the **classic explicit scheme** (FC):

$$\delta_t^+ U_j^n - D\,\delta_x^2 U_j^n = 0. \tag{8.4-2}$$

By applying the definitions of the difference operators and rearranging, we reduce this scheme to the following:

$$U_j^{n+1} = (1 - 2\Gamma D)U_j^n + \Gamma D(U_{j+1}^n + U_{j-1}^n),$$

where $\Gamma := k/h^2$ is the **grid ratio**. Figure 2 shows the stencil for this scheme.

To check consistency, we use Taylor series to compute the truncation error: For any smooth solution $u(t,x)$,

$$\left(\frac{\partial}{\partial t} - \delta_t^+\right) u(t,x) = \mathcal{O}(k); \qquad \left(\frac{\partial^2}{\partial x^2} - \delta_x^2\right) u(t,x) = \mathcal{O}(h^2).$$

Therefore, the classic explicit scheme is consistent with truncation error $\tau_{k,h}(t,x) = \mathcal{O}(k + h^2)$.

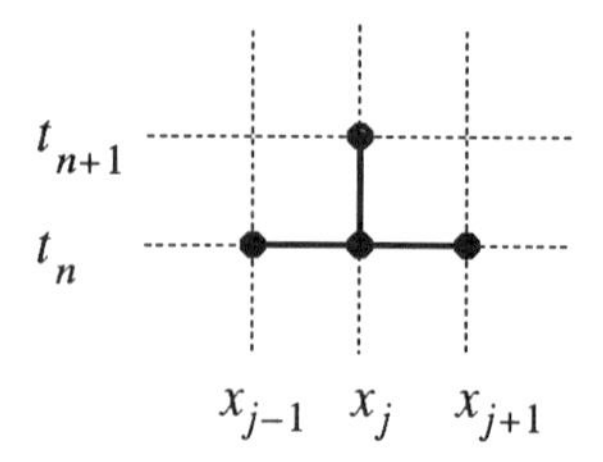

FIGURE 2. *Stencil for the classic explicit approximation to the heat equation.*

To assess the scheme's stability, we employ the von Neumann technique. Substituting $v^n e^{ij\theta}$ for U_j^n in the difference equation, we find that

$$\frac{1}{k}\left(v^{n+1}e^{ij\theta} - v^n e^{ij\theta}\right) = \frac{D}{h^2}\left(v^n e^{i(j+1)\theta} - 2v^n e^{ij\theta} + v^n e^{i(j-1)\theta}\right).$$

Therefore,

$$\begin{aligned} v^{n+1} &= \left[1 + \frac{Dk}{h^2}\left(e^{i\theta} + e^{-i\theta} - 2\right)\right] v^n \\ &= \left[1 + \frac{2Dk}{h^2}(\cos\theta - 1)\right] v^n \\ &= \left(1 - 4D\Gamma\sin^2\frac{\theta}{2}\right) v^n, \end{aligned}$$

so the amplification factor is $A(\theta) = 1 - 4D\Gamma\sin^2(\theta/2)$. Since $|A(\theta)| \leq 1$ for all values of θ precisely when $D\Gamma \leq 1/2$, the classic explicit scheme is stable if

$$k \leq \frac{h^2}{2D}. \tag{8.4-3}$$

This conditional stability parallels that of the explicit scheme FB for the advection equation. However, in the present case the constraint is more severe, since the largest allowable time step $h^2/(2D)$ shrinks more rapidly as we reduce the spatial mesh size h.

In light of these results, Theorem 8.7 allows us to conclude that the scheme (8.4-2) is convergent provided that it meets the stability condition (8.4-3). For this scheme, there is also an elementary direct proof of convergence:

THEOREM 8.8. *If $D\Gamma \leq \frac{1}{2}$ and we use exact initial data ($U_j^0 = f_j$), then the finite-difference scheme (8.4-2) for the initial-value problem (8.4-1) is convergent.*

PROOF: Call $\varepsilon_j^n := u_j^n - U_j^n$. The equations

$$\begin{aligned} u_j^{n+1} &= D\Gamma u_{j-1}^n + (1 - 2D\Gamma)u_j^n + D\Gamma u_{j+1}^n + k\tau_{k,h}(t_n, x_j), \\ U_j^{n+1} &= D\Gamma U_{j-1}^n + (1 - 2D\Gamma)U_j^n + D\Gamma U_{j+1}^n, \end{aligned}$$

imply that

$$\varepsilon_j^{n+1} = \underbrace{D\Gamma}_{\alpha_1} \varepsilon_{j-1}^n + \underbrace{(1 - 2D\Gamma)}_{\alpha_2} \varepsilon_j^n + \underbrace{D\Gamma}_{\alpha_3} \varepsilon_{j+1}^n + \mathcal{O}(k^2 + kh^2).$$

Now, $\alpha_1 + \alpha_2 + \alpha_3 = 1$, and the hypothesis $D\Gamma \le \frac{1}{2}$ guarantees that each of the coefficients α_m is nonnegative. Therefore,

$$\begin{aligned} |\varepsilon_j^{n+1}| &\le \alpha_1 |\varepsilon_{j-1}^n| + \alpha_2 |\varepsilon_j^n| + \alpha_3 |\varepsilon_{j+1}^n| + \mathcal{O}(k^2 + kh^2) \\ &\le \max_j |\varepsilon_j^n| + \mathcal{O}(k^2 + kh^2). \end{aligned}$$

From this inequality we deduce an analog of the maximum principle for the heat equation:

$$\begin{aligned} \max_j |\varepsilon_j^{n+1}| &\le \max_j |\varepsilon_j^n| + \mathcal{O}(k^2 + kh^2) \\ &\le \max_j |\varepsilon_j^{n-1}| + 2\mathcal{O}(k^2 + kh^2) \\ &\vdots \\ &\le \max_j |\varepsilon_j^0| + (n+1)\mathcal{O}(k^2 + kh^2). \end{aligned}$$

But $\max_j |\varepsilon_j^0| = 0$, and $n + 1 = t_{n+1}/k$, so $\max_j |\varepsilon_j^{n+1}| \le t_{n+1}\mathcal{O}(k + h^2)$. ∎

With minor modifications, the same argument establishes the convergence of the classic explicit scheme applied to the initial-boundary-value problem,

$$\begin{aligned} \frac{\partial u}{\partial t} - D\frac{\partial^2 u}{\partial x^2} &= 0, && (t, x) \in (0, T] \times (0, 1), \\ u(0, x) &= f(x), && x \in (0, 1), \\ u(t, 0) = u(t, 1) &= 0, && t \in (0, T]. \end{aligned} \tag{8.4-4}$$

The second finite-difference scheme for the heat equation is the **fully implicit scheme** (BC):

$$\delta_t^- U_j^{n+1} - D\,\delta_x^2 U_j^{n+1} = 0, \tag{8.4-5}$$

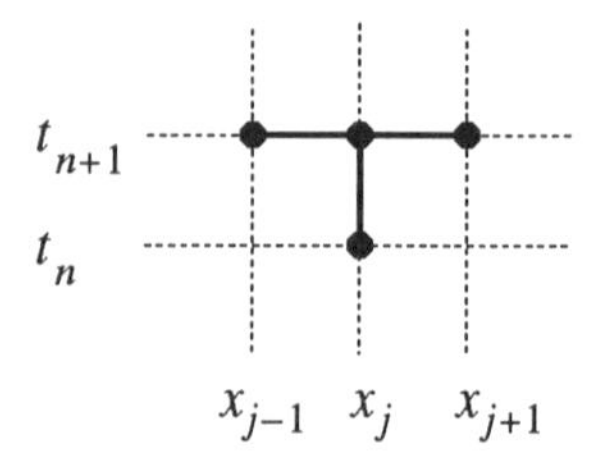

FIGURE 3. *Stencil for the fully implicit approximation to the heat equation.*

which we rewrite as follows:

$$-D\Gamma U_{j-1}^{n+1} + (1 + 2D\Gamma)U_j^{n+1} - D\Gamma U_{j+1}^{n+1} = U_j^n.$$

Figure 3 shows the stencil for this scheme.

Taylor analysis confirms that BC is consistent, having truncation error $\tau_{k,h}(t, x) = \mathcal{O}(k + h^2)$. The von Neumann stability analysis proceeds as follows: Substituting $v^n e^{ij\theta}$ into Equation (8.4-5) and dividing through by $e^{ij\theta}$ yields

$$\frac{v^{n+1} - v^n}{k} = D\frac{v^{n+1}e^{i\theta} - 2v^{n+1} + v^{n+1}e^{-i\theta}}{h^2},$$

that is,

$$v^{n+1} = v^n + \frac{Dk}{h^2}v^{n+1}(2\cos\theta - 2).$$

But $\cos\theta - 1 = -2\sin^2(\theta/2)$, so

$$v^{n+1} = \frac{v^n}{1 + 4D\Gamma\sin^2(\theta/2)}.$$

The amplification factor is $A(\theta) = [1+4D\Gamma\sin^2(\theta/2)]^{-1}$, and thus $|A(\theta)| \leq 1$, independent of θ and Γ. Therefore the fully implicit scheme is unconditionally stable.

We conclude from Theorem 8.7 that the scheme is convergent. An argument analogous to that given for Theorem 8.8 shows that $\max_j |u_j^n - U_j^n| = \mathcal{O}(k + h^2)$.

While unconditional stability is much preferable to the constraint $D\Gamma \leq \frac{1}{2}$ that afflicts the classic explicit scheme, there is a cost associated with the benefit. Namely, one must accommodate the coupling of unknowns associated with time level $n + 1$. In the context of initial-boundary-value problems such as (8.4-4), this coupling leads to the linear systems

$$\begin{bmatrix} 1+2D\Gamma & -D\Gamma & & \\ -D\Gamma & 1+2D\Gamma & -D\Gamma & \\ & & \ddots & \\ & & -D\Gamma & 1+2D\Gamma \end{bmatrix} \begin{bmatrix} U_1 \\ U_2 \\ \vdots \\ U_J \end{bmatrix}^{n+1} = \begin{bmatrix} U_1 \\ U_2 \\ \vdots \\ U_J \end{bmatrix}^{n},$$

to be solved at each time level.

The tridiagonal matrix in this system is symmetric and positive definite, and hence the system has a unique solution at each time level. The need to solve this sequence of tridiagonal systems is not especially onerous from a computational viewpoint. However, in two or three spatial dimensions the analogous implicit schemes generate sparse matrices having larger bandwidth, and efficient solution algorithms become an important practical consideration.

The motivation for the third scheme for the heat equation lies in the observation that, for both the classic explicit and fully implicit schemes, the truncation error is $\mathcal{O}(k+h^2)$. Consequently, reducing the spatial mesh size h improves the accuracy of the schemes faster than reducing the time step k. One can rectify this imbalance by using an averaging approach to center the approximation to $\partial^2 u/\partial x^2$ in time:

$$\delta_t^- U_j^{n+1} - \frac{D}{2}\left(\delta_x^2 U_j^{n+1} + \delta_x^2 U_j^n\right) = 0.$$

This is the **Crank-Nicolson scheme**. Figure 4 shows the stencil for this approximation.

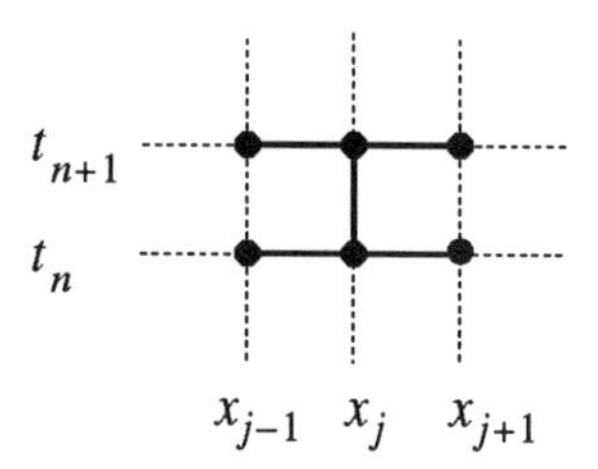

FIGURE 4. *Stencil for the Crank-Nicolson approximation to the heat equation.*

To analyze the consistency of the Crank-Nicolson approximation, we examine Taylor expansions centered at the time $t_{n+1/2} := (n+\frac{1}{2})k$:

$$\delta_t^- u_j^{n+1} = \frac{u_j^{n+1} - u_j^n}{2(k/2)} = \frac{\partial u}{\partial t}(t_{n+1/2}, x_j) + \mathcal{O}(k^2).$$

Also,

$$\begin{aligned}\frac{1}{2}\left(\delta_x^2 u_j^{n+1} + \delta_x^2 u_j^n\right) &= \frac{1}{2}\left[\frac{\partial^2 u}{\partial x^2}(t_{n+1}, x_j) + \frac{\partial^2 u}{\partial x^2}(t_n, x_j) + \mathcal{O}(h^2)\right] \\ &= \frac{\partial^2 u}{\partial x^2}(t_{n+1/2}, x_j) + \mathcal{O}(h^2 + k^2).\end{aligned}$$

Therefore, the scheme has truncation error $\mathcal{O}(k^2 + h^2)$; the temporal approximation is accurate to the same order as the spatial approximation.

We assess the stability of the Crank-Nicolson scheme by thinking of the method in two stages, each having time step $k/2$. The first stage has the form of an explicit scheme:

$$\frac{U_j^{n+1/2} - U_j^n}{k/2} = D\,\delta_x^2 U_j^n.$$

By analogy with the classic explicit scheme, this stage has amplification factor

$$A_1(\theta) = 1 - 4D\frac{\Gamma}{2}\sin^2\frac{\theta}{2}.$$

The second stage has the same form as the fully implicit scheme:

$$\frac{U_j^{n+1} - U_j^{n+1/2}}{k/2} = D\,\delta_x^2 U_j^{n+1}.$$

This stage has amplification factor

$$A_2(\theta) = \frac{1}{1 + 4D(\Gamma/2)\sin^2(\theta/2)}.$$

Taken together, these two stages are equivalent to one step of the Crank-Nicolson scheme, in the sense that one can recover the Crank-Nicolson formula by eliminating $U_j^{n+1/2}$. The combined scheme has amplification factor

$$A(\theta) = A_1(\theta)A_2(\theta) = \frac{1 - 2D\Gamma\sin^2(\theta/2)}{1 + 2D\Gamma\sin^2(\theta/2)}.$$

Since $|A(\theta)| \leq 1$ for all values of θ and Γ, the Crank-Nicolson scheme for the heat equation is unconditionally stable.

The remaining schemes that we discuss for the heat equation are mainly of pedagogical value. The fourth scheme is the **leapfrog scheme** (CC):

$$\delta_t U_j^n - D\,\delta_x^2 U_j^n = 0.$$

Figure 5 shows the stencil for this scheme. A Taylor analysis simpler than the one used for the Crank-Nicolson scheme shows that the leapfrog scheme has

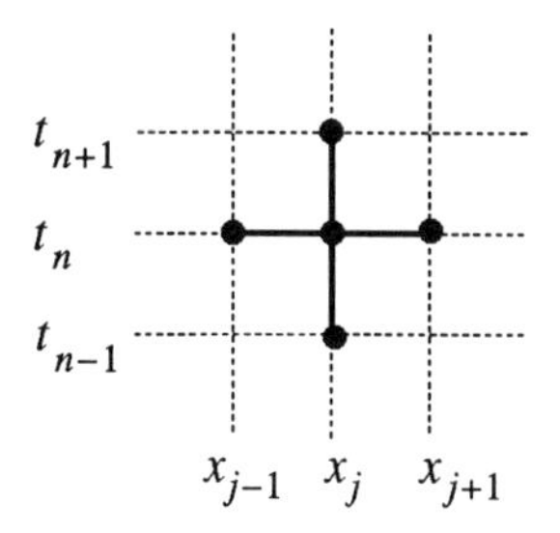

FIGURE 5. *Stencil for the leapfrog and DuFort-Frankel approximations to the heat equation.*

truncation error $\mathcal{O}(k^2 + h^2)$. However, as Problem 8 indicates, the scheme is unstable and therefore useless.

The fifth scheme, the **DuFort-Frankel scheme**, is more curious:

$$\delta_t U_j^n - \frac{D}{h^2}\left[U_{j+1}^n - (U_j^{n+1} + U_j^{n-1}) + U_{j-1}^n\right] = 0.$$

Its stencil appears in Figure 5. We analyze the truncation error for this scheme by substituting Taylor series into the expression

$$\begin{aligned}\tau_{k,h}(t_n, x_j) &= \left[\frac{\partial u}{\partial t}(t_n, x_j) - \delta_t u_j^n\right] \\ &\quad -D\left[\frac{\partial^2 u}{\partial x^2}(t_n, x_j) - \frac{u_{j+1}^n - (u_j^{n+1} + u_j^{n-1}) + u_{j-1}^n}{h^2}\right].\end{aligned}$$

After some manipulation, we find that

$$\begin{aligned}\tau_{k,h}(t_n, x_j) &= 2\left(\frac{k}{h}\right)^2 \frac{\partial^2 u}{\partial t^2}(t_n, x_j) + \frac{k^2}{3}\frac{\partial^3 u}{\partial t^3}(t_n, x_j) \\ &\quad - \frac{h^2}{12}\frac{\partial^4 u}{\partial x^4}(t_n, x_j) + \mathcal{O}(h^3 + k^3).\end{aligned}$$

In this case, $\tau_{k,h} \to 0$ as $h, k \to 0$, provided that $k/h \to 0$.

However, if we shrink k and h in such a manner that $k/h \to c^2$ for some constant c, then the DuFort-Frankel scheme is consistent with the PDE

$$\frac{\partial u}{\partial t} + c^2 \frac{\partial^2 u}{\partial t^2} = D\frac{\partial^2 u}{\partial x^2}.$$

This PDE is a version of the wave equation that includes damping effects; it is hyperbolic, not parabolic. Therefore the DuFort-Frankel scheme is **conditionally consistent**. Problem 8 asks for proof that the scheme is unconditionally stable.

Table 8.2 summarizes properties of finite-difference approximations to the heat equation.

Table 8.2: Properties of difference schemes for the heat equation.

Scheme	Consistency	Stability
Classic explicit (BB)	$\mathcal{O}(k+h^2)$	$k \leq \frac{1}{2}h^2/D$
Fully implicit (BC)	$\mathcal{O}(k+h^2)$	$k < \infty$
Crank-Nicolson	$\mathcal{O}(k^2+h^2)$	$k < \infty$
Leapfrog (CC)	$\mathcal{O}(k^2+h^2)$	unstable
DuFort-Frankel	conditional	$k < \infty$

The Advection-Diffusion Equation

We turn now to an unruly relative of the heat equation, the **advection-diffusion equation**:

$$\frac{\partial u}{\partial t} + v\frac{\partial u}{\partial x} - D\frac{\partial^2 u}{\partial x^2} = 0, \qquad (t,x) \in (0,T) \times (0,L). \tag{8.4-6}$$

We assume that the velocity v and the diffusion coefficient D are both positive constants. This equation arises in mass and heat transfer problems where both advection and diffusion or conduction contribute to the transport.

Although Equation (8.4-6) is parabolic, it exhibits schizophrenic behavior when advection dominates diffusion. In this regime, the equation tends to have solutions in which sharp fronts persist, much as they do for the advection equation studied in Section 8.3. Still, the parabolic nature of the equation ensures that the solution is smooth and obeys a maximum principle,

$$\max_{x\in[0,L]} |u(t_1,x)| \leq \max_{x\in[0,L]} |u(t_2,x)|, \qquad t_1 \leq t_2.$$

Numerical approximations to the PDE tend to suffer pathologic behavior, with the front-preserving and smoothing tendencies competing for prominence. We focus our brief discussion of the advection-diffusion equation on this pathology.

We begin by casting the PDE into a dimensionless form that helps quantify the relative strengths of advection and diffusion. Define dimensionless space and time variables by $\xi := x/L$ and $\vartheta := vt/L$, respectively. The chain rule permits us to convert Equation (8.4-6) to a PDE involving these new independent variables: Since $\partial/\partial t = (v/L)\partial/\partial\vartheta$ and $\partial/\partial x = (1/L)\partial/\partial\xi$,

$$\frac{\partial u}{\partial \vartheta} + \frac{\partial u}{\partial \xi} - P^{-1}\frac{\partial^2 u}{\partial \xi^2} = 0, \qquad (\vartheta,\xi) \in (0, vT/L) \times (0,1).$$

The dimensionless constant $P := vL/D$ is the **Peclet number**; its magnitude indicates the degree to which advection dominates diffusion. Peclet numbers larger than about 100 indicate highly advection-dominated regimes.

Henceforth we consider the following dimensionless problem:

$$\begin{aligned} \frac{\partial u}{\partial t} + \frac{\partial u}{\partial x} - P^{-1}\frac{\partial^2 u}{\partial x^2} &= 0, \qquad (t,x) \in (0,T)\times(0,1), \\ u(0,x) &= 0, \qquad x \in (0,1), \\ u(t,0) &= 1, \qquad t \in (0,T), \\ \frac{\partial u}{\partial x}(t,1) &= 0, \qquad t \in (0,T). \end{aligned} \tag{8.4-7}$$

We examine two finite-difference approximations.

First, consider the explicit, centered-in-space (FC) scheme:

$$\delta_t^+ U_j^n + \delta_x U_j^n - P^{-1}\delta_x^2 U_j^n = 0. \tag{8.4-8}$$

The truncation error for this scheme is $\mathcal{O}(k+h^2)$, so the approximation is consistent. To assess stability, substitute $v^n e^{ij\theta}$ for U_j^n in Equation (8.4-8) and divide through by $e^{ij\theta}$ to get

$$\frac{v^{n+1}-v^n}{k} + \frac{e^{i\theta}-e^{-i\theta}}{2h}v^n - P^{-1}\frac{e^{i\theta}-2+e^{-i\theta}}{h^2}v^n = 0.$$

Rearrangement yields

$$v^{n+1} = \left[1 - 4\Gamma P^{-1}\sin^2(\theta/2) - i(k/h)\sin\theta\right] v^n,$$

where $\Gamma := k/h^2$ as before. Therefore, the amplification factor satisfies the following equation:

$$|A(\theta)|^2 = \underbrace{\left[1 - 4\Gamma P^{-1}\sin^2(\theta/2)\right]^2}_{(\text{I})} + \underbrace{k\Gamma\sin^2\theta}_{(\text{II})}.$$

For the first term on the right we have $(\text{I}) \le 1$ provided that $\Gamma P^{-1} \le 1/2$, paralleling the stability analysis for the classic explicit approximation to the heat equation. For the term arising from advection, we have $(\text{II}) = \mathcal{O}(k^2)$ as $k \to 0$. Hence $|A(\theta)| \le 1 + Mk$ for some positive constant M, provided that $\Gamma P^{-1} \le 1/2$.

How does the $\mathcal{O}(k)$ perturbation to the amplification factor affect stability? The following proposition asserts that the perturbation does not disrupt stability.

PROPOSITION 8.9. *The approximation (8.4-8) to the advection-diffusion equation is stable when $|A(\theta)| \le 1 + Mk$ for some positive constant M.*

PROOF: We sketch the argument here, leaving details for Problem 10. The Fourier transform of the grid function U^n obeys the inequality

$$\|\mathcal{F}U^n\|_{L^2}^2 \leq (1+Mk)^{2n}\|\mathcal{F}U^0\|_{L^2}^2.$$

Hence,

$$\|U^n\|_2 \leq (1+Mk)^{T/k}\|U^0\|_2 \leq e^{MT}\|U^0\|_2,$$

from which stability follows. ∎

Therefore, the scheme (8.4-8) is convergent if $\Gamma P^{-1} \leq \frac{1}{2}$, that is, if $k \leq Ph^2/2$.

Even though FC converges for the advection-diffusion equation, subject to a time-step restriction, it often yields approximate solutions that differ qualitatively from corresponding exact solutions. In particular, the numerical solutions do not always satisfy a discrete version of the maximum principle:

THEOREM 8.10. *Assume that $\Gamma P^{-1} \leq \frac{1}{2}$. All solutions to the finite-difference scheme (8.4-8) satisfy the inequality*

$$\max_j |U_j^{n+1}| \leq \max_j |U_j^n| \tag{8.4-9}$$

if and only if $h \leq 2/P$.

PROOF: First assume that $h \leq 2/P$. According to Equation (8.4-8),

$$\begin{aligned} U_j^{n+1} &= \Gamma P^{-1}\left(1+\tfrac{1}{2}hP\right)U_{j-1}^n + \left(1-2\Gamma P^{-1}\right)U_j^n \\ &\quad +\Gamma P^{-1}\left(1-\tfrac{1}{2}hP\right)U_{j+1}^n. \end{aligned}$$

The inequalities $h \leq 2/P$ and $\Gamma P^{-1} \leq 1/2$ imply that

$$\begin{aligned} |U_j^{n+1}| &\leq \Gamma P^{-1}(1+\tfrac{1}{2}hP)|U_{j-1}^n| + (1-2\Gamma P^{-1})|U_j^n| \\ &\quad +\Gamma P^{-1}(1-\tfrac{1}{2}hP)|U_{j+1}^n| \\ &\leq (\Gamma P^{-1}+1-2\Gamma P^{-1}+\Gamma P^{-1})\max_j |U_j^n| = \max_j |U_j^n|, \end{aligned}$$

establishing the inequality (8.4-9).

For the converse, we prove the contrapositive. Assume that $h > 2/P$, and consider the numerical initial data

$$U_j^0 = \begin{cases} 1, & \text{if } j = 0, 1, \\ 0, & \text{if } j > 1, \end{cases}$$

drawn in Figure 6. Clearly, $\max_j |U_j^0| = 1$. After one time step,

$$\begin{aligned} U_1^1 &= \Gamma P^{-1}\left(1+\tfrac{1}{2}hP\right) + 1 - 2\Gamma P^{-1} \\ &= 1 + \left(\tfrac{1}{2}hP - 1\right)\Gamma P^{-1} > 1. \end{aligned}$$

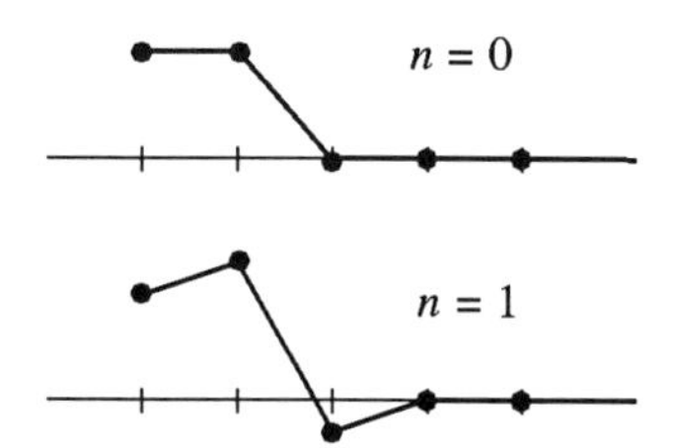

FIGURE 6. *Initial grid function and subsequent solution to the scheme* FC, *violating the discrete version of the maximum principle for the advection-diffusion equation.*

In this case the inequality (8.4-9) fails, as Figure 6 illustrates. ∎

Theorem 8.10 has unpleasant consequences. The condition $h \leq 2/P$ demands that we use extremely fine grids — which require a great deal of computation per time step — when the Peclet number is large. If we violate the condition, then the numerical solution typically exhibits physically spurious wiggles near sharp fronts. Figure 7 shows such a numerical solution for the initial-boundary-value problem (8.4-7), together with the physically realistic exact solution.

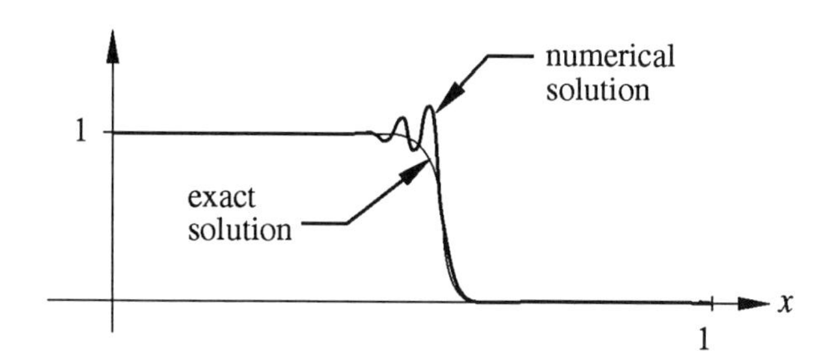

FIGURE 7. *Numerical solution to the advection-diffusion equation generated by the scheme* FC, *showing spurious wiggles near the sharp front.*

There are finite-difference schemes that respect the maximum principle under less restrictive conditions. Perhaps the simplest is the following:

$$\delta_t^+ U_j^n + \delta_x^- U_j^n - P^{-1} \delta_x^2 U_j^n = 0.$$

We denote this scheme as FB; it is explicit in time and uses an approximation to $\partial u / \partial x$ that is **upstream weighted**, in the sense that the difference operator is asymmetric with a bias toward nodes (t_n, x_{j-1}) lying upstream of the central node (t_n, x_j). We can rewrite this scheme in the form

$$
\begin{aligned}
U_j^{n+1} &= \Gamma P^{-1}(1+hP)U_{j-1}^n + \left[1 - 2\Gamma P^{-1}\left(1+\tfrac{1}{2}hP\right)\right] U_j^n \\
&\quad + \Gamma P^{-1} U_{j+1}^n .
\end{aligned}
\tag{8.4-10}
$$

Using the standard methods, one can check that the truncation error for this scheme is $\mathcal{O}(k+h)$ and that the scheme is stable if $\Gamma P^{-1} \leq \frac{1}{2}$, as before. Moreover, one can enforce the maximum principle:

THEOREM 8.11. *Assume that $\Gamma P^{-1} \leq \frac{1}{2}$. The scheme (8.4-10) for the satisfies the maximum principle (8.4-9) if*

$$2\Gamma P^{-1} + C \leq 1, \tag{8.4-11}$$

where $C = k/h$.

PROOF: This is an exercise. ∎

For advection-dominated problems, in which P is very large, the condition (8.4-11) is barely more restrictive than the stability condition.

Figure 8 shows a numerical solution to the advection-diffusion equation generated using FB. As the graph illustrates, there are no spurious wiggles. Instead, the difference scheme smears the sharp front. This smearing mimics the effect of an enhanced diffusion coefficient. We can interpret the smearing by examining the truncation error more closely. Explicitly retaining the terms from the Taylor series that are $\mathcal{O}(h)$, we find that

$$
\begin{aligned}
\delta_t^+ U_j^n &= \frac{\partial u}{\partial t}(t_n, x_j) + \mathcal{O}(k), \\
\delta_x^- U_j^n &= \frac{\partial u}{\partial x}(t_n, x_j) - \frac{h}{2}\frac{\partial^2 u}{\partial x^2}(t_n, x_j) + \mathcal{O}(h^2), \\
\delta_x^2 U_j^n &= \frac{\partial^2 u}{\partial x^2}(t_n, x_j) + \mathcal{O}(h^2).
\end{aligned}
$$

Therefore, up to terms that are $\mathcal{O}(k+h^2)$, the scheme FB is an approximation to the PDE

$$\frac{\partial u}{\partial t} + \frac{\partial u}{\partial x} - \left(\frac{1}{P} + \frac{h}{2}\right)\frac{\partial^2 u}{\partial x^2} = 0.$$

Thus physical diffusion P^{-1} is augmented by a **numerical diffusion** term $h/2$, which vanishes as $h \to 0$ but artificially smears the numerical solution on any realistic spatial grid. Problem 9 calls for an investigation of an implicit scheme that produces similar effects.

The choice between spurious wiggles and numerical diffusion arises in most numerical methods for the advection-diffusion equation, and the literature on this issue is large and clamorous. For two points of view on the dilemma, consult Gresho and Lee [4] and Allen [1]

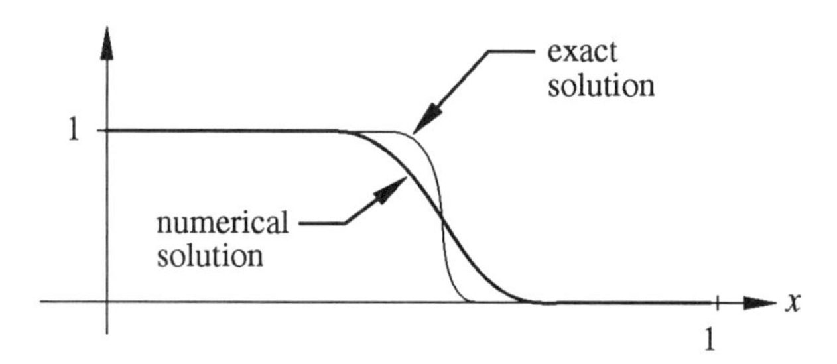

FIGURE 8. *Numerical solution to the advection-diffusion equation generated using the scheme* FB, *showing smearing near the sharp front due to numerical diffusion.*

The Wave Equation

Finally, consider the wave equation, for which the model IVP is

$$\begin{aligned} \frac{\partial^2 u}{\partial t^2} - c^2 \frac{\partial^2 u}{\partial x^2} &= 0, && (t, x) \in (0, T) \times \mathbb{R}, \\ u(0, x) &= f(x), && x \in \mathbb{R}, \\ \frac{\partial u}{\partial t}(0, x) &= g(x), && x \in \mathbb{R}. \end{aligned}$$

This problem has a closed-form solution given by d'Alembert's formula:

$$u(t, x) = \frac{1}{2}[f(x - ct) + f(x + ct)] + \frac{1}{2c} \int_{x-ct}^{x+ct} g(\xi)\, d\xi.$$

By the fundamental theorem of calculus, d'Alembert's formula is equivalent to an equation of the form

$$u(t, x) = \underbrace{F(x - ct)}_{\text{right-running}} + \underbrace{G(x + ct)}_{\text{left-running}}.$$

The component $F(x-ct)$ has a graph whose shape remains constant along the characteristic curves $x-ct =$ constant of the PDE, undergoing pure translation to the right with speed c. Similarly, the component $G(x + ct)$ has a graph whose shape remains constant along the characteristics $x + ct =$ constant, undergoing pure translation to the left with speed c. Figure 9 illustrates these left-running and right-running components. Thus the characteristic curves for this hyperbolic equation play much the same role as those for the advection equation discussed in Section 8.3.

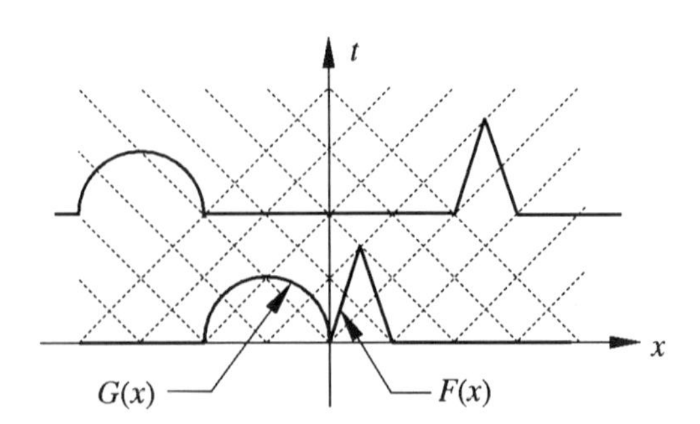

FIGURE 9. *Translation along characteristic curves of left-running and right-running components to a solution of the wave equation.*

Perhaps the most commonly used finite-difference scheme for the wave equation is the **explicit scheme** (CC):

$$\delta_t^2 U_j^n - c^2 \delta_x^2 U_j^n = 0, \tag{8.4-12}$$

or

$$\frac{U_j^{n+1} - 2U_j^n + U_j^{n-1}}{k^2} - c^2 \frac{U_{j+1}^n - 2U_j^n + U_{j-1}^n}{h^2} = 0.$$

This approximation has truncation error $\mathcal{O}(k^2 + h^2)$, so it is consistent.

To assess its stability, substitute $v^n e^{ij\theta}$ for U_j^n in Equation (8.4-12). One obtains

$$v^{n+1} - 2v^n + v^{n-1} = \frac{c^2 k^2}{h^2} \left(e^{i\theta} - 2 + e^{-i\theta}\right) v^n = -4C^2 \sin^2(\theta/2) v^n,$$

where $C := ck/h$. Since $v^n = A(\theta) v^{n-1}$ by definition of the amplification factor,

$$\left[A(\theta) - 2 + \frac{1}{A(\theta)}\right] v^n = -4C^2 \sin^2(\theta/2) A(\theta) v^{n-1},$$

or

$$v^n = \frac{-4C^2 \sin^2(\theta/2) A(\theta)}{A(\theta) - 2 + [A(\theta)]^{-1}} v^{n-1}.$$

Upon equating $A(\theta)$ with the factor multiplying v^{n-1}, rearranging, and taking square roots, we deduce that

$$A(\theta) - 1 = \pm 2iC \sin(\theta/2) \sqrt{A(\theta)}.$$

This relationship furnishes a pair of quadratic equations for $\sqrt{A(\theta)}$, which we can solve to find $A(\theta)$:

$$A(\theta) = \left[\sqrt{1 - C^2 \sin^2(\theta/2)} \pm iC \sin(\theta/2)\right]^2.$$

Observe that $|A(\theta)| \leq 1$ for arbitrary values of θ if and only if $C \leq 1$. Therefore, the scheme (8.4-12) is conditionally stable, having a stability constraint analogous to the CFL condition for the scheme FB applied to the advection equation.

The scheme (8.4-12) has a peculiar feature: We cannot apply it at the initial time step, when $n = 0$, since the values U_j^{-1} are not available. One way to circumvent this problem is to examine Taylor expansions about $t = 0$:

$$\begin{aligned} u(k, x_j) &= u(0, x_j) + k\frac{\partial u}{\partial t}(0, x_j) + \frac{k^2}{2}\frac{\partial^2 u}{\partial t^2}(0, x_j) + \mathcal{O}(k^3) \\ &= f(x_j) + kg(x_j) + \frac{c^2k^2}{2}\frac{\partial^2 u}{\partial x^2}(0, x_j) + \mathcal{O}(k^3). \end{aligned}$$

Replacing $\partial^2 u/\partial x^2$ by a centered difference expression and using the initial condition $u(0, x) = f(x)$, we get

$$u_j^1 = f_j + kg_j + \frac{k^2c^2}{2}\left[\delta_x^2 f_j + \mathcal{O}(h^2)\right] + \mathcal{O}(k^3).$$

A corresponding difference equation for the initial time step results when we neglect the truncation error:

$$U_j^1 = f_j + kg_j + \frac{k^2c^2}{2}\delta_x^2 f_j.$$

It is possible to construct implicit difference schemes for the wave equation. The following formula defines a family of schemes, variably weighted in time:

$$\delta_t^2 U_j^n - c^2\left[\omega\delta_x^2 U_j^{n+1} + (1-2\omega)\delta_x^2 U_j^n + \omega\delta_x^2 U_j^{n-1}\right] = 0.$$

The choice $\omega = 0$ yields the explicit scheme (8.4-12), while the choice $\omega = \frac{1}{2}$ corresponds to a **fully implicit scheme**. For any choice of $\omega \in \left[0, \frac{1}{2}\right]$, the scheme is consistent, having truncation error $\mathcal{O}(k^2 + h^2)$. The von Neumann method shows that the scheme is unconditionally stable if $\omega \in \left[\frac{1}{4}, \frac{1}{2}\right]$.

8.5 Problems

PROBLEM 1. The boundary-value problem

$$\nabla^2 u(x, y) = -\pi^2 \sin(\pi x), \qquad (x, y) \in (0, 1) \times (0, 1),$$

$$u(0, y) = u(1, y) = \frac{\partial u}{\partial y}(x, 0) = \frac{\partial u}{\partial y}(x, 1) = 0,$$

has solution $u(x, y) = \sin(\pi x)$. One finite-difference approximation to this PDE is $(\delta_x^2 + \delta_y^2)U_{j,l} = -\pi^2 \sin(\pi x_j)$, on a grid having nodes $(x_j, y_l) = (jh, lh)$,

$j, l = 0, 1, \ldots, N = 1/h$. To approximate the derivative boundary conditions, one can use the $\mathcal{O}(h^2)$ difference approximations

$$\delta_y U_{j,0} = \delta_y U_{j,N} = 0.$$

This method requires two rows of "fictitious" nodes (x_j, y_{-1}), (x_j, y_{N+1}), together with corresponding unknowns $U_{j,-1}$, $U_{j,N+1}$, for $j = 1, 2, \ldots, N-1$. To balance equations with unknowns, we apply the discrete approximation to the PDE at each node of the form (x_j, y_0) or (x_j, y_N), as well as to the interior nodes. Write a program to solve the resulting system of linear equations using the Gauss-Seidel iterative method on a grid having mesh size $h = 0.1$.

PROBLEM 2. The boundary-value problem

$$\nabla \cdot [e^{xy} \nabla u(x, y)] = 0, \qquad (x, y) \in (0, 1) \times (0, 1) = \Omega,$$

$$u(x, y) = x^2 - y^2, \qquad (x, y) \in \partial\Omega,$$

has solution $u(x, y) = x^2 - y^2$. Formulate and program a finite-difference scheme for this problem. Numerically estimate its convergence rate p (that is, the number p such that $||\text{error}|| = \mathcal{O}(h^p)$) by plotting $\log ||\text{error}||$ versus $\log h$.

PROBLEM 3. Let $\mathcal{L}_h$ be a difference approximation to $-\nabla^2$ having the form

$$\mathcal{L}_h U_{P(0)} = a_0 U_{P(0)} - \sum_{j=1}^{J} a_j U_{P(j)},$$

where $P(0)$ is the index of an arbitrary node in the interior of a grid, and $P(1), P(2), \ldots, P(K)$ are indices of adjacent nodes. Assume that the approximation is **nonnegative**, that is,

$$a_0 > 0, \qquad a_j \geq 0 \quad \text{for} \quad j = 1, 2, \ldots, J,$$

and that it is **weakly diagonally dominant**, that is

$$\sum_{j=1}^{J} |a_j| \leq a_0.$$

Prove that $\mathcal{L}_h$ satisfies the following discrete maximum principle: Whenever $\mathcal{L}_h U_{P(0)} \leq 0$ for all interior grid points $P(0)$, the value of the grid function U at interior grid points is bounded above by the largest value of U on the boundary of the grid. (See Mitchell and Griffiths [7], p. 123.)

PROBLEM 4. Prove Parseval's identity, Equation (8.3-5). (Hint: Expand $||\mathcal{F}U||^2_{L^2([-\pi,\pi])} = \int_{-\pi}^{\pi} |(\mathcal{F}U)(\xi)|^2 \, d\xi$ using the definition of $\mathcal{F}U$, then interchange integration and summation.)

PROBLEM 5. Consider the initial-boundary-value problem

$$\frac{\partial u}{\partial t} + \frac{\partial u}{\partial x} = 0,$$

$$u(t,0) = 1 \quad \text{for} \quad t \geq 0, \qquad u(0,x) = 0, \quad \text{for} \quad x > 0.$$

Compute approximate solutions to this problem at $t = 5$ using (A) the forward-in-time, backward-in-space (FB) scheme with Courant numbers $C < 1$, $C = 1$, and $C > 1$; (B) the backward-in-time, backward-in-space (BB) scheme with $C = 2$; and (C) the forward-in-time, forward-in-space (FF) scheme with any value of $C > 0$ you like.

PROBLEM 6. Consider the advection equation

$$\frac{\partial u}{\partial t} + v\frac{\partial u}{\partial x} = 0, \qquad v > 0.$$

Beginning with the Taylor expansion

$$u(t+k,x) = u(t,x) + k\frac{\partial u}{\partial t}(t,x) + \frac{k^2}{2}\frac{\partial^2 u}{\partial t^2}(t,x) + \mathcal{O}(k^3),$$

substitute the relationships

$$\frac{\partial u}{\partial t} = -v\frac{\partial u}{\partial x}, \qquad \frac{\partial^2 u}{\partial t^2} = v^2\frac{\partial^2 u}{\partial x^2},$$

to arrive at the **Lax-Wendroff scheme**:

$$U_i^{n+1} = U_i^n + \frac{C}{2}(U_{i+1}^n - U_{i-1}^n) + \frac{C^2}{2}(U_{i+1}^n - 2U_i^n + U_{i-1}^n).$$

Here, $C = vk/h$. When is this scheme stable? (Hint: Show that von Neumann stability analysis yields an amplification factor satisfying $|A(\theta)|^2 = 1-4(C^2 - C^4)\sin^4(\theta/2)$.)

PROBLEM 7. Use von Neumann stability analysis to show that the backward-in-time, forward-in-space (BF) scheme $(\delta_t^- + v\delta_x^+)U_i^n = 0$ is a stable approximation to the advection equation with $v > 0$, provided that the Courant number $C \geq 1$.

PROBLEM 8.

(A) Show that the leapfrog scheme for the heat equation is unstable.

(B) Show that the DuFort-Frankel scheme for the heat equation is unconditionally stable.

PROBLEM 9. For the advection-diffusion equation

$$\frac{\partial u}{\partial t}+\frac{\partial u}{\partial x}-P^{-1}\frac{\partial^2 u}{\partial x^2}=0,$$

the approximation $\partial u/\partial x \simeq \delta_x^- U_j^n$ leads to an $\mathcal{O}(h)$ truncation error that acts like an artificial diffusion term. Show that the implicit, centered-in-space scheme

$$\delta_t^- U_j^{n+1}+\delta_x U_j^{n+1}-P^{-1}\delta_x^2 U_j^{n+1}=0$$

also has a truncation error that contributes to a diffusion-like effect. (Hint: Differentiate the PDE with respect to t and x to relate terms in the time truncation error to spatial derivatives.) Verify this conclusion computationally by looking at numerical solutions to the "steep front" problem having the initial condition $u(0,x)=0$, $x\in(0,1)$, and boundary conditions $u(t,0)=1$, $(\partial u/\partial x)(t,1)=0$.

PROBLEM 10. Show that, in von Neumann stability analysis of a difference scheme with time step k, the condition $|A(\theta)|\leq 1+Mk$, where $M>0$ is constant, suffices for stability.

8.6 References

1. M.B. Allen, "Why upwinding is reasonable," in *Proceedings, Fifth International Conference on Finite Elements in Water Resources*, Burlington, Vermont, U.S.A., June, 1984, ed. by J.P. Laible, C.A. Brebbia, W. Gray, and G. Pinder, Springer-Verlag, Berlin, pp. 13–33.

2. M.B. Allen, I. Herrera, and G.F. Pinder, *Numerical Modeling in Science and Engineering*, Wiley, New York, 1988.

3. W.F. Ames, *Numerical Methods for Partial Differential Equations*, 2nd ed., Academic Press, New York, 1977.

4. P.M. Gresho and R.L Lee, "Don't suppress the wiggles — they're telling you something!" in *Finite Elements for Convection-Dominated Flows*, ed. by T.J.R. Hughes, American Society of Mechanical Engineers, New York, 1980, pp. 37–61.

5. E. Isaacson and H.B. Keller, *Analysis of Numerical Methods*, Wiley, New York, 1966.

6. L. Lapidus and G.F. Pinder, *Numerical Solution of Partial Differential Equations in Science and Engineering*, Wiley, New York, 1982.

7. A.R. Mitchell and D.F. Griffiths, *The Finite Difference Method in Partial Differential Equations*, Wiley, New York, 1980.

8. R.D. Richtmyer and K.W. Morton, *Difference Methods for Initial-Value Problems*, 2nd ed., Wiley, New York, 1967.

9. J. Strikwerda, *Finite Difference Schemes and Partial Differential Equations*, Wadsworth, Pacific Grove, CA, 1989.

10. V. Thomée, "Finite difference methods for linear parabolic equations," in *Handbook of Numerical Analysis, Volume I: Finite Difference Methods (Part 1); Solution of Equations in* $\mathbb{R}^n$ *(Part 1)*, ed. by P.G. Ciarlet and J.L. Lions, North-Holland, Amsterdam, 1990, pp. 5–196.

Chapter 9

Introduction to Finite Elements

9.1 Introduction and Background

Finite-element methods have emerged as one of the premier classes of techniques for solving differential equations numerically. While there are strong similarities between finite-element methods and finite-difference methods, the finite-element approach enjoys a much richer mathematical framework, resting on the theory of inner-product spaces. As a consequence, the theory of finite elements has a compelling geometric flavor.

Finite-element methods afford more flexibility than finite differences. For example, the finite-element formulation yields accurate and rigorously based techniques for imposing a variety of boundary conditions. It also admits many geometric forms for approximate solutions, thereby facilitating the discretization of problems posed on oddly shaped domains. Moreover, in some applications — such as solid mechanics — the most commonly used piecewise polynomial approximate solutions enjoy mathematical features that correspond to physical attributes of the structures being modeled. Detailed exploration of these advantages lies beyond this book's scope; interested readers should consult Lapidus and Pinder [3] for a comprehensive overview.

This chapter gives a brief introduction to finite elements in one space dimension. We first outline the basic formulation for a simple boundary-value problem (BVP) involving a second-order ordinary differential equation (ODE), using ideas from the calculus of variations. We then analyze the error associated with piecewise linear trial functions, which constitute the simplest choice in most applications. We also discuss the treatment of various boundary conditions. Then we present a formulation for initial-boundary-value problems based on finite-difference timestepping.

Our discussion aims at simplicity rather than generality. The purpose is to introduce the mechanics of finite elements, sketching some of the more important theoretical aspects. For a more thorough introduction to the theory, we refer to Johnson [2], Strang and Fix [4], and Wait and Mitchell [6].

A Model Boundary-Value Problem

Much of the theory of finite elements rests on ideas from variational formulations of BVPs. To establish terminology and notation, it is useful to examine these ideas. Consider the following BVP: Find a function $u \in C^2([0,1])$ such that

$$\begin{aligned} -[a(x)u'(x)]' &= f(x), \qquad x \in \Omega := (0,1), \\ u(0) = u(1) &= 0. \end{aligned} \tag{9.1-1}$$

Assume that f is a known, continuous, real-valued function defined on $[0,1]$ and that it obeys the bound $|f(x)| \leq L$. Also assume that the coefficient $a(x)$ belongs to $C^1([0,1])$ and that it obeys bounds of the form

$$0 < \alpha \leq a(x) \leq A,$$

for all $x \in [0,1]$. These assumptions guarantee that the problem (9.1-1) has a unique solution in $C^2([0,1])$. This problem is a one-dimensional analog of multidimensional problems involving the elliptic operator $-\nabla \cdot [a(\mathbf{x})\nabla]$.

We cast this model problem in a more general form by the following tactic. First, multiply the ODE in (9.1-1) by **test functions** v, whose nature we specify later. Then integrate the result over Ω, obtaining equations of the form

$$\int_\Omega \Big\{ -[a(x)u'(x)]'v(x) - f(x)v(x) \Big\}\, dx = 0. \tag{9.1-2}$$

This equation roughly says that the residual $R(x) := -[a(x)u'(x)]' - f(x)$ vanishes in a weighted-average sense over the region Ω. In another view, the integral equation asserts that $R(x)$ is orthogonal to the function v with respect to the inner product associated with $L^2(\Omega)$. By analogy with the finite-dimensional Euclidean spaces $\mathbb{R}^n$, we expect $R(x)$ to vanish if we force it to be orthogonal to "enough" test functions $v(x)$.

Another interesting fact about the integral equation (9.1-2) is that integration by parts yields

$$\int_\Omega a(x)u'(x)v'(x)\, dx - a(x)u'(x)v(x)\Big|_0^1 = \int_\Omega f(x)v(x)\, dx. \tag{9.1-3}$$

Consider the demands that the statements (9.1-2) and (9.1-3) make on the functions u and v. For Equation (9.1-2) to make sense, it suffices for v to be square integrable; u, by contrast, must have square integrable derivatives through order 2. For Equation (9.1-3), it is enough to demand that both u and v have square integrable derivatives of order 0 and 1. In other words, from a purely formal point of view, Equation (9.1-3) admits more potential solutions than does Equation (9.1-2). In this sense, Equation (9.1-3) is more general than Equation (9.1-2).

Variational Formulation

Let us explore more deeply the connections between the model problem (9.1-1) and its integral forms (9.1-2) and (9.1-3). We begin by identifying vector spaces of functions that satisfy various degrees of smoothness in the sense of square-integrability:

DEFINITION. *For any nonnegative integer m, the* **Sobolev space** $H^m(\Omega)$ *is*

$$H^m(\Omega) := \left\{ v:\Omega \to \mathbb{R} \; : \; \int_\Omega \left[|v(x)|^2 + |v'(x)|^2 + \cdots + |v^{(m)}(x)|^2 \right] dx < \infty \right\}.$$

In particular,

$$H^0(\Omega) = \left\{ v:\Omega \to \mathbb{R} \; : \; \int_\Omega |v(x)|^2 \, dx < \infty \right\} = L^2(\Omega),$$

$$H^1(\Omega) = \left\{ v:\Omega \to \mathbb{R} \; : \; \int_\Omega \left[|v(x)|^2 + |v'(x)|^2 \right] dx < \infty \right\}.$$

It is straightforward to check that $H^m(\Omega)$ is an inner-product space, with inner products defined as follows:

$$\langle v, w \rangle_0 := \int_\Omega v(x) w(x) \, dx = \langle v, w \rangle,$$

$$\langle v, w \rangle_1 := \int_\Omega \left[v(x) w(x) + v'(x) w'(x) \right] dx,$$

and so forth. These inner products give rise to norms in the usual way:

$$\|v\|_{(0)} := \langle v, v \rangle^{1/2} = \|v\|_{L^2(\Omega)},$$

$$\|v\|_{(1)} := \langle v, v \rangle_1^{1/2},$$

and so forth. For future reference, we record several facts about these inner-product spaces.

PROPOSITION 9.1. *Let m be a nonnegative integer. Then*

(A) $H^{m+1}(\Omega) \subset H^m(\Omega)$.

(B) *If* $v \in H^m(\Omega)$, *then* $\|v\|_{(m)}^2 = \|v\|_{(0)}^2 + \|v'\|_{(0)}^2 + \cdots + \|v^{(m)}\|_{(0)}^2$.

(C) *If* $v \in H^{m+1}(\Omega)$, *then* $\|v\|_{(m+1)} \geq \|v\|_{(m)}$ *and* $\|v\|_{(m+1)} \geq \|v'\|_{(m)}$.

(D) *(Cauchy-Schwarz inequality.) If* $v, w \in H^m(\Omega)$, *then* $|\langle v, w\rangle_m| \leq \|v\|_{(m)} \|w\|_{(m)}$.

(E) *(Triangle inequality.) If* $v, w \in H^m(\Omega)$, *then* $\|v+w\|_{(m)} \leq \|v\|_{(m)} + \|w\|_{(m)}$.

PROOF: This is Problem 1. ∎

With this background, we state the integral form of the model problem (9.1-1) more precisely. Call

$$H_0^1(\Omega) := \left\{ v \in H^1(\Omega) \ : \ v(0) = v(1) = 0 \right\}.$$

This subset of $H^1(\Omega)$ is a vector space in its own right. The **variational form** of the BVP (9.1-1) is as follows: Find a function $u \in H_0^1(\Omega)$ such that

$$\langle au', v'\rangle = \langle f, v\rangle, \tag{9.1-4}$$

for all $v \in H_0^1(\Omega)$. In this formulation, u automatically satisfies the prescribed boundary conditions by belonging to $H_0^1(\Omega)$.

To interpret this form of the BVP, refer to the integral forms discussed earlier. Equation (9.1-4) is equivalent to Equation (9.1-3), since the condition $v \in H_0^1(\Omega)$ implies that the boundary terms arising from integration by parts vanish. Thus the variational form (9.1-4) "almost" demands that $-[a(x)u'(x)]' - f(x)$ be orthogonal to all vectors in the subspace $H_0^1(\Omega)$ of $H^1(\Omega)$. This condition has close connections with the least-squares idea of minimizing the distance to a subspace of a vector space. One can view integration by parts as a device that formally reduces the smoothness requirements on the solution u.

In addition to this geometric interpretation, there are logical connections between the variational form and the original BVP:

THEOREM 9.2. *If u is a solution to the* BVP *(9.1-1), then u is a solution to the variational problem (9.1-4). If u is a solution to the variational problem (9.1-4) and $u \in C^2([0,1])$, then u is a solution to the* BVP *(9.1-1).*

PROOF: This is Problem 2. ∎

For the BVP (9.1-1), the variational form (9.1-4) has yet another, more physical interpretation. The problem models a stationary elastic string held under tension between two fixed endpoints, $x = 0$ and $x = 1$, and subject to a time-independent, transverse applied load $f(x)$. Figure 1 illustrates this configuration. The coefficient $a(x)$ represents the variable coefficient of elasticity along the string, and $u(x)$ stands for the string's transverse displacement.

Guenther and Lee ([1], Section 11-1) show that the total potential energy of a string sustaining a hypothetical displacement $v(x)$ is

$$E(v) := \tfrac{1}{2}\langle av', v'\rangle - \langle f, v\rangle.$$

In the context of the associated mathematical problem (9.1-4), $E(v)$ is the **energy functional**.

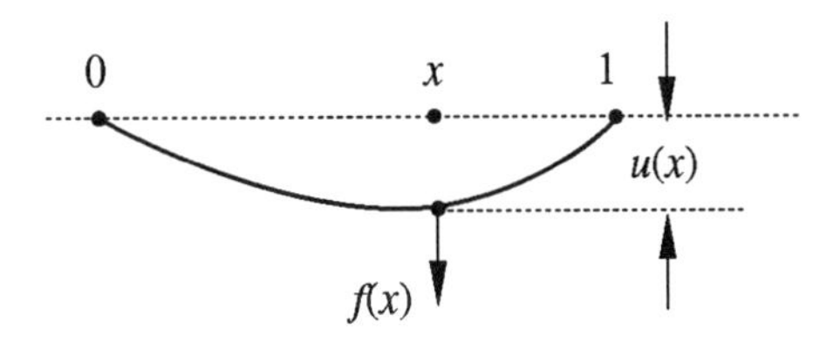

FIGURE 1. *Stationary elastic string, held between fixed ends and undergoing displacement $u(x)$ due to a transverse applied load $f(x)$.*

Physically, the string governed by Equations (9.1-1) assumes the displacement that minimizes potential energy. The variational problem (9.1-4) enjoys a corresponding minimization property:

THEOREM 9.3. *A function $u \in H_0^1(\Omega)$ is a solution to the variational problem (9.1-4) if and only if $E(u) \leq E(v)$ for all $v \in H_0^1(\Omega)$.*

PROOF: First assume that u is a solution to the variational problem, and let $v \in H_0^1(\Omega)$. Denote $w := v - u$. The energy functional is linear, so

$$\begin{aligned} E(v) = E(u+w) &= E(u) + E(w) + \langle au', w'\rangle \\ &= E(u) + E(w) + \langle f, w\rangle \\ &= E(u) + \tfrac{1}{2}\langle aw', w'\rangle, \end{aligned}$$

the last step following from the definition of E. The argument is complete if we show that $\langle aw', w'\rangle \geq 0$. By the hypotheses on $a(x)$,

$$\langle aw', w'\rangle \geq \alpha \int_\Omega (w')^2\, dx \geq 0.$$

Now assume that $E(u) \leq E(v)$ for all $v \in H_0^1(\Omega)$. For arbitrary $v \in H_0^1(\Omega)$, the real-valued function $\Phi(s) := E(u + sv)$ is differentiable, and by hypothesis it has a minimum at $s = 0$. Therefore $\Phi'(0) = 0$. Since

$$\begin{aligned} \Phi(s) &= E(u) + E(sv) + s\langle au', v'\rangle \\ &= E(u) + s\left(\langle au', v'\rangle - \langle f, v\rangle\right) + \tfrac{1}{2}s^2\langle av', v'\rangle, \end{aligned}$$

we have $\Phi'(0) = \langle au', v' \rangle - \langle f, v \rangle$, and the conclusion follows. ∎

Not all problems involving differential equations possess variational forms that are so closely associated with minimum principles. However, the existence of such principles often guides the analysis of finite-element methods and lends them theoretical elegance. In the next section we exploit minimization properties in analyzing the standard Galerkin finite-element formulation of the BVP (9.1-1). In Section 9.3 we examine a finite-element formulation for a time-dependent problem in which minimization principles play only an indirect role.

9.2 A Steady-State Problem

Neither the model problem (9.1-1) nor its variational form (9.1-4) is generally amenable to computation. In the problem (9.1-1), the solution belongs to the vector space $C^2(\Omega)$, while in the variational form u belongs to the larger space $H_0^1(\Omega)$. Both vector spaces are infinite-dimensional. The idea behind the Galerkin finite-element method for Equation (9.1-1) is to solve the variational form (9.1-4) on a finite-dimensional subspace of $H_0^1(\Omega)$, called the **trial space**. Since we need only finitely many degrees of freedom to specify functions in the trial space, it is computationally feasible to determine an approximate solution there.

Several questions arise in this approach. First, does the method have reasonable computational requirements? Second, is the approximate solution, generated by restricting attention to the trial space, reasonably close to the exact solution? Third, does the approach extend readily to more general boundary-value problems? This section presents details of the Galerkin finite-element procedure in a simple setting and examines the most basic techniques for answering these questions.

Construction of the Galerkin Finite-Element Scheme

As with the finite-difference method, begin by constructing a grid on the domain $\Omega = (0,1)$ of the problem. Let $\Delta := \{x_0, x_1, \ldots, x_M\}$, where $x_j := jh$, $h := 1/M$. Associated with this grid is the vector space $\mathcal{M}_0^1(\Delta)$ of piecewise linear polynomials introduced in Section 1.3. Recall that $\mathcal{M}_0^1(\Delta)$ has a basis $\{\ell_0, \ell_1, \ldots, \ell_M\}$, a typical element of which has the form

$$\ell_j(x) = \begin{cases} (x - x_{j-1})/h, & \text{if } \; x_{j-1} \le x \le x_j, \\ (x_{j+1} - x)/h, & \text{if } \; x_j \le x \le x_{j+1}, \\ 0, & \text{otherwise.} \end{cases}$$

Figure 1 illustrates these functions for the case when Ω is a general interval.

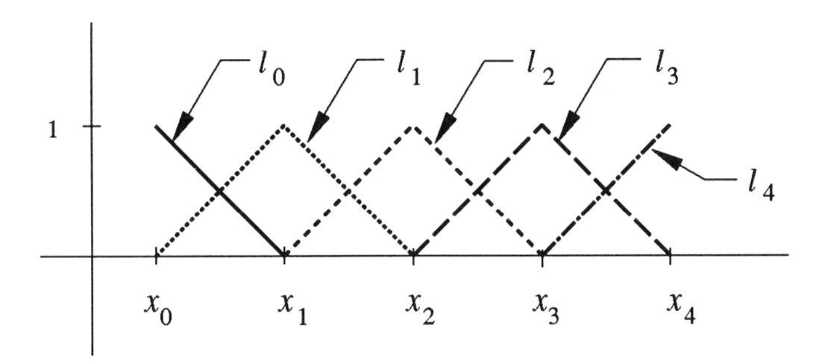

FIGURE 1. *Basis functions $\ell_j(x)$ for the space $\mathcal{M}_0^1(\Delta)$ of piecewise linear functions on a grid Δ.*

To solve the variational problem (9.1-4) numerically, we seek an approximate solution, or **trial function** u_h, that belongs to the following trial space:

$$\mathcal{V}_h := \left\{ v \in \mathcal{M}_0^1(\Delta) \ : \ v(0) = v(1) = 0 \right\}.$$

In other words, u_h is piecewise linear, and it satisfies the prescribed boundary conditions. It has the explicit form

$$u_h(x) = \sum_{j=1}^{M-1} U_j \ell_j(x), \tag{9.2-1}$$

where the coefficients $U_1, U_2, \ldots, U_{M-1}$ are to be determined. One readily checks that $\mathcal{V}_h$ is an $(M-1)$-dimensional subspace of $H_0^1(\Omega)$, the space that contains the exact solution to the variational problem.

To determine the coefficients U_j of the trial function, we impose a finite-dimensional version of the variational problem (9.1-4): We demand that

$$\langle a u_h', v' \rangle = \langle f, v \rangle, \tag{9.2-2}$$

for all $v \in \mathcal{V}_h$. These are the **Galerkin equations**. Since $\mathcal{V}_h$ is finite-dimensional, it suffices to impose Equation (9.2-2) for v ranging over the elements of a basis for $\mathcal{V}_h$. Therefore the Galerkin equations reduce to the following linear system:

$$\begin{aligned} \langle a u_h', \ell_1' \rangle &= \langle f, \ell_1 \rangle, \\ \langle a u_h', \ell_2' \rangle &= \langle f, \ell_2 \rangle, \\ &\vdots \\ \langle a u_h', \ell_{M-1}' \rangle &= \langle f, \ell_{M-1} \rangle. \end{aligned} \tag{9.2-3}$$

Thus there are $M-1$ equations to solve for the unknowns $U_1, U_2, \ldots, U_{M-1}$.

To cast these equations into matrix form, substitute the expression (9.2-1) into the system (9.2-3):

$$\left\langle a \sum_{j=1}^{M-1} U_j \ell_j', \ell_i' \right\rangle = \sum_{j=1}^{M-1} U_j \underbrace{\langle a\ell_j', \ell_i' \rangle}_{a_{i,j}} = \underbrace{\langle f, \ell_i \rangle}_{f_i}, \quad i = 1, 2, \ldots, M-1.$$

By identifying the entries $a_{i,j}$ and f_i as indicated, we have

$$\underbrace{\begin{bmatrix} a_{1,1} & \cdots & a_{1,M-1} \\ \vdots & & \vdots \\ a_{M-1,1} & \cdots & a_{M-1,M-1} \end{bmatrix}}_{\mathsf{A}} \underbrace{\begin{bmatrix} U_1 \\ \vdots \\ U_{M-1} \end{bmatrix}}_{\mathbf{u}} = \underbrace{\begin{bmatrix} f_1 \\ \vdots \\ f_{M-1} \end{bmatrix}}_{\mathbf{f}}. \tag{9.2-4}$$

Several attributes of this linear system are worth noting. First, using Lagrange piecewise linear basis functions forces $a_{i,j} = 0$ whenever $|i - j| > 1$. Therefore the matrix A is tridiagonal. For example, if $a(x) = 1$ in the original boundary-value problem, then

$$\mathsf{A} = \frac{1}{h} \begin{bmatrix} 2 & -1 & & & \\ -1 & 2 & -1 & & \\ & & \ddots & & \\ & & -1 & 2 & -1 \\ & & & -1 & 2 \end{bmatrix}.$$

Second, A is symmetric, since $\langle a\ell_j', \ell_i' \rangle = \langle a\ell_i', \ell_j' \rangle$. Third, A is positive definite. To see this, let $\mathbf{v} = (V_1, V_2, \ldots, V_{M-1})^\top$ be any nonzero vector in $\mathbb{R}^{M-1}$. Corresponding to $\mathbf{v}$ is a function $v \in \mathcal{V}_h$ that has the representation $v(x) = V_1\ell_1(x) + V_2\ell_2(x) + \cdots + V_{M-1}\ell_{M-1}(x)$. Since v is not identically zero,

$$\begin{aligned} 0 < \alpha\langle v', v' \rangle \le \langle av', v' \rangle &= \left\langle a \sum_{i=1}^{M-1} v_i \ell_i', \sum_{j=1}^{M-1} v_j \ell_j' \right\rangle \\ &= \sum_{i=1}^{M-1} \sum_{j=1}^{M-1} v_i \langle a\ell_i', \ell_j' \rangle \, v_j \\ &= \mathbf{v}^\top \mathsf{A} \mathbf{v}. \end{aligned}$$

Hence $0 < \mathbf{v}^\top \mathsf{A} \mathbf{v}$ for arbitrary nonzero $\mathbf{v} \in \mathbb{R}^{M-1}$, that is, A is positive definite.

It follows that the linear system (9.2-4) has a unique solution

$$\mathbf{u} = (U_1, U_2, \ldots, U_{M-1})^\top.$$

Moreover, because the matrix is sparse, symmetric, and positive definite, efficient solution techniques are available to compute $\mathbf{u}$. The sparseness of the matrix is characteristic of the finite-element method; it arises from the use of basis functions that are nonzero only over a small subset of the domain of the problem. Other piecewise polynomial trial spaces yield sparse matrices having different zero structures, as Problem 3 suggests.

A Basic Error Estimate

How close is the finite-element solution u_h to the exact solution u? More specifically, how does the error $\varepsilon_h := u - u_h$ depend upon the mesh size h of the grid Δ? We now show that a finer mesh yields a more accurate solution.

Results from Section 1.3 assert that the piecewise linear interpolant $\hat{u}$ of the exact solution obeys an estimate of the form $\|u - \hat{u}\|_\infty = \mathcal{O}(h^2)$. This observation suggests that an estimate of the form $\|u - u_h\| = \mathcal{O}(h^2)$, in some norm, is the best that we can expect. In this subsection we derive an estimate that is disappointing, since it implies that $\|u - u_h\|_{(0)} = \mathcal{O}(h)$. However, this result plays a crucial role in the subsequent development of more satisfying error estimates.

To streamline the arguments and to make them more readily generalizable, we restate the original Galerkin formulation (9.2-2) in the following, more abstract notation: Find a function $u_h \in \mathcal{V}_h$ such that

$$B(u_h, v) = F(v),$$

for all $v \in \mathcal{V}_h$. Here, $B(\cdot,\cdot)$ is a mapping defined on $H_0^1(\Omega) \times H_0^1(\Omega)$ by the equation

$$B(v, w) := \int_\Omega a(x) v'(x) w'(x)\, dx.$$

This mapping is a **bilinear form**, meaning that it has the following properties:

(i) For any $v, w \in H_0^1(\Omega)$, $B(v, w) = B(w, v)$.

(ii) For any $v, w \in H_0^1(\Omega)$ and any $c \in \mathbb{R}$, $B(cv, w) = cB(v, w)$.

(iii) For any $v_1, v_2, w \in H_0^1(\Omega)$, $B(v_1 + v_2, w) = B(v_1, w) + B(v_2, w)$.

The mapping $F\colon H_0^1(\Omega) \to \mathbb{R}$ is defined by the equation

$$F(v) := \int_\Omega f(x) v(x)\, dx.$$

This mapping is a linear functional: For any functions $v, w \in H_0^1(\Omega)$ and any $c \in \mathbb{R}$, $F(cv + w) = cF(v) + F(w)$.

The following facts about B and F figure prominently in the error analysis.

PROPOSITION 9.4. *The mappings B and F defined above have the following properties:*

(A) *F is* **continuous**, *that is, there exists a constant $L > 0$ such that $|F(v)| \leq L\|v\|_{(1)}$ for all $v \in H_0^1(\Omega)$.*

(B) *B is* **continuous**, *that is, there exists a constant $A > 0$ such that $|B(v,w)| \leq A\|v\|_{(1)}\|w\|_{(1)}$ for all $v, w \in H_0^1(\Omega)$.*

(C) *B is* **coercive**, *that is, there exists a constant $\tilde{\alpha} > 0$ such that $B(v,v) \geq \tilde{\alpha}\|v\|_{(1)}^2$ for all $v \in H_0^1(\Omega)$.*

Part of the proof hinges on a lemma that captures the intuitive notion that differentiation of smooth functions typically leads to wilder behavior:

LEMMA 9.5 (FRIEDRICHS INEQUALITY). *If $v \in H_0^1(\Omega)$, then $\|v\|_{(0)}^2 \leq \|v'\|_{(0)}^2$.*

PROOF: By the fundamental theorem of calculus and the fact that $v(0) = 0$,

$$v(x) = \int_0^x v'(\xi)\,d\xi,$$

for all $x \in [0,1]$. Using the Cauchy-Schwarz inequality, we have

$$\begin{aligned} |v(x)| = \left|\int_0^x v'(\xi)\,d\xi\right| &\leq \int_0^x |v'(\xi)|\,d\xi \\ &\leq \int_0^1 |v'(\xi)|\,d\xi \\ &= \langle |v'|, 1\rangle \leq \|v'\|_{(0)}\|1\|_{(0)} = \|v'\|_{(0)}. \end{aligned}$$

Squaring and integrating yields

$$\int_0^1 |v(x)|^2\,dx \leq \int_0^1 \|v'\|_{(0)}^2 dx,$$

that is, $\|v\|_{(0)}^2 == \|v'\|_{(0)}^2$. ∎

PROOF OF PROPOSITION 9.4: Assertion (A) follows from the Cauchy-Schwarz inequality and the hypothesis, stated in Section 9.1, that f is bounded:

$$|F(v)| = |\langle f, v\rangle| \leq \|f\|_{(0)}\|v\|_{(0)} \leq L\|v\|_{(1)}.$$

Assertion (B) follows similarly:

$$\begin{aligned} |B(v,w)| &= |\langle av', w'\rangle| \\ &\leq \|av'\|_{(0)}\|w'\|_{(0)} \\ &\leq A\|v'\|_{(0)}\|w'\|_{(0)} \\ &\leq A\|v\|_{(1)}\|w\|_{(1)}. \end{aligned}$$

For assertion (c), we reason as follows:

$$B(v,v) = \int_\Omega a(x)v'(x)v'(x)\,dx \;\geq\; \alpha \int_\Omega |v'(x)|^2 dx$$

$$= \alpha \|v'\|_{(0)}^2 = \frac{\alpha}{2}\left(\|v'\|_{(0)}^2 + \|v'\|_{(0)}^2\right).$$

Lemma 9.5 now implies that

$$|B(v,v)| \geq \frac{\alpha}{2}\left(\|v\|_{(0)}^2 + \|v'\|_{(0)}^2\right) = \frac{\alpha}{2}\|v\|_{(1)}^2.$$

Take $\tilde{\alpha} = \alpha/2$. ∎

According to Proposition 9.4, the bilinear form $B(\cdot,\cdot)$ has all the properties needed to guarantee that it is an inner product on the vector space $H_0^1(\Omega)$. It follows that the function $\|\cdot\|_B := \sqrt{B(\cdot,\cdot)}$ constitutes a norm on $H_0^1(\Omega)$. By analogy with the elastic string problem mentioned in Section 9.1, we call this norm the **energy norm** associated with B. The energy norm is equivalent to the Sobolev norm $\|\cdot\|_{(1)}$, since, for any $v \in H_0^1(\Omega)$,

$$\sqrt{\frac{\alpha}{2}}\|v\|_{(1)} \leq \|v\|_B \leq \sqrt{A}\|v\|_{(1)}.$$

We now show that the finite-element error $u - u_h$, measured in the norm $\|\cdot\|_{(1)}$, is comparable to that associated with the best possible approximation to u in the trial space $\mathcal{V}_h$.

THEOREM 9.6. *If u_h is the finite-element approximation to the solution u of the* BVP *(9.1-1), then*

$$\|u - u_h\|_{(1)} \leq \frac{2A}{\alpha}\|u - v\|_{(1)},$$

for every $v \in \mathcal{V}_h$.

PROOF: If $u - u_h = 0$, then the conclusion is trivial. Otherwise, let $\varepsilon_h := u - u_h$. The variational formulation of the boundary-value problem implies that $B(u,w) = F(w)$ for all $w \in \mathcal{V}_h$, since $\mathcal{V}_h \subset H_0^1(\Omega)$. Also, by definition, $B(u_h, w) = F(w)$ for all $w \in \mathcal{V}_h$. The linearity of B thus yields

$$B(\varepsilon_h, w) = 0, \tag{9.2-5}$$

for all $w \in \mathcal{V}_h$. Now choose any function $v \in \mathcal{V}_h$, and let $w := u_h - v$, which belongs to $\mathcal{V}_h$. Using coercivity and Equation (9.2-5), we have

$$\frac{\alpha}{2}\|\varepsilon_h\|_{(1)}^2 \leq B(\varepsilon_h, \varepsilon_h) \;=\; B(\varepsilon_h, \varepsilon_h) + B(\varepsilon_h, w)$$

$$= B(\varepsilon_h, u - u_h + w) = B(\varepsilon_h, u - v).$$

Therefore, by the continuity of B,

$$\frac{\alpha}{2}||\varepsilon_h||_{(1)}^2 \leq A||\varepsilon_h||_{(1)}||u-v||_{(1)},$$

and dividing through by $||\varepsilon_h||_{(1)}$ completes the proof. ∎

This theorem allows us to bound the finite-element error $||u-u_h||_{(1)}$ in terms of the approximating power associated with the piecewise linear trial functions in $\mathcal{V}_h$. Naive estimates yield the following:

COROLLARY 9.7. *With u and u_h as in Theorem 9.6, we have $||u-u_h||_{(1)} = \mathcal{O}(h)$.*

PROOF: Denote by $\hat{u}$ the piecewise linear interpolant of u in $\mathcal{V}_h$. By Theorem 9.6 and the Friedrichs inequality,

$$\begin{aligned} ||u-u_h||_{(1)}^2 &\leq \frac{4A^2}{\alpha^2}||u-\hat{u}||_{(1)}^2 \\ &= \frac{4A^2}{\alpha^2}\left(||u-\hat{u}||_{(0)}^2 + ||u'-\hat{u}'||_{(0)}^2\right) \\ &\leq \frac{8A^2}{\alpha^2}||u'-\hat{u}'||_{(0)}^2. \end{aligned}$$

Therefore,

$$||u-u_h||_{(1)} \leq \frac{\sqrt{8}A}{\alpha}\left[\int_\Omega |u'(x)-\hat{u}'(x)|^2\,dx\right]^{1/2}.$$

By interpolation error estimates developed in Chapter 1 (see Proposition 1.4), the integral on the right is less than or equal to $h||u''||_\infty$. Hence,

$$||u-u_h||_{(1)} \leq \frac{\sqrt{8}A}{\alpha}h||u''||_\infty,$$

which completes the proof. ∎

Before exploring further consequences of Theorem 9.6, we make two remarks. First, Corollary 9.7 gives a **global error estimate** for the Galerkin finite-element method. This estimate measures how well the numerical solution u_h approximates the exact solution u in the sense of the norm $||\cdot||_{(1)}$, which measures distances between functions in an *average* sense. The estimate says nothing explicit about the magnitude $|u(x)-u_h(x)|$ of this distance at any particular point $x \in \Omega$. Second, the corollary asserts that $u_h \to u$, in the norm $||\cdot||_{(1)}$, as $h \to 0$. In other words, the Galerkin finite-element scheme is convergent. The theory presented in the remainder of this section merely sharpens the estimates that establish this basic result.

Optimal-Order Error Estimates

The error estimate of Theorem 9.6 is unsatisfying. It measures the error $u - u_h$ in the norm $\|\cdot\|_{(1)}$, which appeals less to intuition than the more familiar norms $\|\cdot\|_{(0)}$ or $\|\cdot\|_\infty$. Since the norm $\|\cdot\|_{(1)}$ dominates the norm $\|\cdot\|_{(0)}$, Corollary 9.7 trivially implies that $\|u - u_h\|_{(0)} = \mathcal{O}(h)$. However, this estimate seems weak, since it suggests that the finite-element error may be qualitatively larger than the error associated with interpolation of the exact solution u in the trial space $\mathcal{V}_h$. We now develop a sharper estimate. The aim is to show that $\|u - u_h\|_{(0)} \leq C\|u - \hat{u}\|_{(0)}$ for some positive constant C. Since one cannot expect qualitatively better estimates in $\mathcal{V}_h$, we call estimates of this form **optimal-order error estimates**.

Lifting the primitive error estimate of Theorem 9.6 to $\mathcal{O}(h^2)$ requires that we invoke the regularity of solutions to boundary-value problems of the type (9.1-1). If the coefficient $a(x)$ is smooth enough, then we expect the solution u to Equation (9.1-1) to depend continuously upon the forcing function f, in the sense that

$$\|u\|_{(2)} \leq K\|f\|_{(0)}, \tag{9.2-6}$$

for some positive constant K ([4], Section 1.2). In higher-dimensional settings this property is subject to vagaries of geometry: Boundary-value problems posed on domains with nonsmooth boundaries may exhibit weaker regularity.

Central to the arguments are estimates of interpolation error in the norm $\|\cdot\|_{(0)} = \|\cdot\|_{L^2(\Omega)}$. The results of Chapter 1 concern only the norm $\|\cdot\|_\infty$, so we digress briefly to develop estimates in $L^2(\Omega)$.

LEMMA 9.8. *If $u \in H^2(\Omega)$, then*

(A) $\|u - \hat{u}\|_{(0)} \leq (h/\pi)^2\|u''\|_{(0)}$,

(B) $\|u' - \hat{u}'\|_{(0)} \leq (h/\pi)\|u''\|_{(0)}$.

In terms of powers of h, these estimates recall the familiar estimates in $L^\infty(\Omega)$.

PROOF: Consider first a single element $[0, h] = [x_0, x_1]$ of the grid. Define $\eta(x) := u(x) - \hat{u}(x)$, and observe that $\eta(0) = \eta(h) = 0$. Thus η has a uniformly convergent Fourier series:

$$\eta(x) = \sum_{n=1}^{\infty} \eta_n \sin\frac{n\pi x}{h}.$$

By the Parseval identity,

$$\int_0^h \eta^2(x)\,dx = \frac{h}{2}\sum_{n=1}^{\infty} \eta_n^2.$$

Term-by-term differentiation of the series yields

$$\int_0^h [\eta'(x)]^2\, dx = \frac{h}{2}\sum_{n=1}^{\infty}\left(\frac{n\pi}{h}\right)^2 \eta_n^2,$$

$$\int_0^h [\eta''(x)]^2\, dx = \frac{h}{2}\sum_{n=1}^{\infty}\left(\frac{n\pi}{h}\right)^4 \eta_n^2.$$

But

$$\left(\frac{n\pi}{h}\right)^2 \eta_n^2 = \frac{h^2}{n^2\pi^2}\left(\frac{n\pi}{h}\right)^4 \eta_n^2 \leq \frac{h^2}{\pi^2}\left(\frac{n\pi}{h}\right)^4 \eta_n^2,$$

the last step following from the fact that $n \geq 1$. Therefore,

$$\begin{aligned}\int_0^h [\eta'(x)]^2\, dx &\leq \frac{h}{2}\sum_{n=1}^{\infty}\frac{h^2}{\pi^2}\left(\frac{n\pi}{h}\right)^4 \eta_n^2 \\ &= \frac{h^2}{\pi^2}\int_0^h [\eta''(x)]^2\, dx \\ &= \frac{h^2}{\pi^2}\int_0^h [u''(x)]^2\, dx,\end{aligned}$$

the component $\hat{u}''(x)$ vanishing identically on $(0, h)$.

The same reasoning applies to any element $[x_{j-1}, x_j]$ of the grid, so

$$\sum_{j=1}^{M}\int_{x_{j-1}}^{x_j} [\eta'(x)]^2\, dx \leq \frac{h^2}{\pi^2}\sum_{j=1}^{M}\int_{x_{j-1}}^{x_j} [u''(x)]^2\, dx,$$

that is, $||u' - \hat{u}'||_{(0)}^2 \leq (h/\pi)^2 ||u''||_{(0)}^2$. This establishes part (B).

Similarly,

$$\eta_n^2 = \left(\frac{h}{n\pi}\right)^4\left(\frac{n\pi}{h}\right)^4 \eta_n^2 \leq \frac{h^4}{\pi^4}\left(\frac{n\pi}{h}\right)^4 \eta_n^2,$$

so

$$\begin{aligned}\int_0^h \eta^2(x)\, dx \leq \frac{h}{2}\sum_{n=1}^{\infty}\frac{h^4}{\pi^4}\left(\frac{n\pi}{h}\right)^4 \eta_n^2 &= \frac{h^4}{\pi^4}\int_0^h [\eta''(x)]^2\, dx \\ &= \frac{h^4}{\pi^4}\int_0^h [u''(x)]^2\, dx.\end{aligned}$$

Summing over the elements of the grid as before yields

$$||u - u_h||_{(0)}^2 \leq (h/\pi)^4 ||u''||_{(0)}^2,$$

proving part (A). ∎

This lemma has an easy corollary:

COROLLARY 9.9. *Under the hypotheses of Theorem 9.8,*

(A) $\|u - \hat{u}\|_{(1)} \leq \dfrac{h\sqrt{2}}{\pi}\|u''\|_{(0)}$,

(B) $\|u - u_h\|_{(1)} \leq \dfrac{h\sqrt{8}A}{\pi\alpha}\|u''\|_{(0)}$.

PROOF: This is an exercise. ∎

We now have the machinery needed to elevate the H^1 estimate of the finite-element error $u - u_h$ to an optimal-order L^2 estimate. The argument, which is by now standard, involves a clever procedure known as the **Nitsche lift**:

THEOREM 9.10. *If u_h is the piecewise linear Galerkin finite-element solution to the* BVP *(9.1-1), then there exists a positive constant Γ such that $\|u - u_h\|_{(0)} \leq \Gamma h^2 \|u''\|_{(0)}$.*

PROOF: Let $\varepsilon_h := u - u_h$. If $\varepsilon_h = 0$, then there is nothing to prove. Otherwise, the facts that $B(u, v) = F(v)$ and $B(u_h, v) = F(v)$ for all $v \in \mathcal{V}_h$ imply that

$$B(\varepsilon_h, v) = 0, \tag{9.2-7}$$

for all $v \in \mathcal{V}_h$. Let $\varphi: \Omega \to \mathbb{R}$ be a solution to the "dual" BVP,

$$\begin{aligned} -[a(x)\varphi'(x)]' &= \varepsilon_h(x), \qquad x \in \Omega, \\ \varphi(0) = \varphi(1) &= 0. \end{aligned}$$

Using integration by parts, we have

$$\|\varepsilon_h\|_{(0)}^2 = \langle \varepsilon_h, \varepsilon_h \rangle = -\langle \varepsilon_h, (a\varphi')' \rangle = \langle \varepsilon_h', a\varphi' \rangle = B(\varepsilon_h, \varphi).$$

Using the identity (9.2-7) and the fact that the interpolant $\hat{\varphi}$ of φ belongs to $\mathcal{V}_h$, we obtain

$$\begin{aligned} \|\varepsilon_h\|_{(0)}^2 &= B(\varepsilon_h, \varphi - \hat{\varphi}) \\ &= \int_\Omega a(x)\varepsilon_h'(x)[\varphi(x) - \hat{\varphi}(x)]'\, dx \\ &\leq A \int_\Omega \varepsilon_h'(x)[\varphi'(x) - \hat{\varphi}'(x)]dx. \end{aligned}$$

The Cauchy-Schwarz inequality and the results of Corollary 9.9 now yield

$$\begin{aligned} \|\varepsilon_h\|_{(0)}^2 &\leq A\|\varepsilon_h'\|_{(0)}\|\varphi' - \hat{\varphi}'\|_{(0)} \\ &\leq A\|\varepsilon_h\|_{(1)}\|\varphi - \hat{\varphi}\|_{(1)} \\ &\leq A\left(\frac{h\sqrt{8}A}{\pi\alpha}\|u''\|_{(0)}\right)\left(\frac{h\sqrt{2}}{\pi}\|\varphi''\|_{(0)}\right). \end{aligned}$$

Invoking the regularity assumption (9.2-6) to substitute for $||\varphi''||_{(0)}$ gives

$$||\varepsilon_h||_{(0)}^2 \leq \frac{4A^2Kh^2}{\pi^2\alpha}||u''||_{(0)}||\varepsilon_h||_{(0)}.$$

Dividing through by $||\varepsilon_h||_{(0)}$ and identifying $\Gamma := 4A^2K/(\pi^2\alpha)$ completes the proof. ∎

Other Boundary Conditions

The theory just sketched for the BVP (9.1-1) does not specify how to handle boundary conditions other than homogeneous Dirichlet conditions. The remainder of this section indicates, with some geometric motivation, how to treat different boundary conditions. The vehicle for this discussion is the following BVP:

$$\begin{gathered}[-a(x)u'(x)]' = f(x), \qquad x \in \Omega := (0,1), \\ u(0) = \beta_1, \qquad u'(1) = \beta_2. \end{gathered} \tag{9.2-8}$$

Here, the functions a and f are as in Equation (9.1-1), and β_1 and β_2 are arbitrary real numbers.

First, consider the Dirichlet condition $u(0) = \beta_1$. Adopt a piecewise linear trial function

$$u_h(x) = \beta_1\ell_0(x) + \sum_{j=1}^{M} U_j\ell_j(x). \tag{9.2-9}$$

Thus $u_h(0) = \beta_1$ automatically, since $\ell_1(0) = \ell_2(0) = \cdots = \ell_M(0) = 0$. But how should we determine the remaining coefficients $U_1, U_2, \ldots, U_M$?

Let us return for a moment to the model problem (9.1-1). Under homogeneous Dirichlet conditions, we determine the unknown coefficients U_j by restricting the variational formulation,

$$B(u,v) = \langle f, v\rangle \qquad \text{for all} \quad v \in H_0^1(\Omega),$$

to the trial space $\mathcal{V}_h$:

$$B(u_h,v) = \langle f, v\rangle \qquad \text{for all} \quad v \in \mathcal{V}_h.$$

Equivalently, we demand that

$$B(u - u_h, v) = 0 \qquad \text{for all} \quad v \in \mathcal{V}_h. \tag{9.2-10}$$

By exploiting the fact that $B(\cdot,\cdot)$ formally possesses the properties of an inner product and considering an analogous least-squares problem in $\mathbb{R}^n$, one can motivate a geometric strategy for determining the unknown coefficients U_j in Equation (9.2-9). Let us adopt some peculiar but suggestive notation for the least-squares problem in $\mathbb{R}^n$. *Temporarily* denote by $B(\mathbf{u},\mathbf{v})$ the standard inner product $\mathbf{u}\cdot\mathbf{v}$ of two vectors $\mathbf{u},\mathbf{v}\in\mathbb{R}^n$, and use the symbol $\mathcal{V}$ to signify the subspace of $\mathbb{R}^n$ spanned by the first M standard unit basis vectors $\mathbf{e}_1,\mathbf{e}_2,\ldots,\mathbf{e}_M$, where $M<n$.

In this setting, the least-squares analog is as follows: Given a vector $\mathbf{u}\in\mathbb{R}^n$, find a vector $\mathbf{u}_h=U_1\mathbf{e}_1+U_2\mathbf{e}_2+\cdots+U_M\mathbf{e}_M\in\mathcal{V}$ that minimizes the distance $\|\mathbf{u}-\mathbf{u}_h\|$ $(=\sqrt{B(\mathbf{u}-\mathbf{u}_h,\mathbf{u}-\mathbf{u}_h)})$. To solve this problem, we force the error $\boldsymbol{\varepsilon}_h:=\mathbf{u}-\mathbf{u}_h$ to be orthogonal to every vector in the subspace $\mathcal{V}$, as drawn in Figure 2 for the case $M=1$. Equivalently, we demand that

$$B(\mathbf{u}-\mathbf{u}_h,\mathbf{e}_i)=0 \qquad i=1,2,\ldots,M. \tag{9.2-11}$$

This condition gives M equations for $U_1,U_2,\ldots,U_M$. The vector $\mathbf{u}_h$ constructed in this way is the projection of $\mathbf{u}$ on $\mathcal{V}$ using the inner product $B(\cdot,\cdot)$.

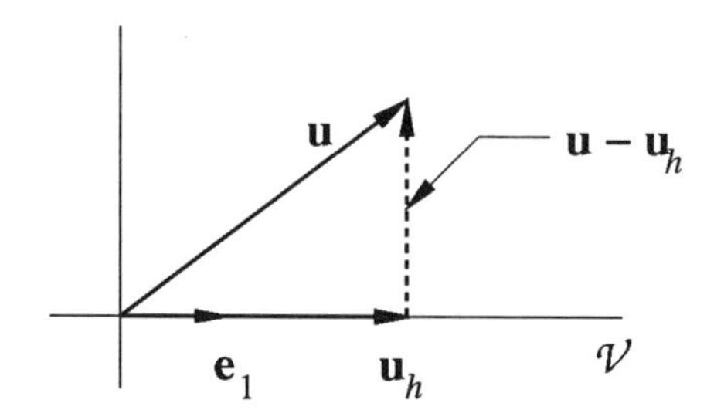

FIGURE 2. *The projection* $\mathbf{u}_h$ *of* $\mathbf{u}\in\mathbb{R}^n$ *onto a subspace* $\mathcal{V}$.

Equations (9.2-11) have the same form as the Galerkin equations (9.2-10). Thus, for the *homogeneous* BVP (9.1-1), the geometry suggests that the finite-element solution u_h is a certain projection of the exact solution u onto the trial space $\mathcal{V}_h$. Because the inner product $B(\cdot,\cdot)$ used in this projection arises from a one-dimensional analog of an elliptic differential operator, we call u_h the **elliptic projection** of u on $\mathcal{V}_h$.

Nonhomogenous Dirichlet conditions lead to a trial function of the form (9.2-9). In this case the geometry is a little different. The analogous problem in $\mathbb{R}^n$ is to approximate a vector $\mathbf{u}$ by a vector $\mathbf{u}_h$ whose coordinate $\mathbf{u}_\partial$ with respect to some basis vector is prescribed. Specifically, we seek an approximating vector from the set

$$\mathbf{u}_\partial+\mathcal{V}:=\left\{\mathbf{u}_\partial+\sum_{j=1}^{M}U_j\mathbf{e}_j\ :\ U_1,U_2,\ldots,U_M\in\mathbb{R}\right\}.$$

Figure 3 illustrates this set for the case $M=1$. The fixed vector $\mathbf{u}_\partial$ is analogous to the fixed component $\beta_1\ell_0(x)$ of the trial function (9.2-9). We call the set $\mathbf{u}_\partial + \mathcal{V}$ an **affine subspace** of $\mathbb{R}^n$; it is not strictly a subspace unless $\mathbf{u}_\partial = \mathbf{0}$. The approximation problem is to find $\mathbf{u}_h \in \mathbf{u}_\partial + \mathcal{V}$ that minimizes the distance $\|\mathbf{u} - \mathbf{u}_h\|$.

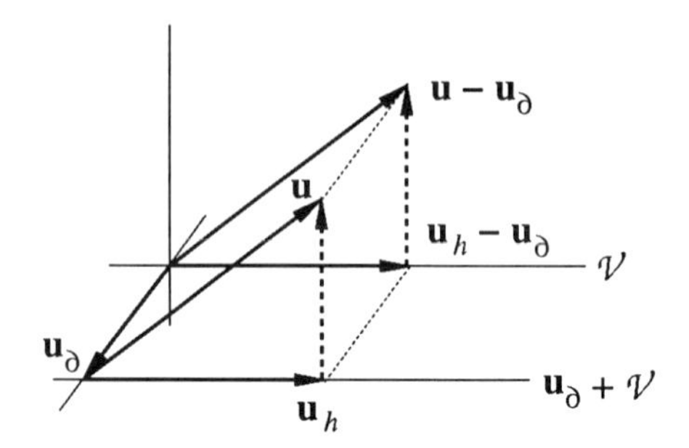

FIGURE 3. *Analog in Euclidean space of the projection corresponding to the Galerkin finite-element approximation u_h to a solution u obeying nonhomogeneous Dirichlet boundary conditions.*

The solution to this problem in $\mathbb{R}^n$ is straightforward: Simply translate the set $\mathbf{u}_\partial+\mathcal{V}$ (a line in Figure 3) by the vector $-\mathbf{u}_\partial$, then solve for $U_1, U_2, \ldots, U_M$ as in Equation (9.2-4). Thus, $\mathbf{u}_h - \mathbf{u}_\partial$ is the projection of $\mathbf{u} - \mathbf{u}_\partial$, using the bilinear form $B(\cdot,\cdot)$:

$$B((\mathbf{u} - \mathbf{u}_\partial) - (\mathbf{u}_h - \mathbf{u}_\partial), \mathbf{e}_i) = B(\mathbf{u} - \mathbf{u}_h, \mathbf{e}_i) = 0, \qquad i = 1, 2, \ldots, M.$$

The weighting vectors $\mathbf{e}_i$ are precisely the basis vectors associated with the unknown coefficients $U_1, U_2, \ldots, U_M$.

We formulate a Galerkin finite-element solution to the BVP (9.2-8) similarly. The affine supspace of $H_0^1(\Omega)$ in this setting is

$$u_\partial + \mathcal{V}_h := \left\{ \underbrace{\beta_1 \ell_0(x)}_{u_\partial} + \sum_{j=1}^{M} U_j \ell_j(x) \ : \ U_1, U_2, \ldots, U_M \in \mathbb{R} \right\}.$$

Any trial function u_h in this set automatically satisfies the nonzero Dirichlet condition. By analogy with the problem in $\mathbb{R}^n$, we solve for $U_1, U_2, \ldots, U_M$ by demanding that $u_h - u_\partial$ be the elliptic projection of $u - u_\partial$ on the subspace $\mathcal{V}_h$. The test functions are precisely the basis functions associated with the unknown coefficients $U_1, U_2, \ldots, U_M$.

There remains the question of how to handle the Neumann condition $u'(1) = \beta_2$. The answer comes from the variational formulation. In terms of the standard inner product on $L^2(\Omega)$, think of the Galerkin equations as requiring that

$$\langle -(au')' + (au_h')', v \rangle = 0,$$

for all $v \in \mathcal{V}_h$. Equivalently,

$$\langle -(au')', \ell_i \rangle = \langle f, \ell_i \rangle, \qquad i = 1, 2, \ldots, M.$$

To reconcile these equations with the smoothness assumptions on $\mathcal{V}_h$, integrate by parts:

$$\underbrace{\langle au_h', \ell_i' \rangle}_{B(u_h, \ell_i)} - a(x) u_h'(x) \ell_i(x) \Big|_0^1 = \underbrace{\langle f, \ell_i \rangle}_{F(\ell_i)}, \quad i = 1, 2, \ldots, M.$$

The boundary terms that result furnish the vehicle for imposing Neumann conditions. By setting $u_h'(1) = \beta_2$ and observing that $\ell_i(0) = 0$ for $i = 1, 2, \ldots, M$, we find that

$$a(x) u_h'(x) \ell_i(x) \Big|_0^1 = \begin{cases} 0, & \text{if } \ i = 1, 2, \ldots, M-1, \\ a(1)\beta_2, & \text{if } \ i = M. \end{cases}$$

Nonhomogeneous boundary conditions affect the matrix equation in a distinctive fashion. For the model problem (9.2-8), we have the following linear system for $i = 1, 2, \ldots, M$:

$$B(u_\partial, \ell_i) + \sum_{j=1}^{M} U_j B(\ell_j, \ell_i) = F(\ell_i) + \begin{cases} 0, & \text{if } \ i = 1, 2, \ldots, M-1, \\ a(1)\beta_2, & \text{if } \ i = M. \end{cases}$$

Call

$$b_i := \beta_1 \int_\Omega a(x) \ell_0'(x) \ell_i'(x)\, dx,$$

which vanishes when $i \neq 1$. As usual, let

$$a_{i,j} := \int_\Omega a(x) \ell_j'(x) \ell_i'(x)\, dx$$

and

$$f_i := \int_\Omega f(x) \ell_i(x)\, dx.$$

The Galerkin equations then yield

$$\begin{bmatrix} a_{1,1} & a_{1,2} & \cdots & a_{1,M} \\ a_{2,1} & a_{2,2} & \cdots & a_{2,M} \\ \vdots & \vdots & & \vdots \\ a_{M,1} & a_{M,2} & \cdots & a_{M,M} \end{bmatrix} \begin{bmatrix} U_1 \\ U_2 \\ \vdots \\ U_M \end{bmatrix} = \begin{bmatrix} f_1 \\ f_2 \\ \vdots \\ f_M \end{bmatrix} - \begin{bmatrix} b_1 \\ 0 \\ \vdots \\ 0 \end{bmatrix} + \begin{bmatrix} 0 \\ \vdots \\ 0 \\ a(1)\beta_2 \end{bmatrix}.$$

The last two vectors on the right account for the boundary conditions.

In summary, for nonhomogeneous Dirichlet conditions it is essential to incorporate the boundary values explicitly into the trial function u_h. Therefore we call these boundary conditions **essential boundary conditions** for the problem (9.2-8). In contrast, we impose Neumann conditions by inserting them into boundary terms that arise naturally from integration by parts. Hence we call Neumann conditions **natural boundary conditions** for this problem. Problem 4 considers the treatment of other boundary conditions.

9.3 A Transient Problem

For time-dependent partial differential equations (PDEs), the most common way to apply the finite-element method is to use Galerkin techniques to discretize the spatial variations. This strategy converts the spatial derivative operators to finite-dimensional approximations but leaves the temporal derivatives intact. One can then obtain fully discrete approximations via methods discussed in Chapter 8.

In this section we elaborate on this approach. For concreteness, we examine an initial-boundary-value problem based upon an extension of the heat equation. Let $\Omega := (0,1)$, and call $J := (0,T]$, where T is some prescribed "final time." Consider the problem

$$\begin{aligned} &\frac{\partial u}{\partial t} - \frac{\partial}{\partial x}\left[a(x)\frac{\partial u}{\partial x}\right] = f(x), && (t,x) \in J \times \Omega, \\ &u(t,0) = u(t,1) = 0, && t \in J, \\ &u(0,x) = g(x), && x \in (0,1). \end{aligned} \tag{9.3-1}$$

We assume that f is continuous on $\overline{\Omega}$ and that a is a continuously differentiable function defined on $\overline{\Omega}$. As in Section 9.2, we also assume that there exist positive constants L, α, and A such that $|f(x)| \leq L$ and

$$0 < \alpha = \inf_{x\in\Omega} a(x) \leq a(x) \leq \sup_{x\in\Omega} a(x) = A,$$

for all $x \in \Omega$. These assumptions guarantee the existence of a solution $u(t,x)$. In the course of the analysis we refer to L^2 norms of various derivatives of u. Assume that the solution u is sufficiently regular that these norms are finite.

We begin by applying the finite-element method to discretize the problem in space. As in Section 9.1, we use a Galerkin method on a piecewise linear trial space to keep the development as simple as possible. We then adopt a finite-difference approximation to convert the resulting semidiscrete approximation to a fully discrete one amenable to computation. Finally, we review a standard error analysis for the scheme. For more details on transient problems, we refer to Thomee [5].

A Semidiscrete Formulation

Corresponding to the classical problem (9.3-1) is a **weak form**, obtained by multiplying the PDE by a test function $v(x)$ and formally integrating by parts over the spatial domain Ω. Explicitly, the weak form is as follows: Find a one-parameter family $u(t,\cdot)$ of functions in $H_0^1(\Omega)$ such that, at every time $t \in J$,

$$\left\langle \frac{\partial u}{\partial t}, v\right\rangle + \left\langle a\frac{\partial u}{\partial x}, \frac{\partial v}{\partial x}\right\rangle = \langle f, v\rangle, \tag{9.3-2}$$

for all $v \in H_0^1(\Omega)$. Consistent with our notation for the steady-state case, we identify the bilinear form

$$B(v, w) := \left\langle a\frac{\partial v}{\partial x}, \frac{\partial w}{\partial x} \right\rangle,$$

and recall that it is continuous and coercive, as shown in Proposition 9.4.

As in the previous section, we formulate Galerkin equations for the weak form (9.3-2) by restricting attention to finite-dimensional subspaces of $H_0^1(\Omega)$. Let $\mathcal{V}_h$ denote the trial space of functions that are piecewise linear on a grid of mesh size h and that vanish at $x = 0$ and $x = 1$. We seek a one-parameter family of functions

$$u_h(t, x) = \sum_{j=1}^{M-1} U_j(t)\ell_j(x)$$

such that $u_h(t, \cdot) \in \mathcal{V}_h$ at each time $t \in J$ and

$$\left\langle \frac{\partial u_h}{\partial t}, v \right\rangle + B(u_h, v) = \langle f, v \rangle, \tag{9.3-3}$$

for all $v \in \mathcal{V}_h$.

This problem reduces to a finite system of ordinary differential equations if we impose Equation (9.3-3) for test functions v ranging over the standard basis for $\mathcal{V}_h$:

$$\left\langle \frac{\partial}{\partial t}\left(\sum_{j=1}^{M-1} U_j\ell_j\right), \ell_i \right\rangle + B\left(\sum_{j=1}^{M-1} U_j\ell_j\,,\, \ell_i\right) = \langle f, \ell_i \rangle, \quad i = 1, 2, \ldots, M-1.$$

In other words,

$$\sum_{j=1}^{M-1} \frac{dU_j}{dt} \underbrace{\langle \ell_j, \ell_i \rangle}_{m_{i,j}} + \sum_{j=1}^{M-1} U_j \underbrace{B(\ell_j, \ell_i)}_{a_{i,j}} = \underbrace{\langle f, \ell_i \rangle}_{f_i}, \quad i = 1, 2, \ldots, M-1.$$

Writing this system in matrix form, we get

$$\begin{bmatrix} m_{1,1} & \cdots & m_{1,M-1} \\ \vdots & & \vdots \\ m_{M-1,1} & \cdots & m_{M-1,M-1} \end{bmatrix} \frac{d}{dt} \begin{bmatrix} U_1 \\ \vdots \\ U_{M-1} \end{bmatrix}$$

$$+ \begin{bmatrix} a_{1,1} & \cdots & a_{1,M-1} \\ \vdots & & \vdots \\ a_{M-1,1} & \cdots & a_{M-1,M-1} \end{bmatrix} \begin{bmatrix} U_1 \\ \vdots \\ U_{M-1} \end{bmatrix} = \begin{bmatrix} f_1 \\ \vdots \\ f_{M-1} \end{bmatrix},$$

or, briefly,

$$\mathsf{M}\frac{d\mathbf{U}}{dt} + \mathsf{A}\mathbf{U} = \mathbf{f}. \tag{9.3-4}$$

Owing to historical connections with mechanics, the matrix M multiplying the time derivative vector is called the **mass matrix**, while the matrix A multiplying the vector $\mathbf{U}$ is the **stiffness matrix**. The stiffness matrix is tridiagonal, symmetric, and positive definite, so it is invertible. One can readily check that the mass matrix M shares these properties. It follows that the system (9.3-4) has a unique solution $\mathbf{U}(t)$, defined for $t \in J$, for any prescribed initial vector $\mathbf{U}(0)$.

A Fully Discrete Scheme

To convert the system (9.3-4) of ODEs to a fully discrete system, we must approximate the time derivatives. Chapter 7 discusses various approaches to this task. In this section we examine a simple finite-difference scheme based on the implicit Euler approximation.

Consider a grid $\Delta_t := \{0, k, 2k, \ldots, Nk\}$ on the time interval $[0, T]$, where $k := T/N$ is the time step. Replace the vector function $\mathbf{U}(t)$ containing time-dependent nodal values of the semidiscrete solution $u_h(t)$, by a grid function $\mathbf{u}\colon \Delta_t \to \mathbb{R}^{M-1}$, whose value at the time level $t = nk$ we denote by $\mathbf{u}^n$. The entries of $\mathbf{u}^n$ are the nodal coefficients of a piecewise linear approximate solution u_h^n to the initial-boundary-value problem (9.3-1).

We approximate the derivative $d\mathbf{U}/dt$ using implicit finite differences:

$$\mathsf{M}\frac{\mathbf{u}^n - \mathbf{u}^{n-1}}{k} + \mathsf{A}\mathbf{u}^n = \mathbf{f}.$$

Rearranging yields

$$(\mathsf{M} + k\mathsf{A})\mathbf{u}^n = k\mathbf{f} + \mathsf{M}\mathbf{u}^{n-1}. \tag{9.3-5}$$

If we know the initial vector $\mathbf{u}^0$, Equation (9.3-5) furnishes a tridiagonal matrix equation for the vector $\mathbf{u}^n$ of unknown nodal values at each time level. Later we discuss an appropriate choice for $\mathbf{u}^0$. Since M and A are symmetric and positive definite, so is the system matrix $\mathsf{M} + k\mathsf{A}$, and it follows that Equation (9.3-5) has a unique solution for $n = 1, 2, \ldots, N$.

Convergence of the Fully Discrete Scheme

Analysis of the time-dependent scheme (9.3-5) is more involved than that of the steady-state scheme (9.2-4). In the transient case, the goal is to show that the error tends to zero, at every time level, as we refine both the spatial and temporal grids. In symbols, we must show that

$$\max_{0 \le n \le N} \|u^n - u_h^n\|_{(0)} \to 0 \qquad \text{as} \quad h, k \to 0.$$

Here, u^n denotes the function $u(nk, \cdot) \in H_0^1(\Omega)$.

The analysis hinges on an error equation. To derive it, recall the weak form of the original problem:

$$\left\langle \frac{\partial u}{\partial t}, v \right\rangle + B(u, v) = \langle f, v \rangle,$$

for all $v \in H_0^1(\Omega)$. It follows from this equation that

$$\left\langle \frac{u^n - u^{n-1}}{k}, v \right\rangle + B(u^n, v) = \langle f, v \rangle + \left\langle \frac{u^n - u^{n-1}}{k} - \frac{\partial u^n}{\partial t}, v \right\rangle, \qquad (9.3\text{-}6)$$

for all $v \in H_0^1(\Omega)$ and for $n = 1, 2, \ldots, N$.

Now consider the fully discrete problem. We seek a sequence $\{u_h^n\}$ of functions in the trial space $\mathcal{V}_h \subset H_0^1(\Omega)$ such that

$$\left\langle \frac{u_h^n - u_h^{n-1}}{k}, v \right\rangle + B(u_h^n, v) = \langle f, v \rangle, \qquad (9.3\text{-}7)$$

for all $v \in \mathcal{V}_h$ and for $n = 1, 2, \ldots, N$. Subtracting Equation (9.3-7) from Equation (9.3-6) and calling $\varepsilon_h^n := u^n - u_h^n$, we arrive at the error equation:

$$\langle \varepsilon_h^n - \varepsilon_h^{n-1}, v \rangle + kB(\varepsilon_h^n, v) = k\langle \tau^n, v \rangle, \qquad (9.3\text{-}8)$$

for all $v \in \mathcal{V}_h$ and for $n = 0, 1, \ldots, N$. Here,

$$\tau^n := \frac{u^n - u^{n-1}}{k} - \frac{\partial u^n}{\partial t}$$

is the truncation error associated with the implicit Euler approximation to the time derivative.

Before proceeding, recall that the bilinear form $B(\cdot, \cdot)$ possesses the properties of an inner product on $H_0^1(\Omega)$. As discussed at the end of Section 9.2, it therefore makes sense to form projections of functions in $H_0^1(\Omega)$ on the trial space $\mathcal{V}_h$ using this inner product. Drawing upon the analogy between the one-dimensional problem (9.1-1) and elliptic problems in higher-dimensional settings, we define the elliptic projection of the solution u on $\mathcal{V}_h$ as the function w_h that satisfies the condition

$$B(u - w_h, v) = 0, \qquad \text{for all } v \in \mathcal{V}_h, \qquad (9.3\text{-}9)$$

at each time $t \in J$. It follows from the error estimates of Theorem 9.10 that there exists a constant $\Gamma > 0$ such that

$$\|u - w_h\|_{(0)} \leq \Gamma h^2 \left\| \frac{\partial^2 u}{\partial x^2} \right\|_{(0)}, \qquad (9.3\text{-}10)$$

at each $t \in J$. Also, since $\partial w_h / \partial t$ is the elliptic projection of $\partial u / \partial t$ (why?),

$$\left\| \frac{\partial u}{\partial t} - \frac{\partial w_h}{\partial t} \right\|_{(0)} \leq \Gamma h^2 \left\| \frac{\partial^3 u}{\partial t \partial x^2} \right\|_{(0)}. \qquad (9.3\text{-}11)$$

Mimicking notation used for u, we denote by w_h^n the function $w_h(nk, \cdot) \in \mathcal{V}_h$.

The strategy for the error analysis is to decompose the error ε_h^n into an elliptic-projection error and a remainder:

$$\varepsilon_h^n = \eta^n + \xi^n, \qquad (9.3\text{-}12)$$

where $\eta^n := u^n - w_h^n$ and $\xi^n := w_h^n - u_h^n$. We substitute this decomposition into the error equation (9.3-8). Rearranging gives an equation that governs the growth of ξ^n, which belongs to the trial space $\mathcal{V}_h$, in terms of the elliptic projection error η^n and the truncation error τ^n. Knowing estimates for η^n and τ^n, we then estimate ξ^n.

To implement this strategy, note that the decomposition (9.3-12) and the error equation (9.3-8) imply that

$$\begin{aligned}\langle \xi^n - \xi^{n-1}, v\rangle + kB(\xi^n, v) &= k\langle \tau^n, v\rangle \\ &\quad -\langle \eta^n - \eta^{n-1}, v\rangle - kB(\eta^n, v),\end{aligned}$$

for all $v \in \mathcal{V}_h$ and for $n = 1, 2, \ldots, N$. By Equation (9.3-9) and the fact that η^{n-1} and η^n are elliptic projection errors, the last term on the right vanishes. Moreover, by specifically choosing $v = \xi^n$, we have

$$\langle \xi^n, \xi^n\rangle - \underbrace{\langle \xi^{n-1}, \xi^n\rangle}_{(\text{I})} + kB(\xi^n, \xi^n) = \underbrace{k\langle \tau^n, \xi^n\rangle}_{(\text{II})} - \underbrace{\langle \eta^n - \eta^{n-1}, \xi^n\rangle}_{(\text{III})}. \quad (9.3\text{-}13)$$

To make further progress, we need estimates for the terms labeled (I), (II), and (III).

LEMMA 9.11. $(\text{I}) \le \frac{1}{2}\langle \xi^{n-1}, \xi^{n-1}\rangle + \frac{1}{2}\langle \xi^n, \xi^n\rangle$.

PROOF: This inequality follows directly from the fact that $\langle v - w, v - w\rangle \ge 0$ for all $v, w \in H_0^1(\Omega)$. ∎

LEMMA 9.12. $(\text{II}) \le \dfrac{k^2}{2}\displaystyle\int_{(n-1)k}^{nk} \left\| \frac{\partial^2 u}{\partial t^2} \right\|_{(0)}^2 dt + \frac{k}{2}\langle \xi^n, \xi^n\rangle$.

PROOF: According to the reasoning in Lemma 9.11,

$$k\langle \tau^n, \xi^n\rangle \le \frac{k}{2}\|\tau^n\|_{(0)}^2 + \frac{k}{2}\langle \xi^n, \xi^n\rangle.$$

It remains to estimate $(k/2)\|\tau^n\|_{(0)}^2$. By the Taylor theorem with integral remainder (see Theorem 6.3),

$$\begin{aligned}\|\tau^n\|_{(0)}^2 &= \left\| \frac{\partial u^n}{\partial t} - \frac{u^n - u^{n-1}}{k} \right\|_{(0)}^2 \\ &= \left\| \frac{1}{k}\int_{(n-1)k}^{nk} [t - (n-1)k]\frac{\partial^2 u}{\partial t^2}\, dt \right\|_{(0)}^2 .\end{aligned}$$

Now use the Cauchy-Schwarz inequality:

$$
\begin{aligned}
\|\tau^n\|_{(0)}^2 &\leq \frac{1}{k^2}\left\|\sqrt{\int_{(n-1)k}^{nk}[t-(n-1)k]^2\,dt}\sqrt{\int_{(n-1)k}^{nk}\left(\frac{\partial^2 u}{\partial t^2}\right)^2 dt}\right\|_{(0)}^2 \\
&= \frac{1}{k^2}\int_\Omega\left\{\underbrace{\int_{(n-1)k}^{nk}[t-(n-1)k]^2\,dt}_{\leq k^3}\int_{(n-1)k}^{nk}\left(\frac{\partial^2 u}{\partial t^2}\right)^2 dt\right\}dx.
\end{aligned}
$$

As indicated, the underbraced integral is bounded in magnitude by k^3. Therefore, an interchange in the order of integration produces the inequality

$$
\|\tau^n\|_{(0)}^2 \leq k\int_\Omega\int_{(n-1)k}^{nk}\left(\frac{\partial^2 u}{\partial t^2}\right)^2 dt\,dx = k\int_{(n-1)k}^{nk}\left\|\frac{\partial^2 u}{\partial t^2}\right\|_{(0)}^2 dt.
$$

Multiplying by $k/2$ completes the proof. ∎

LEMMA 9.13. $|(\text{III})| \leq \dfrac{\Gamma^2 h^4}{2}\displaystyle\int_{(n-1)k}^{nk}\left\|\frac{\partial^3 u}{\partial t \partial x^2}\right\|_{(0)}^2 dt + \frac{k}{2}\langle\xi^n,\xi^n\rangle.$

PROOF: By the fundamental theorem of calculus and the Cauchy-Schwarz inequality,

$$
\begin{aligned}
|\langle\eta^n-\eta^{n-1},\xi^n\rangle| &= \left|\left\langle\int_{(n-1)k}^{nk}\frac{\partial\eta}{\partial t}\,dt\,,\,\xi^n\right\rangle\right| \\
&\leq \left\langle\sqrt{\int_{(n-1)k}^{nk}dt}\sqrt{\int_{(n-1)k}^{nk}\left(\frac{\partial\eta}{\partial t}\right)^2 dt}\,,\,|\xi^n|\right\rangle \\
&= \left\langle\sqrt{\int_{(n-1)k}^{nk}\left(\frac{\partial\eta}{\partial t}\right)^2 dt}\,,\,\sqrt{k}|\xi^n|\right\rangle.
\end{aligned}
$$

Applying the trick used in Lemma 9.11, interchanging the order of integration, and exploiting the inequality (9.3-11), we obtain

$$
\begin{aligned}
|\langle\eta^n-\eta^{n-1},\xi^n\rangle| &\leq \frac{1}{2}\int_\Omega\int_{(n-1)k}^{nk}\left(\frac{\partial\eta}{\partial t}\right)^2 dt + \frac{k}{2}\langle\xi^n,\xi^n\rangle \\
&= \frac{1}{2}\int_{(n-1)k}^{nk}\left\|\frac{\partial\eta}{\partial t}\right\|_{(0)}^2 dt + \frac{k}{2}\langle\xi^n,\xi^n\rangle \\
&\leq \frac{\Gamma^2 h^4}{2}\int_{(n-1)k}^{nk}\left\|\frac{\partial^3 u}{\partial t\partial x^2}\right\|_{(0)}^2 dt + \frac{k}{2}\langle\xi^n,\xi^n\rangle,
\end{aligned}
$$

completing the proof. ∎

Because multiple integrals over time and space appear frequently in what follows, we adopt some new notation. If $v(t,\cdot)$ is a one-parameter family of functions in $H_0^1(\Omega)$, then

$$\|v\|_{L^2(J\times\Omega)}^2 := \int_0^T \int_\Omega |v(t,x)|^2\,dx\,dt.$$

The lemmas just established have the following consequence:

LEMMA 9.14. *Let $k \le 1/4$. Then for $n = 1, 2, \ldots, N$,*

$$\|\xi^n\|_{(0)}^2 \le 2\|\xi^0\|_{(0)}^2 + 2k^2 \left\|\frac{\partial^2 u}{\partial t^2}\right\|_{L^2(J\times\Omega)}^2 + 2h^4\Gamma^2 \left\|\frac{\partial^3 u}{\partial t \partial x^2}\right\|_{L^2(J\times\Omega)}^2 + \sum_{m=0}^{n-1} \|\xi^m\|_{(0)}^2.$$

PROOF: Applying Lemmas 9.11, 9.12, and 9.13, we get for $m = 1, 2, \ldots, n \le N$ the following inequality:

$$\begin{aligned}\frac{1}{2}\langle\xi^m,\xi^m\rangle + kB(\xi^m,\xi^m) \;\le\;& \frac{1}{2}\langle\xi^{m-1},\xi^{m-1}\rangle + \frac{k^2}{2}\int_{(m-1)k}^{mk} \left\|\frac{\partial^2 u}{\partial t^2}\right\|_{(0)}^2 dt \\ &+ \frac{h^4\Gamma^2}{2}\int_{(m-1)k}^{mk} \left\|\frac{\partial^3 u}{\partial t\partial x^2}\right\|_{(0)}^2 dt + k\langle\xi^m,\xi^m\rangle.\end{aligned}$$

But $kB(\xi^m,\xi^m) \ge 0$ by coercivity, so the inequality holds just as well if we neglect this term on the left side. Doing so and computing twice the sum from $m = 1$ to $m = n$ yields

$$\begin{aligned}\|\xi^n\|_{(0)}^2 \;\le\;& \|\xi^0\|_{(0)}^2 + k^2\int_0^{nk} \left\|\frac{\partial^2 u}{\partial t^2}\right\|_{(0)}^2 dt \\ &+ h^4\Gamma^2 \int_0^{nk} \left\|\frac{\partial^3 u}{\partial t\partial x^2}\right\|_{(0)}^2 dt + 2k\sum_{m=1}^{n-1} \|\xi^m\|_{(0)}^2 + 2k\|\xi^n\|_{(0)}^2.\end{aligned}$$

When $k \le 1/4$, $(1-2k) > 1/2$. Also, the integral from 0 to nk of any nonnegative function is bounded above by the integral from 0 to T, so

$$\tfrac{1}{2}\|\xi^n\|_{(0)}^2 \le \|\xi^0\|_{(0)}^2 + k^2\left\|\frac{\partial^2 u}{\partial t^2}\right\|_{L^2(J\times\Omega)}^2 + h^4\Gamma^2\left\|\frac{\partial^3 u}{\partial t\partial x^2}\right\|_{L^2(J\times\Omega)}^2 + 2k\sum_{m=1}^{n-1}\|\xi^m\|_{(0)}^2.$$

But $||\xi^0||^2_{(0)} \geq 0$, so we can add a multiple of it to the right side and preserve the inequality. Doing this and multiplying through by 2 yields

$$||\xi^n||^2_{(0)} \leq 2\left\|\xi^0\right\|^2_{(0)} + 2k^2\left\|\frac{\partial^2 u}{\partial t^2}\right\|^2_{L^2(J\times\Omega)} + 2h^4\Gamma^2\left\|\frac{\partial^3 u}{\partial t\partial x^2}\right\|_{L^2(J\times\Omega)} + 4k\sum_{m=0}^{n-1}||\xi^m||^2_{(0)}.$$

By hypothesis, $4k < 1$, and the proof is complete. ∎

Lemma 9.14 furnishes an inequality that governs the growth of the sequence $||\xi^n||^2_{(0)}$. The following lemma allows us to convert this inequality to a more intelligible one.

LEMMA 9.15 (GRONWALL). *Let $\{\zeta^n\}$ be a sequence that satisfies an inequality of the form*

$$|\zeta^n| \leq \nu + \mu\sum_{m=0}^{n-1}|\zeta^m|, \qquad n = 1, 2, \ldots, N, \tag{9.3-14}$$

for some positive constants ν and μ. Then

$$|\zeta^n| \leq e^{\mu N}(\nu + \mu|\zeta^0|), \qquad n = 1, 2, \ldots, N.$$

(Roughly speaking, sequences bounded in terms of their partial sums grow at worst exponentially.)

PROOF: Define

$$Z^n := \nu + \mu\sum_{m=0}^{n}|\zeta^m|,$$

for $n = 0, 1, \ldots, N$. From the hypothesis (9.3-14) it follows that $Z^n - Z^{n-1} \leq \mu Z^{n-1}$ and hence that $Z^n \leq (1+\mu)Z^{n-1}$, for $n = 1, 2, \ldots, N$. Since $1 + \mu \leq e^\mu$,

$$\begin{aligned} Z^{n-1} \leq (1+\mu)Z^{n-2} \leq \cdots &\leq (1+\mu)^{n-1}Z^0 \\ &\leq e^{\mu N}Z^0 \\ &= e^{\mu N}(\nu + \mu|\zeta^0|). \end{aligned}$$

But the inequality (9.3-14) asserts that $|\zeta^n| \leq Z^{n-1}$, and the conclusion follows. ∎

This lemma applies to the result of the previous one: Letting $\mu = 4k$ and

$$\nu = 2\|\xi^0\|_{(0)}^2 + 2k^2 \left\| \frac{\partial^2 u}{\partial t^2} \right\|_{L^2(J\times\Omega)}^2 + 2h^4\Gamma^2 \left\| \frac{\partial^3 u}{\partial t \partial x^2} \right\|_{L^2(J\times\Omega)}^2,$$

and identifying $T = Nk$, we find that

$$\|\xi^n\|_{(0)}^2 \leq e^{4T} \left(3\|\xi^0\|_{(0)}^2 + 2k^2 \left\| \frac{\partial^2 u}{\partial t^2} \right\|_{L^2(J\times\Omega)}^2 + 2h^4\Gamma^2 \left\| \frac{\partial^3 u}{\partial t \partial x^2} \right\|_{L^2(J\times\Omega)}^2 \right). \tag{9.3-15}$$

(We have used the hypothesis that $k \leq 1/4$.)

Equation (9.3-15) bounds the growth of the error component ξ^n in terms of its initial value ξ^0, the time step k, and the spatial mesh size h. One can control the size of $\xi^0 = w_h^0 - u_h^0$ by initializing the timestepping algorithm correctly. In particular, let us choose the initial function u_h^0 to be the elliptic projection w_h^0 of the exact solution. In other words, to initialize, we solve the following analog of the steady-state problem discussed in Section 9.2:

$$B(u_h^0, v) = B(g, v),$$

for all $v \in \mathcal{V}_h$. Then ξ^0 vanishes, and Equation (9.3-15) collapses to the inequality

$$\begin{aligned} \|\xi^n\|_{(0)}^2 &\leq e^{4T} \left(2k^2 \left\| \frac{\partial^2 u}{\partial t^2} \right\|_{L^2(J\times\Omega)}^2 + 2h^4\Gamma^2 \left\| \frac{\partial^3 u}{\partial t \partial x^2} \right\|_{L^2(J\times\Omega)}^2 \right) \\ &= \mathcal{O}(k^2 + h^4). \end{aligned} \tag{9.3-16}$$

This inequality is a key ingredient in the following convergence proof.

THEOREM 9.16. Consider the fully discrete scheme (9.3-5). If $k < 1/4$ and $u_h^0 = w_h^0$, then $\|\varepsilon_h^n\|_{(0)} = \mathcal{O}(k + h^2)$.

PROOF: Equation (9.3-16) implies that $\|\xi^n\|_{(0)}$ obeys a bound of the form

$$\begin{aligned} \|\xi^n\|_{(0)}^2 \leq a^2k^2 + b^2h^4 &\leq a^2k^2 + 2abkh^2 + b^2h^4 \\ &= (ak + bh^2)^2, \end{aligned}$$

where

$$a^2 := 2e^{4T} \left\| \frac{\partial^2 u}{\partial t^2} \right\|_{L^2(J\times\Omega)}^2, \qquad b^2 := 2e^{4T}\Gamma^2 \left\| \frac{\partial^3 u}{\partial t \partial x^2} \right\|_{L^2(J\times\Omega)}^2.$$

It follows that

$$\begin{aligned} \|\varepsilon_h^n\|_{(0)} &\leq \|\eta^n\|_{(0)} + \|\xi^n\|_{(0)} \leq \sqrt{2}e^{2T} \left\| \frac{\partial^2 u}{\partial t^2} \right\|_{L^2(J\times\Omega)} k \\ &+ \Gamma \left(\max_{0\leq n\leq N} \left\| \frac{\partial^2 u^n}{\partial x^2} \right\|_{(0)} + \sqrt{2}e^{2T} \left\| \frac{\partial^3 u}{\partial t \partial x^2} \right\|_{L^2(J\times\Omega)} \right) h^2. \end{aligned}$$

This concludes the proof. ∎

9.4 Problems

PROBLEM 1. Prove Proposition 9.1.

PROBLEM 2. Prove Proposition 9.2.

PROBLEM 3. One can often gain insight into the computational requirements of a finite-element scheme by looking at the zero structure of the matrix that it generates. For example, for the equation $u'' = f$, the Galerkin method with piecewise linear trial functions yields a matrix whose zero structure is tridiagonal. What are the zero structures that result when the trial functions are in $\mathcal{M}_0^2(\Delta)$, $\mathcal{M}_0^3(\Delta)$, and $\mathcal{M}_1^3(\Delta)$?

PROBLEM 4. Show that Robin boundary conditions, which have the form

$$\alpha(x)u(x) + \beta(x)\frac{\partial u}{\partial x} = \gamma(x),$$

are natural boundary conditions for model problems of the type considered in Section 9.2.

PROBLEM 5. How does the conclusion of Theorem 9.15 change if we use for u_h^0 the piecewise linear interpolant of the initial function g, instead of its elliptic projection?

9.5 References

1. R.B. Guenther and J.W. Lee, *Partial Differential Equations of Mathematical Physics and Integral Equations*, Prentice-Hall, Englewood Cliffs, NJ, 1988.

2. C. Johnson, *Numerical Solution of Partial Differential Equations by the Finite Element Method*, Cambridge University Press, Cambridge, U.K., 1987.

3. L. Lapidus and G.F. Pinder, *Numerical Solution of Partial Differential Equations in Science and Engineering*, Wiley, New York, 1982.

4. G. Strang and G. Fix, *An Analysis of the Finite Element Method*, Prentice-Hall, Englewood Cliffs, NJ, 1973.

5. V. Thomée, *Galerkin Finite Element Methods for Parabolic Problems*, Lecture Notes in Mathematics, Vol. 1054, Springer-Verlag, Berlin, 1984.

6. R. Wait and A.R. Mitchell, *The Finite Element Method in Partial Differential Equations*, Wiley, Chichester, U.K., 1975.

Appendix A

Divided Differences

Divided differences furnish an alternative to the standard Lagrange interpolating bases for the representation of interpolating polynomials. Let $f \in C^{n+1}([a,b])$, and consider a set $\Delta = \{x_0, x_1, \ldots, x_n\}$ of distinct points in $[a,b]$. Let p be the polynomial having degree at most n such that $p(x_i) = f(x_i)$ for $i = 0, \ldots, n$. (Theorem 1.1 guarantees the existence and uniqueness of p.) Can we find coefficients $a_0, a_1, \ldots, a_n \in \mathbb{R}$ for which the representation

$$p(x) = a_0 + a_1(x - x_0) + \cdots + a_n(x - x_0)(x - x_1)\cdots(x - x_{n-1})$$

holds for every $x \in [a,b]$?

The answer is yes. We determine the coefficients $a_0, a_1, \ldots, a_n$ as follows: The condition $p(x_0) = f(x_0)$ implies that

$$a_0 = f[x_0] := f(x_0).$$

Using this result and the condition $p(x_1) = a_0 + a_1(x_1 - x_0) = f(x_1)$, we then obtain

$$a_1 = f[x_0, x_1] := \frac{f[x_1] - f[x_0]}{x_1 - x_0}.$$

Similarly, the condition $p(x_2) = f(x_2)$ yields

$$a_2 = f[x_0, x_1, x_2] := \frac{f[x_1, x_2] - f[x_0, x_1]}{x_2 - x_0},$$

and so forth. The notation adopted in these steps motivates a formal definition.

DEFINITION. *Given a function $f\colon [a,b] \to \mathbb{R}$ and a grid $\Delta = \{x_0, x_1, \ldots, x_n\}$ on $[a,b]$, we define the* **divided differences** *of f on Δ inductively as follows:*

$$f[x_i] := f(x_i),$$

$$f[x_i, x_{i+1}, \ldots, x_{i+k}] := \frac{f[x_{i+1}, \ldots, x_{i+k}] - f[x_i, \ldots x_{i+k-1}]}{x_{i+k} - x_i},$$

where the last line applies for $i = 1, 2, \ldots, n-k$.

With this definition, we can represent the interpolating polynomial p in the form

$$p(x) = f[x_0] + \sum_{j=1}^{n} f[x_0, x_1, \ldots, x_j](x - x_0)(x - x_1) \cdots (x - x_{j-1}). \quad \text{(A-1)}$$

As their difference-quotient form suggests, divided differences have close connections with differentiation.

THEOREM A.1. *If* $f \in C^n([a,b])$ *and* $x_0, x_1, \ldots, x_n \in [a,b]$ *are distinct, then there exists a point* $\zeta \in (a,b)$ *such that*

$$f[x_0, x_1, \ldots, x_n] = \frac{f^{(n)}(\zeta)}{n!}. \quad \text{(A-2)}$$

PROOF: Let p be the polynomial defined in Equation (A-1). Thus $f(x_i) = p(x_i)$ for $i = 0, 1, \ldots, n$, so the function $g := f - p \in C^n([a,b]$ has at least $n+1$ distinct roots in $[a,b]$. Repeated application of Rolle's theorem shows that g' has at least n distinct roots in (a,b), that g'' has at least $n-1$ distinct roots in (a,b), and so forth, until we conclude that $g^{(n)}$ has at least one root $\zeta \in (a,b)$. We now have

$$f^{(n)}(\zeta) - p^{(n)}(\zeta) = f^{(n)}(\zeta) - n!\, f[x_0, x_1, \ldots, x_n] = 0,$$

which establishes Equation (A-2). ∎

By comparing the divided-difference representation (A-1) with the Lagrange form of the interpolating polynomial p, we deduce another fact about divided differences.

THEOREM A.2. *Divided differences are symmetric functions of their arguments, that is, if* $(i_0, i_1, \ldots, i_n)$ *is any permutation of the indices* $(0, 1, \ldots, n)$, *then*

$$f[x_0, x_1, \ldots, x_n] = f[x_{i_0}, x_{i_1}, \ldots, x_{i_n}].$$

PROOF: The Lagrange form for the interpolation polynomial p defined above is

$$p(x) = \sum_{i=0}^{n} f(x_i) L_i(x),$$

where

$$L_i(x) = \frac{(x - x_0) \cdots (x - x_{i-1})(x - x_{i+1}) \cdots (x - x_n)}{(x_i - x_0) \cdots (x_i - x_{i-1})(x_i - x_{i+1}) \cdots (x_i - x_n)}.$$

Comparing the coefficient of x^n in this expression with the coefficient of x^n in (A-1), we see that

$$\begin{aligned} a_n &= f[x_0, x_1, \ldots, x_n] \\ &= \sum_{i=0}^{n} \frac{f(x_i)}{(x_i - x_0)\cdots(x_i - x_{i-1})(x_i - x_{i+1})\cdots(x_i - x_n)}. \end{aligned}$$

This sum is a symmetric function of $x_0, x_1, \ldots, x_n$. ∎

Appendix B

Local Minima

This appendix establishes the connection between the local minima of real-valued functions defined on subsets of $\mathbb{R}^n$ and the zeros of their gradients. This connection often allows one to solve multidimensional optimization problems by the methods discussed in Section 3.7.

The theory generalizes familiar facts about functions $\varphi \in C^2((a,b))$. Recall from elementary calculus that $\varphi'(x^*) = 0$ is a necessary condition for x^* to be a local minimum for φ. Furthermore, given this condition, $\varphi''(x^*) > 0$ is a sufficient condition. In the multidimensional case, the gradient $\nabla\varphi$ plays the role of φ', and the Hessian matrix H_φ plays the role of φ''.

In the following two theorems, $\Omega \subset \mathbb{R}^n$ is open and convex. First we show that $\nabla\varphi(\mathbf{x}^*) = \mathbf{0}$ is a necessary condition for $\mathbf{x}^*$ to be a local minimum.

THEOREM B.1. *Let $\varphi \in C^1(\Omega)$. A point $\mathbf{x}^* \in \Omega$ is a local minimum for φ only if $\nabla\varphi(\mathbf{x}^*) = \mathbf{0}$.*

PROOF: We prove the contrapositive. Assume that $\nabla\varphi(\mathbf{x}^*) \neq \mathbf{0}$, and let $\epsilon > 0$ be any radius small enough to ensure that $\mathcal{B}_\epsilon(\mathbf{x}^*) \subset \Omega$. We construct a point $\mathbf{z} \in \mathcal{B}_\epsilon(\mathbf{x}^*)$ such that $\varphi(\mathbf{z}) < \varphi(\mathbf{x}^*)$. Start by choosing a real number $\alpha > 0$ small enough so that $\mathbf{x}^* - \alpha\nabla\varphi(\mathbf{x}^*) \in \mathcal{B}_\epsilon(\mathbf{x}^*)$. Call $\mathbf{y} := -\alpha\nabla\varphi(\mathbf{x}^*)$, and observe that $\nabla\varphi(\mathbf{x}^*) \cdot \mathbf{y} < \mathbf{0}$. Since the function $\nabla\varphi$ is continuous on Ω, there exists a real number $\overline{\theta} \in [0,1]$ such that $\nabla\varphi(\mathbf{x}^* + \theta\mathbf{y}) \cdot \mathbf{y} < \mathbf{0}$ for all $\theta \in [0,\overline{\theta}]$. The fundamental theorem of calculus now yields

$$\begin{aligned} \varphi(\mathbf{x}^* + \overline{\theta}\mathbf{y}) &= \varphi(\mathbf{x}^*) + \int_0^{\overline{\theta}} \nabla\varphi(\mathbf{x}^* + \theta\mathbf{y}) \cdot \mathbf{y}\, d\theta \\ &< \varphi(\mathbf{x}^*), \end{aligned}$$

since the integrand is negative. But $\mathbf{z} := \mathbf{x}^* + \overline{\theta}\mathbf{y} \in \mathcal{B}_\epsilon(\mathbf{x}^*)$, and $\varphi(\mathbf{z}) < \varphi(\mathbf{x}^*)$ as desired. Since ϵ is an arbitrary positive number, $\mathbf{x}^*$ cannot be a local minimum for φ. ∎

Next we prove that, when $\nabla\varphi(\mathbf{x}^*) = \mathbf{0}$, positive definiteness of the Hessian matrix $\mathsf{H}_\varphi(\mathbf{x}^*)$ suffices to guarantee that $\mathbf{x}^*$ is a local minimum of φ.

THEOREM B.2. *Let $\varphi \in C^1(\Omega)$, and assume that $\nabla\varphi$ is continuously differentiable on Ω. If there is a point $\mathbf{x}^* \in \Omega$ such that $\nabla\varphi(\mathbf{x}^*) = \mathbf{0}$ and $\mathsf{H}_\varphi(\mathbf{x}^*)$ is positive definite, then $\mathbf{x}^*$ is a local minimum for φ.*

PROOF: The continuity of $\mathsf{H}_\varphi(\mathbf{x}^*)$ ensures that there exists a radius $\epsilon > 0$ such that $\mathsf{H}_\varphi(\mathbf{x})$ is positive definite for every $\mathbf{x} \in \mathcal{B}_\epsilon(\mathbf{x}^*)$. (We invite the reader to check this statement carefully.) Pick any $\mathbf{x} \in \mathcal{B}_\epsilon(\mathbf{x}^*)$ with $\mathbf{x} \neq \mathbf{x}^*$. By the Taylor theorem 0.16, there is a point $\boldsymbol{\zeta}$ on the line segment connecting $\mathbf{x}$ and $\mathbf{x}^*$ for which

$$
\begin{aligned}
\varphi(\mathbf{x}) &= \varphi(\mathbf{x}^*) + \nabla\varphi(\mathbf{x}^*)\cdot(\mathbf{x}-\mathbf{x}^*) + \tfrac{1}{2}(\mathbf{x}-\mathbf{x}^*)\cdot\mathsf{H}_\varphi(\mathbf{x}^*)\,(\mathbf{x}-\mathbf{x}^*) \\
&= \varphi(\mathbf{x}^*) + \tfrac{1}{2}(\mathbf{x}-\mathbf{x}^*)\cdot\mathsf{H}_\varphi(\mathbf{x}^*)\,(\mathbf{x}-\mathbf{x}^*) \\
&> \varphi(\mathbf{x}^*).
\end{aligned}
$$

Since $\mathbf{x} \in \mathcal{B}_\epsilon(\mathbf{x}^*)$ is arbitrary, $\mathbf{x}^*$ must be a local minimum for φ. ∎

Appendix C

Chebyshev Polynomials

This appendix reviews some basic properties of the Chebyshev polynomials, which find a variety of applications in classical numerical analysis.

DEFINITION. *The* **Chebyshev polynomials** *are the functions generated by the following recursion:*

$$\begin{aligned} T_0(z) &= 1 \\ T_{n+1}(z) &= 2zT_n(z) - T_{n-1}(z). \end{aligned}$$

This recursion gives rise to several equivalent representations. For example,

$$T_n(z) = \frac{1}{2}\left[\left(z + \sqrt{z^2 - 1}\right)^n + \left(z - \sqrt{z^2 - 1}\right)^n\right], \quad n = 0, 1, 2, \ldots;$$

$$T_n(z) = \cos\left(n \cos^{-1} z\right), \quad -1 \leq z \leq 1.$$

The first few Chebyshev polynomials are as follows:

$$\begin{aligned} T_0(z) &= 1, \\ T_1(z) &= z, \\ T_2(z) &= 2z^2 - 1, \\ T_3(z) &= 4z^3 - 3z. \end{aligned}$$

In general, the nth Chebyshev polynomial has leading coefficient 2^{n-1}.

The nth Chebyshev polynomial T_n has n real zeros. The next proposition gives more specific information:

PROPOSITION C.1. *The Chebyshev polynomial T_n has n zeros in the interval $(-1,1)$ and $n+1$ local extrema in the interval $[-1,1]$. At the local extrema, $|T_n(z)| = 1$.*

PROOF: We use the representation $T_n(z) = \cos(n\cos^{-1} z)$. Notice that $\cos(n\theta)$ vanishes for

$$\theta = \frac{(2N+1)}{n}\frac{\pi}{2},$$

where N ranges over the integers. Letting $\theta = \cos^{-1} z$, we see that $T_n(z) = 0$ for

$$z = \cos\left(\frac{2N+1}{n}\frac{\pi}{2}\right), \quad N = 0, 1, 2, \ldots, n-1.$$

These are the n zeros lying in the interval $[-1,1]$. Also, $\cos(n\theta)$ has local extrema at the points $\theta = N\pi/n$, so setting $\theta = \cos^{-1} z$ shows that $T_n'(z) = 0$ for $z = \cos(N\pi/n)$, $N = 0, 1, 2, \ldots, n$. At these points $T_n(z) = (-1)^N$. ∎

Figure 1 depicts the graph of T_8. Notice that the polynomial is relatively "well behaved" in the interval $[-1,1]$, the function values being confined to the range $[-1,1]$. Intuitively, this controlled behavior inside $[-1,1]$ occurs at the expense of the behavior outside the interval, where the polynomial rapidly shoots off toward infinity.

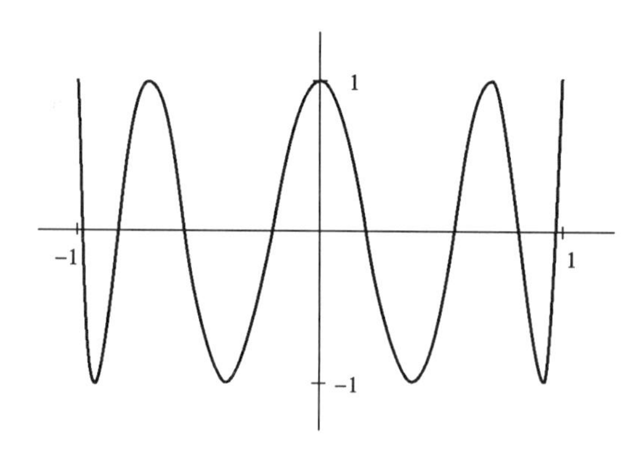

FIGURE 1. *The Chebyshev polynomial $T_8(z)$.*

The next theorem asserts that, in a sense, the controlled behavior inside $[-1,1]$ is the best that we can expect for a polynomial of specified degree.

THEOREM C.2 (MINIMAX PROPERTY). *Of all polynomials p having degree exactly n and leading coefficient 2^{n-1}, T_n possesses the smallest value of $\|p\|_\infty := \sup_{z\in[-1,1]} |p(z)|$.*

(As we have seen, $\|T_n\|_\infty = 1$.)

PROOF: We argue by contradiction. Assume that $p \neq T_n$ is a polynomial having degree exactly n and leading coefficient 2^{n-1} and that $||p||_\infty < ||T_n||_\infty$. Let $z_0, z_1, \ldots, z_n$ denote the extrema of T_n, ordered so that the points $z_0, z_2, z_4, \ldots$ are local maxima and $z_1, z_3, z_5, \ldots$ are local minima. We have

$$p(z_0) < T_n(z_0), \quad p(z_1) > T_n(z_1), \quad p(z_2) < T_n(z_2), \quad \cdots.$$

Thus the nonzero polynomial $p - T_n$ changes signs n times in the interval $(-1, 1)$, which implies that $p - T_n$ has n roots in $(-1, 1)$. It follows that $p - T_n$ has degree at least n. But p and T_n both have degree n and possess the same leading coefficient, so $p - T_n$ has degree at most $n - 1$. This is a contradiction. ∎

Index

PURE AND APPLIED MATHEMATICS

A Wiley-Interscience Series of Texts, Monographs, and Tracts

ADÁMEK, HERRLICH, and STRECKER—Abstract and Concrete Catetories
ADAMOWICZ and ZBIERSKI—Logic of Mathematics
AKIVIS and GOLDBERG—Conformal Differential Geometry and Its Generalizations
ALLEN and ISAACSON—Numerical Analysis for Applied Science
*ARTIN—Geometric Algebra
AZIZOV and IOKHVIDOV—Linear Operators in Spaces with an Indefinite Metric
BERMAN, NEUMANN, and STERN—Nonnegative Matrices in Dynamic Systems
BOYARINTSEV—Methods of Solving Singular Systems of Ordinary Differential Equations
BURK—Lebesgue Measure and Integration: An Introduction
*CARTER—Finite Groups of Lie Type
CHATELIN—Eigenvalues of Matrices
CLARK—Mathematical Bioeconomics: The Optimal Management of Renewable Resources, Second Edition
COX—Primes of the Form $x^2 + ny^2$: Fermat, Class Field Theory, and Complex Multiplication
*CURTIS and REINER—Representation Theory of Finite Groups and Associative Algebras
*CURTIS and REINER—Methods of Representation Theory: With Applications to Finite Groups and Orders, Volume I
CURTIS and REINER—Methods of Representation Theory: With Applications to Finite Groups and Orders, Volume II
*DUNFORD and SCHWARTZ—Linear Operators
Part 1—General Theory
Part 2—Spectral Theory, Self Adjoint Operators in Hilbert Space
Part 3—Spectral Operators
FOLLAND—Real Analysis: Modern Techniques and Their Applications
FRÖLICHER and KRIEGL—Linear Spaces and Differentiation Theory
GARDINER—Teichmüller Theory and Quadratic Differentials
GREENE and KRATZ—Function Theory of One Complex Variable
*GRIFFITHS and HARRIS—Principles of Algebraic Geometry
GROVE—Groups and Characters
GUSTAFSSON, KREISS and OLIGER—Time Dependent Problems and Difference Methods
HANNA and ROWLAND—Fourier Series, Transforms, and Boundary Value Problems, Second Edition
*HENRICI—Applied and Computational Complex Analysis
Volume 1, Power Series—Integration—Conformal Mapping—Location of Zeros
Volume 2, Special Functions—Integral Transforms—Asymptotics—Continued Fractions
Volume 3, Discrete Fourier Analysis, Cauchy Integrals, Construction of Conformal Maps, Univalent Functions
*HILTON and WU—A Course in Modern Algebra
*HOCHSTADT—Integral Equations
JOST—Two-Dimensional Geometric Variational Procedures

*KOBAYASHI and NOMIZU—Foundations of Differential Geometry, Volume I
*KOBAYASHI and NOMIZU—Foundations of Differential Geometry, Volume II
LAX—Linear Algegra
LOGAN—An Introduction to Nonlinear Partial Differential Equations
McCONNELL and ROBSON—Noncommutative Noetherian Rings
NAYFEH—Perturbation Methods
NAYFEH and MOOK—Nonlinear Oscillations
PANDEY—The Hilbert Transform of Schwartz Distributions and Applications
PETKOV—Geometry of Reflecting Rays and Inverse Spectral Problems
*PRENTER—Splines and Variational Methods
RAO—Measure Theory and Integration
RASSIAS and SIMSA—Finite Sums Decompositions in Mathematical Analysis
RENELT—Elliptic Systems and Quasiconformal Mappings
RIVLIN—Chebyshev Polynomials: From Approximation Theory to Algebra and Number Theory, Second Edition
ROCKAFELLAR—Network Flows and Monotropic Optimization
ROITMAN—Introduction to Modern Set Theory
*RUDIN—Fourier Analysis on Groups
SENDOV—The Averaged Moduli of Smoothness: Applications in Numerical Methods and Approximations
SENDOV and POPOV—The Averaged Moduli of Smoothness
*SIEGEL—Topics in Complex Function Theory
Volume 1—Elliptic Functions and Uniformization Theory
Volume 2—Automorphic Functions and Abelian Integrals
Volume 3—Abelian Functions and Modular Functions of Several Variables
STAKGOLD—Green's Functions and Boundary Value Problems
*STOKER—Differential Geometry
*STOKER—Nonlinear Vibrations in Mechanical and Electrical Systems
*STOKER—Water Waves: The Mathematical Theory with Applications
WESSELING—An Introduction to Multigrid Methods
WHITMAN—Linear and Nonlinear Waves
ZAUDERER—Partial Differential Equations of Applied Mathematics, Second Edition

*Now available in a lower priced paperback edition in the Wiley Classics Library.